Edited by
He Tian and Junji Zhang

Photochromic Materials

Edited by He Tian and Junji Zhang

Photochromic Materials

Preparation, Properties and Applications

WILEY-VCH
Verlag GmbH & Co. KGaA

Editors

Dr. He Tian
East China University of Science
and Technology
Key Laboratory for Advanced Materials
200237 Shanghai
China

Dr. Junji Zhang
East China University of Science and
Technology
Key Laboratory for Advanced Materials
200237 Shanghai
China

Cover
Imgram Publishing, UK

Library of Congress Card No.: applied for

British Library Cataloguing-in-Publication Data
A catalogue record for this book is available from the British Library.

Bibliographic information published by the Deutsche Nationalbibliothek
The Deutsche Nationalbibliothek lists this publication in the Deutsche Nationalbibliografie; detailed bibliographic data are available on the Internet at <http://dnb.d-nb.de>.

Print ISBN: 978-3-527-33779-8
ePDF ISBN: 978-3-527-68370-3
ePub ISBN: 978-3-527-68372-7
Mobi ISBN: 978-3-527-68371-0
oBook ISBN: 978-3-527-68373-4

Cover Design Schulz Grafik-Design, Fußgönheim, Germany
Typesetting SPi Global, Chennai, India
Printing and Binding Markono Print Media Pte Ltd, Singapore

Printed on acid-free paper

Contents

List of Contributors

Hari Krishna Bisoyi
Kent State University
Liquid Crystal Institute and
Chemical Physics
Interdisciplinary Program
Kent
OH 44242
USA

Neil R. Branda
Simon Fraser University
4D LABS, Department of
Chemistry
8888 University Drive
Burnaby
BC V5A 1S6
Canada

Liang Chen
Fudan University
Department of Chemistry and
Concerted Innovation Center of
Chemistry for Energy Materials
220 Handan Road
200433 Shanghai
China

Shangjun Chen
Shanghai Normal University
Key Laboratory of Resource
Chemistry of Ministry of
Education
Shanghai Key Laboratory of Rare
Earth Functional Materials
Department of Chemistry
200234 Shanghai
China

Robert Göstl
Humboldt-Universität zu Berlin
Department of Chemistry
Brook-Taylor-Str. 2
12489 Berlin
Germany

Stefan Hecht
Humboldt-Universität zu Berlin
Department of Chemistry
Brook-Taylor-Str. 2
12489 Berlin
Germany

Masahiro Irie
Rikkyo University
Department of Chemistry and Research Center for Smart Molecules
3-34-1 Nishi-Ikebukuro
Toshima-ku
171-8501 Tokyo
Japan

Chi-Chiu Ko
The University of Hong Kong
Institute of Molecular Functional Materials and
Department of Chemistry
Pokfulam Road
Chong Yuet Ming Chemistry Building, 504
Hong Kong
China

and

City University of Hong Kong
Department of Biology and Chemistry
Tat Chee Avenu, Kowloon
Hong Kong
China

Seiya Kobatake
Osaka City University
Department of Applied Chemistry
Graduate School of Engineering
Sugimoto 3-3-138
Sumiyoshi-ku
558-8585 Osaka
Japan

Haichuang Lan
Fudan University
Department of Chemistry and Concerted Innovation Center of Chemistry for Energy Materials
220 Handan Road
200433 Shanghai
China

Quan Li
Kent State University
Liquid Crystal Institute and Chemical Physics Interdisciplinary Program
Kent
OH 44242
USA

Wenlong Li
East China University of Science and Technology
Shanghai Key Laboratory of Functional Materials Chemistry
Key Laboratory for Advanced Materials and Institute of Fine Chemicals
200237 Shanghai
China

Guanglei Lv
Fudan University
Department of Chemistry and Concerted Innovation Center of Chemistry for Energy Materials
220 Handan Road
200433 Shanghai
China

Rémi Métivier
PPSM
ENS Cachan, CNRS
Université Paris-Saclay
61 avenue du Président Wilson
94235 Cachan
France

Masakazu Morimoto
Rikkyo University
Department of Chemistry and Research Center for Smart Molecules
3-34-1 Nishi-Ikebukuro
Toshima-ku
171-8501 Tokyo
Japan

Keitaro Nakatani
PPSM
ENS Cachan CNRS
Université Paris-Saclay
61 avenue du Président Wilson
94235 Cachan
France

Emanuele Orgiu
ISIS and icFRC
Université de Strasbourg and CNRS
Nanochemistry Laboratory
8 allée Gaspard Monge
67000 Strasbourg
France

Jonathan Piard
PPSM
ENS Cachan, CNRS
Université Paris-Saclay
61 avenue du Président Wilson
94235 Cachan
France

Paolo Samorí
ISIS and icFRC
Université de Strasbourg and CNRS
Nanochemistry Laboratory
8 allée Gaspard Monge
67000 Strasbourg
France

Antti Senf
Humboldt-Universität zu Berlin
Department of Chemistry
Brook-Taylor-Str. 2
12489 Berlin
Germany

He Tian
East China University of Science and Technology
Key Laboratory for Advanced Materials and Institute of Fine Chemicals
No. 130 Meilong Road
Shanghai 200237
China

Sheng Wang
Lingnan Normal University
School of Chemistry and Chemical Engineering
Zhanjiang 524048
China

Danielle Wilson
Simon Fraser University
4D LABS, Department of Chemistry
8888 University Drive
Burnaby
BC V5A 1S6
Canada

Vivian Wing-Wah Yam
University of Hong Kong
Institute of Molecular Functional Materials (Areas of Excellence Scheme, University Grants Committee) and Department of Chemistry
Hong Kong
China

Tao Yi
Fudan University
Department of Chemistry and
Concerted Innovation Center of
Chemistry for Energy Materials
220 Handan Road
200433 Shanghai
China

Pei Yu
ICMMO
Université Paris-Sud, CNRS
Université Paris-Saclay
Bâtiment 420
91405 Orsay
France

Junji Zhang
East China University of Science
and Technology
Key Laboratory for Advanced
Materials and Institute of Fine
Chemicals
No. 130 Meilong Road
200237 Shanghai
China

Weihong Zhu
East China University of Science
and Technology
Shanghai Key Laboratory of
Functional Materials Chemistry
Key Laboratory for Advanced
Materials and Institute of Fine
Chemicals
200237 Shanghai
China

1
Introduction: Organic Photochromic Molecules

Keitaro Nakatani, Jonathan Piard, Pei Yu, and Rémi Métivier

1.1
Photochromic Systems

1.1.1
General Introduction

Nowadays, the word "photochromism" (or "photochromic") has been entered in several dictionaries [1]. It stems from the Greek words **φωτός** (photos) and **χρῶμα** (chroma) meaning light and color, respectively. A simple definition of photochromism is the property to undergo a light-induced reversible change of color based on a chemical reaction [2].

Everyone, even without being familiar with this topic, can easily understand that materials possessing such a feature can find useful applications. Generally, using light as a stimulus is extremely attractive for at least two reasons: it can be conveyed to long distances with the "speed of light"; and it is an unlimited energy source although unevenly available in time and space. In addition, the notion of reversible change can be easily connected to objects, useful in everyday life, such as knobs, buttons, dials, handles, and levers, which are used to switch on and off domestic appliance and other devices and machines. Photochromic substances are widely present in glass lenses, initially clear, which turn dark under sunshine [3] (Figure 1.1). They are also present in trendy cosmetics and clothes.

In addition to these objects that have been around for a long time, the digital age has tremendously expanded the fields, where photochromic materials may play a role. The broad and current interest is transmitting, gating, and storing digital data [5]. CD and DVD are among the widely spread storage media, where light writes (and erases) information and optical properties are used to read, just as in photochromic systems. Due their reversible feature, photochromic species match the requirement of the rewritable recording media (CD-RW, DVD-RW), where memory bits have to commute between the two binary states ("0" and "1") upon request. In this domain, there is a race for high-capacity data storage media, where information can be written and erased at high speed. As changes in photochromic

Photochromic Materials: Preparation, Properties and Applications, First Edition.
Edited by He Tian and Junji Zhang.

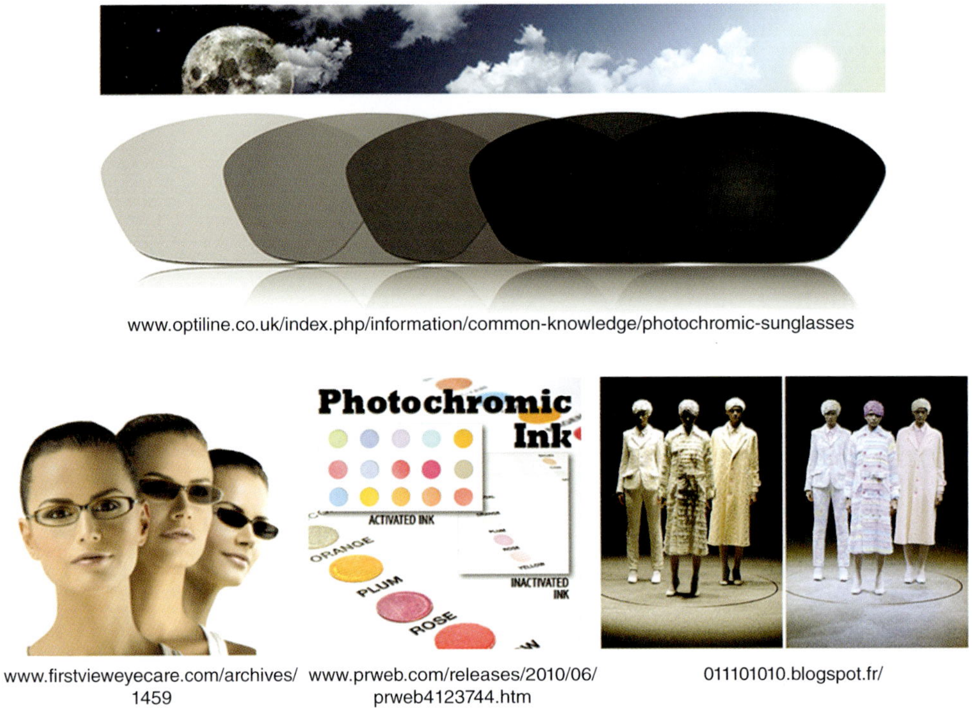

Figure 1.1 Photochromic lenses and clothes: contributions to comfort and to fashion [4].

systems occur in sub-nanosecond timescales, these are suitable for fast switching. Moreover, molecule is the elemental switching unit, comparable to a bit, and occupies less than a cubic nanometer. This means that the memory density can potentially reach a value of more than 10^{18} bit mm^{-3}. Proofs of concept of such media are given in the literature (Figure 1.2a, b), where two-photon phenomena are used to get a high resolution [6].

More recent contributions of photochromism can be found in the topic of fluorescence microscopy imaging, which is spreading very fast in many scientific fields of applications, such as biology, medicine, and materials science. Recent technological progresses have led to faster microscopes with better resolution, along with the development of stable and bright fluorescent probes. However, the optical resolution of conventional microscopy instruments is restricted by the fundamental diffraction limit, whereas the features to be probed are often smaller than 200 nm. To break this severe constraint and limitation, "super-resolution techniques" (or "sub-diffraction imaging methods") were developed and have shown that resolution beyond the diffraction limit was accessible by exploiting controlled optical deactivation processes of fluorescent probes. Among them, microscopy based on photoswitchable fluorophores, such as photochromic fluorescent labels, has been successfully implemented. Figure 1.3a–e shows an example of the comparison between conventional wide-field microscopy and sub-diffractive imaging of

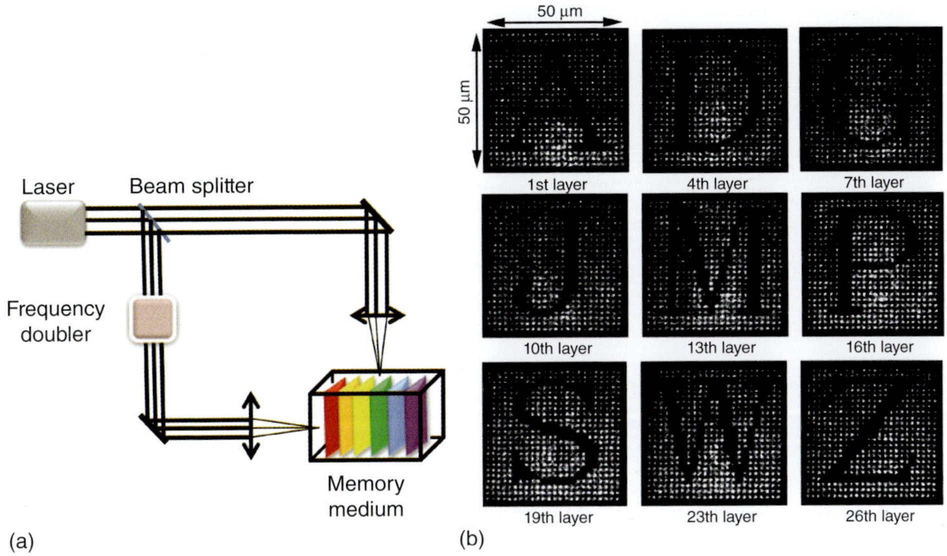

Figure 1.2 Rewritable optical memory medium based on photochromic compounds: (a) general structure of the recording medium [6d] and (b) alphabet letters recorded on the different layers [6b].

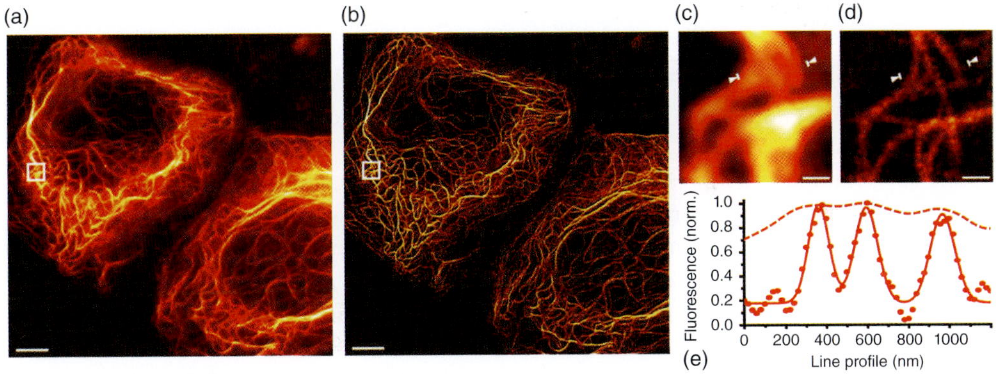

Figure 1.3 Super-resolution image of HeLa cells expressing keratin19 rsCherryRev1.4 by wide-field conventional microscopy (a) and by RESOLFT microscopy (b) (scale bar = 5 μm) and the corresponding magnifications of the highlighted area (c and d, scale bar = 500 nm). Line profiles (e) across the region between the arrows marked in (c) (full line) and (d) (dashed line) [7].

live HeLa cells expressing fluorescent photochromic proteins. No wonder photochromic compounds have entered the bio- and nano-worlds [8].

Light reflection or transmission change, used in the above-mentioned application, is the representative property modified in the photochromic process. As for color, in photochromic systems, traditionally, light is used as a trigger not only to

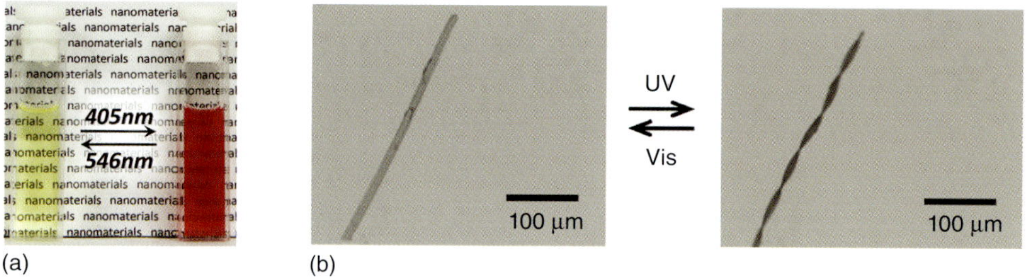

Figure 1.4 (a) Concomitant color and solubility changes of a photochromic solution [13a] and (b) color and shape changes of a photochromic crystal [14c].

induce the change but also to reveal the state of the system at a given moment. Other properties related to light, such as refractive index [9], fluorescence [8c, 10], and even nonlinear optical properties [11], are employed to read out. Indeed, concomitantly to the color, these properties are changed.

Photoswitching other physical or chemical characteristics, such as magnetic, electrical, conductive, or redox properties, is also a matter of interest [12]. Furthermore, one can take advantage of photochromism upon altering or taking the control of features, such as phase, solubility, reactivity, stereochemistry, complexation, or interaction between molecules or ions (Figure 1.4a) [13]. In materials, photochromism can induce shape changes, and opens up a wide field of interest in photo-induced mechanics (Figure 1.4) [14].

1.1.2
Basic Principles

In order to describe photochromism, the most common model introduced is a simple two-way reaction between two molecular species A and B. Although it may sometimes involve other species, the reaction is assumed to be unimolecular (Figure 1.5a).

A and B are separated by a potential barrier (ΔE). If this barrier is low, B is metastable and can revert back spontaneously to A. Previously described photochromic glass lenses operate according to this scheme. Such systems are called *T-type* referring to the thermally induced reaction from B to A. On the contrary, a high barrier features a bistable system. In this case, only photons are able to cause the reaction, and such systems are called *P-type*. In other words, nothing changes in the absence of light.

This last characteristic is important since it makes the difference between photochromic bistable systems and others, such as ferroelectric or (ferro)magnetic systems. In the latter, shuttling between the two states of the bistable system does not follow the same route, displaying the well-known hysteresis, when the polarization or the magnetization is plotted versus the electric or the magnetic field. In photochromic systems, no concept of hysteresis is involved in the rationale of bistability.

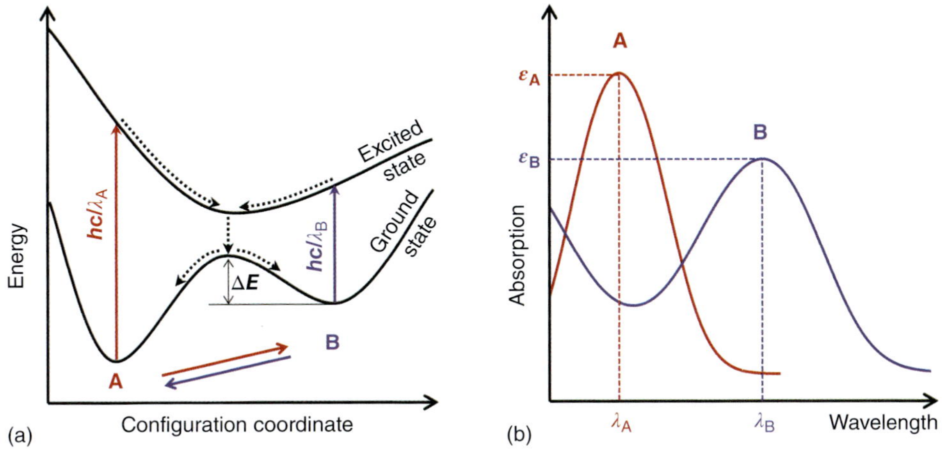

Figure 1.5 Photochromism: a two-way light-induced reaction between two molecules A and B. (a) Potential energy diagram and (b) the related schematic absorption spectra.

In usual photochromic systems, A absorbs in the UV or near-UV, with a characteristic absorption band at wavelength (λ_A). The absorption coefficient of A at this wavelength is ε_A. When a photon at λ_A is absorbed, A is excited from the ground to the excited state. The excited A yields B with a probability of $\phi_{A\to B}$ (see Appendix), known as the *quantum yield*. On the other hand, B reverts back to A, with an analogous pattern, provided that the B is excited at λ_B, where it absorbs. The spectral position of the absorption bands gives an indication of not only the color of light needed to induce the reaction but also the color of the molecule itself (Figure 1.5b). Further quantitative development of this scheme is given in Appendix.

1.1.3 Photochromic Molecules: Some History

The historical reference of photochromism dates back to ancient times and the era of the Alexander the Great (356–323 BC). As King of Macedonia, he got into a vast world conquest. He conquered Asia Minor (now western Turkey) and extended his kingdom to the northwest of India in the east and Egypt to the south. Strategy and carefully coordinated attacks are essential conditions for victory. Thus, Macedonian head warriors were equipped with photochromic bracelets (the compound remains unknown up to now) exhibiting a color change when exposed to sunlight. Such color change was used by all warriors to indicate the right moment to begin the fight [15].

Over 2000 years later in 1867, Fritzsche reported for the first time the following peculiar behavior of tetracene solution: the initial orange color of the solution fades when the sample is irradiated by sunlight but can be recovered as initially when placed in a dark room (Figure 1.6) [16].

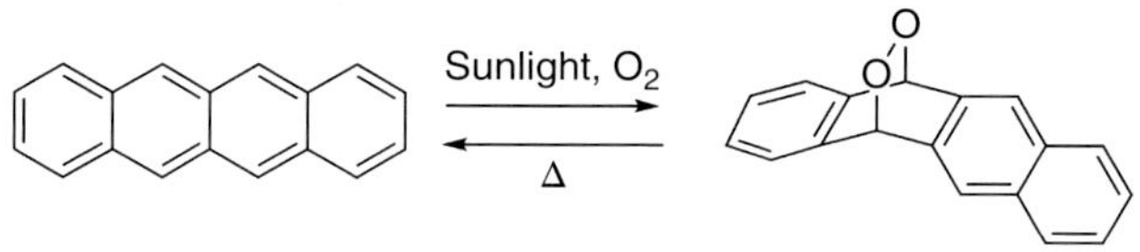

Figure 1.6 Photochromic reaction of tetracene.

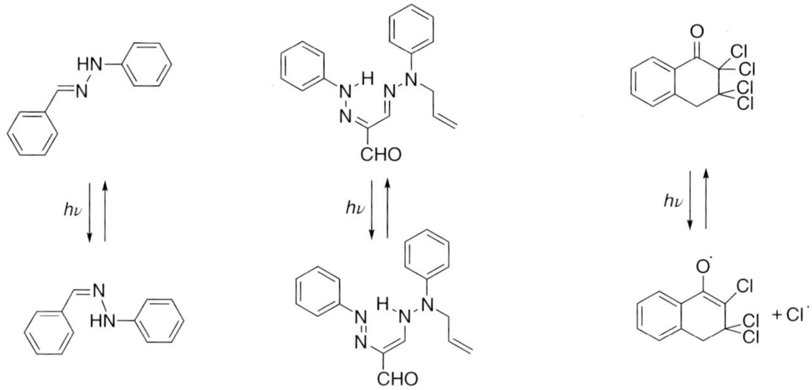

1-Benzylidene-2-phenylhydrazine Mesoaldehyde 1-allyl-1-phenyl-2-phenylosazone Tetrachloro-1,2-ketonaphthalenone

Figure 1.7 Examples of photochromic compounds deriving from phenylhydrazine, phenylosazone, and naphthalenone.

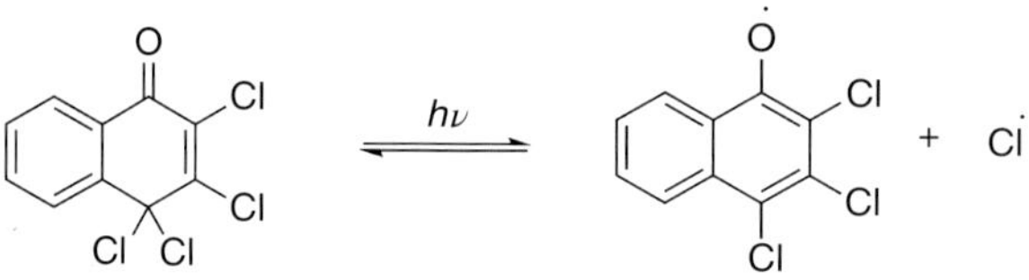

Figure 1.8 Solid-state photochromic reaction of 2,3,4,4-tetrachloronaphthalen-1-(4*H*)-one.

This first observation was followed by some studies [17] on solutions and materials with a similar behavior. Wislicenus noticed the color change of benzalphenylhydrazines (Figure 1.7) [17d]. Later, Biltz confirmed these observations and demonstrated the same behavior for some osazones (Figure 1.7) [18]. Finally, in 1899 Markwald, apart from his work on 1-benzylidene-2-phenylhydrazine (Figure 1.7) and tetrachloro-1,2-ketonaphthalenone (Figure 1.7), discovered the first solid-state photochromic organic compound [19]: the 2,3,4,4-tetrachloronaphthalen-1-(4*H*)-one (Figure 1.8). By that time, he was the first person to recognize this phenomenon as a new reversible photoreaction and gave the name (in German) of "Phototropie."

Other main families of photochromic molecules dating back to this period are fulgides [20], salicylideneanilines (also called *anils*) [21], stilbenes [22], and nitrobenzylpyridines [23].

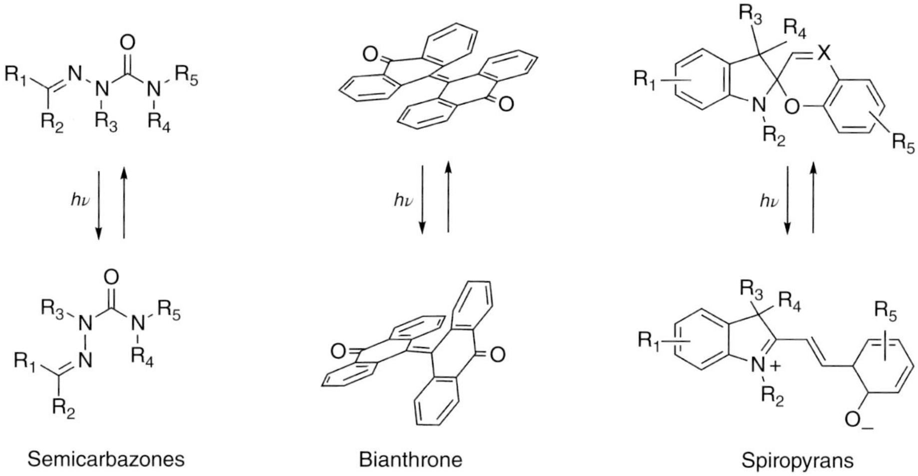

Figure 1.9 Photochromic reactions of semicarbazones, bianthrone, and spiropyrans.

Until the 1920s, much of the work was dedicated to the study of the phenomenon under a practical and descriptive approach than under a deeper scientific approach, that is, on the understanding of mechanisms. Therefore, all the efforts were focused on the synthesis of new molecules and on the optimization of irradiation conditions and fatigue resistant properties [24]. In the 1930s, although attraction for photochromism was low, nevertheless some major advances took place during this period. Indeed, Harris and Gheorghiu were pioneers in mechanistic studies of this phenomenon, respectively, on malachite green [25] and semicarbazones (Figure 1.9) [26].

The 1950s and 1960s probably represent a significant period for photochromic compounds, with the advent of technologies and methods enabling their investigations, especially spectroscopy, contrasting with the period described previously. Many new molecules, both organic and inorganic, were then synthesized and further studies on the mechanism were conducted during this period [15b, 24, 24g, 27]. Among all studies, the work of Hirschberg with the synthesis of the first bianthrone [28] and spiropyrans [29] (Figure 1.9) enabled major advances in the field of photochromism. Also, it was the period when it became usual to call this phenomenon "photochromism."

Although there is a considerable amount of research going on azobenzene derivatives and other photochromic systems such as azine [30] and thioindigoides [31] (Figure 1.10), bottlenecks such as the lack of photoresistance of organic photochromic molecules, leading to degradation, prevented a fast development of applications in the 1960s and the early 1970s. However, during the 1980s, spirooxazines [32], particularly spironaphthoxazines, were developed for their high fatigue resistance, along with chromenes. This marked a significant turning point for photochromism in their use in variable transmission glasses. In such applications, T-type systems are required. In the meantime, compounds such as

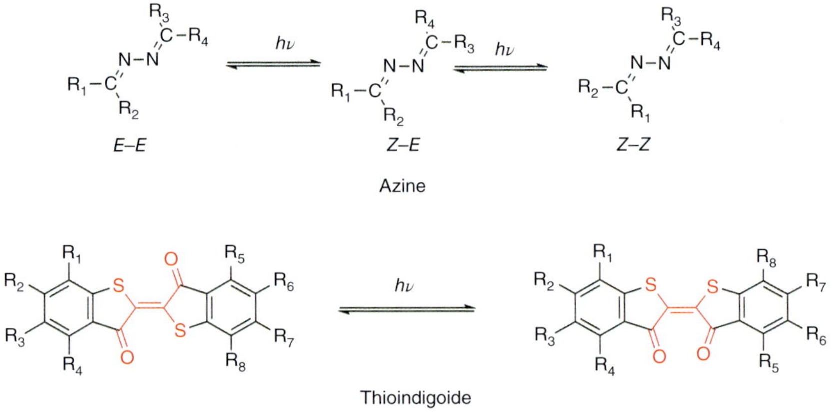

Figure 1.10 General formula and reaction schemes for azines and thioindigoides.

azobenzene derivatives, known for long as dyes and reported to be photochromic half a century earlier [33], were being intensively investigated for their photoswitching properties [34]. Other families of photochromic compounds, such as dihydropyrenes [35], anthraquinones bearing aryloxy groups [36], viologens, based on a photoinduced electron transfer [37], or flavylium [38] and oxazolidines [39], exhibiting both photochromism and acidochromism can be mentioned.

Regarding families of P-type molecules, applications for data storage [6c, 40] and molecular switches emerged in the 1990s. Although molecules such as fulgides have a century-long history as already mentioned, this period corresponds to the discovery of the diarylethene family. This domain certainly contributes to a tremendous increase in the number of publications since the 1990s. More details are given in the following section, which focuses on the most widely investigated photochromic systems.

Photochromism can be considered as being a fast growing domain of research, as it can be substantiated by a simple survey on the evolution of the number of publications in this subject (Figure 1.11). Since the 1990s, a fivefold increase in this number was observed. It is noteworthy to mention that a large number of special issues and review articles appeared during recent years [41] in addition to the references already cited.

1.2 Organic Photochromic Molecules: Main Families

The molecules presented in the previous section made the history of organic photochromism of the twentieth century. Some families of compounds spread, and others almost disappeared from the scientific scene. From the year 2000 to the present, most studies on organic photochromism deal with a group of less than 10 families of compounds. More than 2000 publications concern the diarylethene family. An approaching number of publications is reached when spiropyran,

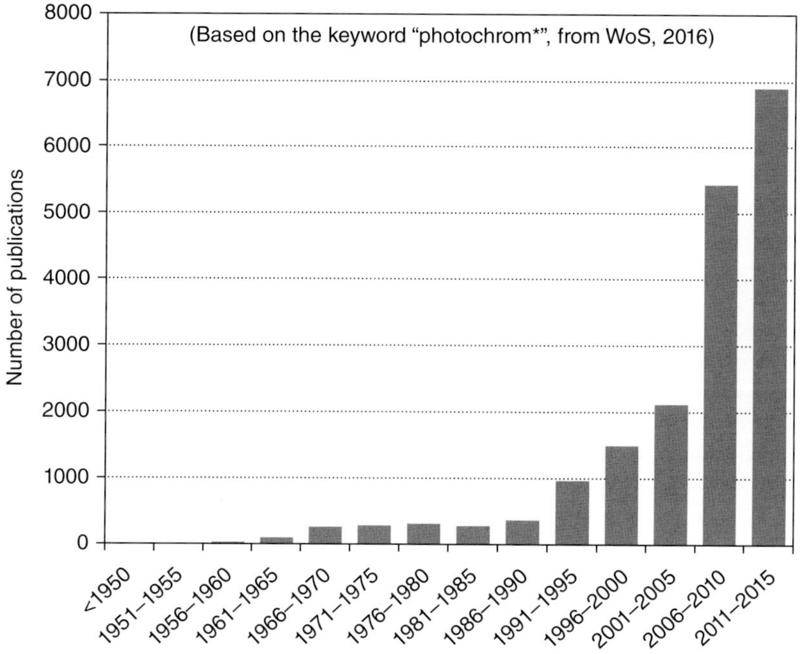

Figure 1.11 Number of publications on photochromism. (*Source*: ISI Web of Science.)

spiroxazine and chromene are added up. Azobenzene derivatives are also widely studied, and these totalize more than 1000 publications during the same period. A few other families of organic photochromic molecules exceed 100 publications, such as salicylideneanilines (anils), fulgides, and hexaarylbiimidazoles (HABIs).

This section mainly focuses on the families that are categorized by the type of chemical reaction involved during the photochromic process.

1.2.1 Proton Transfer

Derivatives of salicylideneaniline are probably the most studied photochromic molecules involving a proton transfer [42]. Also called *anils*, they are mainly substituted salicylideneanilines and variants, where aniline is replaced by groups, such as aminopyridine or aminothiophene. In fact, anils are Schiff bases. Reports on the photochromism of anils date back to the beginning of the twentieth century [21]. However, thorough studies leading to significant characterization started only during the 1960s [43]. Proton transfer reactions are present in different mechanisms involving natural systems [44]. Photochromism of anils can be performed in several media, from solution to inclusions or encapsulated media [6, 411, 45], and even in some cases in the solid state, with medium-dependent behavior.

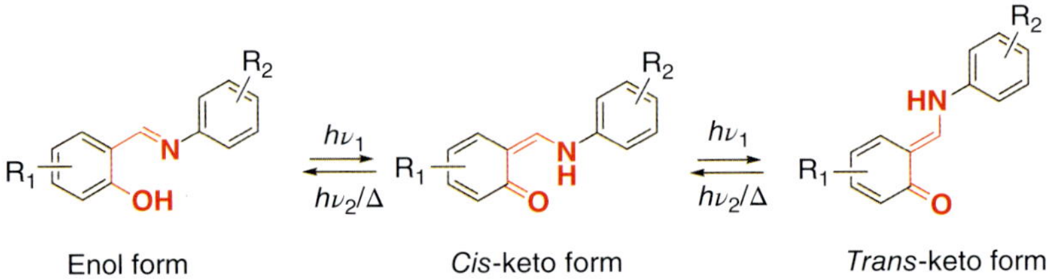

Figure 1.12 Photoreaction (phototautomerism) of anils: the enol to keto reaction.

Under UV irradiation, an intramolecular proton transfer is induced and allows the passage from the enol (in fact a phenol) to the *cis*-keto (ketone) form in the excited state called excited-state intramolecular proton transfer (ESIPT). This transfer is quickly followed by a *cis–trans* isomerization at an excited state to reach a *trans*-keto form (Figure 1.12). Considering the result of this reaction, it may also be called *phototautomerism*. All the process occurs within a few picoseconds in solution, and a few hundred picoseconds even in solid state [46].

Enol form absorbs only in the UV (or near-UV) and is pale yellow in most cases, whereas the keto isomer is usually red. Conjugation is less extended in the latter, so such feature may sound unexpected; however, the bathochromic shift is due to an n–π^* transition evolving in the keto form [47]. Upon UV excitation, a luminescence can be observed (Figure 1.13). Investigations during the last few decades show that this is concomitant to the ESIPT: the deexcitation of the keto state is then responsible for it, leading to a Stokes shift of more than 150 nm [46d, 48].

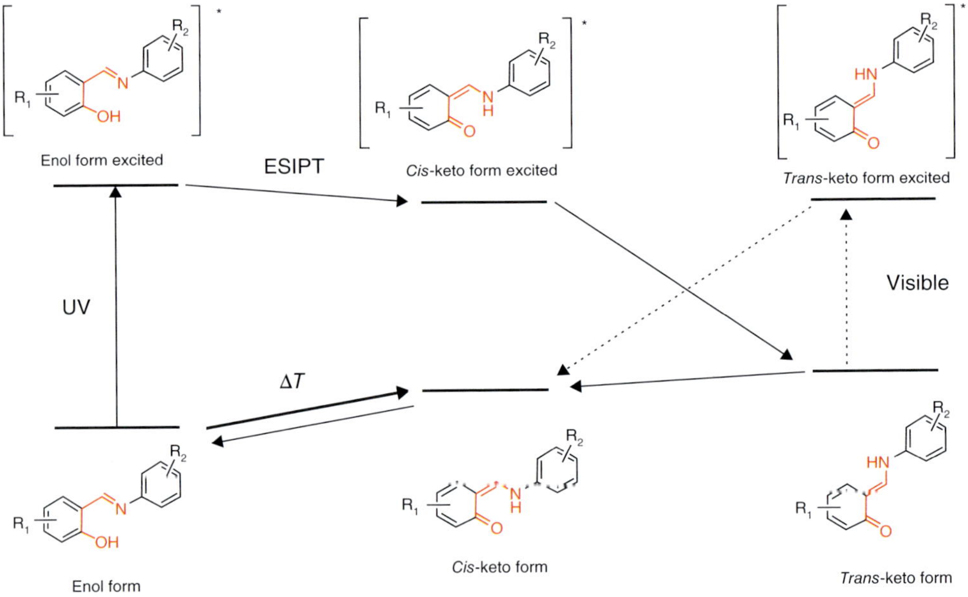

Figure 1.13 Photochromism of anils: schematic diagram illustrating the UV and visible light-induced reactions, and the excited-state intramolecular proton transfer (ESIPT).

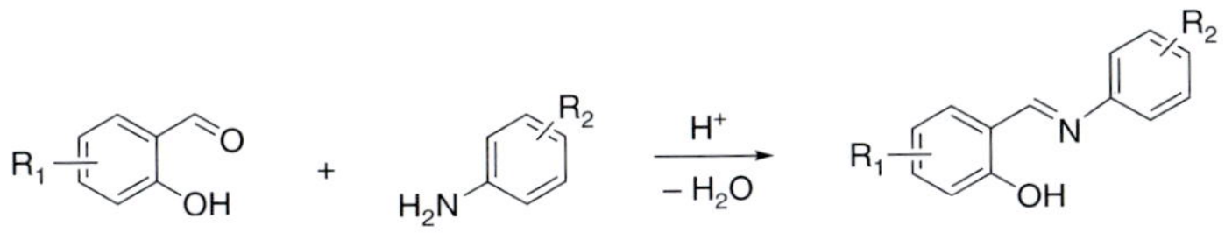

Figure 1.14 Synthesis of anils: basic method by condensation of salicylaldehyde and aniline derivatives.

Anils are of T-type. In a solution, the keto photoproduct reverts back to the enol very quickly, typically within a few milliseconds [47d, 49], whereas in the solid state, it may vary from a few seconds to several months [42b, 50]. A consequence of this is that the photochromism of anils can barely be observed in solution (see Section 1.3.4 for detailed explanations), but only in the solid state.

The synthesis of anils is rather straightforward (Figure 1.14). Based on the condensation of salicylaldehyde and aniline derivatives in an acidic medium, hundreds of molecules with various substituents providing steric and/or electronic features were synthesized.

The second most studied family of compounds with a proton transfer is the dinitrobenzylpyridine (DNBP). Although its color change upon UV irradiation was evidenced in the 1920s [23], unraveling the nature of the species and the reactions involved was possible only several decades later [51]. The colorless CH form (Figure 1.15) is the most stable, and UV excitation yields the OH and NH forms.

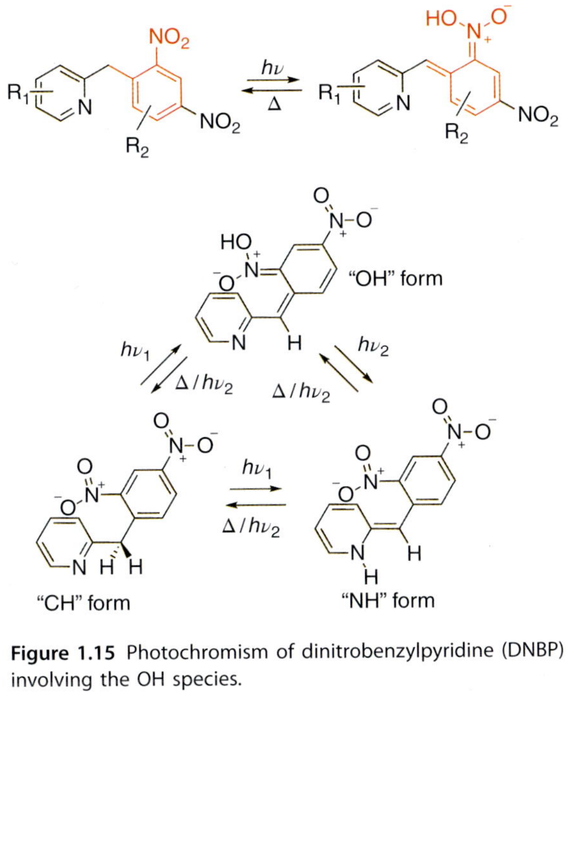

Figure 1.15 Photochromism of dinitrobenzylpyridine (DNBP) between the CH and NH forms, involving the OH species.

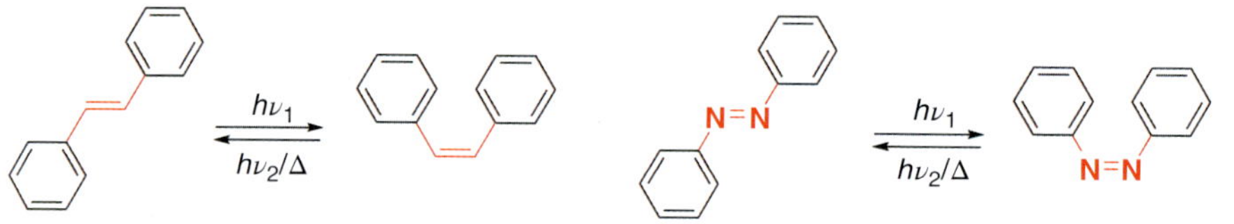

Figure 1.16 *Trans–cis* isomerization of stilbene and azobenzene.

The OH form is the least stable and spontaneously converts to its isomers. DNBP can be obtained by nitration in HNO_3/H_2SO_4 mixture from benzylpyridine.

1.2.2
Trans–Cis Photoisomerization

Most chemists have heard about the isomerization around carbon–carbon double bonds at the early stage of their studies when they learn about *trans–cis* (or *E–Z*) isomerism. Textbook cases are stilbene, as well as azobenzene (Figure 1.16) and their derivatives. These reactions can be induced by light, which circumvents the relatively high-energy barrier in the ground state. Moreover, in many photochromic systems, *trans–cis* reaction is one step of the process, as it is the case for anils and spiropyrans (see Sections 1.2.1 and 1.2.4), and appears as an alternate photoinduced process in fulgides or diarylethenes (see Section 1.2.4).

The knowledge of the development of azobenzene derivatives that has been used as dyes for more than a century is essential; nowadays, they represent the main family of photochromic molecules based on *trans–cis* reaction. It would not be possible to cite the over thousand publications, but there are some reference book chapters and book reviews [34, 52].

For the sake of simplicity, in the following, the derivatives of the azobenzene called *azobenzene(s)* are discussed. The color of azobenzene is generally yellow and its substitution yields a bathochromic shift, providing an orange or red color. Both isomers have two characteristic bands, $\pi-\pi^*$ and $n-\pi^*$, the latter one lying in a lower energy region and being less intense, as it is symmetry forbidden.

Changes in the absorption spectra between the two forms are generally not very pronounced due to a small difference in electron delocalization between the two isomers, and the color change is invisible by the naked eye. In contrast, the photochromic reaction induces very significant change of the free volume of the molecule [53]. Therefore, these molecules are extensively used to induce mass rotation or migration in materials, leading to dichroic and birefringent media, surface relief patterning (Figure 1.17a–c) [11, 54], or mechanical effects [14b,d]. In addition, azobenzenes are introduced for their switching ability when properties not related to color, for example, magnetic, are sought [55].

There are several different methods to synthesize azobenzenes and are thoroughly described in a recent review article [52b]. The most widely used methods

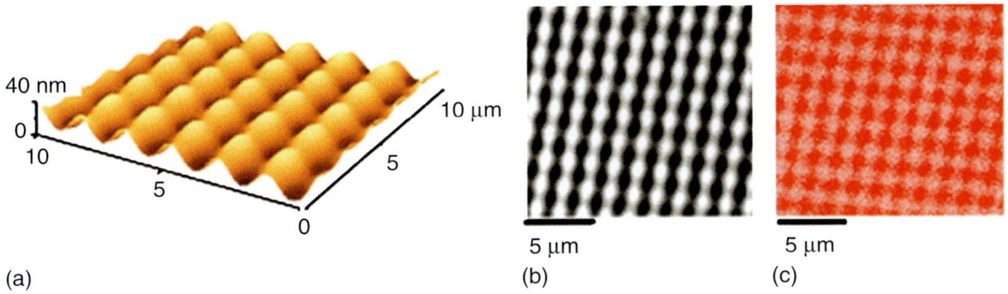

Figure 1.17 Crossed surface relief grating obtained by 2D photopatterning on a thin film of azobenzene derivative: (a) atomic force microscope (AFM) topographic image, (b, c) one- and two-photon transmission images with dark areas corresponding to hills [54b].

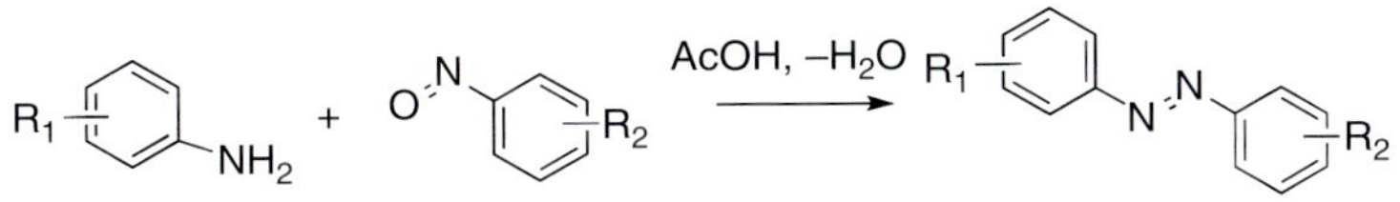

Figure 1.18 Synthesis scheme of the Mills reaction used to obtain azobenzenes.

are the diazo coupling and the Mills reaction between aromatic nitroso and aromatic amine compounds (Figure 1.18).

1.2.3
Homolytic Cleavage

In this class of compounds, HABI discovered by Hayashi and Maeda in the 1960s and its derivatives play an essential role [56]. The homolytic cleavage of the C–N bond between the two imidazole rings involved in this type of compound can be induced by heating, light irradiation, or pressure. It leads to the formation of two radicals (TPIR triphenylimidazolyl radical) as depicted in Figure 1.19. Recombination to revert back to the starting imidazole dimer (also called TPID triphenylimidazolyl dimer) can be carried out thermally and driven by radical diffusion.

TPIR has a large absorption band in the visible light, whereas TPID absorbs only in the UV light and is thus colorless. Compounds of this family exhibit

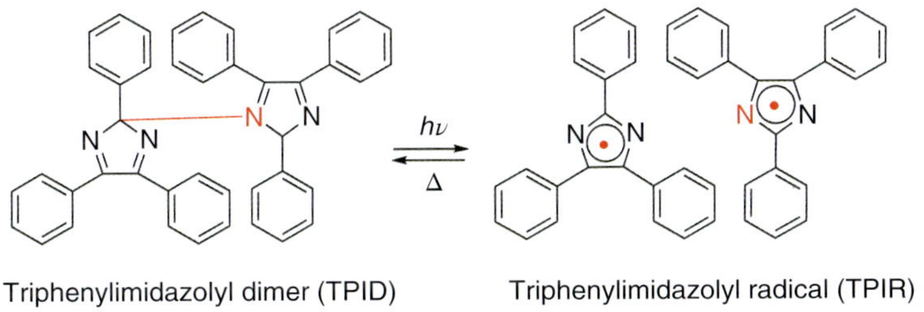

Figure 1.19 Photochromism of hexaarylbiimidazole (HABI) between the triphenylimidazolyl dimer (TPID) and the triphenylimidazolyl radical (TPIR) pair.

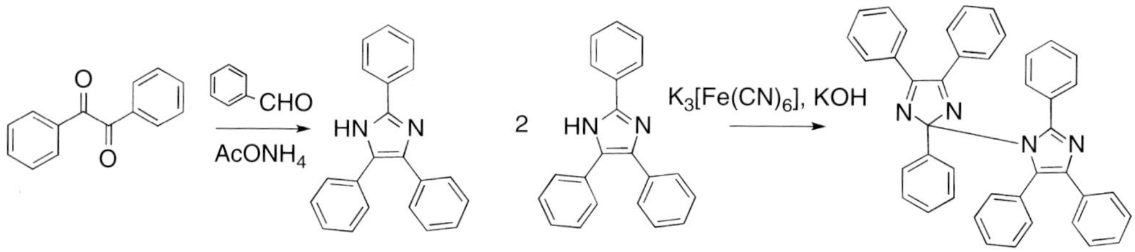

Figure 1.20 Basic synthesis scheme of HABI.

a T-type photochromism. Although the cleavage upon UV irradiation takes place within less than 100 fs, recombination can take up to few minutes at room temperature [57].

The preparation of HABI is rather straightforward (Figure 1.20), at least for the simplest and basic methods. The precursor, triarylimidazole (lophine), known since the end of the nineteenth century, can be obtained by the reaction between benzoin and an aldehyde in the presence of ammonium acetate.

Along with photochromism, HABI also exhibits piezo- or thermochromism. This has to be related to the controversy about the way the two imidazole moieties are connected: in fact, upon the synthesis of the TPID form, several isomers are formed. Studies carried out during several decades lead to the conclusion that C–N bonded ones lead to photochromism, whereas C–C bonded ones lead to piezo- or thermochromism [58]. During the 1970s, HABI was used as industrial photoinitiator for imaging and photoresists. In the process, the TPIR abstracts hydrogen atoms from crystal violet precursor enabling imaging [59].

1.2.4 Cyclization Reaction

In the mostly encountered photochromic systems, the reaction is a cyclization. In most examples, the reaction involves six π electrons delocalized over six different atoms. This is the case for spiropyran, fulgides, and diarylethenes and some related families such as spirooxazine, chromene, and fulgimides. Overall, they represent a largely predominant part of the research carried out on organic photochromism during the last decades.

1.2.4.1 Spiropyrans, Spirooxazines, and Chromenes

As illustrated in Figure 1.21, the most common type of spiropyran is based on indoline and benzopyran. This basic structure can bear substituents (alkyl, alkoxy, nitro, halogeno, etc.), and it can also be extended to larger structures (e.g., naphthopyran instead of benzopyran) or to modified structures (e.g., thioindoline in place of indoline) [29, 60].

Spiropyrans have been extensively studied along the past half-century. For such compounds, UV light irradiation of the colorless closed form (also called the *spiro* or the *N form*, for normal) leads to the carbon–oxygen bond cleavage (open ring

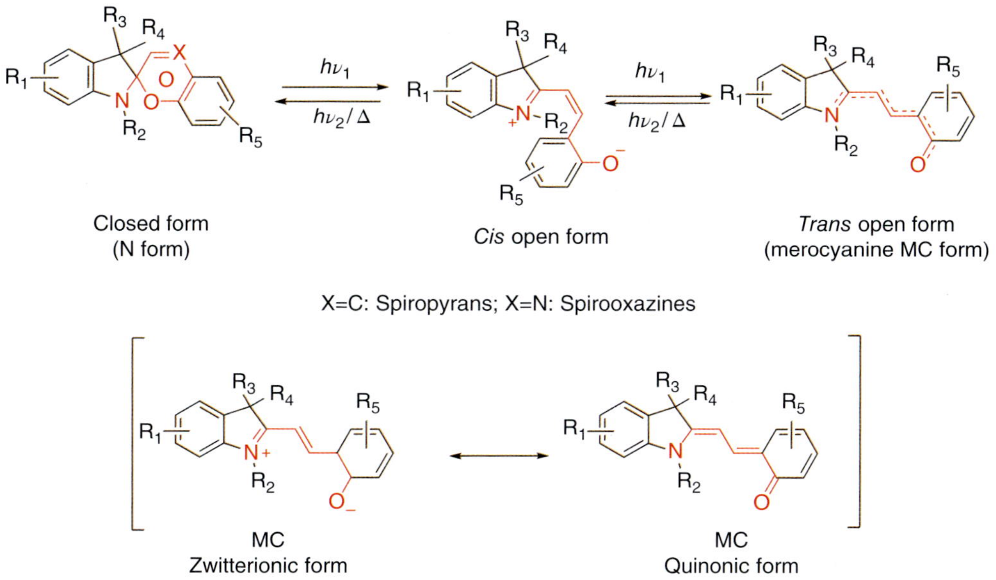

Figure 1.21 Photoisomerization of spiropyrans between the closed and open merocyanine (MC) forms. Resonant zwitterionic and quinonic forms of MC.

reaction) of the pyran ring, followed by a *cis–trans* isomerization to finally reach the colored merocyanine (MC) form. The latter is highly conjugated and has two mesomeric (resonant) forms, zwitterionic and quinonic. This feature can be connected to the presence of an intense absorption band in the visible, and the color of the MC form. On the contrary, in the closed form, conjugation is broken at the spiro carbon atom and the molecule is colorless and absorbs only in the UV region.

Spiropyran is a typical example of T-type compounds family as no P-type molecule has been reported yet. In a majority of cases, the closed form is the more stable one, the MC form being metastable, although a noticeable amount of the MC may exist at equilibrium for some compounds [61] and also in polar media [62]. Due to its electronic structure, the MC form is highly solvatochromic and its stability is strongly solvent dependent: for example, in one representative compound, the so-called 6-nitro-BIPS (1′,3′-dihydro-1′,3′,3′-trimethyl-6-nitrospiro(2*H*-1-benzopyran-2,2′-2*H*-indole)), the rate of the thermal reaction is 300 times slower in ethanol than in benzene and is related to a higher activation barrier of the ground state of the former than that of the latter [63].

In spiropyrans, nonbonding orbitals bearing lone pairs of the nitrogen and oxygen atoms connected to the spiro carbon, respectively, interact with the antibonding orbitals of the C–O (σ^*(C–O)) and the C–N (σ^*(C–N)) bonds. As illustrated in Figure 1.22, the former is stronger than the latter and leads to a weakening of the C–O bond even in the ground state. This C–O bond is then cleaved upon UV irradiation.

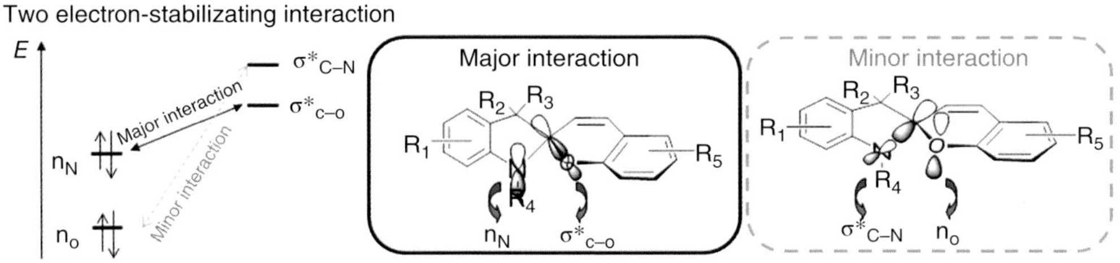

Figure 1.22 Orbital interactions depicting the interactions between the nonbonding and the antibonding orbitals in spiropyrans around the spiro carbon, leading to the weakening of the C–O bond.

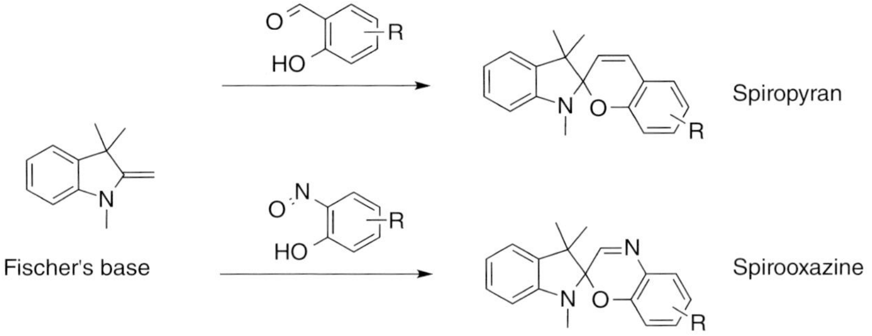

Figure 1.23 Common synthesis scheme for indoline-based spiropyrans and spirooxazines, involving the Fischer's base.

Spirooxazines and chromenes are other families of molecules featuring photochromism under a similar reaction pattern, and also they are of T-type. In spirooxazines, a CH group has been replaced by a nitrogen atom (Figure 1.23) [32b,c, 64]. As shown in this synthesis scheme, the simplest method to obtain the most common indoline-based spiropyran or spirooxazine is a reaction between the Fischer's base (1,3,3-trimethyl-2-methylene-indoline) and the corresponding salicylaldehyde or the nitrosophenol derivative.

When applications are targeted, molecules are introduced in a polymer matrix as dopants, or covalently attached to a polymer chain, the more so because spiropyrans are photochromic in the pure solid state only in a few cases [65]. Due to their high fatigue resistance, spirooxazine with an extended group (naphthopyrans rather than benzopyrans) finds applications in the field of eyewears, in fact more than spiropyrans. Another family of compounds named chromene [64, 66] is also used industrially for the same purpose, and its photochromism is based on a similar reaction (Figure 1.24).

However less common, it is worth mentioning that there are other examples where color is evolved upon light-induced ring opening, with other types of bond cleavage: spirodihydroindolizines where a C–C bond is broken [67] and spiroperimidines [68] based on a C–N bond cleavage.

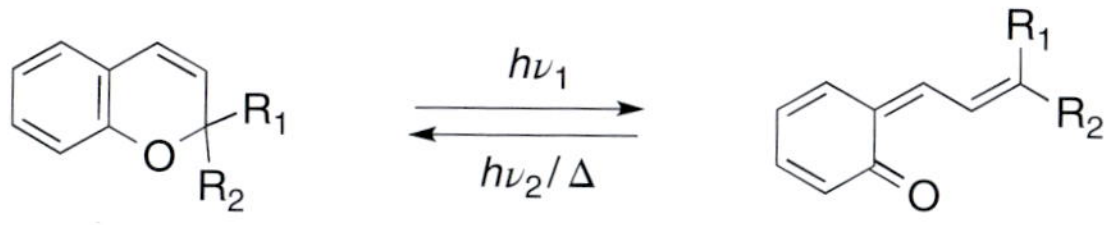

Figure 1.24 Photochromism of chromenes.

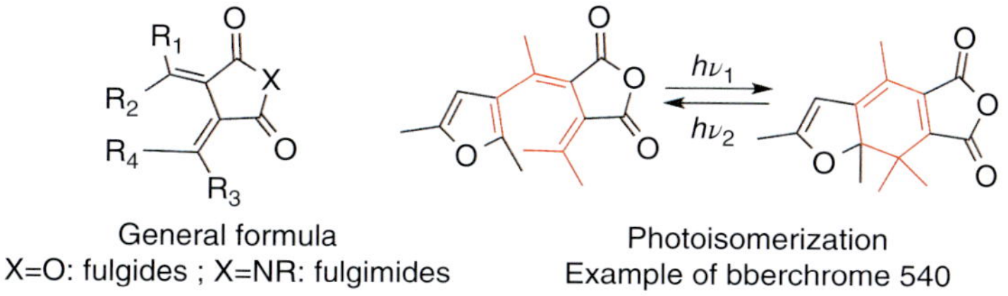

Figure 1.25 General structure of fulgides and fulgimides. The photochromic reaction is written for an example of furylfulgide, Aberchrome 540.

1.2.4.2 **Fulgides and Fulgimides**

Similar to spiropyrans and the related compounds described previously, photochromism in fulgides and fulgimides (Figure 1.25) is based on a cyclization reaction. More specifically, a conversion between π and σ bonds leads to an electrocyclization, and the main difference with the case of spiropyrans is that the open form is colorless and the cyclized form is the colored species.

Fulgides have been known since the early twentieth century and their name comes from the Latin "fulgere" meaning "shine" [69] because the first compounds synthesized by Stobbe in 1905 exhibited a bright character when crystallized [20, 70]. Nevertheless, to have interesting photochromic features, fulgides must have at least one exo-methylene carbon substituted with an aryl group. By this method, a 1,3,5-hexatriene-type structure is constructed and may lead to an electrocyclic reaction [71]. The first fulgides had a significant thermal back-reaction and were T-type photochromic compounds. But during the 1980s, the introduction of heterocycles as aryl groups (e.g., furylfulgides or indolinofulgides) enabled the thermal stabilization of the closed form and the P-type photochromic compound could be obtained [71, 72]. The open form may undergo *trans–cis* reactions, especially the one swapping R_1 and R_2. Although studies on fulgides report on such observations [73], this reaction is less appealing because it induces only a minor change of color and other properties compared to the ring-closure reaction.

In addition to their use in rewritable optical storage devices of information [74], furylfulgides are also well known as *chemical actinometers* [2, 75], particularly the Aberchrome 450 [76] (Figure 1.25). Fulgimides [69c–e, 77] have also been widely studied because of the possibility to attach a broad variety of substituents, without deeply affecting the photochromic properties [77]. Thus, fulgimides have been substituted with not only polymer chains [78] but also fluorophores [79] or proteins [80].

One classical synthesis method of fulgides is mainly based on a double Stobbe condensation, which involves a dialkylsuccinate reacting sequentially in aldol-like

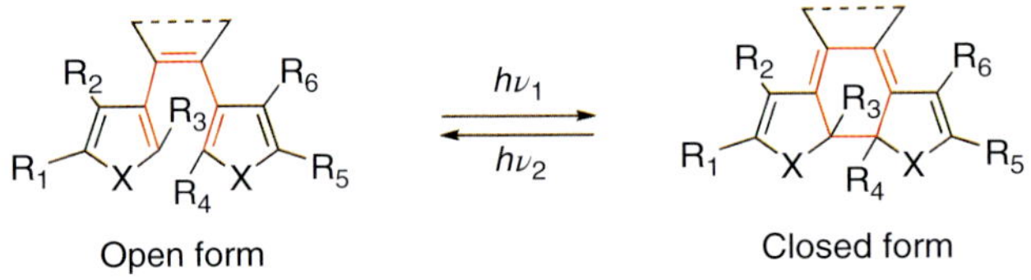

Figure 1.26 Typical reaction scheme of the photochromism of diarylethene when the aryl group is a heterocycle (e.g., X = O, N, S).

reactions with a ketone bearing the R_3 and R_4 moieties, and another one bearing the R_1 and R_2 moieties followed by ester hydrolysis and a dehydration, which enables the formation of the anhydride ring.

1.2.4.3 **Diarylethenes**

The first report on the photochromism of diarylethenes dates back to 1988 by Irie [81]. They represent the most studied family of photochromic compounds due to their very good fatigue resistance and their highly bistable character [82], and hence they were considered as very serious candidates for optoelectronic applications (memories, switches, etc.). Molecular engineering, based on theoretical and experimental basis, to improve the performances of diarylethenes has occupied (and still occupies) an important place in the scientific topics since that time and will be described later.

As for fulgides, diarylethenes undergo a ring-closure reaction between a colorless open form and a colored closed form (Figure 1.26). Although "aryl" may include other aromatic structures, most of the diarylethenes are based on heterocycles. For reasons developed in 1.3.3, the dithienylethene subgroup is nowadays the most widespread group.

All diarylethenes display a 1,3,5-hexatriene structure in the colorless open form. According to Woodward–Hoffman rules [82] based on the symmetry of π orbitals, electrocyclization reactions on such structures follow a conrotatory mechanism under photochemical control. Under UV irradiation, a 1,3-cyclohexadiene (colored closed form) is evolved. The reverse reaction (cycloreversion) is induced by irradiation in the visible range. Extension of the delocalization of the π system explains the color of the closed form.

A previously introduced compound in the part dealing with *trans–cis* photoisomerization, stilbene, is in fact a type of diarylethene, as the two phenyl groups are linked by an ethene bridge. In addition to the *trans–cis* isomerization, its *cis* isomer is known to undergo a light-induced cyclization reaction, leading to the formation of dihydrophenanthrene [83] (Figure 1.27). However, it oxidizes irreversibly to yield phenanthrene by the elimination of hydrogen. Introducing substituents on the reacting carbon atoms can avoid this undesirable reaction, thus hydrogen atoms are replaced in diarylethenes (R_3 and R_4 positions, Figure 1.26) by a large variety of groups, the most common ones being alkyl and alkoxy.

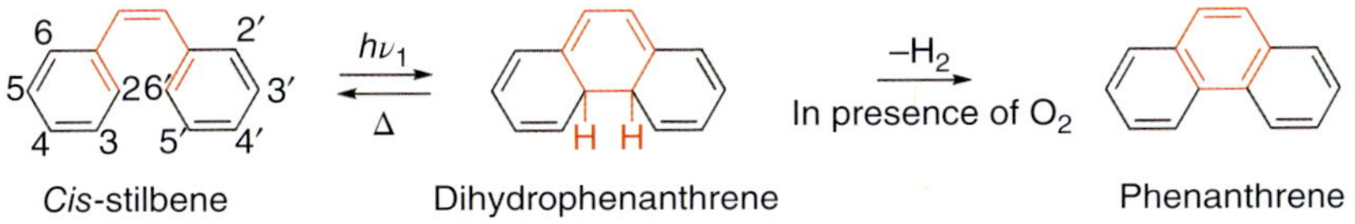

Figure 1.27 Photocyclization reaction of *cis*-stilbene and the subsequent oxidation to phenanthrene.

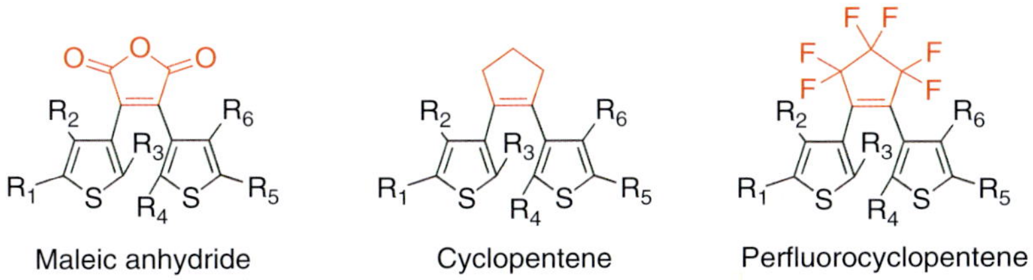

Figure 1.28 Typical examples of rings including the ethene bridge in diarylethenes (more specifically, dithienylethenes).

The analogy with *cis*-stilbene leads us to care about the *trans–cis* reaction, which competes with the cyclization reaction. Since the former induces less change than the latter for most properties, it is usually considered to be less interesting and is even avoided. In order to do so, in many molecules, the ethene bridge is included in a cyclic group (Figure 1.28).

At the time diarylethenes emerged in the world of photochromism, spiropyran and the related compounds, as previously described, were occupying the front scene for T-type compounds. Diarylethenes quickly showed some potential as P-type systems, and much effort was invested in the molecular engineering to increase the stability of both forms of the molecule (see Section 1.3.3).

Various synthetic methods have been developed for the preparation of diarylethenes [64]. Diarylethenes with hexafluorocyclopentene as ethene bridge can be, in general, synthesized by the reaction of appropriate heteroaryl organolithium reagent with octafluorocyclopentene (Figure 1.29a). However, Pd-catalyzed cross-coupling, Suzuki cross-coupling, for example, is another quite general synthetic approach to diarylethenes (Figure 1.29b) [84].

It would be unrealistic to list all the diarylethene derivatives that have been developed based on dithienylethenes, even if one focuses only on compounds with dithienylperfluorocyclopentene bridges. A lot of modifications in diarylethene structure were done in order to modify their photochromic properties. Beyond dithienylethenes, variants of diarylethenes with thiazolyl cycles (see Section 1.3.5), terarylenes [85], or arylbutadienes [86] can be mentioned as other photochromic systems.

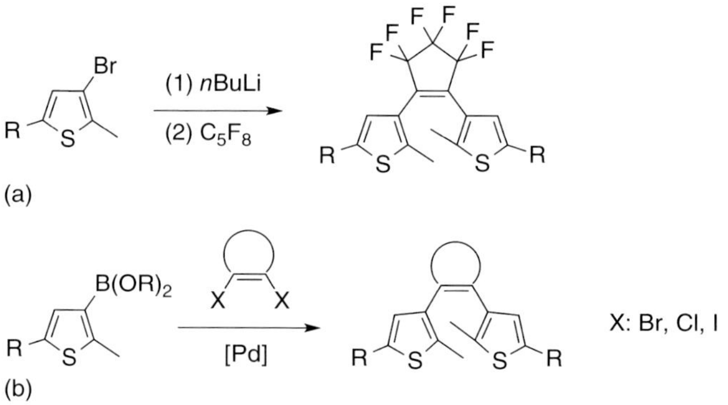

Figure 1.29 Synthesis of diarylethenes: typical reaction binding the ethane bridge with the aryl cycles (here thienyl).

1.3
Molecular Design to Improve the Performance

1.3.1
Figures of Merit

In order to fulfill the requirements as photoswitches, several characteristics are necessary and a lot of research has been carried out to improve them.

Some of them are rather obvious and "universal":

1) The number of cycles that a photochromic system is capable to complete should be as high as possible. In other words, fatigue resistance is an important feature.
2) The switching should not be greedy in energy, in order to have an efficient conversion in terms of extent and speed. To be under such conditions, the quantum yield of the photoreaction should be high (see Appendix).

There are also other important characteristics; however, the expectations may vary according to the targeted application. For instance:

3) The absence of a thermal reaction leading to a complete bistability (the composition remains the same as in the absence of light) is essential in P-type systems, whereas for a T-type system, the reversion from B to A state is desired to occur in a glance.
4) Usually, a large absorption spectrum shift between A and B states is sought: not only this means a large color change but it is also linked to a large change of other properties. However, in some applications, performing as much A–B cycles as possible is necessary, and a large overlap of their spectra can be advantageous.

Finally, in terms of most applications, photochromic systems should be capable to operate in the solid state, and structural properties of the molecules have an important influence on their ability to undergo reactions in the solid state.

In the following, for a selection of figures of merits, such as fatigue resistance, bistability for P-type systems, or speediness of the back-reaction for T-type systems, molecular design related to their improvement is presented.

1.3.2
Fatigue Resistance: Increasing the Number of Operating Cycles

To work as a switch, photochromic systems should be able to undergo a large number of cycles. For example, it would be inconceivable to change a glass panel on a building or a car windshield every year. In applications where a 10-year product lifetime would be reasonable, one should consider not only that the material should withstand the 3650 day/night cycles but also that the A – B photoreaction is permanently occurring during daytime, even at photostationary state (PSS) where the global composition of the medium remains constant (see Appendix).

Unfortunately, as in all reactive media, fatigue is also present in photochromic systems. A few decades ago, investigation on side products and degradation mechanism was intensively conducted on spiropyrans [87]. Electronic effects of the substituents showed that the introduction of electron-donating groups (EDGs) at almost all positions improved the fatigue resistance, whereas the introduction of electron-withdrawing groups (EWGs) had the opposite effect [88]. In addition, comparative studies between spiropyrans and analogous spirooxazines were carried out and showed that the latter had a better fatigue resistance [89]. A difference of deactivation processes with molecular oxygen between the two compound families was reported: in spirooxazines, the mechanism is more photophysical than photochemical.

Diarylethenes are reputed for their good fatigue resistance, owing to many molecular engineering works, although not totally exempt from irreversible side reactions under some conditions [84c, 90]. In Section 1.2.4, issues related to *trans*–*cis* photoisomerization and oxidation of a tricyclic compound were discussed in the case of stilbene. Substituting hydrogen wherever this atom can be involved in a side reaction, is an answer to avoid it. The same design is applied in fulgides [91].

Back to diarylethenes, the extent of side reactions depends on both the molecular structure of the switch and experimental conditions, leading to diarylethenes with cycling ability ranging from a few tens to hundreds of thousands cycles [90a]. Despite the abundance of literature data on diarylethenes, it is still impossible to predict the photoresistance of a diarylethene, as there are several different methods of degradation [92]. Nevertheless, some empirical correlations are observed between the fatigue behavior of a diarylethene and its molecular structure. When the reacting carbon atoms are not substituted, a possible side reaction is an oxidation reaction concomitant to the cyclization, analogously to the one described for stilbene (see Section 1.2.4, Figure 1.27).

First, all else being equal, hexafluorocyclopentene as ethenic bridge confers, in general, higher fatigue resistance to diarylethenes than cyclopentene does [93].

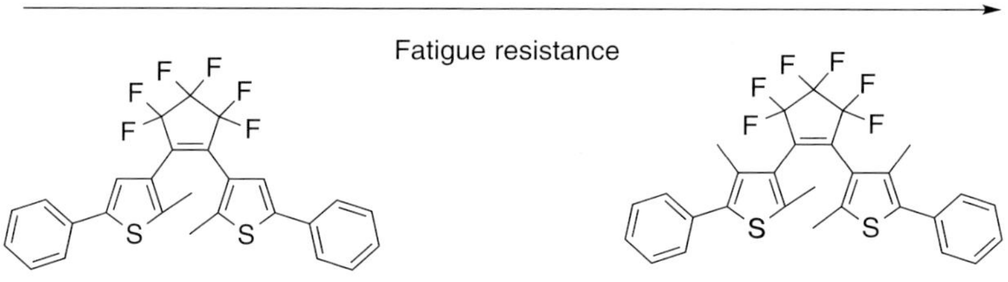

Figure 1.30 Effect of the substitution by methyl groups on thiophene rings on fatigue resistance.

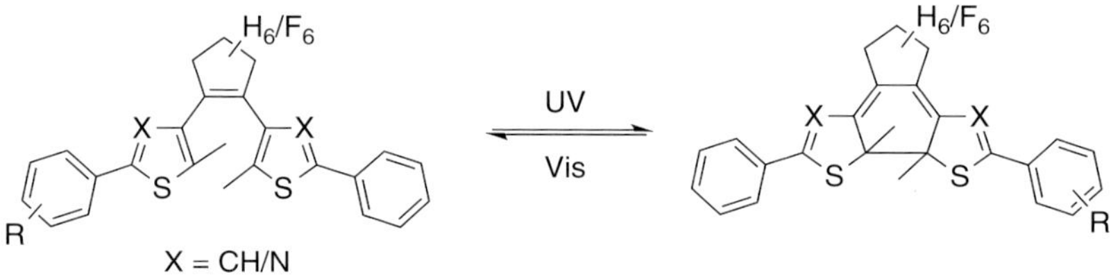

Figure 1.31 Substituted dithienylethenes (X = CH) and dithiazolylethenes (X = N).

Note that maleic anhydride or maleimide as ethenic bridge offers comparable photoresistance to hexafluorocyclopentene but has lower chemical stability [90a].

Second, diarylethenes with benzo[*b*]thiophenes as heteroaryl are, in general, more fatigue resistant than those with thiophenes. Within the thiophene-based diarylethenes, the fatigue resistance is very sensitive to small structure changes. For instance, a simple methyl substitution of thiophene ring was found to bring about a spectacular improvement in its photoresistance (Figure 1.30) [90a]. Note also that diarylethenes with thiazole or oxazole, which are more electron deficient than thiophene and benzothiophene, are also known for their high fatigue resistance [90a, 94].

Finally, the fatigue resistance of a diarylethene is sensitive not only to the nature of the two heteroaryl groups directly involved in the photochromic reactions but also to peripheral structure changes. Indeed, a clear correlation has been established between the fatigue resistance and the peripheral substitution of phenyl groups in two families of diarylethenes, dithienylethenes and dithiazolylethenes (Figure 1.31) [84c].

Contrary to spiropyrans, substitution of phenyl rings by EWG was clearly found to have beneficial effect on their photoresistance, while that of EDG produced the opposite effect. The most spectacular improvement of the fatigue resistance was observed when the phenyl rings are substituted at 3- and 5-positions by chemically inert and strongly electron-withdrawing trifluoromethyl group (CF_3) or pentafluorosulfanyl group (SF_5). More importantly, this enhancement of fatigue resistance was achieved without altering their photoreactivity [84c].

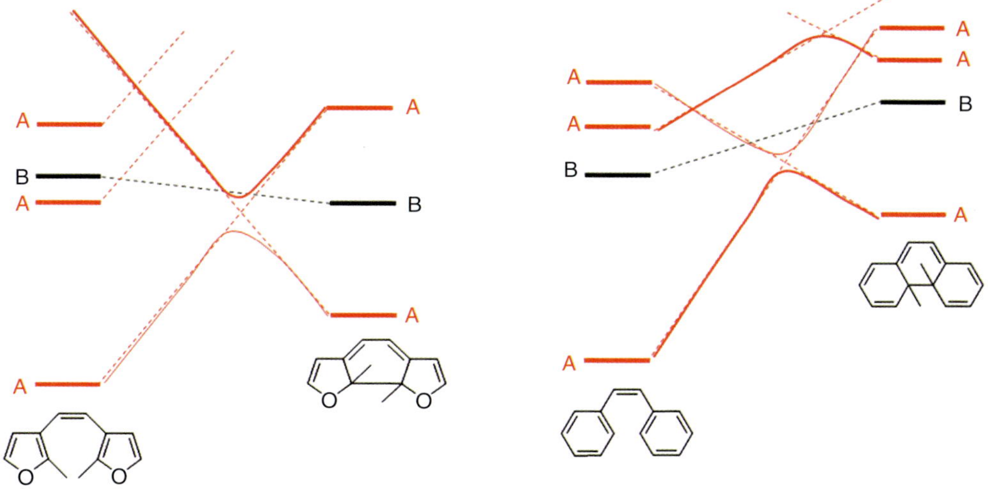

Figure 1.32 Correlation diagram showing the influence of the ground-state energy difference between B and A forms on the potential barrier for the B form.

1.3.3
Bistability: Avoiding Unwanted Thermal Back-Reaction in the Dark

In some applications, the required system should undergo switching only during the application of the light stimulus. For example, in memories, recorded information should not be altered in the dark. To reach this goal, P-type systems are sought where the ground-state potential barrier is particularly high for the B form (ΔE in Figure 1.5).

In the case of diarylethenes, a comparative study between different types of aromatic groups leads to the result: the lower the aromaticity of the aryl group, the smaller the ground-state energy difference between B and A forms, and thus, the higher the ΔE value [82, 90a]. Figure 1.32 shows the comparison between diphenylethene and difurylethene where the ground-state energy differences are, respectively, 27.3 and 9.2 kcal mol^{-1}. For dithienylethene, this energy difference is −3.3 kcal. Consequently, for aryl moieties with a low aromatic stabilization energy, such as furans, thiophenes, and thiazoles, the B form is thermally stable compared to phenyl, pyrrole, or indole [90a]. This is the reason why a large majority of diarylethenes bears thiophene (or benzothiophene) moieties, although molecules based on other groups, such as thiazole and oxazole, are now growing in importance.

It is also important to note that one can also play with the substitution of these heteroaryl groups as well as the nature of the ethenic bridge to tune the thermal stability of the colored B form of such a diarylethene, as exemplified below.

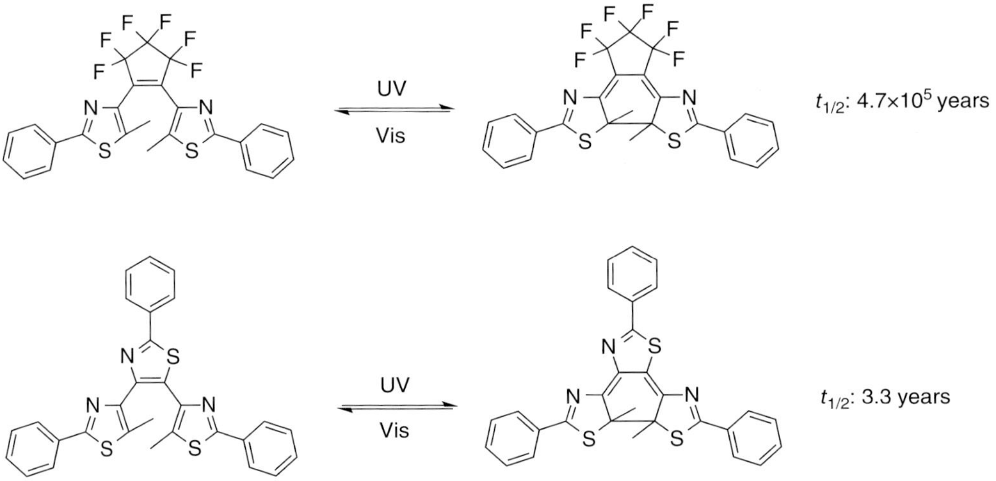

Figure 1.33 Comparison between hexafluorocyclopentene- and phenylthiazole-bridged diarylethenes: effect of the bridge on the thermal stability of the B form.

1.3.3.1 **Influence of Ethenic Bridge on the Thermal Stability of the B Form**

With hexafluorocyclopentene as the ethene bridge, the B form is extremely thermally stable, with a half-lifetime estimated to be 4.7×10^5 years at 30 °C [95], while that of the analog with a phenylthiazole bridge is only about 3.3 years at 20 °C [96] (Figure 1.33). The main reason for such a huge difference is that the cyclization results in a larger energy loss in the case of a phenylthiazole bridge than with hexafluorocyclopentene due to an extra loss of aromatic stabilization energy associated with the central thiazole. In other words, the ground-state energy difference between A and B forms is larger with phenylthiazole than that with hexafluorocyclopentene, leading to a smaller energy barrier for the thermal back-reaction, therefore a lower thermal stability of the B form with a phenylthiazole bridge.

Introducing six-member rings to bridge the aryl groups can be advantageous to develop new molecular structures and functionalities and also to increase the quantum yield of $\phi_{A \to B}$ to some extent [97]. However, concomitantly, it induces a high aromaticity, thus leading to T-type systems, unless EWG such as benzobisthiadiazoles are introduced [98].

1.3.3.2 **Impact of the Heteroaryl Substituents on the Thermal Stability of the B Form**

Figure 1.34 describes two situations, where the stability of the colored form B is influenced by the strength of the C–C single bond created upon the A to B reaction. Indeed, the B form is destabilized by a weakening of this bond, which can stem from either a larger steric hindrance (Figure 1.34a) when passing from methyl to isopropyl substituent on the reactive carbon atoms [99], or a stronger electron-withdrawing effect of the substituent at 5-position of thiophene moiety (Figure 1.34b) when going from aldehyde to dicyanoethene group [12]. In both

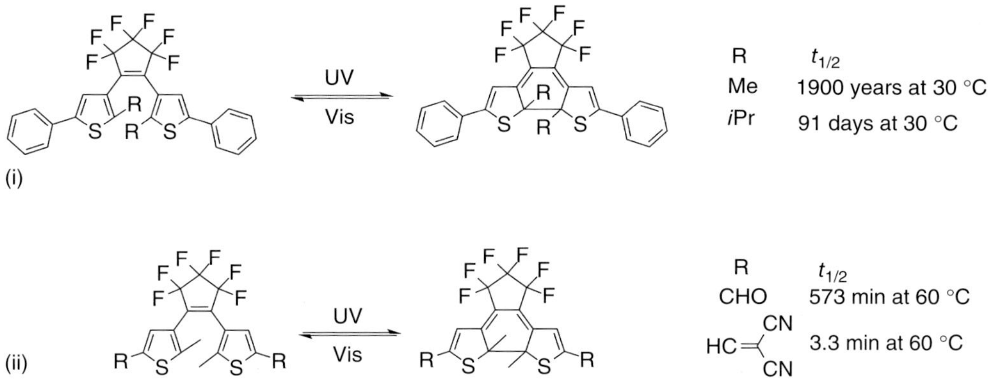

Figure 1.34 Substituent effect on the thermal stability of the B form in diarylethenes.

cases, a weaker C–C bond means a higher energy difference between B and A, thus leading to a lower energy barrier to the thermal back-reaction.

1.3.4
Fast Photochromic Systems: Reverting Back Spontaneously to the Colorless State in a Glance

In 1.3.3, the design to reach molecules with a perfect stability of both A and B forms was described, thus leading to systems that are not affected in the absence of light. These are based on P-type photochromic molecules for applications such as memories. In the contrary, another interesting development of photochromic compounds is to target systems, which revert back to the initial state when light is switched off. For example, ophthalmic applications require such a behavior. T-type molecules have appropriate characteristics, provided that the reaction occurs in a glance from A to B when light is switched on, and the other way around when it is switched off.

Evidently, a high thermal B to A reaction rate (k) is required for a rapid color fading when light is switched off. However, as described in Appendix, a high k value would inconveniently lower the conversion extent, $\alpha_B(\lambda_{irr})$, under UV irradiation. The consequence would be that only a low absorbance in the visible spectrum is reached under UV irradiation. Hence, a compromise for the value of k is necessary. Since the eye's reflex rate is at the order of some 10 ms, the trade-off value is some $10\ s^{-1}$, considering applications where the human's vision is involved. In addition, high UV light intensity and/or strong absorption coefficient of the B form in the visible spectrum help the performance of the photochromic system.

Many efforts are carried out to control k by a careful design of molecules for several families of molecules. Within the literature on T-type molecule families, such as spiropyrans and azobenzene, thermal back-reaction rate is probably among the most discussed properties in feature chapters and reviews [29, 34, 52, 60]. For the

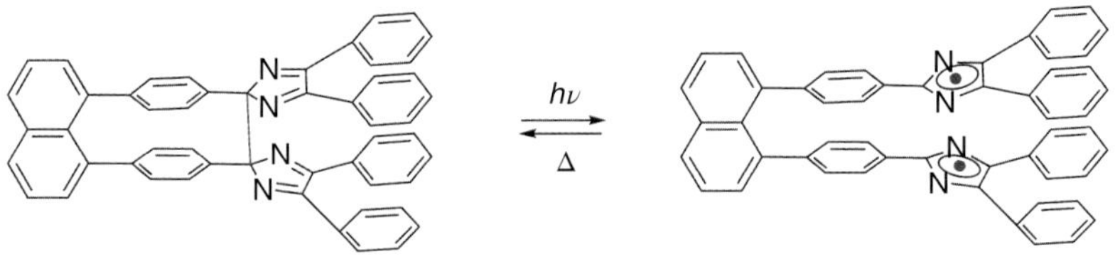

1.8-TPID-naphthalene 1.8-bis-TPIR-naphthalene

Figure 1.35 Example of fast HABI: naphthalene moiety avoids the separation in two molecules upon bond cleavage, leading to a fast thermal back-reaction [57].

latter, substituents such as amino groups or further conjugated push–pull structures speed up the thermal back-reaction, and recent work reports on the capability of fluorine substituents in slowing down the back-reaction rate drastically[100].

As mentioned previously, HABI undergoes a photochromic reaction between a TPID dimeric form and a pair of radicals TPIR (see Section 1.2.3, Figure 1.19). Once colored, the fading of the color of TPIR lasts for several minutes in solution [56]. In the last decade, an important work was carried out, leading to a series of fast HABI molecules. To speed up the thermal back-reaction, Abe has developed a new type of HABI structure where a naphthalene or cyclophane moiety bridges the two parts of the molecule and prevents the two radicals from leaving apart [101] (Figure 1.35). Diffusion of TPIR is thus annihilated [102]. The back-reaction is considerably accelerated and takes place in 180 ms. Other compounds, structured identically, have been the subject of several studies to modulate photochromic properties [103] or to understand the photodissociation mechanism [104].

In addition to the molecular structure, the environment can drastically influence *k* for a given molecule. The case of spiropyrans has been presented previously (see Section 1.2.4), where the stability of the colored MC is strongly solvent dependent. In addition, the case of anils can be pointed out: the thermal back-reaction ranges from a few milliseconds in solution, where the photochromism is merely visible due to an excessive value of *k* (see Appendix), to several days, in some cases even years, in the bulk [41l, 50b, 105].

1.3.5
Gaining Efficiency of the Photoreaction: the Example of Diarylethenes

Traditionally, clear and strong color change in photochromic systems used to be the target in many cases. In such context, colorability, which is the product of ε_B in the visible light spectrum and the quantum yield $\phi_{A\to B}$ [106], is one indicator to measure the efficiency of a photochromic system. The advantage of colorability is that it can be determined directly from the experiment, knowing the experimental conditions, even if the value of ε_B is unknown. This is useful in the case for T-type systems, where the B form is metastable and difficult to characterize. However, when properties other than color change are the main interests, the relevant figure

Table 1.1 Quantum yield ($\phi_{B\rightarrow A}$): effect of substituents on 1- and 5-positions of 1,2-bis(2,6-dimethyl-3-thienyl)perfluorocyclopentene (Figure 1.26, $-R_2=-R_6=-CH_3$).

Substituent at R_1 and R_5 positions	$\phi_{A\rightarrow B}$	$\phi_{B\rightarrow A}$
–H	0.21	0.13
–CN	0.44	7.5×10^{-2}
–Ph	0.46	1.5×10^{-2}
$-Ph-NEt_2$	0.37	2.5×10^{-3}
–Th–CN	0.12	1.3×10^{-3}
–Th–Th–CN	0.12	1.3×10^{-4}

Ph and Th stand for phenyl and thienyl, respectively.

would rather be the quantum yield(s). As it appears clearly in the expression of the conversion extent (see Appendix), the values of the quantum yields $\phi_{A\rightarrow B}$ and $\phi_{B\rightarrow A}$ count.

Coming back to diarylethenes, the bridging ethylene is included in a five-membered ring. This is the result of a compromise between the color change and the quantum yield, $\phi_{A\rightarrow B}$ [82, 90a]. Molecules of interest have a $\phi_{A\rightarrow B}$ value of a few 10^{-1}, whereas $\phi_{B\rightarrow A}$ can vary in a wide range of several decades. Regarding $\phi_{B\rightarrow A}$, it is lowered by the extension of the conjugation. For example, on 1,2-bis(2,6-dimethyl-3-thienyl)perfluorocyclopentene (Figure 1.26, $-R_2=-R_6=-CH_3$), $\phi_{B\rightarrow A}$ lowers from 0.13 to 7.5×10^{-2}, 1.5×10^{-2}, and 2.5×10^{-3} when, respectively, cyano, phenyl, and aminophenyl substituents are introduced at 1- and 5-positions (Table 1.1) [107]. Even the B to A photoreaction is annihilated when substituents belong to carotenoids family [108].

In addition, substituents linked to reactive carbon atoms ($-R_3$ and $-R_4$) affect $\phi_{B\rightarrow A}$ [109]: bulky groups and EDG lower its value, whereas EWG increase it, as illustrated in Table 1.2 [13e, 109b, 110].

From the data in the tables, it is difficult to draw any clear tendency about the influence of the substituent's properties on the quantum yield of the forward reaction, $\phi_{A\rightarrow B}$.

Furthermore, conformational analysis of diarylethenes shows that two conformations of the A form exist: antiparallel (ap) and parallel (p). The former

Table 1.2 Quantum yield ($\phi_{B\rightarrow A}$): effect of substituents on 3- and 4-positions of 1,2-bis(1,5-diphenyl-3-thienyl)perfluorocyclopentene (Figure 1.26, $-R_1=-R_5=-Ph$).

Substituent at R_3 and R_4 positions	$\phi_{A\rightarrow B}$	$\phi_{B\rightarrow A}$
$-CH_3$	0.59	1.3×10^{-2}
$-C_2H_5$	0.52	8.1×10^{-3}
–CN	0.42	0.41
$-OCH_3$	0.44	$<2\times10^{-5}$

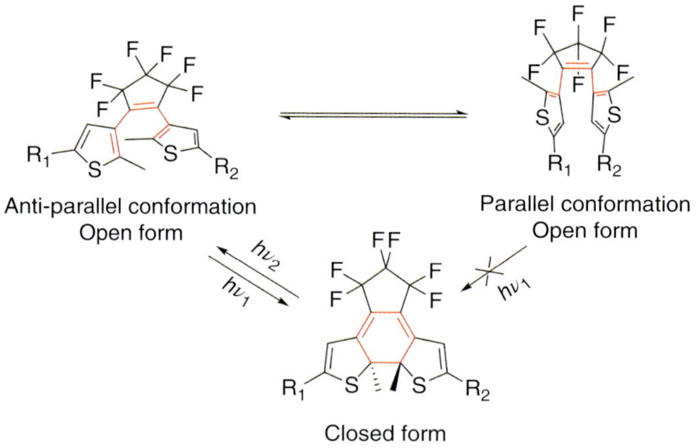

Figure 1.36 Antiparallel (ap) and parallel (p) conformations of the open form of diarylethenes. Only the ap conformer is reactive.

is symmetric with respect to 180° rotation about a C_2 axis, and the latter is symmetric with respect to reflection in a mirror plane. As it is rationalized by Woodward–Hoffmann's rule, only the ap conformer can undergo a photochemical ring-closure reaction (Figure 1.36) [82, 90a]. Usually for conformers, both spectra overlap and it is practically impossible to irradiate selectively one of them. In this context, obtaining a diarylethene with 100% ap conformation is sought after because a 50 : 50 ratio between the conformations would fix an upper limit of $\phi_{A\rightarrow B}$ to 50%.

Assuming these considerations, the proportion of ap form, and therefore the efficiency, can be increased by placing the dithienylethenes in a confined space, such as in crystalline phase [111], in doped polymer [112] or in inclusion material like cyclodextrin [113]. In contrast, parallel conformation can be largely favored, for example, by the formation of a rigid bridge between the two aromatics arms through hydrogen bonds. Such structure freezing leads to the unresponsiveness of the compound under light irradiation [114]. Freezing/defreezing can be achieved in specific cases by acid–base reaction [115].

In addition, several strategies in the molecular design have been endeavored to increase the proportion of the ap form.

One possibility is the introduction of bulky groups. Literature reports on an example where substitution of methyl by bulkier isopropyl groups on the reacting carbon atoms increases significantly the ap:p ratio (94 : 6 instead of 65 : 35), leading to more favorable $\phi_{A\rightarrow B}$ value (0.52 compared to 0.35) [99a]. Bulkiness can also be introduced elsewhere: a high $\phi_{A\rightarrow B}$ value of 0.83 was reached by introducing four methyl groups on the five-membered bridging ring [116] (Figure 1.37a).

Also, linking the two aryl groups with an additional bridge at *ortho*-positions of the reacting carbon atoms can favor the ap conformation [117] (Figure 1.37b).

During the past 5 years, several research groups have been working on the intramolecular interactions between the aryl groups and the bridging ring,

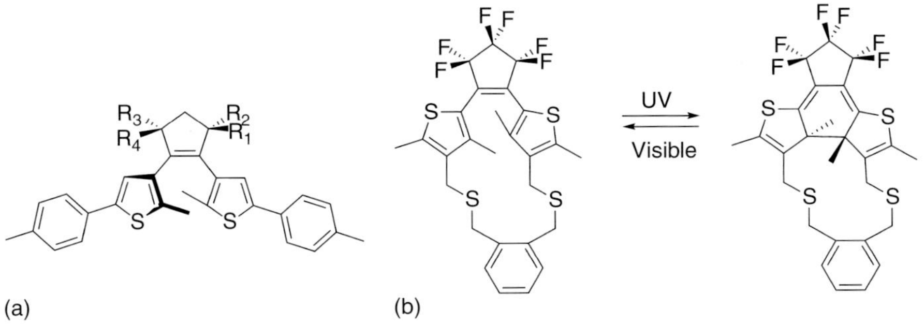

Figure 1.37 Antiparallel conformation (ap) favored by introducing bulkiness in the bridging ring (a) [116] or constraint with an additional bridge (b) [117b].

	Hexane	Methanol
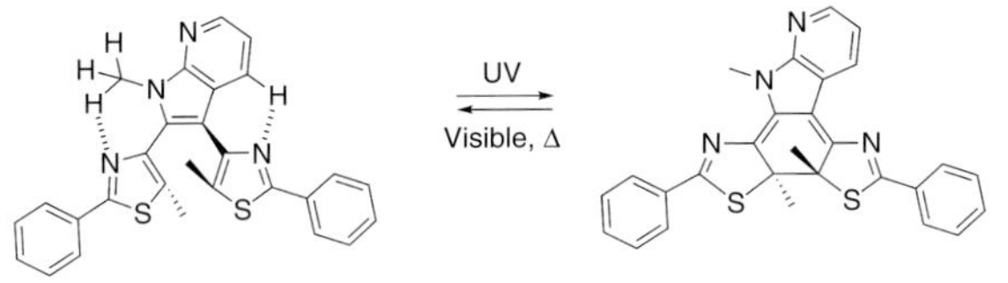	0.90	0.90
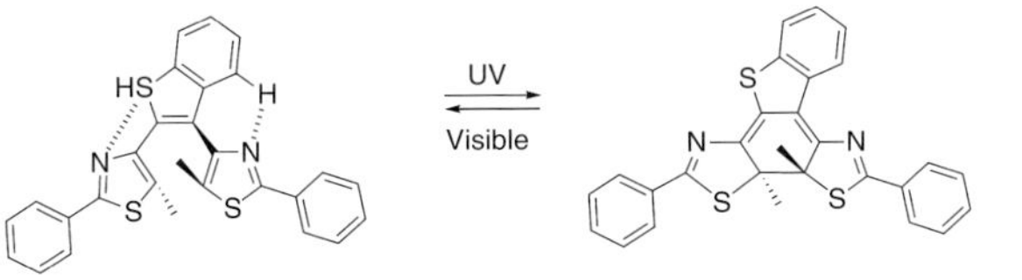	0.98	0.54

Figure 1.38 Intramolecular interactions favoring the ap conformation and solvent effect on the $\phi_{A\rightarrow B}$ value [118b].

in order to introduce some rigidity and block the structure within the ap conformation. Representative results were obtained on a series of teraryl-type diarylethenes, where intramolecular interactions play a crucial role [118]. As a result, $\phi_{A\rightarrow B}$ reaches high values of 0.90 and 0.98 (Figure 1.38) in hexane. In methanol, the S–N bond, present in one of the molecules, apparently weakens and $\phi_{A\rightarrow B}$ is lowered, thus showing the importance of these intramolecular interactions. This is also illustrated in another example where the methanol solvent plays a crucial role in the intramolecular interactions, fixing the conformation to ap and increasing $\phi_{A\rightarrow B}$ compared with what is observed in hexane (Figure 1.39) [119].

Blocking the conformation, in order to prevent the interconversion between ap and p conformers, has also been achieved in a diarylethene with thiazole groups and in a bridging six-membered benzo(thiadiazole) ring. A quantum yield as high as 0.91 was obtained for the cyclization of the ap form [120]. The energy barrier between the ap and p conformers was calculated to be more than 140 kJ mol^{-1}.

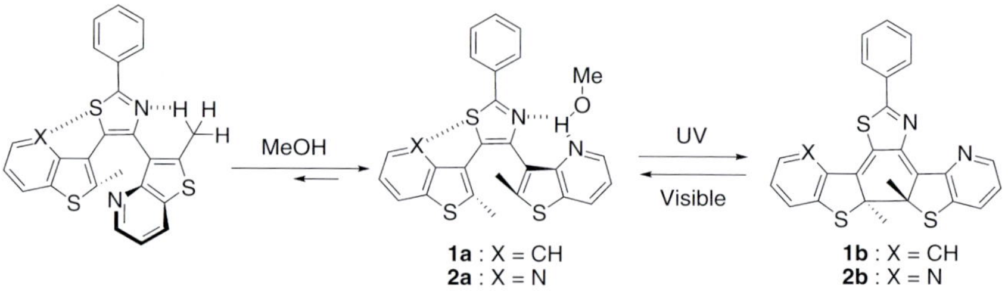

Molecule	Hexane	Methanol
1a	0.90	0.90
2a	0.98	0.54

Figure 1.39 Solvent–solute hydrogen bond favoring the ap conformation and increasing the $\phi_{A\to B}$ value [119].

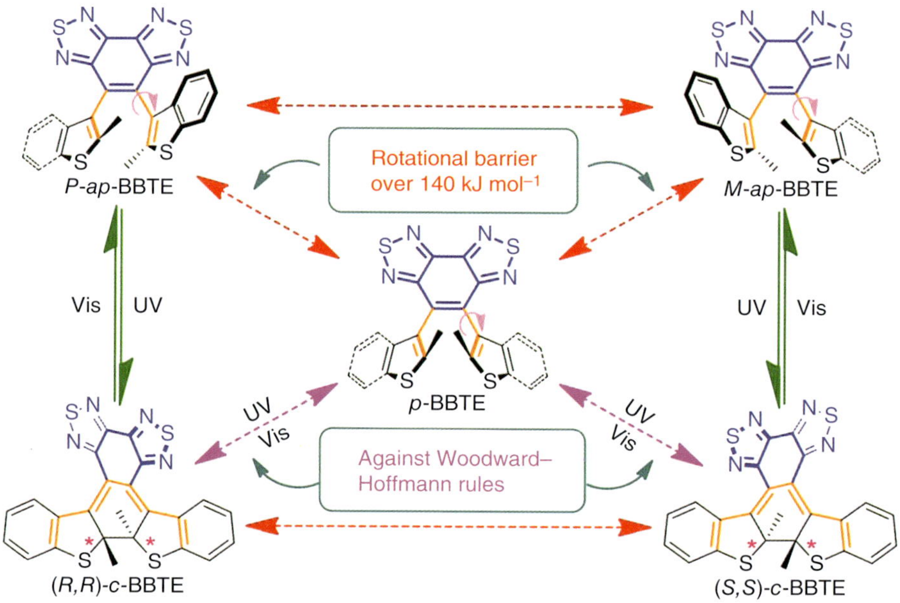

Figure 1.40 Enantiospecific photochromism resulting from a high-energy barrier between the open form conformations [121].

Such a design gives the possibility to isolate not only the ap and p conformers but also the pair of enantiomers of the p conformation. This feature was exploited further to undergo stereospecific reactions, where each enantiomer of the open form leads to a specific closed-form enantiomer [121] (Figure 1.40). Another example with a chiral center leads to a 100% diastereoselective process with a $\phi_{A\to B}$ value of 85% (Figure 1.41) [122].

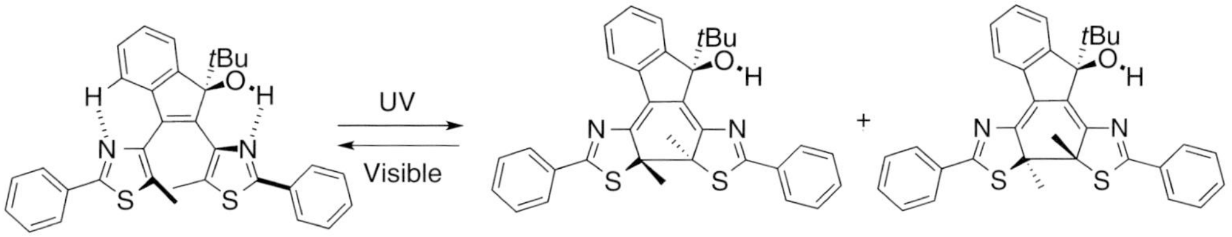

Figure 1.41 Stereospecific photochromism resulting from the control of the conformation of the open form by intramolecular hydrogen bonds [122].

1.4 Conclusion

The world of organic photochromism is full of variety and versatility because of the infinite possibilities of molecular engineering. Different geometrical and electronic considerations lead to different properties, and much effort has been done to tune these. In addition to what has been described in this chapter, obtaining molecules active in the solid state is another issue, where the design of the molecule has some influence, although this property involves the material. Criteria on intramolecular distances [101] or angles [42b] were established, and strategies such as introducing bulky groups [9b] were followed in order to ensure reactivity in the bulk state.

Appendix

General Kinetics of a Photochromic Reaction and Determination of the Conversion Extent $\alpha_B(\lambda_{irr})$

When a homogeneous sample (under constant stirring), containing a given photochromic species that exists in its two isomeric states A and B, is exposed to light irradiation (at a wavelength λ_{irr}, see Figure 1.A.1), the composition of the mixture evolves with the following general differential equation:

$$\frac{dC_A(t)}{dt} = -\frac{dC_B(t)}{dt} = \varphi_{B\to A} I_B^{abs}(\lambda_{irr}, t) - \varphi_{A\to B} I_A^{abs}(\lambda_{irr}, t) + k_{B\to A} C_B(t) \quad (A1)$$

with t the time; $C_A(t)$ and $C_B(t)$, respectively, the concentrations of A and B; $\varphi_{A\to B}$ and $\varphi_{B\to A}$ the photochromic quantum yields of the A → B and B → A reactions; $k_{B\to A}$ the thermal back-reaction rate from B to A (kinetics is assumed to be of first order); and $I_A^{abs}(\lambda_{irr}, t)$ and $I_B^{abs}(\lambda_{irr}, t)$ the intensities of the irradiation light absorbed by A and B, expressed as follows:

$$I_A^{abs}(\lambda_{irr}, t) = \frac{[1 - 10^{-Abs(\lambda_{irr}, t)}]}{Abs(\lambda_{irr}, t)} I_0(\lambda_{irr}) \varepsilon_A(\lambda_{irr}) \ell C_A(t) \quad (A2)$$

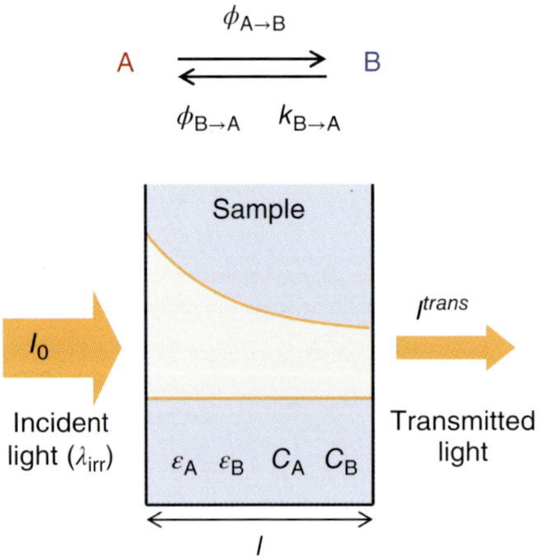

Figure 1.A.1 Definition of the system under irradiation and its variables.

$$I_B^{abs}(\lambda_{irr}, t) = \frac{[1 - 10^{-Abs(\lambda_{irr},t)}]}{Abs(\lambda_{irr}, t)} I_0(\lambda_{irr})\varepsilon_B(\lambda_{irr})\ell C_B(t) \tag{A3}$$

where $\varepsilon_A(\lambda_{irr})$ and $\varepsilon_B(\lambda_{irr})$ designate the molar absorption coefficients of the A and B species at the wavelength λ_{irr}, ℓ the optical path of the sample, $I_0(\lambda_{irr})$ the irradiation light intensity per volume unit of the sample, and Abs(λ_{irr}, t) the absorbance of the sample, which follows the Beer–Lambert law:

$$Abs(\lambda_{irr}, t) = \varepsilon_A(\lambda_{irr})\ell C_A(t) + \varepsilon_B(\lambda_{irr})\ell C_B(t) \tag{A4}$$

Often, the term *photokinetic factor* is employed to designate the nonlinear quantity $[1 - 10^{-Abs(\lambda_{irr},t)}]/Abs(\lambda_{irr}, t)$.

In the general situation, in the absence of any approximation, the differential equation (A1) shows no analytic solution. Indeed, the "photokinetic factor," and thus $I_A^{abs}(\lambda_{irr}, t)$ and $I_B^{abs}(\lambda_{irr}, t)$, vary upon time, as it can be seen from Eqs. (A2) and (A3).

Irradiation at a Specific Wavelength: Isosbestic Point

When irradiation is performed at the isosbestic point, the molar absorption coefficients of both A and B species are equal to the same value ε, and the absorption of the sample Abs(λ_{irr}, t) holds a constant value during the irradiation process (Abs(λ_{irr})), as well as the "photokinetic factor." Then, we can define the constant K:

$$K = \frac{I_0(\lambda_{irr})}{C_{tot}}[1 - 10^{-Abs(\lambda_{irr})}] \tag{A5}$$

where $C_{tot} = C_A(t) + C_B(t)$. As a consequence, the general differential kinetic equation simplifies on

$$\frac{dC_A(t)}{dt} = (K\varphi_{B\to A} + k_{B\to A})C_{tot} - (K(\varphi_{A\to B} + \varphi_{B\to A}) + k_{B\to A})C_A(t) \tag{A6}$$

Then, the time evolution of the molar fractions $x_A(t)$ and $x_B(t)$ of, respectively, A and B takes the expression of an exponential function:

$$\begin{aligned} x_A(t) = {} & \left(x_A^0 - \frac{K\varphi_{B\to A} + k_{B\to A}}{K\left(\varphi_{A\to B} + \varphi_{B\to A}\right) + k_{B\to A}} \right) \\ & \times \exp[-(K(\varphi_{A\to B} + \varphi_{B\to A}) + k_{B\to A})t] \\ & + \frac{K\varphi_{B\to A} + k_{B\to A}}{K(\varphi_{A\to B} + \varphi_{B\to A}) + k_{B\to A}} \end{aligned} \tag{A7}$$

$$\begin{aligned} x_B(t) = {} & -\left(\frac{K\varphi_{A\to B}}{K\left(\varphi_{A\to B} + \varphi_{B\to A}\right) + k_{B\to A}} - x_B^0 \right) \\ & \times \exp[-(K(\varphi_{A\to B} + \varphi_{B\to A}) + k_{B\to A})t] \\ & + \frac{K\varphi_{A\to B}}{K(\varphi_{A\to B} + \varphi_{B\to A}) + k_{B\to A}} \end{aligned} \tag{A8}$$

where x_A^0 and x_B^0 represent the initial molar fraction of, respectively, A and B.

Then, the conversion extent $\alpha_B(\lambda_{irr})$ at the PSS, corresponding to the molar fraction of B isomer in the sample after an infinite time of irradiation, can be easily expressed:

$$\alpha_B(\lambda_{irr}) = \frac{K\varphi_{A\to B}}{K(\varphi_{A\to B} + \varphi_{B\to A}) + k_{B\to A}} \tag{A9}$$

Coming back to the general case (irradiation at any wavelength), the photochromic kinetics does not follow an exponential behavior. In the following, we focus on the conversion extent of the B isomer at the PSS ($\alpha_B(\lambda_{irr})$) and consider the case of P-type and T-type photochromic systems.

This situation is obtained either by extrapolating the expression of $x_B(t)$ to infinite time (as it was done with Eq. (A8) to obtain Eq. (A9) in the case of the irradiation at the isosbestic point) or by writing that $\frac{dC_A(t)}{dt} = 0$ in Eq. (A1).

Case A: When the Thermal Back-Reaction is Negligible Compared to the Photochemical Reaction (Typically P-type)

In this situation, $k_{B\to A}C_B \ll \varphi_{B\to A}I_B^{abs}$ and the last term of the general kinetic equation is suppressed. In the following, the PSS is simply characterized by

$$\alpha_B(\lambda_{irr}) = \frac{\varepsilon_A(\lambda_{irr})\varphi_{A\to B}}{\varepsilon_A(\lambda_{irr})\varphi_{A\to B} + \varepsilon_B(\lambda_{irr})\varphi_{B\to A}} \tag{A10}$$

It is worth noticing that the composition at PSS does not depend on the irradiation conditions. Only the kinetics to reach PSS does.

Table 1.A.1 Values of $\alpha_B(\lambda_{irr})$ as a function of the thermal back-reaction rate $k_{B\rightarrow A}$.

$k_{B\rightarrow A}$ (s^{-1})	10	1	0.1
$\alpha_B(\lambda_{irr})$	0.01	0.10	0.54

Case B: When the Thermal Back-Reaction is More Efficient than the Photochemical *B* → *A* Reaction (Typically T ype)

In this situation, $k_{B\rightarrow A}C_B \gg \varphi_{B\rightarrow A}I_B^{abs}$ and the first term of Eq. (A1) is suppressed. In order to have a simple analytic resolution of the equation, we need to consider the case where the absorption of the sample at the irradiation wavelength is low enough ($\mathrm{Abs}(\lambda_{irr}) \ll 1$) to allow the linearization of the differential equation with ($[1 - 10^{-\mathrm{Abs}(\lambda_{irr})}] = \ln 10 \times \mathrm{Abs}(\lambda_{irr})$). In this case, Eq. (A1) becomes

$$\frac{dC_A(t)}{dt} = -\ln 10 \times I_0(\lambda_{irr})\ell\varepsilon_A(\lambda_{irr})\varphi_{A\rightarrow B}C_A(t) + k_{B\rightarrow A}C_B(t) \quad (A11)$$

and the conversion extent at the PSS $\alpha_B(\lambda_{irr})$ is dependent on the irradiation intensity $I_0(\lambda_{irr})$:

$$\alpha_B(\lambda_{irr}) = \frac{1}{1 + \frac{k_{B\rightarrow A}}{\ln 10 \times I_0(\lambda_{irr})\ell\varepsilon_A(\lambda_{irr})\varphi_{A\rightarrow B}}} \quad (A12)$$

From Eq. (A12), we can deduce that the limit of $\alpha_B(\lambda_{irr})$ approaches 0 when $k_{B\rightarrow A}$ is very high.

Let us assume

- a sample of 1 ml, $\ell = 1$ cm,
- containing the isomer A, which has the following characteristics: $\varepsilon_A(\lambda_{irr}) = 10\,000\ \mathrm{mol^{-1}\ l\ cm^{-1}}$ and $\varphi_{A\rightarrow B} = 0.5$,
- under an irradiation of 3 mW at $\lambda_{irr} = 400$ nm, which corresponds to an intensity per unit volume of $I_0(\lambda_{irr}) = 10^{-5}\ \mathrm{mol\ l^{-1}\ s^{-1}}$.

Table 1.A.1 shows that, under the above-mentioned conditions, if the thermal reaction rate is close to 1 s^{-1}, the conversion extent is 10%, whereas if it is 10 times faster, only 1% of A is transformed to B. This explains why a fast thermal back-reaction can be an obstacle to visualize photochromism.

References

1. Stevenson, A. and Lindberg, C. A. (eds). (2010) *New Oxford American Dictionary*, 3rd edn, Oxford University Press, p. 2096.
2. Bouas-Laurent, H. and Durr, H. (2001) Organic photochromism. *Pure Appl. Chem.*, **73** (4), 639–665.
3. Higgins, S. (2003) Photochromics – colourful molecular athletes. *Chim. Oggi-Chem. Today*, **21** (1-2), 63–67.
4. www.optiline.co.uk/index.php/information/common-knowledge/photochromic-sunglasses http://

011101010.blogspot.fr (24 October 2015).

5. Szacilowski, K. (2008) Digital information processing in molecular systems. *Chem. Rev.*, **108** (9), 3481–3548.
6. (a) Corredor, C.C., Huang, Z.L., Belfield, K.D., Morales, A.R., and Bondar, M.V. (2007) Photochromic polymer composites for two-photon 3D optical data storage. *Chem. Mater.*, **19** (21), 5165–5173; (b) Toriumi, A., Kawata, S., and Gu, M. (1998) Reflection confocal microscope readout system for three-dimensional photochromic optical data storage. *Opt. Lett.*, **23** (24), 1924–1926; (c) Kawata, S. and Kawata, Y. (2000) Three-dimensional optical data storage using photochromic materials. *Chem. Rev.*, **100** (5), 1777–1788; (d) Parthenopoulos, D.A. and Rentzepis, P.M. (1989) 3-dimensional optical storage memory. *Science*, **245** (4920), 843–845; (e) Ogawa, K. (2014) Two-photon absorbing molecules as potential materials for 3D optical memory. *Appl. Sci.*, **4** (1), 1–18.
7. Lavoie-Cardinal, F., Jensen, N.A., Westphal, V., Stiel, A.C., Chmyrov, A., Bierwagen, J., Testa, I., Jakobs, S., and Hell, S.W. (2014) Two-color RESOLFT nanoscopy with green and red fluorescent photochromic proteins. *ChemPhysChem*, **15** (4), 655–663.
8. (a) Zhang, J., Wang, J., and Tian, H. (2014) Taking orders from light: progress in photochromic bio-materials. *Mater. Horiz.*, **1** (2), 169–184; (b) Szymanski, W., Beierle, J.M., Kistemaker, H.A., Velema, W.A., and Feringa, B.L. (2013) Reversible photocontrol of biological systems by the incorporation of molecular photoswitches. *Chem. Rev.*, **113**, 6114–6178; (c) Tian, Z., Wu, W., and Li, A.D. (2009) Photoswitchable fluorescent nanoparticles: preparation, properties and applications. *ChemPhysChem*, **10** (15), 2577–2591; (d) Klajn, R., Stoddart, J.F., and Grzybowski, B.A. (2010) Nanoparticles functionalised with reversible molecular and supramolecular switches. *Chem. Soc. Rev.*, **39** (6), 2203–2237; (e) Feczko, T. (2013) Organic nanoparticulate photochromes. *Curr. Org. Chem.*, **17** (16), 1771–1789.
9. (a) Chauvin, J., Kawai, T., and Irie, M. (2001) Refractive index change of an amorphous bisbenzothienylethene. *Jpn. J. Appl. Phys. Part 1*, **40** (4A), 2518–2522; (b) Kawai, T., Fukuda, N., Groschl, D., Kobatake, S., and Irie, M. (1999) Refractive index change of dithienylethene in bulk amorphous solid phase. *Jpn. J. Appl. Phys. Part 2*, **38** (10B), L1194–L1196; (c) Kim, E., Choi, Y.K., and Lee, M.H. (1999) Photoinduced refractive index change of a photochromic diarylethene polymer. *Macromolecules*, **32** (15), 4855–4860; (d) Kim, M.S., Maruyama, H., Kawai, T., and Irie, M. (2003) Refractive index changes of amorphous diarylethenes containing 2,4-diphenylphenyl substituents. *Chem. Mater.*, **15** (24), 4539–4543; (e) Bertarelli, C., Bianco, A., Castagna, R., and Pariani, G. (2011) Photochromism into optics: opportunities to develop light-triggered optical elements. *J. Photochem. Photobiol., C*, **12** (2), 106–125.
10. (a) Fukaminato, T. (2011) Single-molecule fluorescence photoswitching: design and synthesis of photoswitchable fluorescent molecules. *J. Photochem. Photobiol., C*, **12** (3), 177–208; (b) Yun, C., You, J., Kim, J., Huh, J., and Kim, E. (2009) Photochromic fluorescence switching from diarylethenes and its applications. *J. Photochem. Photobiol., C*, **10** (3), 111–129; (c) Cusido, J., Deniz, E., and Raymo, F.M. (2009) Fluorescent switches based on photochromic compounds. *Eur. J. Org. Chem.*, **2009** (13), 2031–2045; (d) Yildiz, I., Deniz, E., and Raymo, F.M. (2009) Fluorescence modulation with photochromic switches in nanostructured constructs. *Chem. Soc. Rev.*, **38** (7), 1859–1867; (e) Raymo, F.M. and Tomasulo, M. (2005) Electron and energy transfer modulation with photochromic switches. *Chem. Soc. Rev.*, **34** (4), 327–336.
11. Delaire, J.A. and Nakatani, K. (2000) Linear and nonlinear optical properties of photochromic molecules

and materials. *Chem. Rev.*, **100** (5), 1817–1845.

12. (a) Gilat, S.L., Kawai, S.H., and Lehn, J.M. (1995) Light-triggered molecular devices – photochemical switching of optical and electrochemical properties in molecular wire type diarylethene species. *Chem. Eur. J.*, **1** (5), 275–284; (b) Kawai, T., Kunitake, T., and Irie, M. (1999) Novel photochromic conducting polymer having diarylethene derivative in the main chain. *Chem. Lett.*, (9), 905–906; (c) Matsuda, K. and Irie, M. (2000) A diarylethene with two nitronyl nitroxides: photoswitching of intramolecular magnetic interaction. *J. Am. Chem. Soc.*, **122** (30), 7195–7201; (d) Tanifuji, N., Irie, M., and Matsuda, K. (2005) New photoswitching unit for magnetic interaction: diarylethene with 2,5-bis(arylethynyl)-3-thienyl group. *J. Am. Chem. Soc.*, **127** (38), 13344–13353; (e) Matsuda, K., Yamaguchi, H., Sakano, T., Ikeda, M., Tanifuji, N., and Irie, M. (2008) Conductance photoswitching of diarylethene-gold nanoparticle network induced by photochromic reaction. *J. Phys. Chem. C*, **112** (43), 17005–17010; (f) Tsujioka, T. and Irie, M. (2010) Electrical functions of photochromic molecules. *J. Photochem. Photobiol., C*, **11** (1), 1–14.
13. (a) Patra, A., Métivier, R., Brisset, F., and Nakatani, K. (2012) Photochromic one-dimensional nanostructures based on dithienylethene: fabrication by light-induced precipitation and reversible transformation in the nanoparticle state. *Chem. Commun.*, **48** (19), 2489–2491; (b) Takeshita, M. and Irie, M. (1998) Photoresponsive tweezers for alkali metal ions. Photochromic diarylethenes having two crown ether moieties. *J. Org. Chem.*, **63** (19), 6643–6649; (c) Natali, M. and Giordani, S. (2012) Molecular switches as photocontrollable "smart" receptors. *Chem. Soc. Rev.*, **41** (10), 4010–4029; (d) Yamamoto, S., Matsuda, K., and Irie, M. (2003) Absolute asymmetric photocyclization of a photochromic diarylethene derivative in single crystals. *Angew. Chem. Int. Ed.*, **42** (14), 1636–1639; (e) Kodani, T., Matsuda, K., Yamada, T., Kobatake, S., and Irie, M. (2000) Reversible diastereoselective photocyclization of a diarylethene in a single-crystalline phase. *J. Am. Chem. Soc.*, **122** (40), 9631–9637; (f) Yamaguchi, T., Uchida, K., and Irie, M. (1997) Asymmetric photocyclization of diarylethene derivatives. *J. Am. Chem. Soc.*, **119** (26), 6066–6071.
14. (a) Kobatake, S., Takami, S., Muto, H., Ishikawa, T., and Irie, M. (2007) Rapid and reversible shape changes of molecular crystals on photoirradiation. *Nature*, **446** (7137), 778–781; (b) Yu, Y.L., Nakano, M., and Ikeda, T. (2003) Directed bending of a polymer film by light – miniaturizing a simple photomechanical system could expand its range of applications. *Nature*, **425** (6954), 145–145; (c) Kitagawa, D., Nishi, H., and Kobatake, S. (2013) Photoinduced twisting of a photochromic diarylethene crystal. *Angew. Chem. Int. Ed.*, **52** (35), 9320–9322; (d) Koshima, H., Ojima, N., and Uchimoto, H. (2009) Mechanical motion of azobenzene crystals upon photoirradiation. *J. Am. Chem. Soc.*, **131** (20), 6890–6891.
15. (a) Wöhrle, D., Tausch, M.W., and Stroher, W.-D. (1998) *Photochemie*, Wiley-VCH, Weinheim; (b) Dessauer, R. and Paris, J.P. (1963) Photochromism. *Adv. Photochem.*, **1**, 275–278.
16. Fritzsche, J. (1867) Note sur les carbures d'hydrogène solides, tirés du goudron de houille, *C.R. Acad. Sci.*, **69**, 1035.
17. (a) Phipson, T.L. (1881) *Chem. News*, **43**, 283; (b) Ter Meer, E. (1876) *Annu. Chem.*, **181** (1); (c) Orr, J.B. (1881) *Chem. News*, **44**, 12; (d) Wislicenus, W. (1893) *Annu. Chem.*, (277), 366.
18. (a) Biltz, H. (1899) *Annu. Chem.*, (305), 170; (b) Biltz, H. (1899) *Phys. Chem.*, (30), 527; (c) Biltz, H. and Wienands, A (1899) *Annu. Chem.*, (308), 1.
19. Markwald, W. (1899) *Phys. Chem.*, **30**, 140.
20. Stobbe, H. (1905) The colour of the 'fulgenic acid' and 'fulgide'. *Ber Dtsch. Chem. Ges.*, **38**, 3673–3682.

21. (a) Senier, A. and Shepheard, F.G. (1909) Studies in phototropy and thermotropy. Part I. Arylidene- and Naphtylidene-amines. *J. Chem. Soc.*, 1943–1955; (b) Senier, A., Shepheard, F.G., and Clarke, R. (1912) Studies in phototropy and thermotropy. Part III. Arylideneamines. *J. Chem. Soc.*, 1950–1958.
22. Stobbe, H. and Mallison, H. (1913) Phototropie-Erscheinungen bei Stilben-Derivaten. *Ber Dtsch. Chem. Ges.*, **46** (2), 1226–1238.
23. Tschitschibabin, A.E., Kuindshi, B.M., and Benewolenskaja, S.W. (1925) Über Nitrierungsprodukte des α- und des γ-Benzyl-pyridins. *Ber Dtsch. Chem. Ges. (A and B Ser.)*, **58** (8), 1580–1587.
24. (a) Stobbe, H. (1922) *Verhandl*, Sachs Akademie, Leipzig, p. 161; (b) Stobbe, H. (1923) *Chem, Abstr.*, (17), 3020; (c) Chakley, L. (1929) *Chem. Rev.*, **6**, 217; (d) Bhatanagar, S.S., Kapur, P.L., and Hashmi, M.S. (1938) *J. Indian Chem. Soc.*, (15), 573; (e) Overbeck, J.V. (1939) *Bot. Rev.*, (5), 655; (f)Brown, G.H. (1959) Wright Air Development Center; (g) Brown, G.H. and Shaw, W.G. (1961) Phototropism, *Rev. Pure Appl. Chem.*, **11**, 2.
25. Harris, L., Kaminsky, J., and Simard, R.G. (1935) The absorption spectrum of malachite green leucocyanide and the mechanism of the dark reaction after photolysis, *J. Am. Chem. Soc.*, (57), 1151.
26. (a) Georghiu, C.V. and Arrventien, B. (1930) *Bull. Soc. Chim.*, **47**, 105; (b) Georghiu, C.V. (1934) *Bull. Soc. Chim.*, **1**, 97; (c) Georghiu, C.V. and Matei, V. (1939) *Bull. Soc. Chim.*, **6**, 1324.
27. (a) Douzou, P. and Wippler, C. (1961) *Science*, **28**, 70; (b) Schwab, H. and Bertelson, R.C. (1964) Photochromism, state of the art review, Unconventional Photographic Systems Symposium; (c) Luck, W. and Sand, H. (1964) Phenomenon of phototropy, *Angew. Chem. Int. Ed.*, **3**, 570; (d) Exelby, R. and Grinter, R. (1965) Phototropy (or photochromism), *Chem. Rev.*, **65**, 247; (e) Dorion, G.H. and Weissbein, L. (1963) *Discovery*; (f) Inoue, E., Kodado, H., Shimizu, I., and Yoshida, K. (1967) Photoreversible photographic system, *Photogr. Sci. Eng.*, **11**, 181; (g) Cohen, S.D. and Newman, G.H. (1967) Inorganic photochromism, *J. Photogr. Sci.*, **15**, 290.
28. Hirsberg, Y. (1950) Photochromie dans la série de la bianthone. *Compte Rendu*, (7), 116.
29. Fischer, E. and Hirshberg, Y. (1952) Formation of coloured forms of spirans by low-temperature irradiation. *J. Chem. Soc.*, 4522–4524.
30. (a) Appenroth, K., Reichenbacher, M., and Paetzold, R. (1980) Thermochromism and photochromism of aryl-substituted acyclic azines. 2. Photokinetics. *J. Photochem.*, **14** (1), 39–50; (b) Appenroth, K., Reichenbacher, M., and Paetzold, R. (1980) Thermochromism and photochromism of aryl-substituted acyclic azines. 3. Investigations on the mechanism of photochemical isomerization. *J. Photochem.*, **14** (1), 51–60.
31. Fukunishi, K., Tatsuma, M., Rahman, S., Kuwabara, M., Yamanaka, H., and Nomura, M. (1990) Cis trans isomerization of thioindigo derivatives adsorbed on silica-gel modified with octadecyl groups. *Bull. Chem. Soc. Jpn.*, **63** (12), 3701–3703.
32. (a) Chu, N.Y.C. (1983) Photochromism of spiroindolinonaphthoxazine. I. Photophysical properties. *Can. J. Chem.*, **61** (2), 300–305; (b) Chu, N.Y.C. (1990) in *Photochromism, Molecules, and Systems* (eds H. Dürr and H. Bouas-Laurent), Elsevier, Amsterdam, pp. 493–509; (c) Maeda, S. (2002) in *Organic Photochromic and Thermochromic Compounds* (eds J.C. Crano and R.J. Guglielmetti), Kluwer Academic Publishers, New York, pp. 85–109.
33. Hartley, G.S. (1937) The Cis-form of azobenzene. *Nature*, **140** (3537), 281–281.
34. Rau, H. (1990) in *Photochemistry and Photophysics*, vol. **2** (ed J.F. Rabek), CRC Press, Boca Raton, FL 119–141.
35. (a) Mitchell, R.H. (1999) The metacyclophanediene-dihydropyrene photochromic pi switch. *Eur. J. Org. Chem.*, (11), 2695–2703;

(b) Roldan, D., Kaliginedi, V., Cobo, S., Kolivoska, V., Bucher, C., Hong, W., Royal, G., and Wandlowski, T. (2013) Charge transport in photoswitchable dimethyldihydropyrene-type single-molecule junctions. *J. Am. Chem. Soc.*, **135** (16), 5974–5977.

36. (a) Barachevsky, V.A. (1994) Photochromism of Aryloxyquinones. *Mol. Cryst. Liq. Cryst. Sci. Technol., Sect. A: Mol. Cryst. Liq. Crys.*, **246** (1), 95–102; (b) Gritsan, N.P. and Klimenko, L.S. (1994) Photochromism of quinoid compounds. *Mol. Cryst. Liq. Cryst. Sci. Technol., Sect. A: Mol. Cryst. Liq. Crys.*, **246** (1), 103–106.
37. Kamogawa, H., Masui, T., and Nanasawa, M. (1980) Photochemically induced reduction of viologens in solid polar aprotic polymer matrices. *Chem. Lett.*, (9), 1145–1148.
38. Pina, F., Melo, M.J., Laia, C.A., Parola, A.J., and Lima, J.C. (2012) Chemistry and applications of flavylium compounds: a handful of colours. *Chem. Soc. Rev.*, **41** (2), 869–908.
39. (a) Mancois, F., Sanguinet, L., Pozzo, J.L., Guillaume, M., Champagne, B., Rodriguez, V., Adamietz, F., Ducasse, L., and Castet, F. (2007) Acido-triggered nonlinear optical switches: benzazolo-oxazolidines. *J. Phys. Chem. B*, **111** (33), 9795–9802; (b) Sevez, G., Gan, J., Delbaere, S., Vermeersch, G., Sanguinet, L., Levillain, E., and Pozzo, J.L. (2010) Photochromic performance of a dithienylethene-indolinooxazolidine hybrid. *Photochem. Photobiol. Sci.*, **9** (2), 131–135; (c) Sanguinet, L., Pozzo, J.L., Rodriguez, V., Adamietz, F., Castet, F., Ducasse, L., and Champagne, B. (2005) Acido- and phototriggered NLO properties enhancement. *J. Phys. Chem. B*, **109** (22), 11139–11150.
40. Irie, M. (2000) Photochromism: memories and switches – introduction. *Chem. Rev.*, **100** (5), 1683–1683.
41. (a) Irie, M. (2000) *Chem. Rev.*, **100** (5); (b) Irie, M. (2010) *Photochem. Photobiol. Sci.*, **9** (2); (c) (2009) *New J. Chem.*, **33** (6); (d) (2007) *J. Phys. Org. Chem.*, **20** (11); (e) (2011) *J. Photochem. Photobiol. C*, **12** (2-3); (f) Zhang, J., Zou, Q., and Tian, H. (2013) Photochromic materials: more than meets the eye. *Adv. Mater.*, **25** (3), 378–399; (g) Feringa, B.L. (2007) The art of building small: from molecular switches to molecular motors. *J. Org. Chem.*, **72** (18), 6635–6652; (h) Irie, M., Fukaminato, T., Matsuda, K., and Kobatake, S. (2014) Photochromism of diarylethene molecules and crystals: memories, switches, and actuators. *Chem. Rev.*, **114** (24), 12174–12277; (i) Tian, H. and Yang, S.J. (2004) Recent progresses on diarylethene based photochromic switches. *Chem. Soc. Rev.*, **33** (2), 85–97; (j) Matsuda, K. and Irie, M. (2004) Diarylethene as a photoswitching unit. *J. Photochem. Photobiol., C*, **5** (2), 169–182; (k) Irie, M. (2010) Photochromism of diarylethene single molecules and single crystals. *Photochem. Photobiol. Sci.*, **9** (12), 1535–1542; (l) Amimoto, K. and Kawato, T. (2005) Photochromism of organic compounds in the crystal state. *J. Photochem. Photobiol., C*, **6** (4), 207–226; (m) Andreasson, J. and Pischel, U. (2010) Smart molecules at work – mimicking advanced logic operations. *Chem. Soc. Rev.*, **39** (1), 174–188; (n) Tamaoki, N. and Kamei, T. (2010) Reversible photo-regulation of the properties of liquid crystals doped with photochromic compounds. *J. Photochem. Photobiol., C*, **11** (2-3), 47–61.
42. (a) Hadjoudis, E. (1995) Photochromic and thermochromic anils. *Mol. Eng.*, **5**, 301–337; (b) Hadjoudis, E. and Mavridis, I.M. (2004) Photochromism and thermochromism of Schiff bases in the solid state: structural aspects. *Chem. Soc. Rev.*, **33** (9), 579–588; (c) Hadjoudis, E., Chatziefthimiou, S., and Mavridis, I. (2009) Anils: photochromism by H-transfer. *Curr. Org. Chem.*, **13** (3), 269–286.
43. Cohen, M.D., Schmidt, G.M.J., and Flavian, S. (1964) Topochemistry. Part VI.* Experiments on photochromy and thermochromy of crystalline anils of salicylaldehydes. *J. Chem. Soc.*, 2041–2051.
44. Ritter, E., Przybylski, P., Brzezinski, B., and Bartl, F. (2009) Schiff bases in

biological systems. *Curr. Org. Chem.*, **13** (3), 241–249.

45. (a) Destro, R., Gavezzotti, A., and Simonetta, M. (1978) Salicylideneaniline. *Acta Crystallogr., Sect. B*, **34**, 2867–2869; (b) Sliwa, M., Naumov, P., Choi, H.J., Nguyen, Q.T., Debus, B., Delbaere, S., and Ruckebusch, C. (2011) Effects of a self-assembled molecular capsule on the ultrafast photodynamics of a photochromic salicylideneaniline guest. *ChemPhysChem*, **12** (9), 1669–1672.
46. (a) Mitra, S. and Tamai, N. (2003) Dynamics of photochromism in salicylideneaniline: a femtosecond spectroscopic study. *Phys. Chem. Chem. Phys.*, **5** (20), 4647–4652; (b) Tamai, N. and Miyasaka, H. (2000) Ultrafast dynamics of photochromic systems. *Chem. Rev.*, **100** (5), 1875–1890; (c) Ziolek, M., Kubicki, J., Maciejewski, A., Naskrecki, R., and Grabowska, A. (2004) An ultrafast excited state intramolecular proton transfer (ESPIT) and photochromism of salicylideneaniline (SA) and its "double" analogue salicylaldehyde azine (SAA). A controversial case. *Phys. Chem. Chem. Phys.*, **6** (19), 4682–4689; (d) Sliwa, M., Mouton, N., Ruckebusch, C., Aloïse, S.P., Poizat, O., Buntinx, G., Métivier, R.M., Nakatani, K., Masuhara, H., and Asahi, T. (2009) Comparative investigation of ultrafast photoinduced processes in salicylidene-aminopyridine in solution and solid state†. *J. Phys. Chem. C*, **113** (27), 11959–11968.
47. (a) Cohen, M.D. and Schmidt, G.M.J. (1962) Photochromy and thermochromy of ANILS1. *J. Phys. Chem.*, **66** (12), 2442–2446; (b) Becker, R.S. and Richey, W.F. (1967) Photochromic anils. Mechanisms and products of photoreactions and thermal reactions. *J. Am. Chem. Soc.*, **89** (6), 1298–1302; (c) Ledbette, J.W. (1966) Spectroscopic evidence for enol imine-keto enamine tautomerism of N-(o- and p-hydroxybenzylidene) anils in solution. *J. Phys. Chem.*, **70** (7), 2245; (d) Hadjoudis, E. and Hayon, E. (1970) Flash photolysis of some photochromic N-benzylideneanilines. *J. Phys. Chem.*, **74**, 3184.
48. Mitra, S. and Tamai, N. (1999) A combined experimental and theoretical study on the photochromism of aromatic anils. *Chem. Phys.*, **246** (1-3), 463–475.
49. Nakagaki, R., Kobayashi, T., Nakamura, J., and Nagakura, S. (1977) Spectroscopic and kinetic studies of photochromism of N-Salicylideneanilines and related compounds. *Bull. Chem. Soc. Jpn.*, **50** (8), 1909–1912.
50. (a) Hadjoudis, E., Vittorakis, M., and Moustakali-Mavridis, I. (1987) Photochromism and thermochromism of schiff bases in the solid state and in rigid glasses. *Tetrahedron*, **43** (7), 1345–1360; (b) Sliwa, M., Letard, S., Malfant, I., Nierlich, M., Lacroix, P.G., Asahi, T., Masuhara, H., Yu, P., and Nakatani, K. (2005) Design, synthesis, structural and nonlinear optical properties of photochromic crystals: toward reversible molecular switches. *Chem. Mater.*, **17** (18), 4727–4735.
51. Naumov, P., Sakurai, K., Ishikawa, T., Takahashi, J., Koshihara, S.Y., and Ohashi, Y. (2005) Intramolecular nitro-assisted proton transfer in photoirradiated 2-(2′,4′-dinitrobenzyl)pyridine: polarized optical spectroscopic study and electronic structure calculations. *J. Phys. Chem. A*, **109** (32), 7264–7275.
52. (a) Bandara, H.M. and Burdette, S.C. (2012) Photoisomerization in different classes of azobenzene. *Chem. Soc. Rev.*, **41** (5), 1809–1825; (b) Merino, E. (2011) Synthesis of azobenzenes: the coloured pieces of molecular materials. *Chem. Soc. Rev.*, **40** (7), 3835–3853; (c) Rau, H. (1990) in *Photochromism, Molecules, and Systems* (eds H. Dürr and H. Bouas-Laurent), Elsevier, Amsterdam, pp. 165–192.
53. Zhao, Y. and Ikeda, T. (2008) *Smart Light-Responsive Materials: Azobenzene-Containing Polymers and Liquid Crystals*, Wiley-VCH Verlag GmbH, Hoboken, NJ.
54. (a) Ichimura, K. (2000) Photoalignment of liquid-crystal systems. *Chem. Rev.*, **100** (5), 1847–1873; (b) Ishow, E., Brosseau, A., Clavier, G., Nakatani,

K., Pansu, R.B., Vachon, J.J., Tauc, P., Chauvat, D., Mendonca, C.R., and Piovesan, E. (2007) Two-photon fluorescent holographic rewritable micropatterning. *J. Am. Chem. Soc.*, **129** (29), 8970–8971; (c) Okano, K., Tsutsumi, O., Shishido, A., and Ikeda, T. (2006) Azotolane liquid-crystalline polymers: huge change in birefringence by photoinduced alignment change. *J. Am. Chem. Soc.*, **128** (48), 15368–15369; (d) Sobolewska, A. and Miniewicz, A. (2008) On the inscription of period and half-period surface relief gratings in azobenzene-functionalized polymers. *J. Phys. Chem. B*, **112** (15), 4526–4535; (e) Ryabchun, A., Bobrovsky, A., Sobolewska, A., Shibaev, V., and Stumpe, J. (2012) Dual photorecording on cholesteric azobenzene-containing LC polymer films using helix pitch phototuning and holographic grating recording. *J. Mater. Chem.*, **22** (13), 6245.

55. Venkataramani, S., Jana, U., Dommaschk, M., Sonnichsen, F.D., Tuczek, F., and Herges, R. (2011) Magnetic bistability of molecules in homogeneous solution at room temperature. *Science*, **331** (6016), 445–448.
56. Hayashi, T. and Maeda, K. (1960) Preparation of a new phototropic substance. *Bull. Chem. Soc. Jpn.*, **33** (4), 565–566.
57. Satoh, Y., Ishibashi, Y., Ito, S., Nagasawa, Y., Miyasaka, H., Chosrowian, H., Taniguchi, S., Mataga, N., Kato, D., Kikuchi, A., and Abe, J. (2007) Ultrafast laser photolysis study on photodissociation dynamics of a hexaarylbiimidazole derivative. *Chem. Phys. Lett.*, **448** (4–6), 228–231.
58. (a) Delbaere, S., Orio, M., Berthet, J., Sliwa, M., Hatano, S., and Abe, J. (2013) Insights into the recombination of radical pairs in hexaarylbiimidazoles. *Chem. Commun.*, **49** (52), 5841–5843; (b) Goto, T., Tanino, H., and Kondo, T. (1980) Photochemical and thermal interconversions of three dimers of 2,4,5-triphenylimidazolyl (lophyl). *Chem. Lett.*, (4), 431–434.
59. Dessauer, R. (2006) *Photochemistry History and Commercial Applications of Heaarylbiimidazoles*, Elsevier.
60. (a) Guglielmetti, R. (1990) in *Photochromism, Molecules, and Systems* (eds H. Dürr and H. Bouas-Laurent), Elsevier, Amsterdam, pp. 314–466; (b) Bertelson, R.C. (2002) in *Organic Photochromic and Thermochromic Compounds* (eds J.C. Crano and R.J. Guglielmetti), Kluwer Academic Publishers, New York, pp. 11–83.
61. (a) Guiliano, M., Davin-Pretelli, E., Mille, G., Chouteau, J., and Guglielmetti, R. (1978) Etudes spectroscopiques infrarouge et Raman de spiropyrannes et mérocyanines benzothiazoliniques. *Helv. Chim. Acta*, **61** (3), 1072–1085; (b) Samat, A.M., Guglielmetti, R.J., and Martin, G.J. (1976) Etude des formes fermées et ouvertes de spiropyrannes benzothiazoliniques en RMN13C et1H par transformée de Fourier. *Org. Magn. Reson.*, **8** (2), 62–73.
62. (a) Schaudel, B., Guermeur, C., Sanchez, C., Nakatani, K., and Delaire, J.A. (1997) Spirooxazine- and spiropyran-doped hybrid organic-inorganic matrices with very fast photochromic responses. *J. Mater. Chem.*, **7** (1), 61–65; (b) Koelsch, C.F. and Workman, W.R. (1952) Some thermochromic spirans. *J. Am. Chem. Soc.*, **74** (24), 6288–6289.
63. Flannery, J.B. (1968) Photo- and thermochromic transients from substituted 1′,3′,3′-trimethylindolinobenzospiropyrans. *J. Am. Chem. Soc.*, **90** (21), 5660–5671.
64. Berkovic, G., Krongauz, V., and Weiss, V. (2000) Spiropyrans and spirooxazines for memories and switches. *Chem. Rev.*, **100** (5), 1741–1753.
65. (a) Benard, S. and Yu, P. (2000) New spiropyrans showing crystalline-State photochromism. *Adv. Mater.*, **12** (1), 48–50; (b) Benard, S. and Yu, P. (2000) A spirooxazine showing crystalline state photochromism. *Chem. Commun.*, (1), 65–66; (c) Spagnoli, S., Block, D., Botzung-Appert, E., Colombier, I., Baldeck, P.L., Ibanez, A., and Corval,

A. (2005) Photochromism of spiropyran nanocrystals embedded in sol-gel matrices. *J. Phys. Chem. B*, **109** (18), 8587–8591.

66. (a) Van Gemert, B. (2002) in *Organic Photochromic and Thermochromic Compounds* (eds J.C. Crano and R.J. Guglielmetti), Kluwer Academic Publishers, New York, pp. 111–140; (b) Becker, R.S. and Michl, J. (1966) Photochromism of synthetic and naturally occurring 2H-chromenes and 2H-pyrans. *J. Am. Chem. Soc.*, **88** (24), 5931–5933.
67. (a) Dürr, H. (2002) in *Organic Photochromic and Thermochromic Compounds* (eds J.C. Crano and R.J. Guglielmetti), Kluwer Academic Publishers, New York, pp. 223–266; (b) Durr, H. and Kranz, C. (1994) A new-type of photochromic spirodihydroindolizines and their ability of cation-binding. *Mol. Cryst. Liq. Cryst. Sci. Technol. Sect. A*, **246**, 135–138.
68. Norikane, Y., Davis, R., and Tamaoki, N. (2009) Photochromism of a spiroperimidine compound in polymer matrices. *New J. Chem.*, **33** (6), 1327.
69. (a) Whittall, J. (1990) in *Photochromism, Molecules, and Systems* (eds H. Dürr and H. Bouas-Laurent), Elsevier, Amsterdam, pp. 467–492; (b) Heller, H.G. (1995) in *Handbook of Organic Photochemistry and Photobiology* (eds W.M. Horspool and P.-S. Song), CRC Press, Boca Raton, FL; (c) Whittall, J. (1992) in *Applied Photochromic Polymer Systems* (ed C.B.M. Ardle), Blackie, p. 80; (d) Hart, R.J., Heller, H.G., and Salisbur, K. (1968) Photochemical rearrangements of some photochromic fulgimides. *Chem. Commun.*, (24), 1627–1628; (e) Goldschmidt, S., Riedle, R., and Reichardt, A. (1957) Uber die Bisdiphenylenfulgide und die Spaltung der Bisdiphenylenfulgensaure in Optisch Aktive Komponenten. *Ann. Chem.*, **604** (1-3), 121–132.
70. (a) Stobbe, H. and Kullenberg, A. (1905) Amino-triphenyl fulgen acid (13. Treatise on butadiene compounds). *Ber. Dtsch. Chem. Ges.*, **38**, 4087–4090; (b) Stobbe, H. and Kullenberg, A. (1905) Yellow nitro triphenyl fulgen acids and their red fulgides – (12 Treatise on butadiene compounds). *Ber. Dtsch. Chem. Ges.*, **38**, 4081–4087; (c) Stobbe, H. and Eckert, R. (1905) A parallel of coloured furyl- and phenyl-fulgide. (1 Treatise on butadiene compounds). *Ber. Dtsch. Chem. Ges.*, **38**, 4075–4081; (d) Stobbe, H. and Leuner, K. (1905) Two dimethyl-cumyl-fulgen acid and a third isomere. (10 Paper on butadiene connections). *Ber. Dtsch. Chem. Ges.*, **38**, 3897–3903; (e) Stobbe, H. (1905) Yellow monoaryl fulgide – (9 Paper on butadiene connections). *Ber. Dtsch. Chem. Ges.*, **38**, 3893–3897; (f) Stobbe, H. and Leuner, K. (1905) Colourless alkylfulgide (8. essay on the butadien compounds). *Ber. Dtsch. Chem. Ges.*, **38**, 3682–3685.
71. Heller, H.G. and Langan, J.R. (1981) Photochromic heterocyclic fulgides. 3. The use of (E)-alpha-(2,5-dimethyl-3-furylethylidene) (isopropylene) succinic anhydride as a simple convenient chemical actinometer. *J. Chem. Soc., Perkin Trans.*, **2** (2), 341–343.
72. Heller, H.G. and Oliver, S. (1981) Photochromic heterocyclic fulgides. 1. Rearrangement reactions of (E)-alpha-3-furylethylidene (isopropylidene) succinic anhydride. *J. Chem. Soc., Perkin Trans.*, **1** (1), 197–202.
73. (a) Yokoyama, Y., Goto, T., Inoue, T., Yokoyama, M., and Kurita, Y. (1988) Fulgides as efficient photochromic compounds – role of the substituent on furylalkylidene moiety of furylfulgides in the photoreaction. *Chem. Lett.*, (6), 1049–1052; (b) Yokoyama, Y., Iwai, T., Kera, N., Hitomi, I., and Kurita, Y. (1990) Steric effect of alkylidene groups of furylfulgides on the photochromic behavior. *Chem. Lett.*, (2), 263–264.
74. Matsui, F., Taniguchi, H., Yokoyama, Y., Sugiyama, K., and Kurita, Y. (1994) Application of photochromic 5-dimethylaminoindolylfulgide to photon-mode erasable optical memory media with nondestructive readout ability based on wavelength dependence of bleaching quantum yield. *Chem. Lett.*, (10), 1869–1872.

75. (a) Uhlmann, E. and Gauglitz, G. (1996) New aspects in the photokinetics of Aberchrome 540. *J. Photochem. Photobiol., A*, **98** (1-2), 45–49; (b) Guo, Z.X., Wang, G.J., Tang, Y.W., and Song, X.Q. (1995) Photokinetic study on the photochromic reaction of Aberchrome-540(TM) – a further comment about the use of aberchrome-540(TM) in chemical actinometry. *J. Photochem. Photobiol., A*, **88** (1), 31–34.
76. Darcy, P.J., Heller, H.G., Strydom, P.J., and Whittall, J. (1981) Photochromic heterocyclic fulgides. 2. Electrocyclic reactions of (e)-alpha-2,5-dimethyl-3-furylethylidene (alkyl-substituted methylene)-succinic anhydrides. *J. Chem. Soc. Perkin Trans.*, **1** (1), 202–205.
77. Yokoyama, Y. (2000) Fulgides for memories and switches. *Chem. Rev.*, **100** (5), 1717–1739.
78. (a) Deblauwe, V. and Smets, G. (1988) Quantum yields of the photochromic reactions of heterocyclic fulgides and fulgimides. *Makromol. Chem.-Macromol. Chem. Phys.*, **189** (10), 2503–2512; (b) Cabrera, I., Dittrich, A., and Ringsdorf, H. (1991) Thermally irreversible photochromic liquid-crystal polymers. *Angew. Chem. Int. Ed.*, **30** (1), 76–78.
79. Walz, J., Ulrich, K., Port, H., Wolf, H.C., Wonner, J., and Effenberger, F. (1993) Fulgides as switches for intramolecular energy-transfer. *Chem. Phys. Lett.*, **213** (3-4), 321–324.
80. (a) Willner, I., Liondagan, M., Rubin, S., Wonner, J., Effenberger, F., and Bauerle, P. (1994) Photoregulation of alpha-chymotrypsin activity in organic media – effects of bioimprinting. *Photochem. Photobiol.*, **59** (4), 491–496; (b) Willner, I., Rubin, S., Wonner, J., Effenberger, F., and Bauerle, P. (1992) Photoswitchable binding of substrates to proteins – photoregulated binding of alpha-D-mannopyranose to concanavalin-a modified by a thiophenefulgide dye. *J. Am. Chem. Soc.*, **114** (8), 3150–3151.
81. Irie, M. and Mohri, M. (1988) Thermally irreversible photochromic systems – reversible photocyclization of diarylethene derivatives. *J. Org. Chem.*, **53** (4), 803–808.
82. Nakamura, S. and Irie, M. (1988) Thermally irreversible photochromic systems – a theoretical study. *J. Org. Chem.*, **53** (26), 6136–6138.
83. Waldeck, D.H. (1991) Photoisomerization dynamics of stilbenes. *Chem. Rev.*, **91** (3), 415–436.
84. (a) Hiroto, S., Suzuki, K., Kamiya, H., and Shinokubo, H. (2011) Synthetic protocol for diarylethenes through Suzuki-Miyaura coupling. *Chem. Commun.*, **47** (25), 7149–7151; (b) Shorunov, S.V., Krayushkin, M.M., Stoyanovich, F.M., and Irie, M. (2006) A convenient synthesis of 3,4-diaryl(hetaryl)-substituted maleimides and maleic anhydrides. *Russ. J. Org. Chem.*, **42** (10), 1490–1497; (c) Herder, M., Schmidt, B.M., Grubert, L., Patzel, M., Schwarz, J., and Hecht, S. (2015) Improving the fatigue resistance of diarylethene switches. *J. Am. Chem. Soc.*, **137** (7), 2738–2747.
85. (a) Nakashima, T., Atsumi, K., Kawai, S., Nakagawa, T., Hasegawa, Y., and Kawai, T. (2007) Photochromism of thiazole-containing triangle terarylenes. *Eur. J. Org. Chem.*, **19**, 3212–3218; (b) Kawai, S., Nakashima, T., Atsumi, K., Sakai, T., Harigai, M., Imamoto, Y., Kamikubo, H., Kataoka, M., and Kawai, T. (2007) Novel photochromic molecules based on 4,5-dithienyl thiazole with fast thermal bleaching rate. *Chem. Mater.*, **19**, 3479–3483; (c) Jeong, Y.C., Gao, C., Lee, I.S., Yang, S.I., and Ahn, K.H. (2009) The considerable photostability improvement of photochromic terarylene by sulfone group. *Tetrahedron Lett.*, **50** (37), 5288–5290; (d) Kutsunugi, Y., Kawai, S., Nakashima, T., and Kawai, T. (2009) Photochromic properties of terarylene derivatives having a pi-conjugation unit on central aromatic ring. *New J. Chem.*, **33** (6), 1368–1373.
86. (a) Shrestha, S.M., Nagashima, H., Yokoyama, Y., and Yokoyama, Y. (2003) Synthesis of novel thermally irreversible photochromic 1-aryl-1,3-butadiene derivatives. *Bull. Chem. Soc. Jpn.*, **76**

(2), 363–367; (b) Peters, A., Vitols, C., McDonald, R., and Branda, N.R. (2003) Novel photochromic compounds based on the 1-thienyl-2-vinylcyclopentene backbone. *Org. Lett.*, **5** (8), 1183–1186; (c) Takami, S. and Irie, M. (2004) Synthesis and photochromic properties of novel yellow developing photochromic compounds. *Tetrahedron*, **60** (29), 6155–6161; (d) Kochi, J., Ubukata, T., and Yokoyama, Y. (2010) Photochromism of novel Arylbutadienes. ISOP 2010, Yokoyama National University, Yokohama, Japan.

87. Malatesta, V. (1999) Photodegradation of organic photochromes, in *Organic Photochromic and Thermochromic Compounds* (eds J.C. Crano and R.J. Guglielmetti), Plenum, New York.
88. Gautron, R. (1968) Photochromism of indolinospiropyrans. 4. Study of degradation by physical methods. Relationship with structure. *Bull. Soc. Chim.*, (8), 3190.
89. Gautron, R., Eloy, D., Escaffre, P., Guglielmetti, R., Pottier, E., and Tardieu, P. (1991) Colorability and Photodegradation of some spiro[indoline-oxazines] through repeated flashes photolysis. *Bull. Soc. Chim. Belg.*, **100** (4), 315–328.
90. (a) Irie, M. (2000) Diarylethenes for memories and switches. *Chem. Rev.*, **100** (5), 1685–1716; (b) Liu, H.-H. and Chen, Y. (2012) Selective photoconversion of photochromic diarylethenes and their properties. *New J. Chem.*, **36** (11), 2223; (c) Shoji, H. and Kobatake, S. (2013) Thermal bleaching reactions of photochromic diarylethenes with thiophene-S,S-dioxide for a light-starting irreversible thermosensor. *Chem. Commun.*, **49** (23), 2362–2364.
91. Suzuki, H., Tomoda, A., Ishizuka, M., Kaneko, A., Furui, M., and Matsushima, R. (1989) Photochromism of 4-oxazolyl fulgides. *Bull. Chem. Soc. Jpn.*, **62** (12), 3968–3971.
92. (a) Taniguchi, H., Shinpo, A., Okazaki, T., Matsui, F., and Irie, M. (1992) Photodegradation mechanism of photochromic diarylethene derivatives. *Nippon Kagaku Kaishi*, **10**, 1138–1140; (b) Erko, F.G., Berthet, J., Patra, A., Guillot, R., Nakatani, K., Métivier, R., and Delbaere, S. (2013) Spectral, conformational and photochemical analyses of photochromic dithienylethene:cis-1,2-dicyano-1,2-bis(2,4,5-trimethyl-3-thienyl)ethene revisited. *Eur. J. Org. Chem.*, **2013** (34), 7809–7814; (c) Irie, M., Lifka, T., Uchida, K., Kobatake, S., and Shindo, Y. (1999) Fatigue resistant properties of photochromic dithienylethenes: by-product formation. *Chem. Commun.*, (8), 747–748.
93. de Jong, J.J.D., Lucas, L.N., Hania, R., Pugzlys, A., Kellogg, R.M., Feringa, B.L., Duppen, K., and van Esch, J.H. (2003) Photochromic properties of perhydro- and perfluorodithienylcyclopentene molecular switches. *Eur. J. Org. Chem.*, **2003** (10), 1887–1893.
94. Takami, S., Kuroki, L., and Irie, M. (2007) Photochromism of mixed crystals containing bisthienyl-, bisthiazolyl-, and bisoxazolylethene derivatives. *J. Am. Chem. Soc.*, **129** (23), 7319–7326.
95. Takami, S., Kobatake, S., Kawai, T., and Irie, M. (2003) Extraordinarily high thermal stability of the closed-ring isomer of 1,2-Bis(5-methyl-2-phenylthiazol-4-yl)perfluorocyclopentene. *Chem. Lett.*, **32** (10), 892–893.
96. Nakashima, T., Atsumi, K., Kawai, S., Nakagawa, T., Hasegawa, Y., and Kawai, T. (2007) Photochromism of thiazole-containing triangle terarylenes. *Eur. J. Org. Chem.*, **2007** (19), 3212–3218.
97. (a) Yam, V.W.W., Ko, C.C., and Zhu, N.Y. (2004) Photochromic and luminescence switching properties of a versatile diarylethene-containing 1,10-phenanthroline ligand and its rhenium(I) complex. *J. Am. Chem. Soc.*, **126** (40), 12734–12735; (b) Suzuki, K., Ubukata, T., and Yokoyama, Y. (2012) Dual-mode fluorescence switching of photochromic bisthiazolylcoumarin. *Chem. Commun.*, **48** (5), 765–767.
98. Zhu, W., Yang, Y., Métivier, R., Zhang, Q., Guillot, R., Xie, Y., Tian, H., and Nakatani, K. (2011) Unprecedented stability of a photochromic

bisthienylethene based on benzobisthiadiazole as an ethene bridge. *Angew. Chem. Int. Ed.*, **50** (46), 10986–10990.

99. (a) Uchida, K., Tsuchida, E., Aoi, Y., Nakamura, S., and Irie, M. (1999) Substitution effect on the coloration quantum yield of a photochromic bisbenzothienylethene. *Chem. Lett.*, (1), 63–64; (b) Kobatake, S., Uchida, K., Tsuchida, E., and Irie, M. (2000) Photochromism of diarylethenes having isopropyl groups at the reactive carbons. Thermal cycloreversion of the closed-ring isomers. *Chem. Lett.*, (11), 1340–1341.

100. Knie, C., Utecht, M., Zhao, F., Kulla, H., Kovalenko, S., Brouwer, A.M., Saalfrank, P., Hecht, S., and Bleger, D. (2014) ortho-Fluoroazobenzenes: visible light switches with very long-Lived Z isomers. *Chem. Eur. J.*, **20** (50), 16492–16501.

101. Abe, J. (2013) in *New Frontiers in Photochromism* (eds M. Irie, Y. Yokoyama, and T. Seki), Springer, Tokyo, pp. 161–181.

102. Fujita, K., Hatano, S., Kato, D., and Abe, J. (2008) Photochromism of a radical diffusion-inhibited hexaarylbiimidazole derivative with intense coloration and fast decoloration performance. *Org. Lett.*, **10** (14), 3105–3108.

103. (a) Kishimoto, Y. and Abe, J. (2009) A fast photochromic molecule that colors only under UV light. *J. Am. Chem. Soc.*, **131** (12), 4227–4229; (b) Hatano, S. and Abe, J. (2008) Activation parameters for the recombination reaction of intramolecular radical pairs generated from the radical diffusion-inhibited HABI derivative. *J. Phys. Chem. A*, **112** (27), 6098–6103; (c) Mutoh, K., Hatano, S., and Abe, J. (2010) an efficient strategy for enhancing the photosensitivity of photochromic [2.2]paracyclophane-bridged imidazole dimers. *J. Photopolym. Sci. Technol.*, **23** (3), 301–306; (d) Harada, Y., Hatano, S., Kimoto, A., and Abe, J. (2010) Remarkable acceleration for back-reaction of a fast photochromic molecule. *J. Phys. Chem. Lett.*, **1** (7), 1112–1115.

104. Miyasaka, H., Satoh, Y., Ishibashi, Y., Ito, S., Nagasawa, Y., Taniguchi, S., Chosrowjan, H., Mataga, N., Kato, D., Kikuchi, A., and Abe, J. (2009) Ultrafast photodissociation dynamics of a hexaarylbiimidazole derivative with pyrenyl groups: dispersive reaction from femtosecond to 10 ns time regions. *J. Am. Chem. Soc.*, **131** (21), 7256–7263.

105. Patra, A., Métivier, R., Piard, J., and Nakatani, K. (2010) SHG-active molecular nanorods with intermediate photochromic properties compared to solution and bulk solid states. *Chem. Commun.*, **46** (34), 6385–6387.

106. Appriou, P., Guglielmetti, R., and Garnier, F. (1978) Study on photochemical process involved in benzopyranring-opening of photochromic spiropyrans. *J. Photochem.*, **8** (3), 145–165.

107. (a) Irie, M., Eriguchi, T., Takada, T., and Uchida, K. (1997) Photochromism of diarylethenes having thiophene oligomers as the aryl groups. *Tetrahedron*, **53** (36), 12263–12271; (b) Uchida, K. and Irie, M. (1995) A photochromic dithienylethene that turns yellow by UV irradiation. *Chem. Lett.*, (11), 969–970.

108. Bens, A.T., Frewert, D., Kodatis, K., Kryschi, C., Martin, H.D., and Trommsdorff, H.P. (1998) Coupling of chromophores: carotenoids and photoactive diarylethenes – photoreactivity versus radiationless deactivation. *Eur. J. Org. Chem.*, (11), 2333–2338.

109. (a) Morimitsu, K., Kobatake, S., and Irie, M. (2005) Control of cycloreversion quantum yields of diarylethenes by introduction of substituents at the reactive carbons. *Mol. Cryst. Liq. Cryst.*, **431**, 451–454; (b) Morimitsu, K., Kobatake, S., Nakamura, S., and Irie, M. (2003) Efficient photocycloreversion reaction of diarylethenes by introduction of cyano substituents to the reactive carbons. *Chem. Lett.*, **32** (9), 858–859.

110. Irie, M., Lifka, T., Kobatake, S., and Kato, N. (2000) Photochromism of 1,2-bis(2-methyl-5-phenyl-3-thienyl)perfluorocyclopentene in a single-crystalline phase. *J. Am. Chem. Soc.*, **122** (20), 4871–4876.

111. (a) Spangenberg, A., Metivier, R., Gonzalez, J., Nakatani, K., Yu, P., Giraud, M., Leaustic, A., Guillot, R., Uwada, T., and Asahi, T. (2009) Multiscale approach of photochromism: synthesis and photochromic properties of a diarylethene in solution, in nanoparticles, and in bulk crystals. *Adv. Mater.*, **21** (3), 309–313; (b) Kobatake, S., Matsumoto, Y., and Irie, M. (2005) Conformational control of photochromic reactivity in a diarylethene single crystal. *Angew. Chem. Int. Ed.*, **44** (14), 2148–2151.

112. Stellacci, F., Bertarelli, C., Toscano, F., Gallazzi, M.C., Zotti, G., and Zerbi, G. (1999) A high quantum yield diarylethene-backbone photochromic polymer. *Adv. Mater.*, **11** (4), 292–295.

113. (a) Takeshita, M., Choi, C.N., and Irie, M. (1997) Enhancement of the photocyclization quantum yield of 2,2′-dimethyl-3,3′-(perfluorocyclopentene-1,2-diyl)bis(benzo[b]-thiophen e-6-sulfonate) by inclusion in a cyclodextrin cavity. *Chem. Commun.*, (23), 2265–2266; (b) Takeshita, M., Kato, N., Kawauchi, S., Imase, T., Watanabe, J., and Irie, M. (1998) Photochromism of dithienylethenes included in cyclodextrins. *J. Org. Chem.*, **63** (25), 9306–9313; (c) Takeshita, M. and Irie, M. (1999) Photoreversible circular dichroism change of 1,2-bis(1-benzothiophen-3-yl)perfluorocycpentane modified cyclodextrin. *Tetrahedron Lett.*, **40** (7), 1345–1348.

114. Irie, M., Miyatake, O., Uchida, K., and Eriguchi, T. (1994) Photochromic diarylethenes with intralocking arms. *J. Am. Chem. Soc.*, **116** (22), 9894–9900.

115. Massaad, J., Micheau, J.C., Coudret, C., Sanchez, R., Guirado, G., and Delbaere, S. (2012) Gated photochromism and acidity photomodulation of a diacid dithienylethene dye. *Chem. Eur. J.*, **18** (21), 6568–6575.

116. Gostl, R., Kobin, B., Grubert, L., Patzel, M., and Hecht, S. (2012) Sterically crowding the bridge of dithienylcyclopentenes for enhanced photoswitching performance. *Chem. Eur. J.*, **18** (45), 14282–14285.

117. (a) Takeshita, M., Nagai, M., and Yamato, T. (2003) A photochromic thiophenophan-1-ene. *Chem. Commun.*, (13), 1496–1497; (b) Hossain, M.K., Takeshita, M., and Yamato, T. (2005) Synthesis, structure, and photochromic properties of dithia-(dithienylethena)phane derivatives. *Eur. J. Org. Chem.*, (13), 2771–2776.

118. (a) Fukumoto, S., Nakashima, T., and Kawai, T. (2011) Photon-quantitative reaction of a dithiazolylarylene in solution. *Angew. Chem. Int. Ed.*, **50** (7), 1565–1568; (b) Fukumoto, S., Nakashima, T., and Kawai, T. (2011) Intramolecular hydrogen bonding in a triangular dithiazolyl-azaindole for efficient photoreactivity in polar and nonpolar solvents. *Eur. J. Org. Chem.*, **2011** (26), 5047–5053.

119. Nakashima, T., Fujii, R., and Kawai, T. (2011) Regulation of folding and photochromic reactivity of terarylenes through a host-guest interaction. *Chem. Eur. J.*, **17** (39), 10951–10957.

120. Li, W., Jiao, C., Li, X., Xie, Y., Nakatani, K., Tian, H., and Zhu, W. (2014) Separation of photoactive conformers based on hindered diarylethenes: efficient modulation in photocyclization quantum yields. *Angew. Chem. Int. Ed.*, **53** (18), 4603–4607.

121. Li, W., Li, X., Xie, Y., Wu, Y., Li, M., Wu, X.Y., Zhu, W.H., and Tian, H. (2015) Enantiospecific photoresponse of sterically hindered diarylethenes for chiroptical switches and photomemories. *Sci. Rep.*, **5**, 9186.

122. Ogawa, H., Takagi, K., Ubukata, T., Okamoto, A., Yonezawa, N., Delbaere, S., and Yokoyama, Y. (2012) Bisarylindenols: fixation of conformation leads to exceptional properties of photochromism based on 6pi-electrocyclization. *Chem. Commun.*, **48** (97), 11838–11840.

2
Photochromic Transitional Metal Complexes for Photosensitization

Chi-Chiu Ko and Vivian Wing-Wah Yam

2.1
Introduction

Photochemical reactions of most organic compounds (except those with carbonyl substituent) occur predominantly at the lowest-lying singlet excited state (S_1). This is attributed to the very inefficient intersystem crossing (ISC) from the singlet to the triplet state due to its spin-forbidden nature. In contrast, triplet excited-state photochemistry of carbonyl-containing organic compounds are well reported because the singlet–triplet energy splitting of these types of compounds are much smaller, leading to the effective population of their triplet excited state on light absorption [1]. To tackle the inefficient ISC, the photoreactivity of an organic compound derived from the triplet state can be achieved by using a triplet photosensitizer, through which light is absorbed by the photosensitizer and energy is transferred from its triplet excited state to the acceptor organic molecule to populate the triplet excited state (Figure 2.1). To achieve photosensitization, which bypasses the reactivity from singlet excited states, the photosensitizer must possess long-lived and higher-lying triplet (T_1) energy than the acceptor organic molecule for efficient energy transfer but lower-lying singlet state (S_1) for selective photoexcitation of the photosensitizer (Figure 2.1).

In the organic photochromic systems, even though the photochromism is originated from different reaction mechanisms such as reversible *trans–cis* photoisomerizations as well as the reversible photoinduced ring-opening and ring-closing reactions, most of these reactions are derived from the excited singlet state [2]. Studies on the photochromic reactivity of azo compounds [3], stilbenes [3], spiropyrans [4], and spirooxazines [5] via the triplet excited-state pathway achieved by intermolecular photosensitization with organic photosensitizers have also been reported in the literature [3–5]. Along with the rapid development of phosphorescent metal polypyridine complexes in the 1970s, diffusion-controlled intermolecular sensitization of photochromic reactions of stilbenes with the triplet excited state of tris(bipyridyl)ruthenium(II) [6] and tricarbonyl rhenium(I) bipyridine [8] complexes has been demonstrated. The

Photochromic Materials: Preparation, Properties and Applications, First Edition.
Edited by He Tian and Junji Zhang.

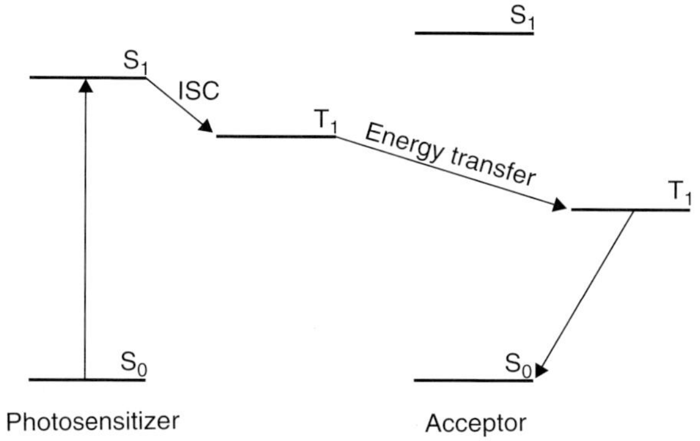

Figure 2.1 Schematic representation of photosensitization via triplet–triplet energy transfer process (ISC: intersystem crossing).

use of inorganic/organometallic triplet photosensitizers provides a number of advantages over the organic counterparts. These include the much more efficient ISC due to the spin–orbit coupling in the presence of heavy metal atom and the ease of systematic variations of the triplet state energy of the transition metal complexes. However, the efficiency of the intermolecular triplet–triplet energy transfer processes is limited by diffusion. Such limitation can be overcome by an intramolecular process to result in highly efficient photosensitization processes.

In recent years, wide varieties of ligands containing different families of photochromic moieties have been designed and coordinated to transition metal complex systems to achieve photosensitization and/or to afford metal complexes with photoswitchable functional properties [8]. This section is only focused on the photosensitization of photochromic ligands with different transition metal complex systems. The rational design of ligands, selection of metal complex systems, and the effects of metal coordination on their photophysical and photochromic properties will also be discussed.

2.2 Photosensitization of Stilbene- and Azo-Containing Ligands

The pioneering study of intramolecular photosensitization of 4-stilbazole (STPY, Figure 2.2), which is a stilbene-like ligand, from the triplet metal-to-ligand charge transfer (^{3}MLCT) excited state was reported in 1973 by Whitten and coworkers [9]. This work was based on their earlier investigation of the intermolecular sensitization of stilbene with the tris(bipyridyl)ruthenium(II) complex [6]. The *trans–cis* photoisomerization of STPY in $[Ru(bpy)_2(STPY)_2]^{2+}$ has been carried out with excitation into the charge transfer absorption band of the ruthenium complex [10]. The energy transfer process from the ^{3}MLCT excited state to the triplet ligand-centered (^{3}LC) or intraligand (^{3}IL) state of the stilbene-containing ligand was first proposed on the basis of the short-lived phosphorescence

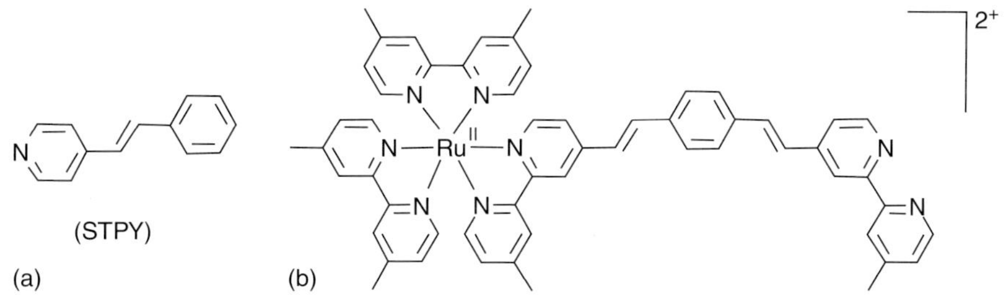

Figure 2.2 Schematic drawing of (a) STPY and (b) the tris(bipyridyl)ruthenium(II) complex with styryl-bridged bipyridine ligand.

together with the observation of a much longer-lived transient absorption in the mononuclear and dinuclear tris(bipyridyl)ruthenium complexes with the styryl-bridged bipyridine ligand (Figure 2.2) [10].

On the other hand, Yam and coworkers have designed a series of stilbene- and azo-containing pyridine ligands (Figure 2.3) and incorporated these ligands into the tricarbonyl rhenium(I) diimine complex system [11]. As the triplet state of the *trans*-isomer of these ligands is lower lying than the ^{3}MLCT state of these rhenium(I) complexes, triplet *trans*-to-*cis* photoisomerization has been photosensitized by the intramolecular energy transfer process from the ^{3}MLCT excited state. Such intramolecular energy transfer process from the ^{3}MLCT state to the ^{3}LC(*trans*-isomer) state and the mechanism for the photosensitized photoisomerization (Figure 2.3b) have recently been elucidated by ultrafast time-resolved absorption, infrared, and resonance Raman spectroscopic studies on $[Re(CO)_3(bpy)(STPY)]^+$ by Vlček and coworkers [12]. The mechanism was further supported by computational studies [13].

Due to the efficient intramolecular quenching of the ^{3}MLCT state, the typically observed MLCT phosphorescence of the *trans*-isomer in the rhenium complex system has become very weak. On conversion to the *cis*-isomer, the ^{3}LC state became higher lying due to the reduced π electron delocalization [14], and thus the intramolecular energy transfer process from the ^{3}MLCT state to ^{3}LC state would be disfavored. Consequently, the MLCT phosphorescence of the rhenium(I) complex was restored upon conversion from the *trans*-isomer to the *cis*-isomer (Figure 2.3c). Thus, these complexes not only demonstrated the photosensitization of the photochromic stilbene- or azo-containing ligands but also led to the development of photoswitchable luminescent materials. Later on, other stilbene-like pyridine ligands, such as 1,3,5-tris(4-ethenylpyridyl)benzene, 1,2,4,5-tetrakis(4-ethenylpyridyl)benzene, 1,2-di-(4-pyridyl)ethylene, octadecyl-substituted styrylpyridine, azacrown-containing styrylpyridine, and 4-(1-naphthylvinyl)pyridine have been designed and incorporated into the rhenium(I) complex systems to achieve photosensitized photochromism and to develop complexes with photoswitchable luminescent properties [15]. The potential use of these photoswitchable luminescent complexes as cell imaging probe has also been illustrated [16g].

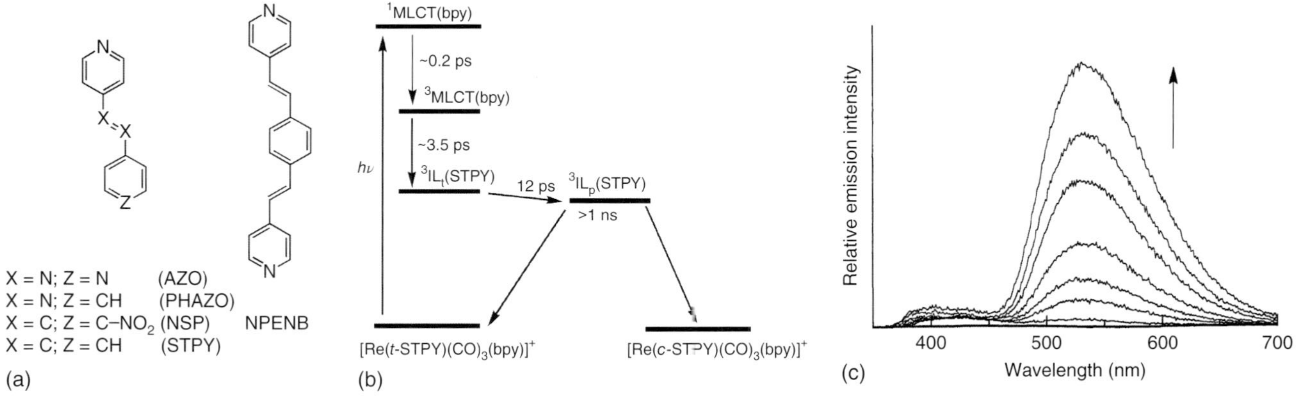

Figure 2.3 (a) Schematic drawing of stilbene and azo-containing pyridine ligands. (b) Excited-state dynamics of $[Re(CO)_3(bpy)(STPY)]^+$. (From [12]. Reproduced with the permission from American Chemical Society.) (c) Emission spectral traces of $[Re(CO)_3(phen)(NSP)]^+$ upon irradiation at $\lambda = 330$ nm in degassed CH_2Cl_2. (From [12]. Reproduced with the permission from Royal Society of Chemistry.)

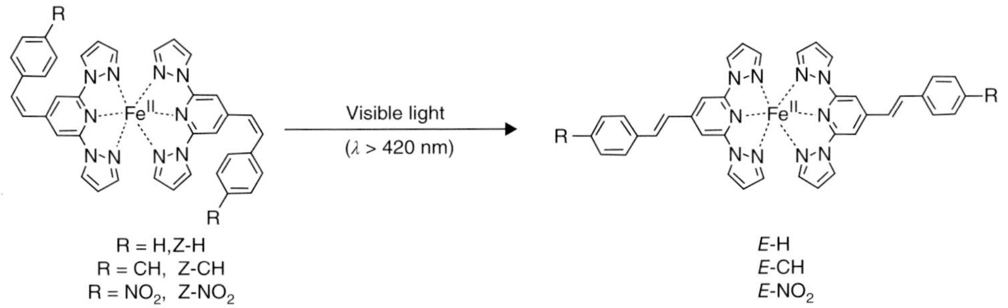

Figure 2.4 Visible light-induced *cis*-to-*trans* isomerization of substituted [Fe(STPPY)$_2$]. (From [17c]. Reproduced with the permission from American Chemical Society.)

Apart from using the phosphorescent MLCT excited state to sensitize the photoisomerization of the C=C and N=N double bonds, excitation into the MLCT absorption of the iron(II) complexes [16] such as [Fe(STPY)$_4$X$_2$] and [Fe(STPPY)$_2$] (STPPY = 2,6-di(1*H*-pyrazol-1-yl)-4-styrylpyridine) (Figure 2.4) has also been shown to induce *cis*-to-*trans* isomerization of the C=C bond in the ligands. However, the detailed mechanisms including the excited state dynamics for these MLCT-induced photoisomerizations have not been elucidated and would require further investigations.

Other than photosensitization with the MLCT excited states of the metal complexes, excitation into the ligand-to-ligand (interligand) charge transfer (LL′CT) transition [π(thiolate) → π^*(diimine)] in the ditholato-bipyridine platinum(II) complexes (Figure 2.5) could also induce efficient *cis*-to-*trans* isomerization of the azo-containing diimine (Figure 2.5a) and the azo-containing benzene-1,2-dithiolato (Figure 2.5b) ligands [17]. This extended the excitation energy to much lower energy in the visible region when compared to that in azobenzene. By incorporating both the azo-containing diimine and the azo-containing benzene-1,2-dithiolato ligands into the platinum complexes, the *cis*-to-*trans* isomerizations of both ligands could be probed by excitation into the LL′CT transition. Because the *trans*-to-*cis* isomerizations of the two different ligands in the platinum complex could be induced by different wavelengths of excitations, a photocontrollable ternary photochromic system with three different stable states has been obtained (Figure 2.5c).

2.3
Photosensitization of Spirooxazine-Containing Ligands

The photochromism of spirooxazines is attributed to the reversible photoinduced ring-opening reaction from the cleavage of the spiro-C–O bond, which results in the planarization of the molecule and the extension of the π-conjugation in the colored photomerocyanine conformer [2]. Unlike the *trans*–*cis* photoisomerization, in which the triplet reaction pathway has been well documented [1, 3], the photochromic ring-opening reaction of spirooxazines

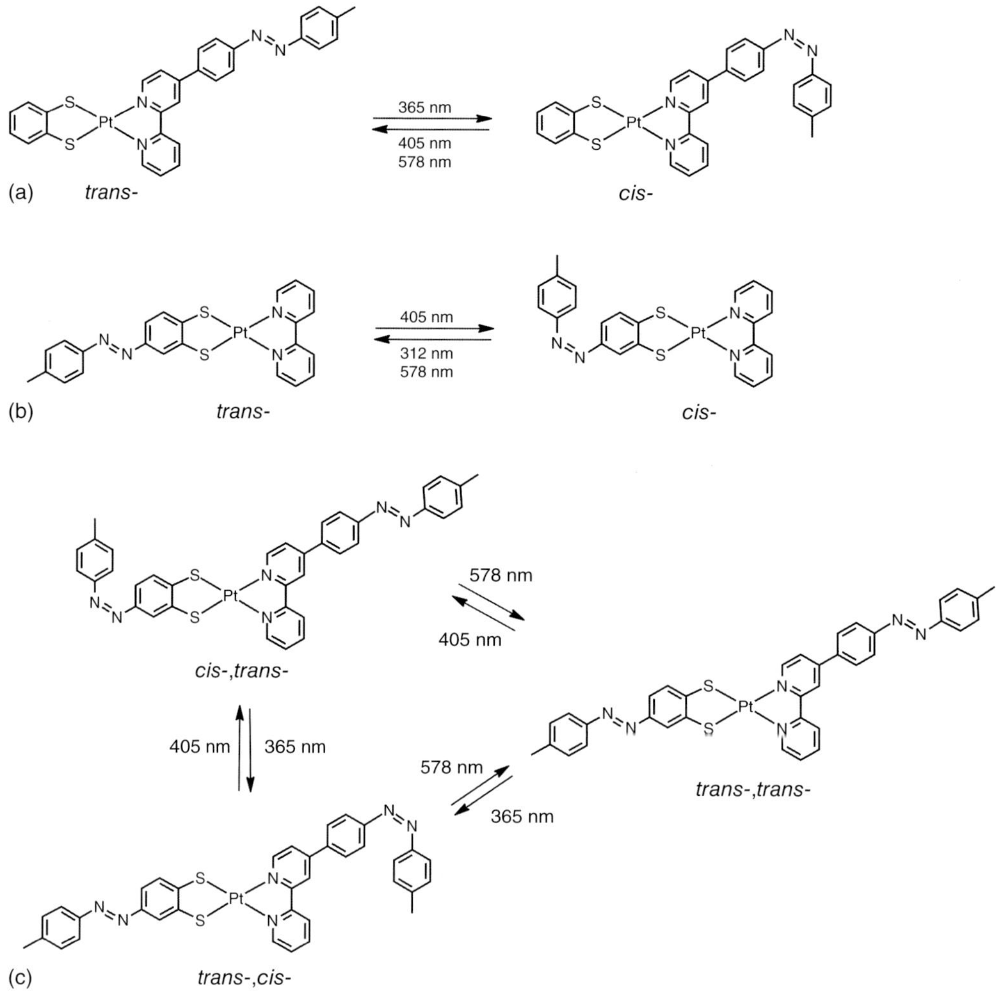

Figure 2.5 Dithiolato-bipyridine platinum(II) complexes with (a) azo-containing bipyridine ligand, (b) azo-containing benzene-1,2-dithiolato ligand, and (c) photocontrollable ternary photochromic platinum(II) complex with two different azo-containing ligands. (From [18b]. Reproduced with the permission from Wiley-VCH.)

from the triplet excited state was only revealed by Wilkinson and coworker [5] in 1996 with the use of organic triplet sensitizers. By the introduction of nicotinyl substituent, a spirooxazine-containing pyridine ligand (**SOPY**) was developed. The coordination of this ligand to rhenium(I) tricarbonyl diimine system, $[Re(CO)_3(N–N)(\textbf{SOPY})]^+$ (N–N = ^{t}Bu$_2$bpy, Me$_2$bpy, phen) (Figure 2.6) with sufficient triplet-state energy has led to successful photosensitization of the photochromic ring-opening reaction [18]. The successful generation of photomerocyanine (open form) with the photoexcitation into the MLCT

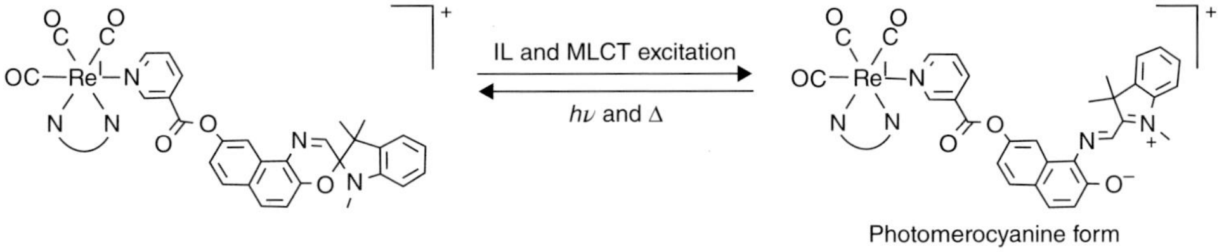

Figure 2.6 Photosensitized photochromism of [Re(CO)$_3$(N−N)(**SOPY**)]$^+$ (N−N = ^{t}Bu$_2$bpy, Me$_2$bpy, phen). (From [19]. Reproduced with the permission from American Chemical Society.)

absorption band of these complexes, where the free ligand (**SOPY**) does not absorb, suggested an intramolecular photosensitization of the spirooxazine by the MLCT excited state [19]. As all these complexes only show LC phosphorescence instead of the typical MLCT phosphorescence, the photochromic ring-opening reaction was proposed to be initiated with an efficient intramolecular energy transfer from the ^{3}MLCT state to the reactive ^{3}SOPY state, followed by ring-opening reaction. With the triplet photosensitization in the complexes, the excitation wavelength for the photochromism could be extended from the UV region in the free ligand to the lower-energy visible region in the metal complexes.

Extension of this work to a series of spirooxazine-containing bipyridine ligands (**L1**−**L4**) and their tricarbonyl rhenium(I) complexes (Figure 2.7a) was reported [19]. However, the photochromism of the spirooxazine-containing bipyridine ligands could not be sensitized by the ^{3}MLCT excited state of the rhenium(I) complexes as the ^{3}MLCT state to the ^{3}LC state triplet−triplet energy transfer was disfavored by the lower-lying ^{3}MLCT state (166−188 kJ mol^{-1}, estimated from the phosphorescence of these complexes) compared to the ^{3}LC state of the spirooxazine (210−225 kJ mol^{-1}) [20]. On the contrary, the lower-lying ^{3}MLCT excited states of these complexes have been shown to lead to quenching of the triplet photochromic pathway (Figure 2.7b), which resulted in lower quantum efficiency for the photochromic ring-opening reaction [9, 20]. The quenching of the reactive ^{3}LC state by the ^{3}MLCT state was supported by the photochromic ring-opening reaction quantum yields, which have been found to be in line with the energies of the MLCT excited states ([Re(CO)$_3$(**L1**)Cl] (0.15) < [Re(CO)$_3$(**L2**)Cl] (0.18) < [Re(CO)$_3$(**L3**)Cl] (0.48) < [Re(CO)$_3$(**L4**)Cl] (0.65)). Similarly, the significant decrease of photocyclization quantum yields of the fulgimide-containing photochromic ligand in the tris(bipyridyl)osmium(II) and tris(bipyridyl)ruthenium(II) complexes (Figure 2.7c) due to intramolecular quenching of the reactive ^{3}LC state by the lower-lying ^{3}MLCT excited state have also been proposed and supported by transient absorption spectroscopic studies [20].

Through the incorporation of spirooxazine-containing bipyridine ligand (**L2**) to the bis-arylalkynylplatinum(II) complexes (Figure 2.8), which possess higher ^{3}LL′CT/^{3}MLCT energy, the photochromism of **L2** can be photosensitized with the LL′CT/MLCT excitation [22]. With the photochromic

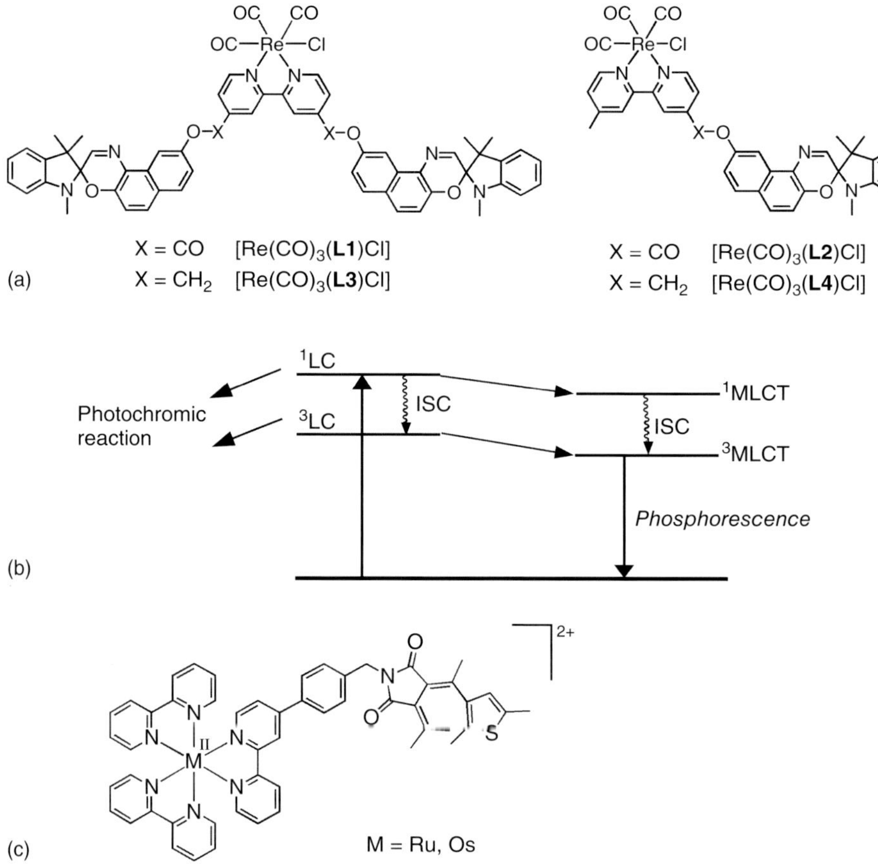

Figure 2.7 (a) Rhenium(I) tricarbonyl complexes with spirooxazine-containing bipyridine ligands. (b) Qualitative energy state diagram for the quenching of the photochromic reaction with the lower lying MLCT excited state. (From [20]. Reproduced with the permission from Wiley-VCH.)(c) Fulgimide-containing tris(bipyridyl) osmium(II) and tris(bipyridyl) ruthenium(II) complexes.

spirooxazine-containing ligand together with the gelator-functionalized alkynyl ancillary ligands, some of these complexes not only exhibit sensitized photochromism with emission reversibly switched by the photochromic reactions but also show liquid-crystalline and gelation properties (Figure 2.8) [21].

2.4
Photosensitization of Diarylethene-Containing Ligands

The study of the photodegradation of the photochromic stilbene, which was attributed to the photocyclization of the *cis*-isomer to give the dihydrophenanthrene [2, 22], has led to the development of the new families of photochromic diarylethenes. With a cyclic group, such as cyclopentenes,

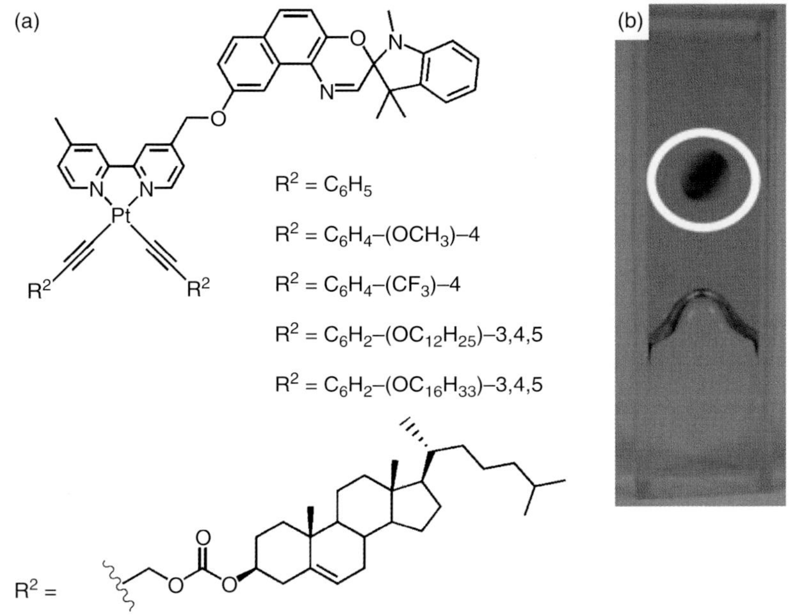

Figure 2.8 (a) Schematic diagram for [Pt(**L2**)(C≡CR^2)$_2$]. (b) Photochromic gel formed by the cholesteryl platinum(II) bipyridine complex. (From [21]. Reproduced with the permission of Wiley-VCH.)

perfluorocyclopentenes, maleic anhydrides, and maleimides, fixing the *cis*-conformation of the aryl groups in the diarylethenes, the diarylthenes would undergo reversible photocyclization and photocycloreversion reaction [23]. The study of Irie and coworkers revealed that both photochromic isomers are thermally stable when the thiophene derivatives with low aromatic stabilization energy are used as the aryl groups in the diarylethenes [23, 24]. These studies have provided the important foundation for the development of thermally stable photochromic diarylethenes.

The triplet photochromic reaction pathway for diarylethenes was not discovered until the study of the diarylethene-containing transition metal complexes (Figure 2.9) [25]. The earliest design of diarylethene-containing ligand and its application for the photomodulation of luminescent properties of the metal complexes have been studied by Lehn and coworker [26]. In 2004, the photosensitizations of the photochromism of the diarylethenes were reported independently by two different research groups at almost the same time. These include the photosensitization of diarylethene-containing phenanthroline ligand in its tricarbonylrhenium(I) complex (Figure 2.8a) reported by Yam and coworkers [26a] and the photosensitization of diarylethene-containing bridging bipyridine ligand in the dinuclear tris(bipyridyl)ruthenium(II) complex (Figure 2.8b) reported by Belser, De Cola, and coworkers [26b]. The designs of these ligands represent two very different approaches and thus have led to the development of different types

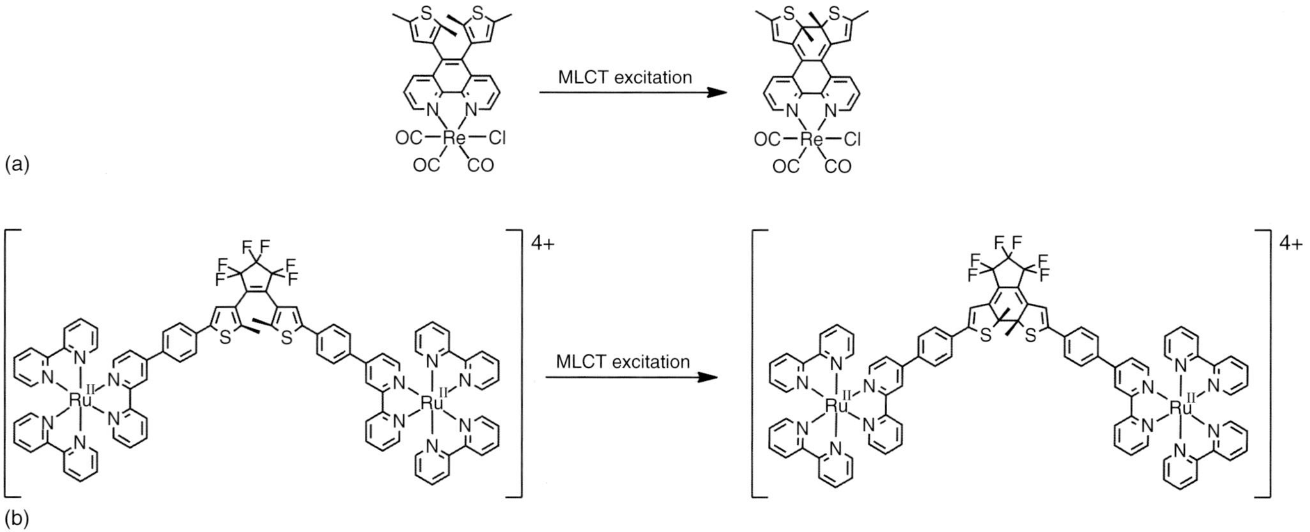

Figure 2.9 Triplet photocyclization of (a) diarylethene-containing phenanthroline ligand in the tricarbonyl rhenium(I) complex and (b) diarylethene-containing bridging bipyridine ligand in the dinuclear tris(bipyridyl)ruthenium(II) complex.

of diarylethene-containing ligands and transition metal complexes. Similar to the earliest ligand design reported by Lehn and coworkers [27], the diarylethene-containing bridging bipyridine ligand was designed by linking the thiophene moieties of the diarylethene unit to two bipyridine ligand through the phenyl spacer [26b]. The diarylethene-containing phenanthroline ligand represented a new and different design involving the ligand itself directly forming part of the diarylethene framework [26a], which should possess a greater perturbation and interaction between the metal center and the photochromic moiety. Excitation into the MLCT absorption of the rhenium(I) complex led to the photocyclization reaction of the diarylethene unit. Subsequent ultrafast transient absorption studies confirmed the triplet-state photochromism and photosensitization [27].

In the dinuclear ruthenium(II) complex (Figure 2.10a), the photochromic cyclization reaction would similarly be carried out with the excitation into the MLCT absorption band of the ruthenium(II) complex. The energy transfer from the ^{3}MLCT state to the ^{3}IL state followed by photocyclization was proposed [26]. With the reversible photochromic reaction, the phosphorescence of the ruthenium(II) complexes could be switched on and off (Figure 2.10b). It is worth to mention that the photochromism of the same ligands in the dinuclear osmium(II) analog could not be photosensitized by the MLCT excitation and was largely suppressed. This has been rationalized by the fact that the energy of the ^{3}MLCT state in the osmium(II) complexes would be considerably lower than that of the ruthenium(II) counterpart as well as the ^{3}LC state of the diarylethene. As a result, the photochromic reaction was inhibited due to the quenching of the reactive LC excited state of the dithienylethene by the lower-lying MLCT state (Figure 2.7b). Due to similar reason, a significant slowing down of the photocyclization of another diarylethene-containing bipyridine ligand upon coordination to iron(II) metal center (Figure 2.10c) has been reported [28], which was in contrast to successful photosensitization of the photochromism in the ruthenium(II) complex analog. These two examples illustrated the effect of the excited-state energy of the metal complexes as the photosensitizers.

The drastic decrease of the photocyclization quantum yields due to the quenching of the photochromic reactive excited state in the presence of the lower-lying MLCT/LL′CT excited state of a platinum(II) complex with two diarylethenes has also been reported [29]. The systematic variation of excited-state energy by modifying the π-conjugation of the diarylethenes not only suppressed the quenching pathways but also led to stepwise/multistage photochromism [30]. By the control of energy transfer processes through the rational choice of suitable metal complex linkers and diarylethene-containing ligands, other multistage photochromic transition metal complexes with more than two diarylethene moieties have also been reported [30].

Other diarylethene-containing ligands, which connected the diarylethene moiety to various ligands through a spacer unit, have been designed to construct a supramolecular system linking the inorganic/organometallic photosensitizers with different types of diarylethenes [29–36]. For the photochromic bis(2-methylbenzothien-3-yl)maleimide-coupled phenanthroline ligands (**L5** and **L6**)

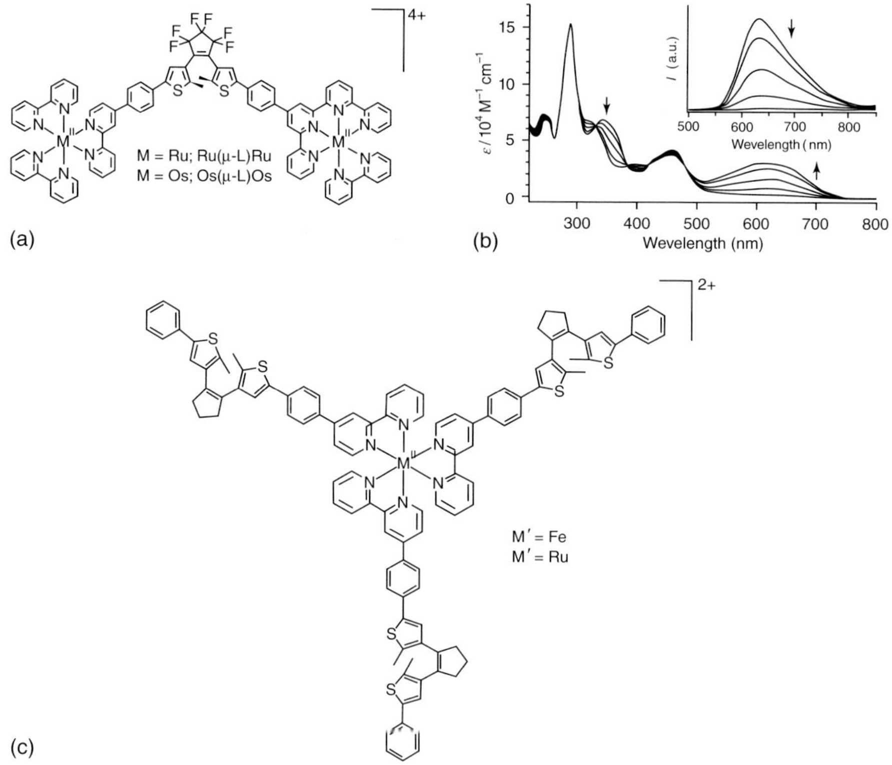

Figure 2.10 (a) Dithienylethene-bridged dinuclear tris(bipyridyl)ruthenium(II) and osmium(II) complexes. (b) UV–vis spectral changes accompanying the photocyclization of the dinuclear ruthenium complex. Inset: the corresponding decrease in emission intensity of the dinuclear ruthenium complex. Conditions: MeCN, 293 K. (From [26b]. Reproduced with the permission from American Chemical Society.) (c) Tris(bipyridyl) iron(II) and ruthenium(II) complexes with three dithienylcyclopentenes.

reported by Scandola and coworkers [31], their photochromism could also be sensitized by the ^{3}MLCT excited state in their ruthenium(II) complexes, $[Ru(bpy)_2(\mathbf{L5})]^{2+}$ and $[Ru(bpy)_2(\mathbf{L6})]^{2+}$ (Figure 2.11a). The singlet and triplet photochromic cyclization pathway of the free ligands and the MLCT-sensitized photochromism in the complexes were investigated in detail by ultrafast spectroscopic and computational studies. Their study revealed that the rates of photocyclization from the ^{3}LC excited state (~10 μs in deaerated solvent) are much slower than that from the ^{1}LC excited state (~1 ps) and the triplet photocyclization is oxygen sensitive (Figure 2.11b). Moreover, the internal conversion processes including the ^{3}MLCT-^{3}LC triplet–triplet energy transfer would be significantly slowed down in **L6** in the presence of the additional CH_2 unit in the spacer. This showed the importance of the linker in the ligand design for efficient photosensitization.

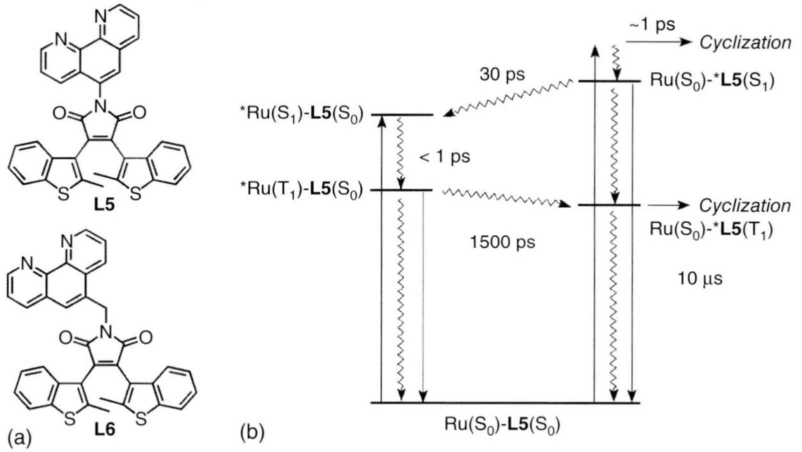

Figure 2.11 (a) Bis(2-methylbenzothien-3-yl)maleimide-coupled polypyridine ligands. (b) Qualitative energy-level diagram and photocyclization mechanism for $[Ru(bpy)_2(\mathbf{L5})]^{2+}$. (From [32]. Reproduced with the permission of American Chemical Society.)

Linker-dependent photosensitization of diarylethenes with the ^{3}MLCT/^{3}LL′CT state of platinum(II) terpyridine system (Figure 2.12) was also reported [32]. The lengthening together with the introduction of a nonconjugated group ($CH_2{-}OC_6H_4$) in the linkage (Figure 2.12) would result in reduction of orbital overlap between the metal-based and LC states and thus reduce the efficiency of the energy transfer process. The reduced energy transfer efficiency not only would affect the kinetics of the internal conversion processes but would also lead to substantial decrease of the quantum yields of the corresponding MLCT/LL′CT photosensitized cyclization reaction (Figure 2.12) [33b]. Apart from the successful intramolecular sensitization with the triplet excited state of tris(bipyridyl)ruthenium(II) [26–31, 33] and terpyridylplatinum(II) [33] photosensitizers, intramolecular sensitizations by the triplet excited state of diphosphino dialkynyl platinum(II) complexes [34], bis(pyridyl)copper(II) complexes [35], and perylenebisimide organic photosensitizers [36] have also been reported.

On top of the ligand design derived from connecting the ligating group with photochromic units as pendants, a novel ligand design involving the ligand itself directly forming part of the diarylethene framework was first reported by Yam and coworkers in 2004 [26a]. The functionalization of 1,10-phenanthroline with the thienyl groups at 5- and 6-positions has led to the development of a series of photochromic 5,6-dithienyl-1,10-phenanthroline ligands (Figure 2.13), which have the phenanthroline ligating moiety in the photochromic diarylethene framework [26a]. In these photochromic dithienylphenanthroline ligands, the parallel (photochemically inactive) and antiparallel (photochemically active) forms are not readily interconvertible at room temperature as the rotations of the thienyl rings are restricted by the steric hindrance of the 1,10-phenanthroline ring. The conrotatory cyclization processes of these ligands as predicted by the Woodward–Hoffman rule have been corroborated by NMR spectroscopic study

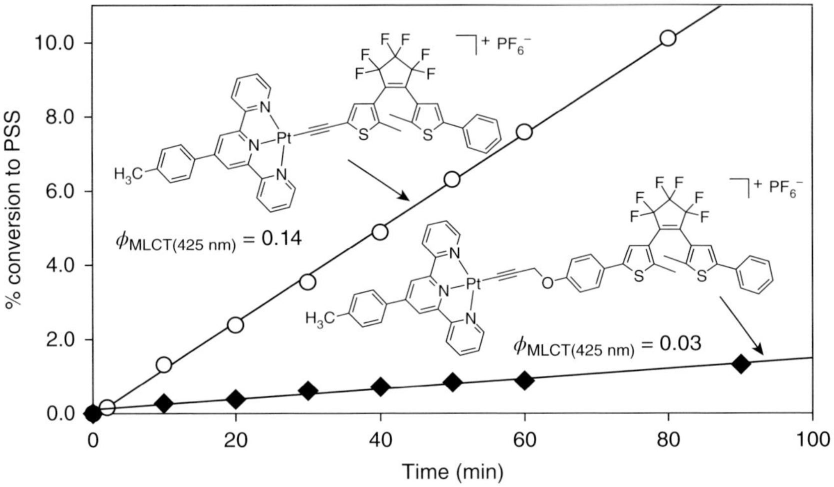

Figure 2.12 Percent conversion to the photostationary states as a function of time for CH_3CN solutions of the diarylethene-containing platinum(II) terpyridine complexes upon irradiation into their MLCT absorption band. (From [33b]. Reproduced with the permission from American Chemical Society.)

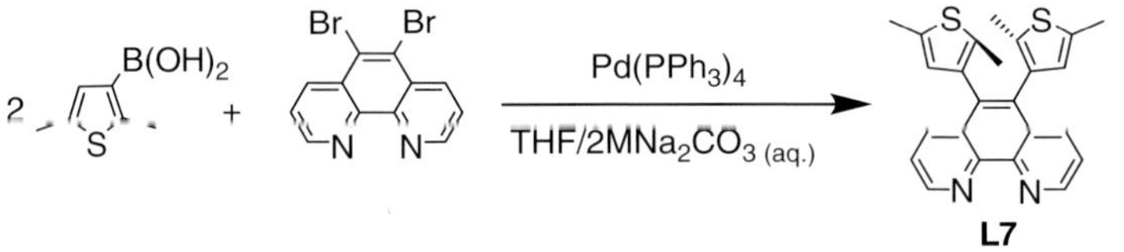

Figure 2.13 Synthetic routes to photochromic 5,6-dithienyl-1,10-phenanthroline ligand. (From [26a]. Reproduced with the permission from American Chemical Society.)

through the monitoring of the variations of the two conformations of the open form and closed form during the photocyclization reaction [28].

The photosensitization of **L7** was first reported in the tricarbonylrhenium(I) complexes [26a, 28]. In the rhenium(I) complex, the photocyclization of **L7** could be carried out with the photoexcitation into the LC and MLCT absorptions. With the MLCT-sensitized photochromism, the excitation wavelength could be extended from the high-energy UV light ($\lambda \leq 340$ nm), corresponding to the LC absorption in the free ligand, to the MLCT absorption band with $\lambda \leq 480$ nm in the visible region. The sensitized photochromism involving an intramolecular energy transfer from the ^{3}MLCT state to the ^{3}IL state prior to the cyclization (Figure 2.14) has been suggested [26a] and confirmed by the ultrafast transient absorption spectroscopy [28]. Based on the reversible photochromic reactivities, the phosphorescence energy and origins of the complex could also be reversibly switched (Figure 2.14c). The photochromism of these ligands has also been shown to be sensitized by the ^{3}MLCT excited state of different metal complex systems including the bis(alkynyl)platinum(II) [37] and tris(bipyridyl)ruthenium(II)

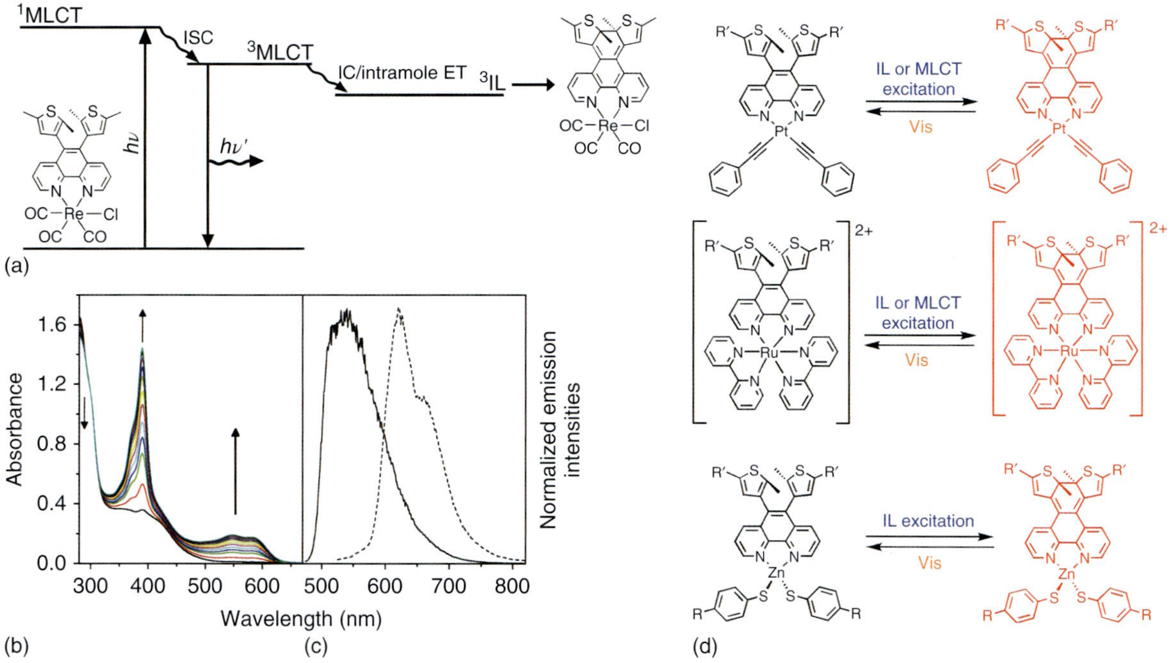

Figure 2.14 (a) Proposed qualitative energy-level diagram for the photosensitized photochromism of [Re(CO)$_3$(**L7**)Cl] by MLCT excitation. (b) UV–vis absorption spectral changes of [Re(CO)$_3$(**L7**)Cl] in benzene (7.16×10^{-5} M) upon MLCT excitation at 440 nm. (c) Overlaid normalized corrected emission spectra of the open form (——) and the closed form (- - -) of [Re(CO)$_3$(**L7**)Cl] in EtOH–MeOH glass (4 : 1 v/v) at 77 K. (d) Photochromism of different complexes of **L7**. (From [28]. Reproduced with the permission from Wiley-VCH.)

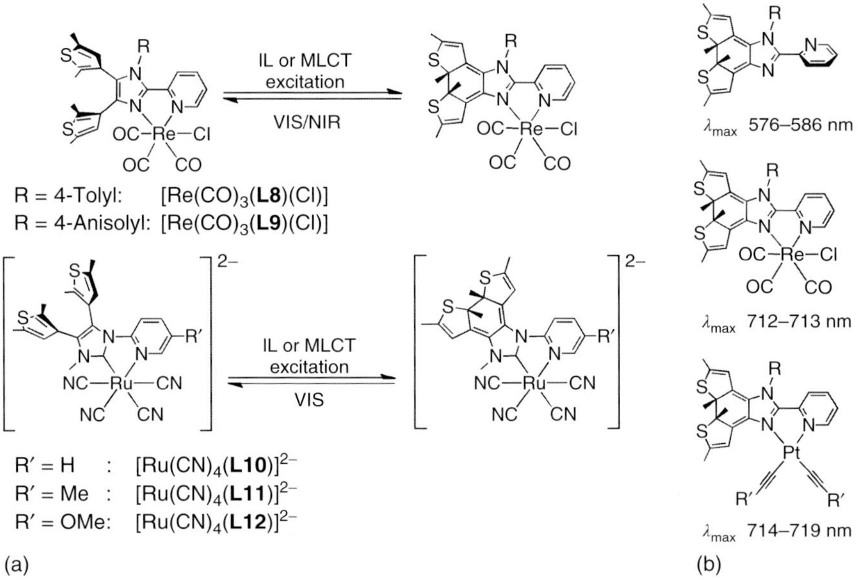

Figure 2.15 (a) Photochromism of the tricarbonylrhenium(I) and the tetracyanoruthenate(II) complexes with different dithienyl-substituted 2-pyridylimidazole ligands. (b) Absorption maxima of the closed forms of the dithienyl-substituted 2-(2-pyridyl)imidazole ligands and their metal complexes.

complexes (Figure 2.14d) [38]. However, it could not be sensitized by the ^{3}LL′CT excited state of the bis(thiolate)zinc(II) complexes [39].

Subsequently, similar design of the diarylethene-containing diimine (**L8**–**L9**) and pyridyl *N*-heterocyclic carbene (**L10**–**L12**) ligands based on the 2-pyridylimidazoles has also been reported [40, 41]. Similar to **L7**, the photocyclization of **L8**–**L9** in the tricarbonylrhenium(I) complexes and **L10**–**L12** in the tetracyanoruthenate(II) complexes could be photosensitized with the ^{3}MLCT excited state of the complexes (Figure 2.15a) [41, 42]. Importantly, the absorptions of the closed forms in the rhenium(I) complexes were found to be significantly red-shifted compared to those of their free ligands (710 nm vs 580 nm) (Figure 2.15b). This was attributed to the more extensive π-conjugation due to the metal-assisted planarization of the 2-pyridylimidazole moiety and the perturbation of the metal center [41]. Such effects have also been observed when these ligands were incorporated into the dialkynylplatinum(II) complexes [42]. This provides a simple and effective strategy in the design of the NIR photochromic diarylethenes.

Along with the design of the photochromic diimine ligands, various diarylethene-containing cyclometalated 2-thienylpyridine ligands have also been developed [43, 44]. Photosensitization of these cyclometalated ligands and the significant red shift of their closed-form absorption in their cyclometalated iridium(III) [44] and platinum(II) [45] complexes have also been demonstrated (Figure 2.16a). The photochromic behaviors and the absorption properties of the

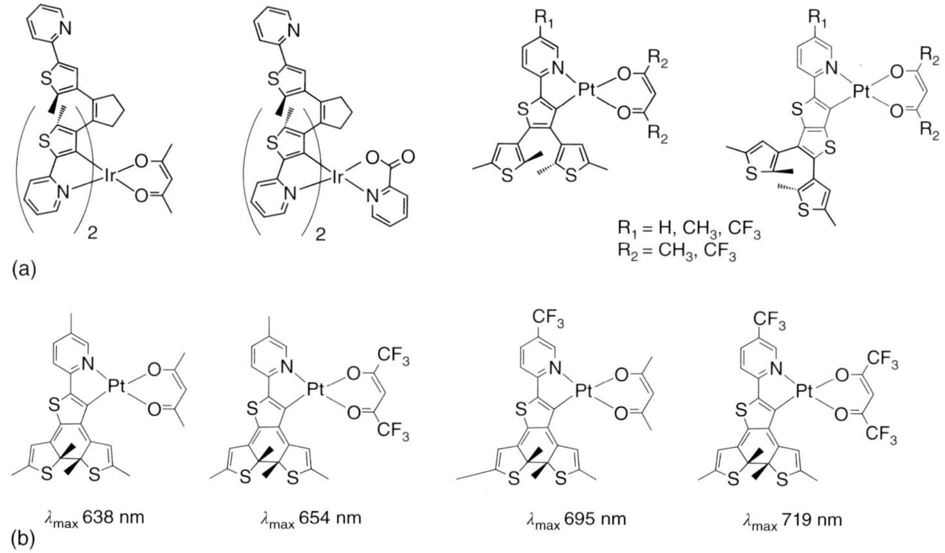

Figure 2.16 (a) Diarylethene-containing cyclometalated iridium(III) and platinum(II) complexes. (b) Effect of the substituent on the diarylethene-containing ligand and the ancillary ligand on the closed-form absorption.

closed forms of these cyclometalated complexes were also found to vary with the nature of the ancillary ligands of the complexes (Figure 2.16b) [44c, 45], which has offered an alternative way to fine-tune the photochromism. In the cyclometalated iridium(III) complexes, their luminescent properties were shown to be photoswitchable by the photochromic reactivities. These photoswitchable properties of these complexes have been demonstrated to be useful as photomarkers in living cells [44c].

Other than the above examples, a wide variety of diarylethene-containing ligands [30, 31b, 45] (Figure 2.17) have been designed and used to develop transition metal complexes with photomodulating luminescent properties, photoswitchable electronic communication, photoswitchable nonlinear optical properties, and electrochromism. By coordinating these ligands to a suitable transition metal complex system with a sufficient triplet excited state energy, it is anticipated that their photochromism could also be photosensitized by the excited state of the transition metal complex, leading to new sets of perturbed photochromic properties.

2.5 Photosensitization of Photochromic N^C-Chelate Organoboranes

The photochromism of N^C-chelate organoboranes has recently been reported [46]. By introduction of an ethynyl ligating substituent on the N^C chelate of $BMes_2$ (Figure 2.18), it has been incorporated into the tricarbonylrhenium(I) bipyridine, diphosphinoplatinum(II), and phosphinogold(I) complexes [46c]. In

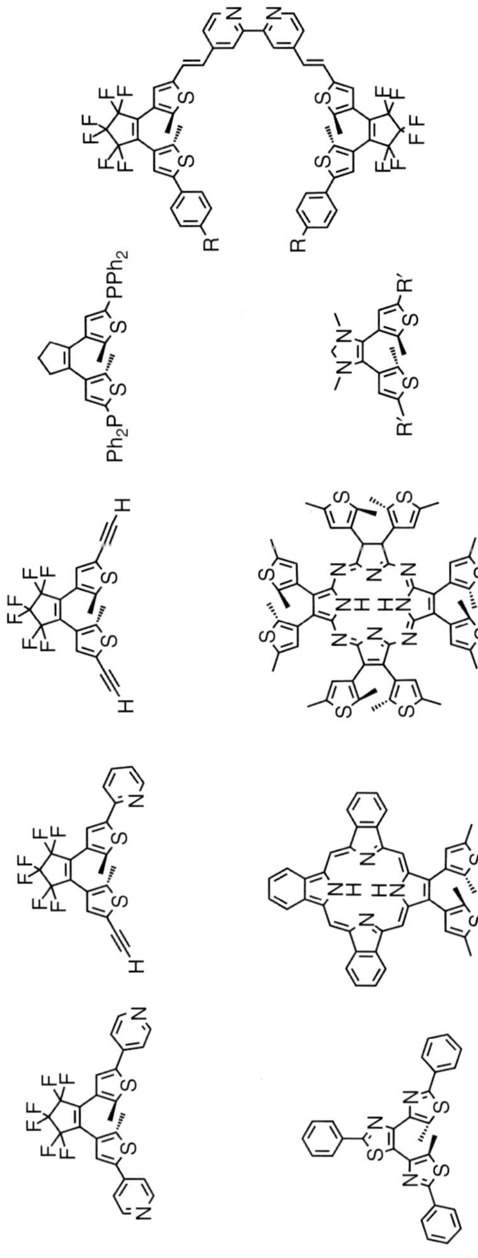

Figure 2.17 Selected examples of different types of diarylethene-containing ligands.

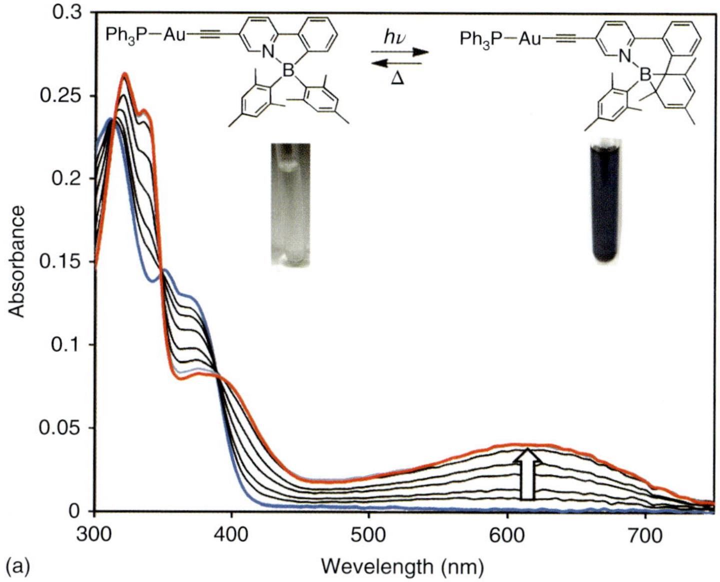

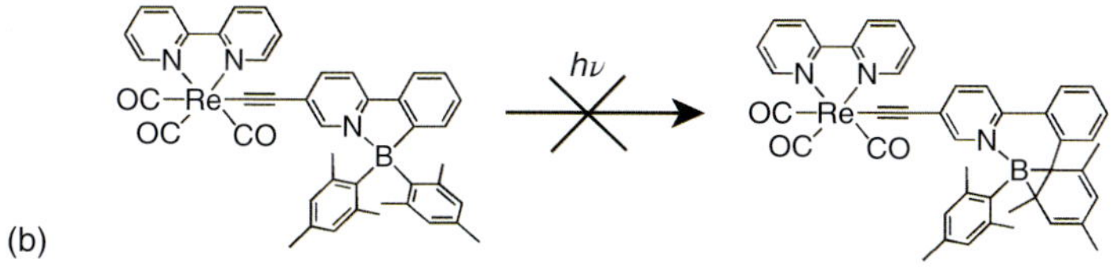

Figure 2.18 (a) UV–vis absorption spectral change of phosphino gold(I) complexes with photochromic N^C chelate of $BMes_2$ (10 μM) in toluene upon excitation at 365 nm. (b) Inhibited photochromic reaction in the rhenium complex with the same N^C chelate of $BMes_2$. (From [46c]. Reproduced with the permission from Wiley-VCH.)

the rhenium(I) complexes, the photochromic activity was completely inhibited. Similar to those observed in other photochromic systems [21, 27b, 30, 31], the inactivity was attributed to the presence of the lower-lying ^{1}MLCT and ^{3}MLCT excited state, which quenched the reactive excited state. In contrast, triplet-sensitized photochromism of the N^C chelate of $BMes_2$ was observed in the gold(I) complex [46c]. The energy transfer processes in these complexes have also been supported from the calculated triplet energies of rhenium(I), gold(I), and boron(III) moieties [46c].

2.6 Conclusion

As illustrated from the above-mentioned examples, the photochromic reactivity of many different families of photochromic compounds could proceed through their singlet and triplet excited states. With judicious choice of a triplet photosensitizer with sufficient energy for energy transfer process, the photochromism can occur through triplet-sensitized reaction pathway. The efficiency of the triplet-sensitized pathway can be enhanced by the heavy atom effect when inorganic/organometallic triplet sensitizers are employed. The intramolecular photosensitization of the photochromic ligands, designed by coupling the photochromic moiety with a ligating group or incorporating a ligand framework into the photochromic moiety, can be achieved by the coordination of an inorganic/organometallic system with long-lived and higher-lying triplet excited state. To improve the efficiency of the photosensitization, the distance, conjugation of the linkage, and energy difference between the triplet donor (photosensitizer) and acceptor (photochromic unit) should be optimized. This not only extends the photochromism to lower-energy excitation but also provides a simple approach in tuning the photochromic properties, including quantum yields, photoreaction kinetics, and absorption properties. In addition to sensitized photochromism, the incorporation of photochromic ligand into the transition metal complexes has also led to the development of transition metal complexes with photoswitchable functional properties.

References

1. Turro, N.J., Ramamurthy, V., and Scaiano, J.C. (2010) *Modern Molecular Photochemistry of Organic Molecules*, University Science Books, Sausalito, CA.
2. Dürr, H. and Bouas-Laurent, H. (1990) *Photochromism: Molecules and Systems*, Elsevier, Amsterdam.
3. (a) Hammond, G.S., Saltiel, J., Lamola, A.A., Turro, N.J., Bradshaw, J.S., Cowan, D.O., Counsell, R.C., Vogt, V., and Dalton, C. (1964) Mechanisms of photochemical reactions in solution. XXII. 1 Photochemical *cis-trans* isomerization. *J. Am. Chem. Soc.*, **86** (16), 3197–3217; (b) Saltiel, J. (1967) Perdeuteriostilbene. The role of phantom states in the *cis-trans* photoisomerization of stilbenes. *J. Am. Chem. Soc.*, **89** (4), 1036–1037; (c) Lee, E.K.C., Denschlag, H.O., and Haninger, G.A. Jr., (1968) *Cis–trans* isomerization of 2-butenes in the gas phase: benzene photosensitization at 2537 Å. *J. Chem. Phys.*, **48** (10), 4547–4555; (d) Saltiel, J., Neuberger, K.R., and Wrighton, M. (1969) Nature of the intermediates in the sensitized *cis–trans* photoisomerization of alkenes. *J. Am. Chem. Soc.*, **91** (13), 3658–3659.
4. Becker, R.S. and Roy, J.K. (1965) The spectroscopy and photosensitization of various photochromic spiropyrans. *J Phys. Chem.*, **69** (4), 1435–1436.
5. Hobley, J. and Wilkinson, F. (1996) Photochromism of naphthoxazine-spiro-indolines by direct excitation and following sensitisation by triplet-energy donors. *J. Chem. Soc., Faraday Trans.*, **92** (8), 1323–1330.
6. Whitten, D.G. and Zarnegar, P.P. (1971) Photochemistry of ruthenium complexes. Ligand isomerization via orbitally different excited states. *J. Am. Chem. Soc.*, **93** (15), 3776–3777.
7. (a) Wrighton, M. and Morse, D.L. (1974) Nature of the lowest excited state in tricarbonylchloro-1,10-phenanthrolinerhenium(I) and related complexes. *J. Am. Chem. Soc.*, **96** (4), 998–1003; (b) Wrighton, M.S., Morse, D.L. and Pdungsap, L. (1975) Intraligand lowest excited states in tricarbonylhalobis(styrylpyridine)-rhenium(I) complexes. *J. Am. Chem. Soc.*, **97** (8), 2073–2079.
8. (a) Nishihara, H. (2005) Combination of redox- and photochemistry of azo-conjugated metal complexes. *Coord.*

Chem. Rev., **249** (13-14), 1468–1475; (b) Kume, S. and Nishihara, H. (2008) Photochrome-coupled metal complexes: molecular processing of photon stimuli. *Dalton Trans.*, **2005** (25), 3260–3271; (c) Ko, C.C. and Yam, V.W.W. (2010) Transition metal complexes with photochromic ligands—photosensitization and photoswitchable properties. *J. Mater. Chem.*, **20** (11), 2063–2070; (d) Guerchais, V., Ordronneau, L., and Le Bozec, H. (2010) Recent developments in the field of metal complexes containing photochromic ligands: Modulation of linear and nonlinear optical properties. *Coord. Chem. Rev.*, **254** (21-22), 2533–2545; (e) Hasegawa, Y., Nakagawa, T., and Kawai, T. (2010) Recent progress of luminescent metal complexes with photochromic units. *Coord. Chem. Rev.*, **254** (21-22), 2643–2651; (f) Akita, M. (2011) Photochromic organometallics, a stimuli-responsive system: an approach to smart chemical systems. *Organometallics*, **30** (1), 43–51.

9. Zarnegar, P., Bock, C.R., and Whitten, D.G. (1973) Photoreactions of transition metal complexes. Ligand reactivity as a probe for excited-state characterization. *J. Am. Chem. Soc.*, **95** (13), 4367–4372.
10. Shaw, J.R., Webb, R.T., and Schmehl, R.H. (1990) Intersystem crossing to both ligand-localized and charge-transfer excited states in mononuclear and dinuclear ruthenium(II) diimine complexes. *J. Am. Chem. Soc.*, **112** (3), 1117–1123.
11. Yam, V.W.W., Lau, V.C.Y., and Cheung, K.K. (1995) Synthesis, photophysics and photochemistry of novel luminescent rhenium(I) photoswitchable materials. *J. Chem. Soc., Chem. Commun.*, 259–261.
12. Busby, M., Matousek, P., Towrie, M., and Vlček, A. Jr., (2005) Ultrafast excited-state dynamics preceding a ligand *trans–cis* isomerization of *fac*-$[Re(Cl)(CO)_3(t\text{-}4\text{-styrylpyridine})_2]$ and *fac*-$[Re(t\text{-}4\text{-styrylpyridine})(CO)_3(2,2'\text{-bipyridine})]^+$. *J. Phys. Chem. A*, **109** (13), 3000–3008.
13. Bossert, J. and Daniel, C. (2006) *Trans–cis* photoisomerization of the styrylpyridine ligand in $[Re(CO)_3(2,2'\text{-bipyridine})(t\text{-}4\text{-styrylpyridine})]^+$: role of the metal-to-ligand charge-transfer excited states. *Chem. Eur. J.*, **12** (18), 4835–4843.
14. Murov, S.L. (1973) *Handbook of Photochemistry*, Marcel Dekker, New York.
15. (a) Lewis, J.D., Pertuz, R.N., and Moore, J.N. (2000) Proton-controlled photoisomerization: rhenium(I) tricarbonyl bipyridine linked to amine or azacrown ether groups by a styryl pyridine bridging ligand. *Chem. Commun.*, 1865–1866; (b) Sun, S.S. and Lees, A.J. (2002) Synthesis, photophysical properties, and photoinduced luminescence switching of trinuclear diimine rhenium(I) tricarbonyl complexes linked by an isomerizable stilbene-like ligand. *Organometallics*, **21** (1), 39–49; (c) Dattelbaum, D.M., Itokazu, M.K., Iha, N.Y.M., and Meyer, T.J. (2003) Mechanism of metal-to-ligand charge transfer sensitization of olefin *trans*-to-*cis* isomerization in the *fac*-$[Re^I(phen)(CO)_3(1,2\text{-bpe})]^+$ cation. *J. Phys. Chem. A*, **107** (20), 4092–4095; (d) Wenger, O.S., Henling, L.M., Day, M.W., Winkler, J.R., and Gray, H.B. (2004) Photoswitchable luminescence of rhenium(I) tricarbonyl diimines. *Inorg. Chem.*, **43** (6), 2043–2048; (e) Patrocinio, A.O.T., Brennaman, M.J., Meyer, T.J., and Iha, N.Y.M. (2010) Excited-state dynamics in *fac*-$[Re(CO)_3(Me_4phen)(L)]^+$. *J. Phys. Chem. A*, **114** (46), 12129–12137; (f) Lin, J.L., Chen, C.W., Sun, S.S., and Lees, A.J. (2011) Photoswitching tetranuclear rhenium(I) tricarbonyl diimine complexes with a stilbene-like bridging ligand. *Chem. Commun.*, **47** (21), 6030–6032; (g) Sathish, V., Babu, E., Ramdass, A., Lu, Z.Z., Chang, T.T., Velayudham, M., Thanasekaran, P., Lu, K.L., Li, W.S., and Rajagopal, S. (2013) Photoswitchable alkoxy-bridged binuclear rhenium(I) complexes – a potential probe for biomolecules and optical cell imaging. *RSC Adv.*, **3** (40), 18557–18566.
16. (a) Tissot, A., Boillot, M.-L., Pillet, S., Codjovi, E., Boukheddaden, K., and Daku, L.M.L. (2010) Unidirectional photoisomerization of styrylpyridine for switching the magnetic behavior of an iron(II) complex: a MLCT pathway in crystalline solids. *J. Phys. Chem. C*, **114** (49), 21715–21722; (b) Hasegawa, Y., Takahashi, K., Kume,

S., and Nishihara, H. (2011) Complete solid state photoisomerization of bis(dipyrazolylstyrylpyridine)iron(II) to change magnetic properties. *Chem. Commun.*, **47** (24), 6846–6848; (c) Takahashi, K., Hasegawa, Y., Sakamoto, R., Nishikawa, M., Kume, S., Nishibori, E., and Nishihara, H. (2012) Solid-state ligand-driven light-induced spin change at ambient temperatures in bis(dipyrazolylstyrylpyridine)iron(II) complexes. *Inorg. Chem.*, **51** (9), 5188–5198.

17. (a) Sakamoto, R., Murata, M., Kume, S., Sampei, H., Sugimoto, M., and Nishihara, H. (2005) Photo-controllable tristability of a dithiolato-bipyridine-Pt(II) complex molecule containing two azobenzene moieties. *Chem. Commun.*, 1215–1217; (b) Sakamoto, R., Kume, S., Sugimoto, M., and Nishihara, H. (2009) *Trans–cis* photoisomerization of azobenzene-conjugated dithiolato-bipyridine platinum(II) complexes: extension of photoresponse to longer wavelengths and photocontrollable tristability. *Chem. Eur. J.*, **15** (6), 1429–1439.
18. Yam, V.W.W., Ko, C.C., Wu, L.X., Wong, K.M.C., and Cheung, K.K. (2000) Syntheses, crystal structure, and photochromic properties of rhenium(I) complexes containing the spironaphthoxazine moiety. *Organometallics*, **19** (10), 1820–1822.
19. Ko, C.C., Wu, L.X., Wong, K.M.C., Zhu, N., and Yam, V.W.W. (2004) Synthesis, characterization and photochromic studies of spirooxazine-containing 2,2′-bipyridine ligands and their rhenium(I) tricarbonyl complexes. *Chem. Eur. J.*, **10** (3), 766–776.
20. Jukes, R.T.F., Kühni, J., Salluce, N., Belser, P., De Cola, L., and Hartl, F. (2009) Photochemical, photophysical and redox properties of novel fulgimide derivatives with attached 2,2′-bipyridine (bpy) and $[M(bpy)_3]^{2+}$ (M = Ru and Os) moieties. *Dalton Trans.*, 3993–4002.
21. Kellogg, R.M., Groen, M.B., and Wynberg, H. (1967) Photochemically induced cyclization of some furyl- and thienylethenes. *J. Org. Chem.*, **32** (10), 3093–3100.
22. Li, Y.G., Tam, A.Y.Y., Wong, K.M.C., Li, W., Wu, L.X., and Yam, V.W.W. (2011) Synthesis, characterization, and the photochromic, luminescence, metallogelation and liquid-crystalline properties of multifunctional platinum(II) bipyridine complexes. *Chem. Eur. J.*, **17** (29), 8048–8059.
23. Irie, M. (2000) Diarylethenes for memories and switches. *Chem. Rev.*, **100** (5), 1685–1716.
24. Nakamura, S. and Irie, M. (1988) Thermally irreversible photochromic systems. A theoretical study. *J. Org. Chem.*, **53** (26), 6136–6138.
25. (a) Yam, V.W.W., Ko, C.C., and Zhu, N. (2004) Photochromic and luminescence switching properties of a versatile diarylethene-containing 1,10-phenanthroline ligand and its rhenium(I) complex. *J. Am. Chem. Soc.*, **126** (40), 12734–12735; (b) Jukes, R.T.F., Adamo, V., Hartl, F., Belser, P., and De Cola, L. (2004) Photochromic dithienylethene derivatives containing Ru(II) or Os(II) metal units. sensitized photocyclization from a triplet state. *Inorg. Chem.*, **43** (9), 2779–2792.
26. (a) Fernández-Acebes, A. and Lehn, J.M. (1998) Optical switching and fluorescence modulation in photochromic metal complexes. *Adv. Mater.*, **10** (18), 1519–1522; (b) Fernández-Acebes, A. and Lehn, J.M. (1999) Optical switching and fluorescence modulation properties of photochromic metal complexes derived from dithienylethene ligands. *Chem. Eur. J.*, **5** (11), 3285–3292.
27. Ko, C.C., Kwok, W.M., Yam, V.W.W., and Phillips, D.L. (2006) Triplet MLCT photosensitization of the ring-closing reaction of diarylethenes by design and synthesis of a photochromic rhenium(I) complex of a diarylethene-containing 1,10-phenanthroline ligand. *Chem. Eur. J.*, **12** (22), 5840–5848.
28. Zhong, Y.W., Vilà, N., Henderson, J.C., and Abruña, H.D. (2009) Transition-metal tris-bipyridines containing three dithienylcyclopentenes: synthesis, photochromic, and electrochromic properties. *Inorg. Chem.*, **48** (15), 7080–7085.

29. Li, B., Wen, H.M., Wang, J.Y., Shi, L.X., and Chen, Z.N. (2013) Modulating stepwise photochromism in platinum(II) complexes with dual dithienylethene–acetylides by a progressive red shift of ring-closure absorption. *Inorg. Chem.*, **52** (21), 12511–12520.
30. (a) Roberts, M.N., Carling, C.J., Nagle, J.K., Branda, N.R., and Wolf, M.O. (2009) Successful bifunctional photoswitching and electronic communication of two platinum(II) acetylide bridged dithienylethenes. *J. Am. Chem. Soc.*, **131** (46), 16644–16645; (b) Li, B., Wang, J.Y., Wen, H.M., Shi, L.X., and Chen, Z.N. (2012) Redox-modulated stepwise photochromism in a ruthenium complex with dual dithienylethene-acetylides. *J. Am. Chem. Soc.*, **134** (38), 16059–16067; (c) Li, B., Wu, Y.H., Wen, H.M., Wang, J.Y., Shi, L.X., and Chen, Z.-N. (2012) Gold(I)-coordination triggered multistep and multiple photochromic reactions in multi-dithienylethene (DTE) systems. *Inorg. Chem.*, **51** (3), 1933–1942; (d) Hervault, Y.M., Ndiaye, C.M., Norel, L., Lagrost, C., and Rigaut, S. (2012) Controlling the stepwise closing of identical DTE photochromic units with electrochemical and optical stimuli. *Org. Lett.*, **14** (17), 4454–4457; (e) Wen, H.M., Li, B., Wang, J.Y., Shi, L.X., Chen, C.N., and Chen, Z.N. (2013) Multistate photochromism in a ruthenium complex with dithienylethene–acetylide. *Organometallics*, **32** (6), 1759–1765.
31. Indelli, M.T., Carli, S., Ghirotti, M., Chiorboli, C., Ravaglia, M., Garavelli, M., and Scandola, F. (2008) Triplet pathways in diarylethene photochromism: photophysical and computational study of dyads containing ruthenium(II) polypyridine and 1,2-bis(2-methylbenzothiophene-3-yl)maleimide units. *J. Am. Chem. Soc.*, **130** (23), 7286–7299.
32. (a) Roberts, M.R., Nagle, J.K., Finden, J.G., Branda, N.R., and Wolf, M.O. (2009) Linker-dependent metal-sensitized photoswitching of dithienylethenes. *Inorg. Chem.*, **48** (1), 19–21; (b) Roberts, M.R., Nagle, J.K., Majewski, M.B., Finden, J.G., Branda, N.R., and Wolf, M.O. (2011) Charge transfer and intraligand excited state interactions in platinum-sensitized dithienylethenes. *Inorg. Chem.*, **50** (11), 4956–4966.
33. He, B. and Wenger, O.S. (2012) Ruthenium-phenothiazine electron transfer dyad with a photoswitchable dithienylethene bridge: flash-quench studies with methylviologen. *Inorg. Chem.*, **51** (7), 4335–4342.
34. Brayshaw, S.K., Schiffers, S., Stevenson, A.J., Teat, S.J., Warren, M.R., Bennett, R.D., Sazanovich, I.V., Buckley, A.R., Weinstein, J.A., and Raithby, P.R. (2011) Highly efficient visible-light driven photochromism: developments towards a solid-state molecular switch operating through a triplet-sensitised pathway. *Chem. Eur. J.*, **17** (16), 4385–4395.
35. Munakata, M., Han, J., Maekawa, M., Suenaga, Y., Kuroda-Sowa, T., Nabei, A., and Ebisu, H. (2007) A MLCT-switched photochromic copper(II) coordination polymer with 1,2-bis(2′-methyl-5′-(4″-pyridyl)-3′-thienyl)perfluorocyclopentene in crystalline phase. *Inorg. Chim. Acta*, **360** (8), 2792–2796.
36. Fukaminato, T., Doi, T., Tanaka, M., and Irie, M. (2009) Photocyclization reaction of diarylethene–perylenebisimide dyads upon irradiation with visible (>500 nm) light. *J. Phys. Chem. C*, **113** (27), 11623–11627.
37. Lee, J.K.W., Ko, C.C., Wong, K.M.C., Zhu, N., and Yam, V.W.W. (2007) A photochromic platinum(II) bis(alkynyl) complex containing a versatile 5,6-dithienyl-1,10-phenanthroline. *Organometallics*, **26** (1), 12–15.
38. Belser, P., de Cola, L., Hartl, F., Adamo, V., Bozic, B., Chriqui, Y., Iyer, V.M., Jukes, R.T.F., Kühni, J., Querol, M., Roma, S., and Salluce, N. (2006) Photochromic switches incorporated in bridging ligands: a new tool to modulate energy-transfer processes. *Adv. Funct. Mater.*, **16** (2), 195–208.
39. Ngan, T.W., Ko, C.C., Zhu, N., and Yam, V.W.W. (2007) Syntheses, luminescence switching, and electrochemical studies of photochromic dithienyl-1,10-phenanthroline zinc(II) bis(thiolate) complexes. *Inorg. Chem.*, **46** (4), 1144–1152.

40. Lee, P.H.M., Ko, C.C., Zhu, N., and Yam, V.W.W. (2007) Metal coordination-assisted near-infrared photochromic behavior: a large perturbation on absorption wavelength properties of N,N-donor ligands containing diarylethene derivatives by coordination to the rhenium(I) metal center. *J. Am. Chem. Soc.*, **129** (19), 6058–6059.
41. Duan, G.P. and Yam, V.W.W. (2010) Syntheses and photophysical properties of N-pyridylimidazol-2-ylidene tetracyanoruthenates(II) and photochromic studies of their dithienylethene-containing derivatives. *Chem. Eur. J.*, **16** (42), 12642–12649.
42. Wong, H.L., Zhu, N., and Yam, V.W.W. (2014) Photochromic alkynylplatinum(II) diimine complexes containing a versatile dithienylethene-functionalized 2-(2′-pyridyl)imidazole ligand. *J. Organomet. Chem.*, **751**, 430–437.
43. (a) Tan, W., Zhang, Q., Zhang, J., and Tian, H. (2009) Near-infrared photochromic diarylethene iridium (III) complex. *Org. Lett.*, **11** (1), 161–164; (b) Li, X., Zhang, Q., Tu, Y., Ågren, H., and Tian, H. (2010) Modulation of iridium(III) phosphorescence via photochromic ligands: a density functional theory study. *Phys. Chem. Chem. Phys.*, **12** (41), 13730–13736; (c) Tan, W., Zhou, J., Li, F., Yi, T., and Tian, H. (2011) Visible light-triggered photoswitchable diarylethene-based iridium(III) complexes for imaging living cells. *Chem. Asian J.*, **6** (5), 1263–1268.
44. Chan, J.C.H., Lam, W.H., Wong, H.L., Zhu, N., Wong, W.T., and Yam, V.W.W. (2011) Diarylethene-containing cyclometalated platinum(II) complexes: tunable photochromism via metal coordination and rational ligand design. *J. Am. Chem. Soc.*, **133** (32), 12690–12705.
45. (a) Chen, B.Z., Wang, M.Z., Wu, Y.Q., and Tian, H. (2002) Reversible near-infrared fluorescence switch by novel photochromic unsymmetrical-phthalocyanine hybrids based on bisthienylethene. *Chem. Commun.*, 1060; (b) Luo, Q.F., Chen, B.Z., Wang, M.Z., and Tian, H. (2003) Mono-bisthienylethene ring-fused versus multi-bisthienylethene ring-fused photochromic hybrids. *Adv. Funct. Mater.*, **13** (3), 233–239; (c) Tian, H., Chen, B.Z., Tu, H.Y., and Müllen, K. (2002) Novel bisthienylethene-based photochromic tetraazaporphyrin with photoregulating luminescence. *Adv. Mater.*, **14** (12), 918–923; (d) Aubert, V., Guerchais, V., Ishow, E., Hoang-Thi, K., Ledoux, I., Nakatani, K., and Le Bozec, H. (2008) Efficient photoswitching of the nonlinear optical properties of dipolar photochromic zinc(II) complexes. *Angew. Chem. Int. Ed.*, **47** (3), 577–580; (e) Nakagawa, T., Hasegawa, Y., and Kawai, T. (2008) Photoresponsive europium(III) complex based on photochromic reaction. *J. Phys. Chem. A*, **112** (23), 5096–5103; (f) Liu, Y., Lagrost, C., Costuas, K., Tchouar, N., Le Bozec, H., and Rigaut, S. (2008) A multifunctional organometallic switch with carbon-rich ruthenium and diarylethene units. *Chem. Commun.*, 6117–6119; (g) Yam, V.W.W., Lee, J.K.W., Ko, C.C., and Zhu, N. (2009) Photochromic diarylethene-containing ionic liquids and N-heterocyclic carbenes. *J. Am. Chem. Soc.*, **131** (3), 912–913; (h) Liu, Y., Ndiaye, C.M., Lagrost, C., Costuas, K., Choua, S., Turek, P., Norel, L., and Rigaut, S. (2014) Diarylethene-containing carbon-rich ruthenium organometallics: tuning of electrochromism. *Inorg. Chem.*, **53** (15), 8172–8188.
46. (a) Rao, Y.L., Amarne, H., Zhao, S.B., McCormick, T.M., Martic, S., Sun, Y., Wang, R., and Wang, S. (2008) Reversible intramolecular C–C bond formation/breaking and color switching mediated by a N,C-chelate in (2-ph-py)$BMes_2$ and (5-$BMes_2$-2-ph-py)$BMes_2$. *J. Am. Chem. Soc.*, **130** (39), 12898–12900; (b) Rao, Y.L., Amarne, H., and Wang, S. (2012) Photochromic four-coordinate N,C-chelate boron compounds. *Coord. Chem. Rev.*, **256** (5-8), 759–770; (c) Wang, N., Ko, S.B., Lu, J.S., Chen, L.D., and Wang, S. (2013) Tuning the photoisomerization of a N^C-chelate organoboron compound with a metal-acetylide unit. *Chem. Eur. J.*, **19** (17), 5314–5323.

3
Multi-addressable Photochromic Materials

Shangjun Chen, Wenlong Li, and Weihong Zhu

3.1
Molecular Logic Gates

Information processing for molecular computing devices has inspired scientists to design and construct a variety of systems that mimic the semiconductor logic functions. Exactly, photochromic systems with two states between open and closed forms are highly desirable for construction of logic elements, in which each isomer can represent "0" and "1" of a digital binary code. Up to date, not only simple logic gates (e.g., AND, INHIBIT, and NOT) but also more complex devices (e.g., Adders and Subtractors, Multiplexers/Demultiplexers, Encoders/Decoders, Keypad Locks, and Multivalued logic devices) have been successfully developed with multi-addressable photochromic materials using redox, chemicals, or photons as inputs and optical properties as outputs [1–7]. Among a variety of photochromic systems, we focus on multi-addressable photochromic diarylethenes that have received much attention owing to the thermal irreversibility and excellent fatigue resistance [8–15]. Indeed, thermal stability is very critical for information storage, and fatigue resistance allows molecular logic gates to proceed multiple times and to be reset conventionally [16–19]. We illustrate molecular logic gates, data storage, and molecular memory, especially with the multi-addressable unimolecular photochromic platform.

3.1.1
Two-Input Logic Gates

Two-input logic gates such as AND, INHIBIT, and NOT are very important in information processing since they are the fundamental logic elements for complex devices. Generally, the AND gate gives an output of logical "1" only when all the inputs are in the logical "1" state. If any one of the inputs is in the logical "0" state, the output is also "0." For instance, Lu *et al.* reported a dual-fluorescent donor–acceptor dyad system consisting of a tercarbazole donor and photoswitchable imide acceptor (**1**, Figure 3.1) [20]. Interestingly, the emission

Photochromic Materials: Preparation, Properties and Applications, First Edition.
Edited by He Tian and Junji Zhang.

(a)

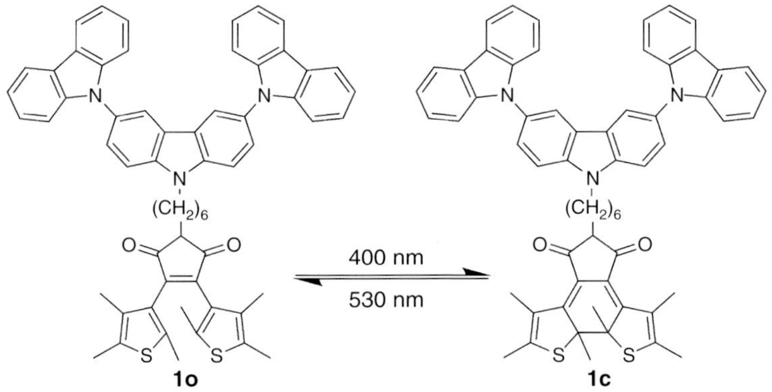

(b)

Input1	Input2	Output
0	0	0
1	0	0
0	1	0
1	1	1

(c)

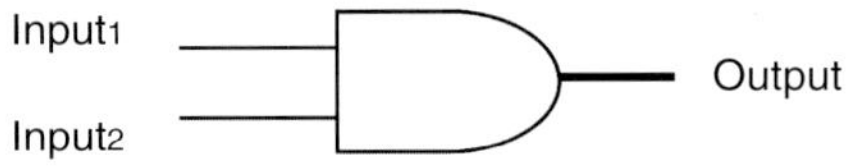

Figure 3.1 (a) Photochromic reaction between dual-fluorescent donor–acceptor dyad **1o** and **1c**, (b) truth table, and (c) symbol of two-input AND logic gate. (Reprinted with permission from [20]. Copyright 2007, American Chemical Society.)

intensity could be regulated by either photoisomeric reaction of diarylethene unit (*Photochromism*) or the polarities of solvents (*Solvatochromism*). Exactly, the AND gate was successfully demonstrated with the emission intensity at 500 nm (Output), which was seriously quenched by irradiation with 400 nm light (Input$_1$) or addition of high-polar solvent (Input$_2$), resulting in a typical AND logic (Figure 3.1).

Concatenation of NOT gate with other logic gates results in the inverted logic gates. For example, a concatenation of a NOT gate with an OR gate results in a NOR, and a concatenation of an AND and a NOT gates gives an INHIBIT (INH) logic gate. Comparing other gates with concatenated NOT, the INHIBIT logic gate does not concern the output but one of the inputs. Tian *et al.* reported a combined NOR and INHIBIT logic gate, whereby the fluorescence arising from the naphthalimide chromophore were exploited as outputs [21]. Since naphthalimides are well-known fluorescent chromophores with high efficiency and chemical stability, incorporation of such an unprecedented six-membered aryl fluorescent moiety as ethene bridge (**2**) exhibits typical photochromism, in which the fluorescence wavelength and intensity arising from the bridged naphthalimide moiety could be well controlled by photochromism and solvatochromism. It is desirable for constructing logic gates. Irradiation at 365 nm for the photocyclization was defined as Input$_1$ that quenched the fluorescence of **2o** in either cyclohexane (420 nm,

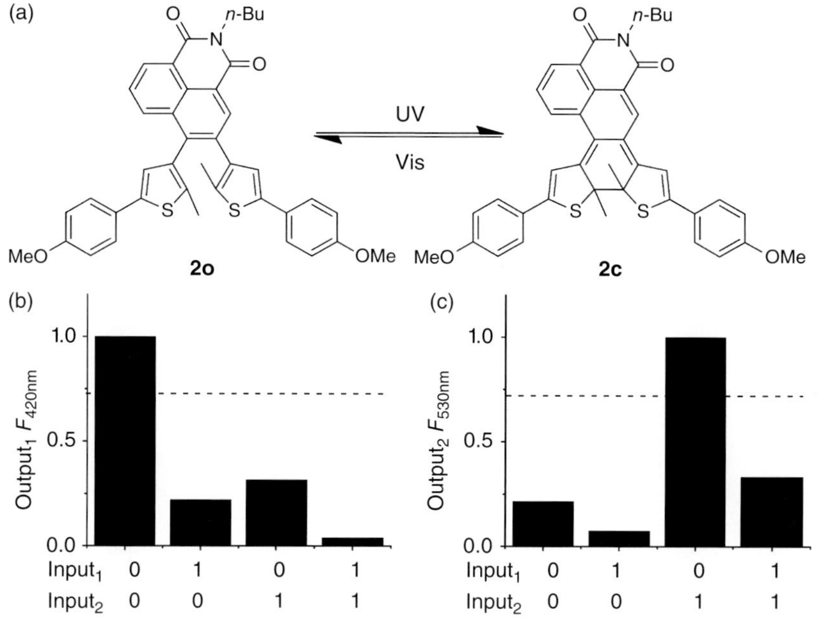

(d)

Input$_1$	Input$_2$	Output
UV	Solvent	F_{420}
0	0	1
1	0	0
0	1	0
1	1	0

(e)

Input$_1$	Input$_2$	Output
UV	Solvent	F_{530}
0	0	0
1	0	0
0	1	1
1	1	0

Figure 3.2 (a) Photochromic reaction between **2o** and **2c**, (b) NOR and (c) INHIBIT gate function, and truth table of (d) NOR and (e) INHIBIT. (Reprinted with permission from [21]. Copyright 2008, American Chemical Society.)

Output$_1$) or acetone (530 nm, Output$_2$), giving an Output 0 regardless of other inputs. The polar solvent by adding acetone into the solution of **2o** in cyclohexane was utilized as Input$_2$. The role of the second input is to redshift the emission band of **2o** from 420 to 530 nm, which also means that Input$_2$ switches Output$_1$ off and switches Output$_2$ on. In binary language, when either Input$_1$ or Input$_2$ is at its 1 state, Output$_1$ is 0. And only when Input$_2$ is at its on state, Output$_2$ is on. So the fluorescence behaviors at 420 and 530 nm are coincided with the Boolean logic NOR and INHIBIT (Figure 3.2), respectively. Reset of the logic gate could be achieved by irradiation with visible light or removing acetone solvent.

3.1.2
Combinatorial Logic Systems

3.1.2.1 Half-Adder and Half-Subtractor

Simple logic gates only perform basic logic operations. Considering multi-addressable photochromic bisthienylethenes, molecular approaches to logic operations have focused especially on the development of more complex molecular devices and on simple electronic circuits. A key requirement for digital computing at the molecular level is to perform arithmetic addition or subtraction operation, which requires the combination of several basic logic gates into more complex circuits, such as half-adder and half-subtractor [22, 23]. The half-adder, a combination of an AND gate and an XOR (exclusive OR) gate that share inputs, carries out binary addition. In contrast, the half-subtractor carrying out binary addition comprises an INHIBIT gate and an XOR gate. A chemically and optically driven binary half-adder based on unimolecular platform was reported by Zhu *et al.* [24]. Due to the existence of pyridine unit, photochromic molecule **3o** (Figure 3.3) can not only respond to light irradiation but also respond to ions (Cu^{2+} and Hg^{2+}) and protons, leading to a multi-addressable system with different switchable absorbance and fluorescence. In tetrahydrofuran (THF) solution, **3o** exhibited a sharp absorption peak at 305 nm and an intense fluorescence

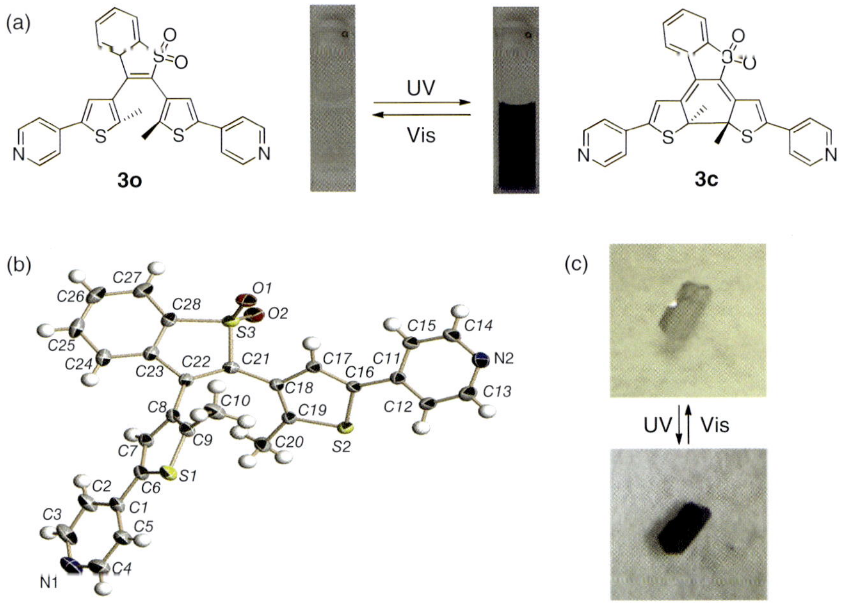

Figure 3.3 (a) Photochromism and photographic images between **3o** and **3c** under the alternative irradiation with UV and visible light in THF, (b) ORTEP representation of the crystal structure of **3o** with displacement ellipsoids shown at the 50% probability level, and (c) color changes of **3o** in the crystalline state. (Reprinted with permission from [24]. Copyright 2012, Royal Society of Chemistry.)

peak at 467 nm upon excitation at 347 nm. Upon irradiation at 365 nm, two new absorption peaks centered at 391 and 575 nm in the visible region emerged, indicative of the formation of closed isomer. Meanwhile, the fluorescence at 467 nm was also quenched. Addition of ions and protons induced bathochromic absorption shift for both the open and closed isomers, that is, the absorbance change from 305 to 345, 345, and 320 nm for the open isomer and from 575 to 638, 640, and 620 nm for the closed isomer upon addition of protons Hg^{2+} and Cu^{2+}, respectively. At the same time, the fluorescence of **3o** was increased by nearly 100% upon adding 2.0 equiv. of Hg^{2+} or H^+, while it was almost completely quenched upon adding 1.0 equiv. of Cu^{2+}. As a molecular half-adder based on **3o**, irradiation at 365 nm (180 s) and addition of Hg^{2+} (2.0 equiv.) were defined as In_1 and In_2, respectively. $Output_1$ (AND) and $Output_2$ (XOR) were the absorptions at 640 and 370 nm. As discussed earlier, the absorption band at 640 nm appeared only when two inputs were applied. So the performance of $Output_1$ worked as an AND logic gate (Figure 3.4). The XOR logic operation was demonstrated by $Output_2$. The absorbance at 370 nm was increased upon neither irradiation of light nor addition of ions, while the value of the band is under the threshold when both the inputs were present or absent. Upon operating with the same inputs and two different output channels, integration of the two molecular logic gates resulted in a half-adder. Meanwhile, molecule **3o** can also be employed as a molecular half-subtractor gate under the same inputs as the half-adder gate. INHIBIT gates give an output of 1 only when one particular input (not the other or both) is applied, which was demonstrated by the performance of the absorption at 345 nm. The value of this band was above the threshold only when addition of Hg^{2+} was applied. On the other hand, the XOR gate used in the half-adder still works (just changing the threshold value). In this way, both the INHIBIT and XOR gates share the same inputs for constructing a half-subtractor (Figure 3.4).

Molecules consisting of different photoswitchable chromophores are promising candidates for applications in molecular logic systems. They may exhibit multiple accessible states via controlling the different photoisomer reactions of each photochromic unit upon excitation. Andréasson *et al.* reported an all-photonic half-adder based on the multi-addressable system **4** (Figure 3.5) comprising a dithienylethene (DTE) and two fulgimides (FGs) [25]. Upon irradiation with 397 nm light, the FG photochrome was photoisomerized independently into the cyclic closure state FGc, while the 302 nm light-induced DTE unit was photoisomerized into the closed forms (DTEc). Thus, there existed four constitutionally isomeric forms (FGo-DTEo, FGc-DTEo, FGo-DTEc, and FGc-DTEc) with total different optical properties. The AND and XOR gates for the half-adder share the same inputs: irradiation at 397 nm (Input a) or at 302 nm (Input b). The absorbance peak at 535 nm corresponding to the isomer FGc-DTEc acts as the output of the AND gate. The all-closed isomer FGc-DTEc can be generated only when both inputs are applied. Therefore, this system can be worked as a photonic AND gate. The absolute value (modulus) in the absorbance change at 393 nm, $|\Delta A|$, was chosen as the output of the XOR gate. The absorption

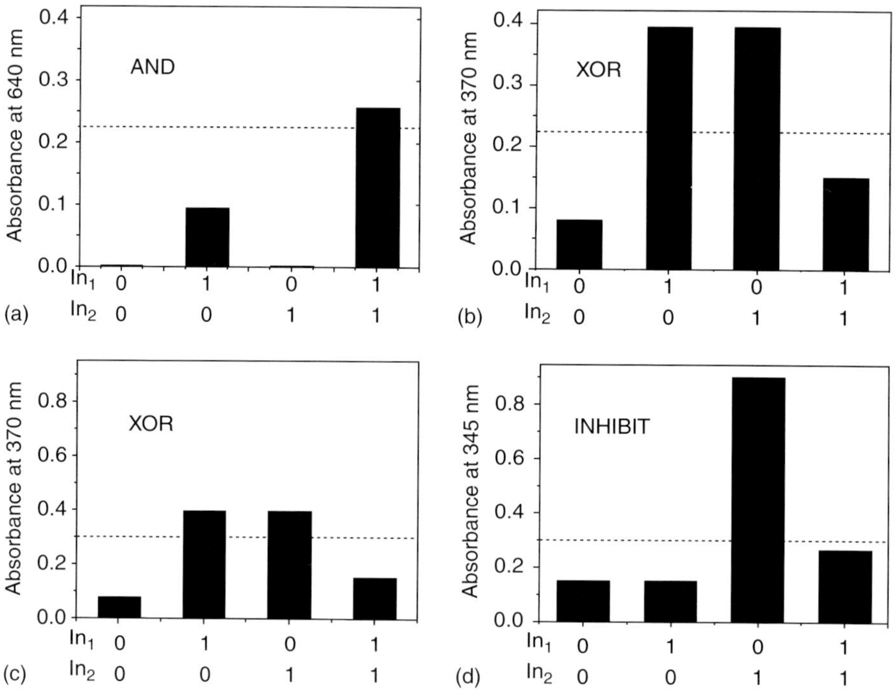

Figure 3.4 Performance of **3o** working as a half-adder or half-subtractor. (Reprinted with permission from [24]. Copyright 2012, Royal Society of Chemistry.)

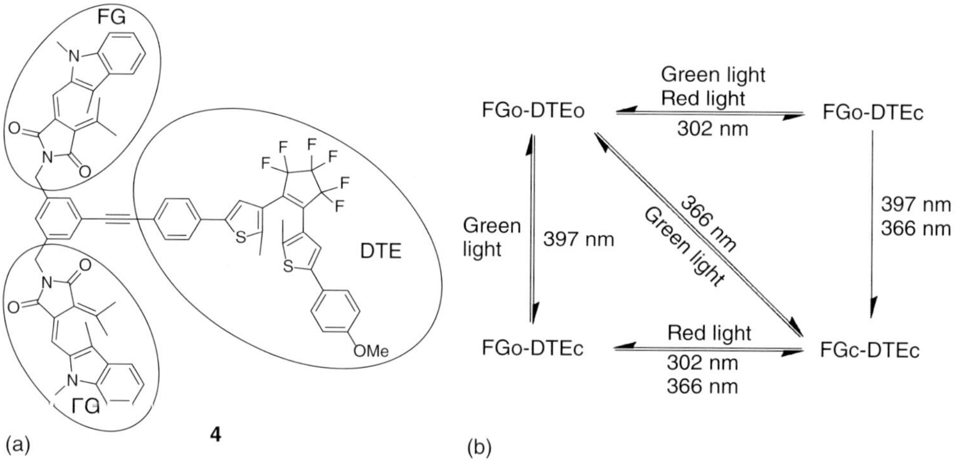

Figure 3.5 (a) Chemical structure of **4** (FGo-DTEo) and (b) photochemical interconversions among four photoisomeric structures. (Reprinted with permission from [25]. Copyright 2011, Copyright 2007, American Chemical Society.)

band at 393 nm in the isomer FGc-DTEo is much high, while that of the isomer FGo-DTEc is much low. On the other hand, the absorbance of FGo-DTEo and FGc-DTEc is essentially identical to this wavelength, that is, the output is turn on only when a single input is applied. In this way, integration of AND and XOR gates in one system results in the all-photonic half-adder. Actually, the FG-DTE triad is also capable of performing binary subtraction (half-subtractor) using the two same inputs in the half-adder gate. Under this condition, just the INHIBIT gate is needed to be constructed and the isomer FGo-DTEc acts as such a gate. FGo-DTEc, characterized by strong absorption at 393 nm (output of the INHIBIT gate), can be generated only from the closed form of the DTE unit. Photochemical closure of DTE unit can proceed only upon UV irradiation with 302 nm. Therefore, this system can be regarded as a photonic INHIBIT gate. It should be noted that the half-adder and half-subtractor can also be constructed in the above two systems by carefully choosing the different input and output signals.

3.1.2.2 **Keypad Locks**

Molecular keypad locks are capable of authorizing password entries, thus providing new opportunities to protect information on the molecular scale. The main keypad characteristic is that it gives an output signal (opens) only when given the correct inputs in the correct order [26, 27]. Although several molecular keypad locks have been constructed, only a few DAE-based molecular keypad locks were involved.

A light-driven, all-photonic molecular two-input keypad lock with fluorescence output was developed by Andréasson *et al.* in 2009 [28]. The system (**5**, Figure 3.6) consists of a photochromic FG and a DTE unit linked with a central tetraarylporphyrin (P). Each photochromic unit can be reversibility interconverted between the open (FGo, DTEo) and closed states (FGc, DTEc) upon irradiation at different wavelength. As the principle of keypad lock, the fluorescence from porphyrin moiety can be modulated with the energy transfer between photochromic isomers and tetraarylporphyrin unit. When both FG and DTE are in the open state, the molecule does not absorb significantly at $\lambda = 470$ nm, and the emission of P at 650 nm is weak. Exposing the corresponding solution to red light ($\lambda > 585$ nm) for 15 min (input B) results in no isomerization, and the emission keeps the same. In contrast, irradiation at 365 nm light for 2 min (Input A) results in the photocyclization for both photochromes. In this case, the fluorescence of P was strongly quenched as a result of the efficient energy transfer between DTEc unit and porphyrin moiety. Further irradiation of the closed isomer with red light transforms the DTEc unit into the open form (DTEo) and causes no effect to the FGc. Only at this state, the molecule absorbs light significantly at $\lambda = 470$ nm, along with intensive emission at 650 nm. In other words, the emission intensity is high enough to open the lock only upon the "A followed by B" input sequence (Figure 3.6). Moreover, this keypad lock could be easily reset by green light and can be used numerous times without any photodegradation.

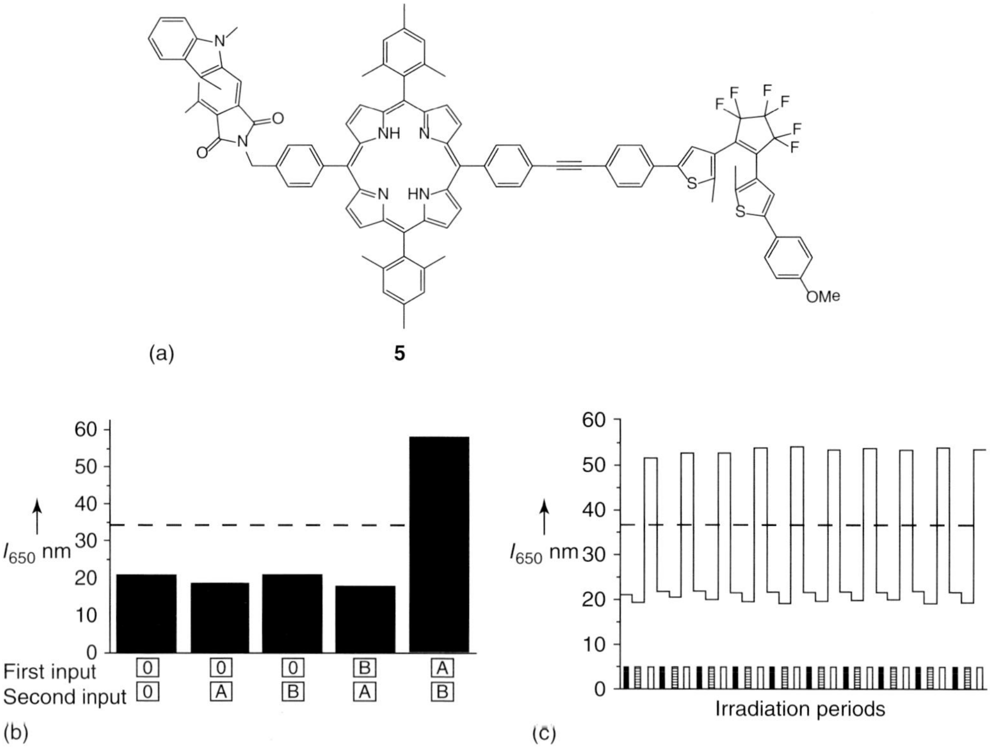

Figure 3.6 (a) Chemical structure of **5**, (b) fluorescence emission intensities ($\lambda = 650$ nm upon excitation at $\lambda = 470$ nm) following various input sequences, and (c) photocycling of **5**.

In order to further boost the password in practical applications, it is still a great challenge to design more complicated keypad locks using three or more inputs at a unimolecular platform for the purpose of practical applications. A novel three-input keypad lock based on an unsymmetrical diarylethene was developed in the laboratory of Tian [29]. Compound **6o** contains a phenol Schiff base derivative as interaction site and a naphthalene moiety as a fluorophore, in which the fluorescence could be modulated by the inputs of UV–vis irradiation (Input U), Cu^{2+} (Input C), and CN^- (Input T). Out of six possible input combinations, that is, UCT, UTC, CUT, CTU, TUC, TCU, only the UCT input combination (i.e., it is the password) gives birth to an instinct fluorescent output signal (Figure 3.7).

Recently, Zhu *et al.* has reported a keypad lock with three inputs constructed on a multi-addressable photochromic bisthienylethene [30]. Incorporation of two phenyl-substituted imidazole units into the framework of a photochromic DAE results in a novel photochromic switch (**7**), which exhibits sequence-dependent responses via efficient interaction of the specific imidazole unit with protons and Ag^+ (Figure 3.8). Upon alternating addition of proton and base, the absorption spectra of **7** in THF display reversible changes due to the protonation–deprotonation of the imidazole moiety. However, the photocyclization of **7** is

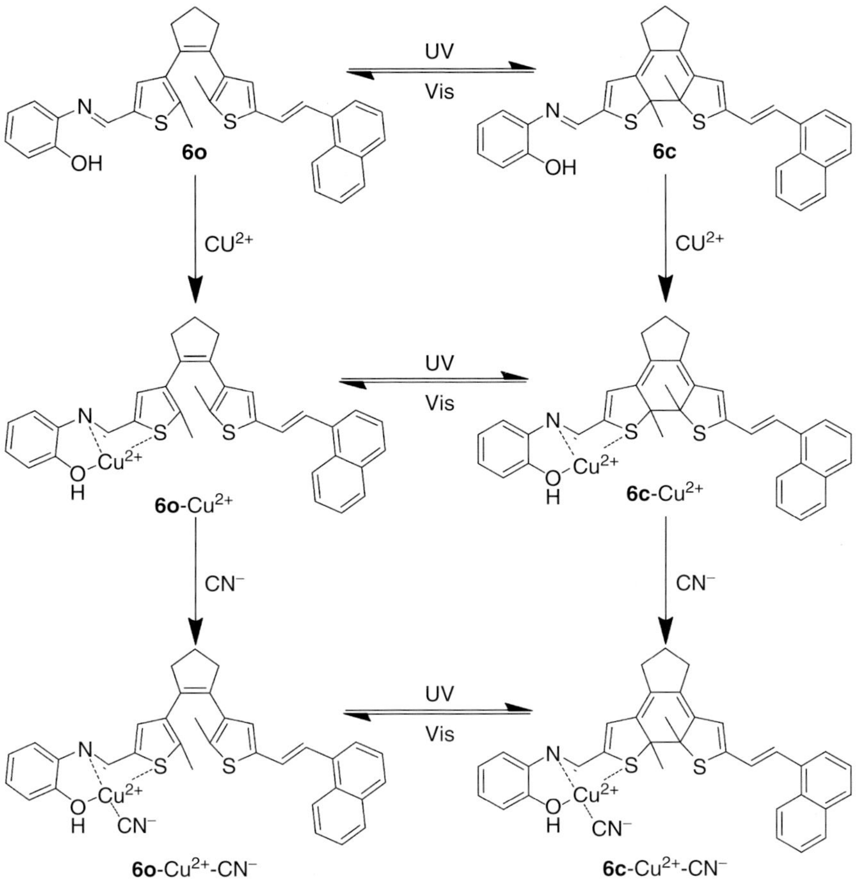

Figure 3.7 Proposed sensing processes of compounds **6o** with Cu^{2+} and then CN^-, and the photochromic processes responding to light stimuli.

suppressed to some extent in the presence of Ag^+. Compound 7 underwent normal photochromic reactions with characteristic absorption maximum at 610 nm under UV light in the absence of Ag^+. Addition of 2.0 equiv. of Ag^+ to the THF solution of 7 induced a slightly decrease of the absorption band at about 334 nm. However, upon further UV irradiation, absorption band at 610 nm increased very slightly until the photostationary state (PSS) was reached. The value of absorbance intensity (A_{610nm}) was much smaller than that of closed isomer of the free 7. Here the remarkable decrease of A_{610nm} could be possibly attributed to the efficient interaction between Ag^+ and 7, which was confirmed by the distinct ^{1}H NMR signal change in the hydrogens on the imidazole and thiophene units. More interestingly, the absorbance at 610 nm exhibited sequential dependence on combinational inputs of protons (P), Ag^+ (A), and UV irradiation (U). Only the correct inputs order, "AUP" (Ag^+ was firstly added, followed by irradiation of light with 365 nm, the protons were added finally) gives small absorption

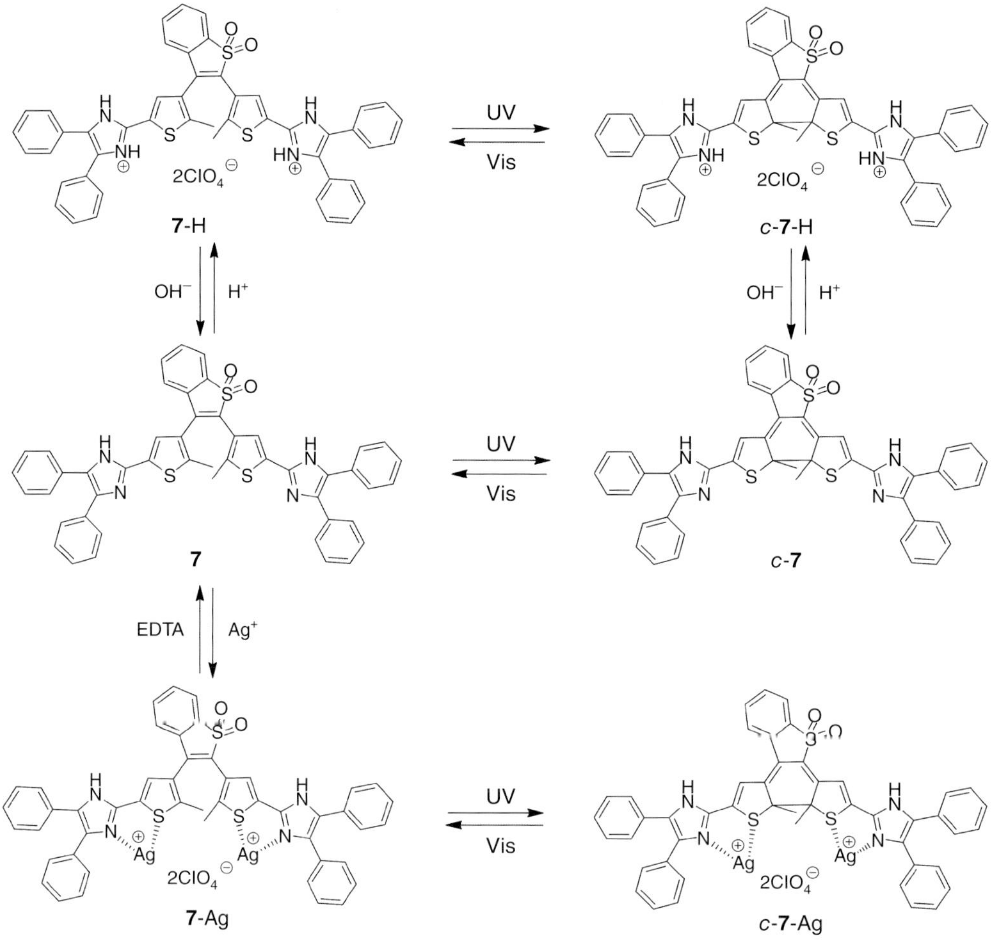

Figure 3.8 Proposed sensing processes of compounds **7** with H^+ and Ag^+ and the photochromic processes responding to light stimuli. (Reprinted with permission from [30]. Copyright 2013, American Chemical Society.)

intensity value at 610 nm. Moreover, the established sequence-dependent system can stand or persevere more than 4 h. Under specific conditions (threshold was set as $0.05 < A_{610nm} < 0.2$), only the input order "AUP" results in the "true" output signal (A_{610nm}), while the output signals from all other cases such as A, U, P, AU, UA, AP, PA, UP, PU, APU, PAU, PUA, UPA, UAP, or no inputs are all out of the threshold region. Obviously, a three-input keypad lock could be constructed with the output of the absorbance at 610 nm under this condition.

A similar fluorescent-dependent keypad lock with three inputs of UV irradiation (U), Cu^{2+} (C), and Hg^{2+} (H) was also reported by Zeng and coworkers (Figure 3.9) [31]. The fluorescence of the compound **8o** is strongly dependent on the combinational three inputs, that is, only the sequence of three input

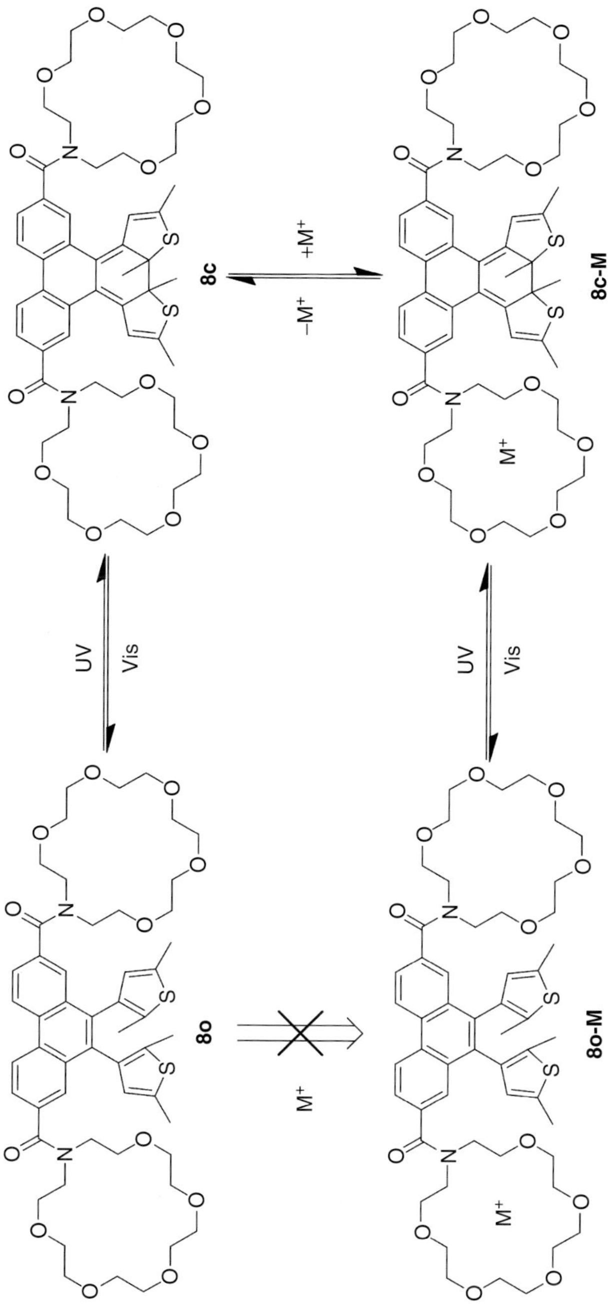

Figure 3.9 Proposed sensing processes of compounds **8** with Cu^{2+} and Hg^{2+} and the photochromic processes responding to light stimuli.

keys CHU gives birth to a strong fluorescent intensity at 530 nm, along with a characteristic keypad lock. However, all other sequences with the resulting weak fluorescence outputs fail to open the lock.

3.1.2.3 Digital Encoder and Decoder

An important function in data processing is to compress information for transmission or storage. A digital encoder can convert data into a code. For example, a 4-to-2 encoder converts four bits of data into two bits. On the other hand, a decoder recovers the information in its original form, for example, a 2-to-4 decoder converts two coded inputs to four readable outputs [31]. Actually, with the appropriate combination of chemical and photonic inputs and outputs, compound **3o** as mentioned earlier performs as both a single-bit 4-to-2 encoder and 2-to-4 decoder, respectively. Briefly, the encoder inputs are light at 365 nm (Input 1), Cu^{2+} (Input 2), H^+ (Input 3), and Hg^{2+} (Input 4). The outputs are fluorescence at 467 nm (Output 1) and absorption at 335 nm (Output 2), respectively. The initial state is the open isomer of **3o**. As described earlier, when Input 3 and Input 4 were applied, the fluorescence intensity at 467 nm was increased, while the fluorescence was completely quenched in the presence of Input 1 and Input 2. Meanwhile, both Input 2 and Input 4 induce a strong absorbance at 335 nm, while Input 1 and Input 3 can only result in a weak absorbance at 335 nm. This means that both Input 1 and Input 2 can turn Output 1 off, Input 3 and Input 4 can turn Output 1 on, while Output 2 can be switched on by Input 2 and Input 4, and switched off by Input 1 and Input 3. Thus, the truth table of the 4-to-2 encoder is represented.

Input 1 and Input 2 can also be used as the inputs of the decoder, and the outputs are fluorescence at 467 nm and the absorptions at 391, 341, and 412 nm. By a suitable choice of the thresholds, a 2-to-4 decoder can be easily constructed. At the initial state, **3o** exhibits strong fluorescence at 467 nm, while it has no absorption band in the visible range. When Input 1 is applied, the absorption at 391 nm is increased strongly, while the fluorescence is decreased. Alternatively, applying Input 2 induces the absorbance at 341 nm above the threshold, giving an on response for Output 3. Finally, applying both Input 1 and Input 2 causes remarkable absorption at 412 nm, which results in switching Output 4 on. Thus, the decoder function of **3o** has been demonstrated.

There are also complicated molecular logic gates such as multiplexer/demultiplexer and transfer gates, parity generator/checker for error detection [32–34] can be constructed based on DTE systems (e.g., the compound FG-DTE). The critical point is the careful screening of the optical properties of the photochromic systems to fulfill the truth tables of the corresponding logic gates.

There are very few chemical logic systems that can be described as elementary multi-input logic gates. A single photochromic DTE switch with four optical outputs corresponding to four inputs was proposed by Tian *et al.* [35]. The logic gate is based on a very simple cyclopentene derivative containing pyridine unit (**9o**). Its absorption, fluorescent intensity, and emission peak can be reversibly modulated by UV/vis light irradiation, Zn^{2+}, and protons (Figure 3.10). The solution

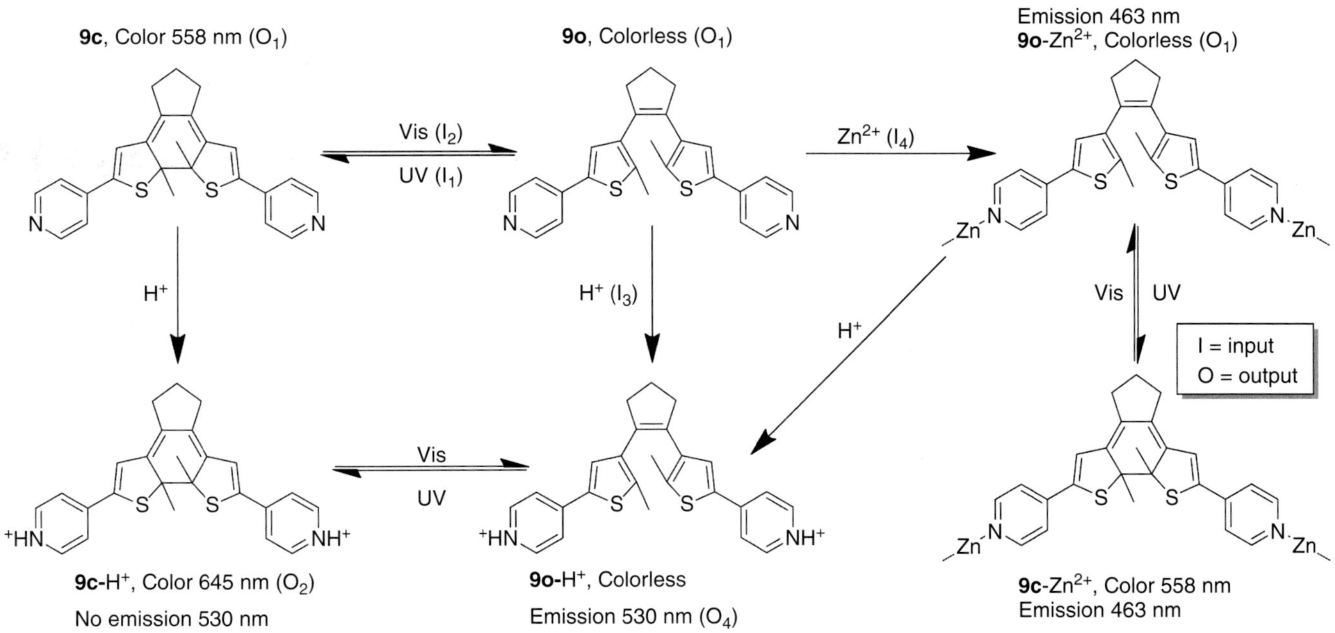

Figure 3.10 Proposed sensing processes of compounds **9** with Zn^{2+} and H^+ and the photochromic processes responding to light stimuli.

of **9o** was colorless. Irradiating the solution with UV light (254 nm) generated a new absorption band at 558 nm, resulting in purple color. The colored solution was completely bleached by further irradiation with visible light at a wavelength of 570 nm. Protonation of **9** induced the emission peak shift from 415 to 530 nm. Furthermore, UV light irradiation resulted in absorption redshift at 645 nm. Such a remarkable spectral redshift may be due to the increased conjunction system or changed charge-transfer state when the pyridine unit was protonated. Upon addition of Zn^{2+}, a new emission peak at 463 nm was observed. And the fluorescence intensity was increased along with the amount of Zn^{2+}. The fluorescence at 541 or at 463 nm can be completely quenched by irradiation with UV light (254 nm). It is should be noted that both the closed isomer of **9** and the complexes with protons or Zn^{2+} are thermal stable. In other words, the molecule has a memory effect. Actually, a complicated four inputs (254 nm, 570 nm, proton, and Zn^{2+}) logic gate with four outputs (absorption at 558 and 645 nm, emission at 463 and 530 nm) was realized on the unimolecular platform. Although the present system is in solution and inputs are not all optical, the concept shown here may be useful to design some kind of "wet" computer that work much more like our brain.

3.2
Data Storage and Molecular Memory

In the 1950s, Hirshberg first put forward the concept that photochromism could apply in computer data storage [36]. Since then, photochromic molecular memories have been widely explored, emerging in a large number of reports and patents. In the well-established photochromic systems nowadays, the color changes are commonly caused by photoinduced interconversion between two chemical species (isomers) through rearranging either chemical bonds or electronic distribution. Scheme 3.1 illustrates a typical photochromic reaction process. A stable isomer **A** can transform to a relative high-energy isomer **B**, overcoming certain energy barrier with the assistance of photon excitation. Apparently, such process is controlled by chemical kinetics, and a spontaneous back-reaction inevitably happens when the exciting source is removed. Thus, the reversal rate is related to its activation barrier, in other words, the thermal stability of isomer **B** depends on the activation barrier. Accordingly, photochromic systems can be separated into two categories:

P-type (photochemically reversible): Isomer **B** is thermally stable enough that only photon excitation can lead to back-transformation of isomer **A**.
T-type (thermally reversible): Isomer **B** is thermally unstable, and thus the reversal reaction can be triggered by heat or photostimulus.

Obviously, T-type system is not suitable for the practical application as molecular data storage since the written storage data would be gradually erased after a certain time. Thus, only P-type compounds are selected as the candidates, endowing the change in photoswitched state between two thermally

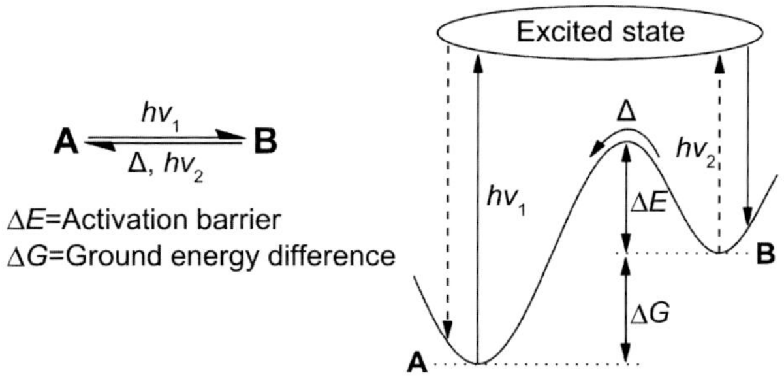

Scheme 3.1 Illustration of photoisomerization.

stable isomers. Compared with T-type compounds such as azobenzenes and spiropyrans, P-type compounds are really rare, which generally consist of two main systems as fulgides and diarylethenes. Over the past 20 years, diarylethenes obviously earned much more attentions due to their durable persistency up to $\sim 10^4$ times, which is another indispensable requirement for molecular memories.

When the bistable written progress is settled, another important issue, nondestructive readout, should also be carefully fixed. Generally for photochromic memory materials, common reading information by the UV–vis spectrophotometer inevitably induces the molecular photoexcitation, leading to data falsification after certain times of accumulation. It means that the reading progress becomes potentially destructive, and thus we should seek on other detecting method for avoiding the interference to the stored states. Therefore, the following illustrations mainly focus on diarylethene system with nondestructive readability based on fluorescence, infrared, and optical rotation spectroscopy.

3.2.1 Fluorescence Spectroscopy

The fluorescence signals can be readily and sensitively recognized while the weak and color-tuned excitation light scarcely ever erases the recorded signal during the fluorescence readout process. Thus, fluorescent diarylethenes are promising to realize erasable optical memory with high-contrast fluorescence recording and nondestructive readout of signals.

Since diarylethenes are structurally nonfluorescent, early works mainly focused on introducing fluorophores to the system. The fluorescence can be modulated by Förster resonance energy transfer (FRET) from the fluorophore to the closed form. Figure 3.11 shows a representative system **10** [37, 38] composed of a photochromic diarylethene unit, a rigid adamantyl spacer, and a fluorescent bis(phenylethynyl)anthracene chromophore. When anthracene is excited by irradiation at 488 nm, the excited state tends to transfer the energy to the closed form of diarylethene, causing fluorescence quenching with perfect efficiency of 99.9%. However, the transfer to the open isomer with higher level energy hardly takes place, and thus the molecule shows strong green emission. The fluorescence

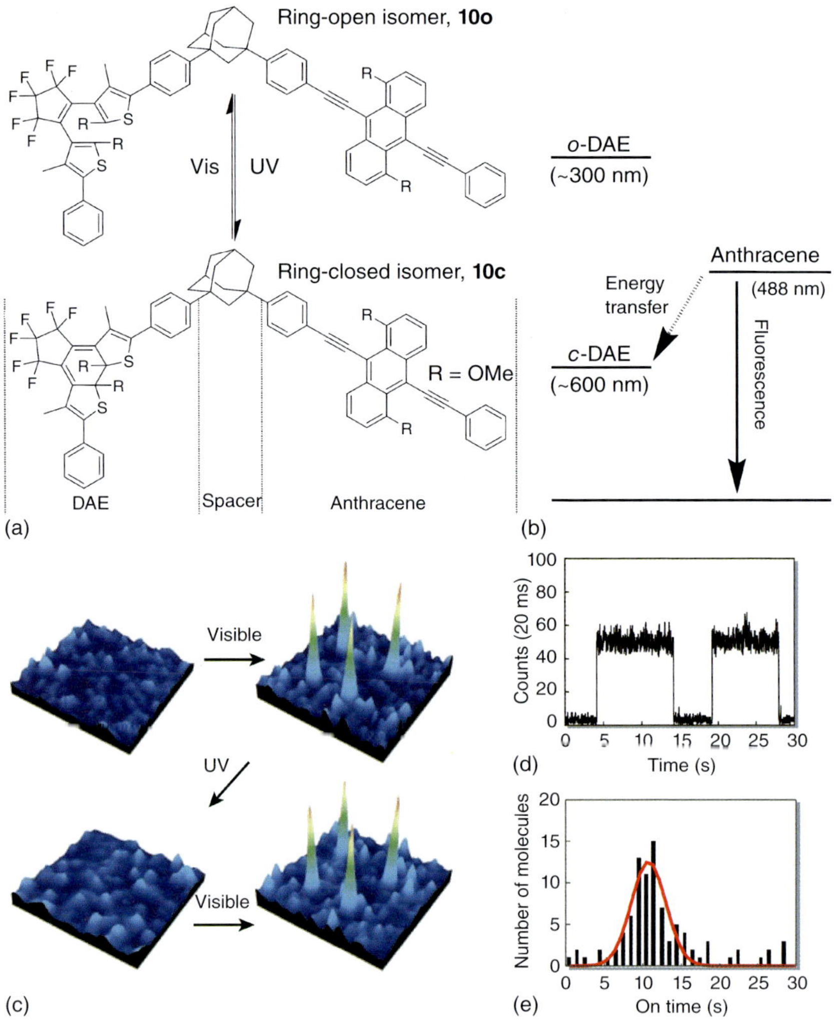

Figure 3.11 (a) Schematic diagrams of fluorescent diarylethene **10**, (b) involved FRET mechanism, (c) fluorescence images, (d) time trace, and (e) histogram of the time of single photoswitching molecules (**10**) upon alternate irradiation with 488 and 325 nm light. (Reprinted with permission from [38]. Copyright 2002, Rights Managed by Nature Publishing Group.)

quantum yields of closed and open forms in the system are 0.001 and 0.73, respectively. Digital on/off switching between the two distinct states can be observed even at a single-molecule level, as shown in Figure 3.11. When novel techniques to address each molecule by photon are developed, such fluorescent diarylethene derivatives can be potentially utilized for ultrahigh-density erasable optical data storage (1 bit per molecule, ~petabit per square inch).

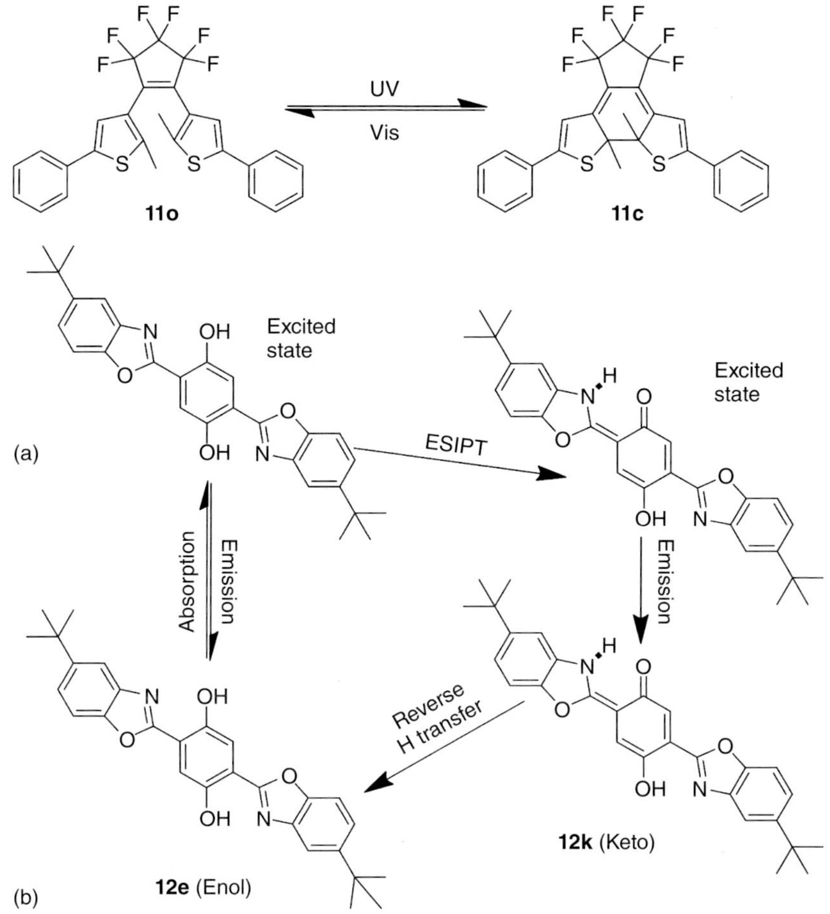

Figure 3.12 (a) Chemical structure and photochromic reaction of diarylethene **11** and (b) four-level photochemical and photophysical ESIPT process of dye **12**. (Reprinted with permission from [39]. Copyright 2006, American Chemical Society.)

A similar delicate example [39] was presented by loading a simple diarylethene (**11**) and another fluorescent dye (**12**) on polymer films (Figure 3.12). The carefully selected dye undergoes excited-state intramolecular proton-transfer (ESIPT) process with the large Stokes' shifted keto emission, matching the absorption band of the closed form of **BP-BTE**. The dye also shows commendable aggregation-induced enhanced emission (AIEE; $\Phi_F^{powder} = 10\%$, $\Phi_F^{soln} = 2\%$ in chloroform), which can fix the common "concentration quenching" problem in the highly loaded polymer films. In this way, the emission of the polymer film can be reversibly modulated by UV/vis light (on/off fluorescence switching ratio over 290). Moreover, the fluorescence can be excited at a wavelength where both open and closed forms show little absorbance. Thus, excellent readout can be almost achieved (over 125 000 shots).

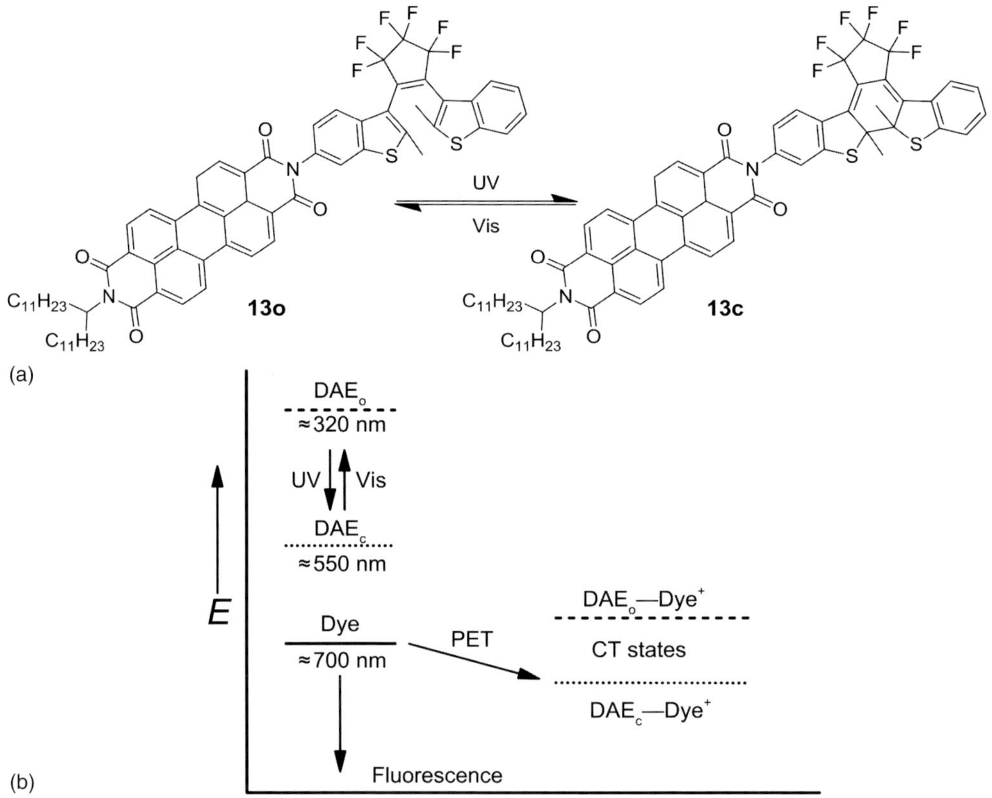

Figure 3.13 (a) Photochromic reaction between **13o** and **13c** and (b) schematic energy diagram for the fluorescent open form (dashed line) and the nonfluorescent closed form (dotted line) of a diarylethene (DAE) and CT states of the DAE–dye conjugate.

Although high-contrast fluorescence modulation can be realized employing FRET, the energy transfer to closed isomer of a diarylethene, raising its energy level, would potentially cause the cycloreversion during the readout step, and thus destroys the recorded memory. An alternative solution (Figure 3.13a, compound **13**) is employing photoinduced electron transfer (PET) if the open and closed forms of the photochromic moieties were to possess different redox properties. The electron transfer involves oxidation of a fluorescent dye group and reduction of a DAE unit, and the solvent-dependent driving force for PET is exergonic for the DAE_C^-–dye$^+$, but not for the DAE_O^-–dye$^+$ charge-transfer (CT) state. Furthermore, absorption bands of both DAE isomers should be shorter than the fluorescence spectrum of the dyes. Figure 3.13b illustrates the process [40] when certain solution with specific polarity is selected, the fluorescence of open and closed forms can be controlled following such mechanism, leading to nondestructive readout optical memories.

The most excellent example is Figure 3.14a, compound **14** [41], which consists of a diarylethene with shorter absorption band in both forms and a perylene bisimide

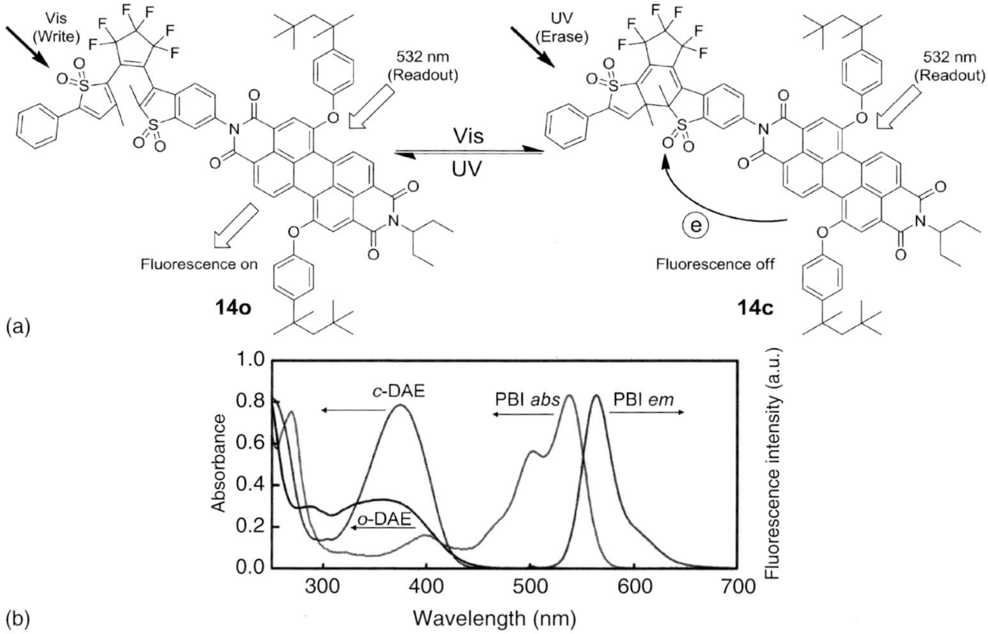

Figure 3.14 (a) Molecular structure of diarylethene **14o** and (b) absorption and fluorescence spectra of each component in 1,4-dioxane. Absorption spectra of the open-ring isomer of diarylethene unit, the closed-ring isomer of diarylethene unit, and the PBI unit, and the fluorescence spectrum of the PBI unit. (Reprinted with permission from [41]. Copyright 2001, American Chemical Society.)

(PBI) dye with long emission. The fluorescence quantum yield of the closed form dramatically decreases when the dielectric constant of the solvent increases above 5, while that of the open isomer remains almost constant with increasing dielectric constant (Figure 3.14b). Therefore, in polar solvents, the fluorescence can be reversibly switched upon alternate irradiation with visible and UV light. Such principle can also apply to single-molecule fluorescence with nondestructive readout ability.

Alternatively, employing other methods that selectively excite the fluorophore part other than the diarylethene unit can also reach the goal of nondestructive readout. A diarylethene coordinated with $[Eu(hfa)_3(H_2O)_2]$ (hfa = 1,1,1,5,5,5-hexafluoroacetylacetonate) to give a photochromic Eu(III) complex [42] (Figure 3.15, compound **15**). The fluorescence of the complex can be modulated by UV/vis irradiation due to the different coordination environment of open and closed forms. In this complex, the absorption of diarylethene is completely separated with that of Eu(III) center. It means that exciting the complex at its own absorption region would not affect the diarylethene core. This facilitates writing, erasing, and reading data with light sources of three different wavelengths without destructive readout.

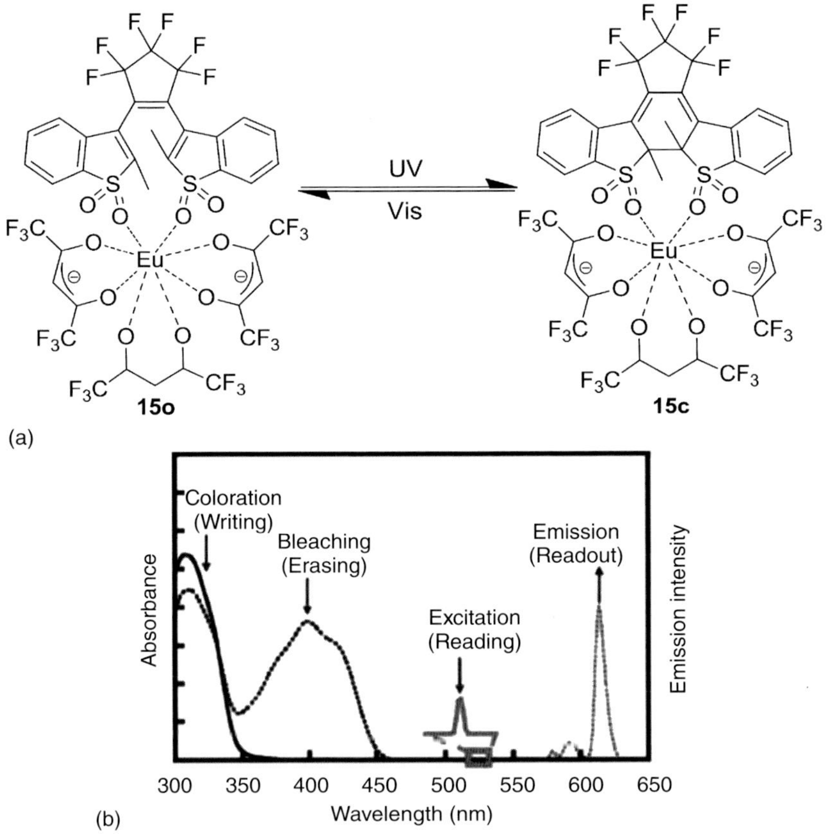

Figure 3.15 (a) Molecular structures of photochromic complexes **15** and (b) schematic representation of transitions of the diarylethene (black line) and the Eu(III) complex (red line). (Reprinted with permission from [42]. Copyright 2009, Royal Society of Chemistry.)

3.2.2
Infrared Spectroscopy

Another promising method for nondestructive readout is detecting infrared spectral changes of memory materials. Infrared spectroscopy exploits the fact that molecules absorb specific frequencies that are characteristic of their structure. These absorptions are resonant frequencies, that is, the frequency of the absorbed radiation should match the transition energy of the bond or group that vibrates. Indeed, exciting such vibration energy would not cause the rearrangement of electrons, thus limiting the structural changes during the reading progress.

For diarylethene **16** [43], the infrared intensity of the closed form is stronger than that of the open form in the IR region at 1500–1700 cm^{-1}. Such IR spectral changes can be used to read out the image written on a polymer film containing the diarylethene derivative. As shown in Figure 3.16, the IR images can

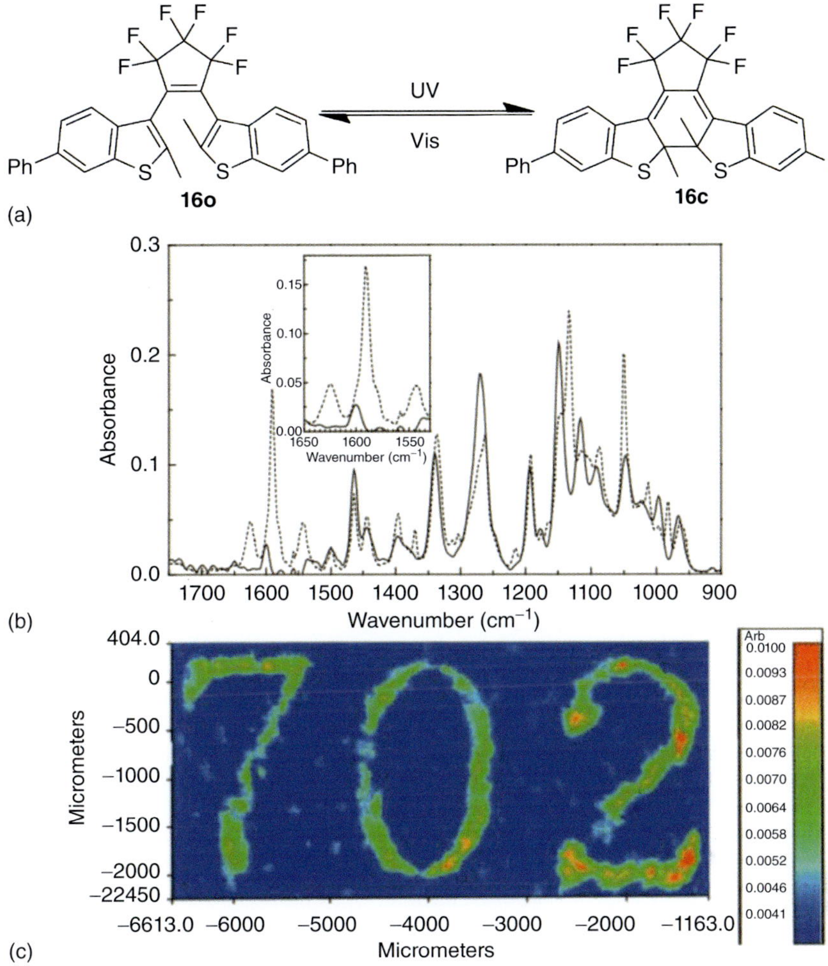

Figure 3.16 (a) Molecular structure of the open and closed forms of diarylethene **16**, (b) IR spectra of open form (solid line) and closed form (broken line) of diarylethene **16**, and (c) IR images of the recorded film containing 4 wt% diarylethene **16**. (Reprinted with permission from [43]. Copyright 2003, WILEY-VCH Verlag GmbH & Co. KGaA, Weinheim.)

tell the difference between the letters and the background. Moreover, even after prolonged readout of the image using IR light, no decrease in the signal-to-noise (S/N) ratio was observed, suggesting the potential for memory materials with non-destructive readout ability.

As a matter of fact, the use of long-wavelength IR light is disadvantageous for high-density data storage since the light spot diameter cannot be focused shorter than the wavelength of the writing light. In contrast, multi-addressable recording system may solve the problem by multiplying the data in the same spot. Three

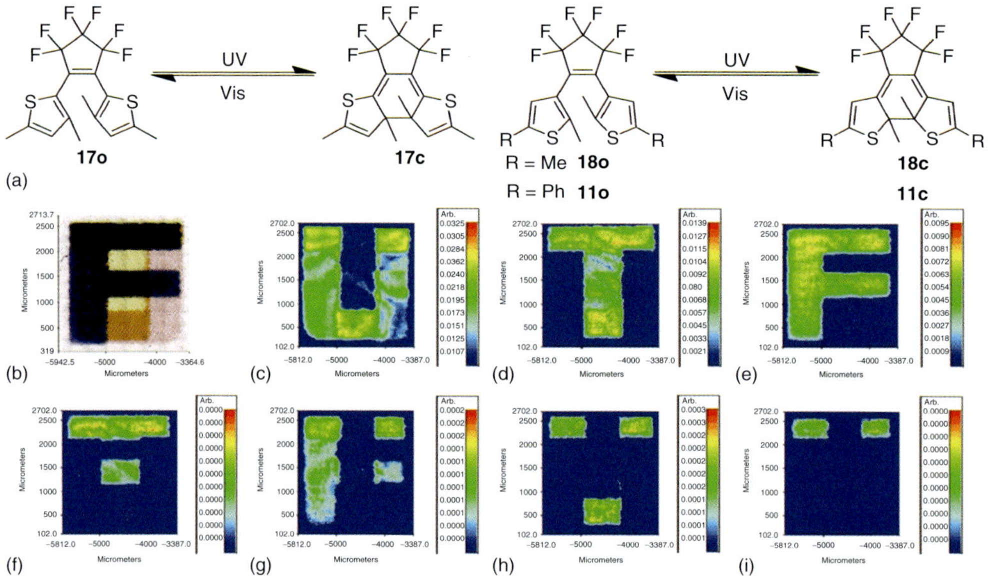

Figure 3.17 (a) Molecular structure of the open and closed forms of diarylethene **17**, **18**, and **11** and (b) visible and IR images of the recorded film containing the three diarylethenes. IR image detected at 1549 cm^{-1} (c), 1655 cm^{-1} (d), 1527 cm^{-1} (e), both 1655 and 1527 cm^{-1} (f), both 1549 and 1527 cm^{-1} (g), both 1655 and 1549 cm^{-1} (h), and all three wavelengths (i). (Reprinted with permission from [44]. Copyright 2005, WILEY-VCH Verlag GmbH & Co. KGaA, Weinheim.)

bits of eight-state information from triple frequency (Figure 3.17) can be recorded on the polymer film doped with three diarylethenes **17**, **18**, and **11** with separate absorption [44]. The UV/vis lighted states can also be well controlled and read out nondestructively by appropriate wavenumber of IR light.

3.2.3
Optical Rotation

Optical rotation can be detected in the region outside the electronic absorption bands, eliminating photoexcitation of molecules in the preferable mode. Thus, it is another promising candidate for realizing nondestructive readout. To achieve the goal, asymmetric (chiral) structures must be induced to the system to cause the light polarization.

A closer look at the photoswitching process of DAEs reveals not only a structural change from the ring-open to the ring-closed isomers but also a chirality transformation from the axial helicity of central hexatriene moiety to the central asymmetry of two reactive stereogenic centers (Scheme 3.2). Since the flexible ring-open isomers show rapid rotation of aryl groups, the accompanied loss in inherent chirality takes place inevitably. Accordingly, for common DAEs, their chirality changes cannot be directly employed. Chemical bonding the aryl groups

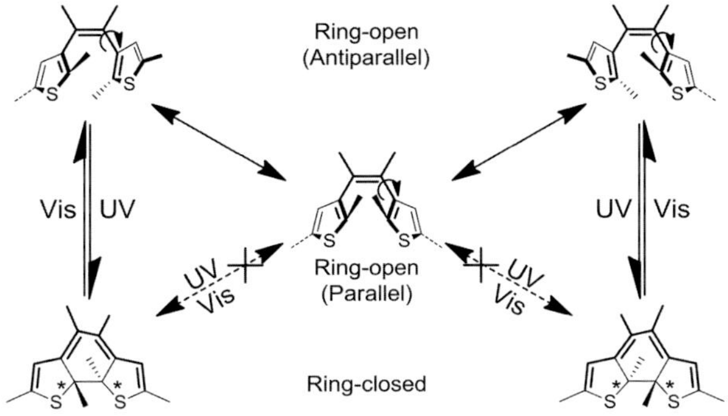

Scheme 3.2 Illustration of the photochromic reaction in diarylethene.

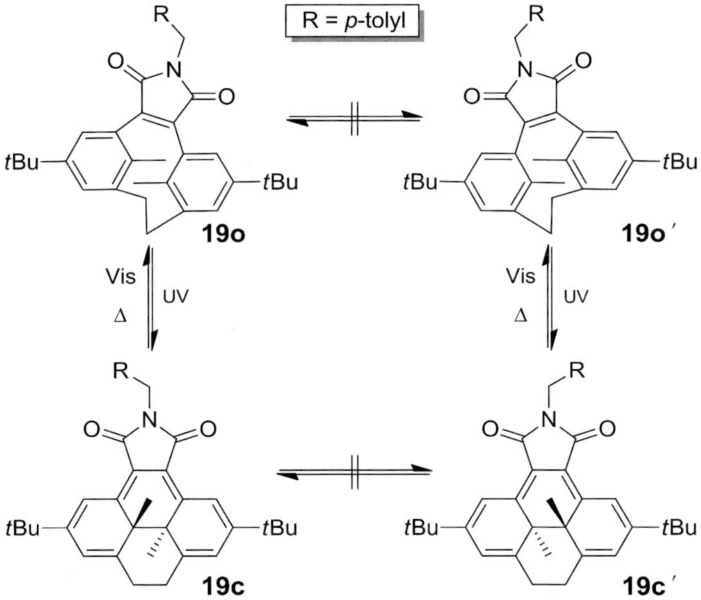

Figure 3.18 Enantiospecific photochromic reaction between **19o** and **19c**.

can eliminate the undesired racemism. Interestingly, chiral **19o** exhibits enantiospecific transformation during the photochromic reaction [45] (Figure 3.18), and its reversible optical rotation changes at 750 nm indicate the potential for nondestructive readout [46]. However, such system has a severe disadvantage for photomemory due to the thermal reversible feature of its colored form.

Another approach is to induce chiral substituents, and diastereomers form predominantly because of the energy differences in each isomer. Allylic 1,3-strain [47, 48] has been used to investigate diastereoselective photochromic reactions with electronic repulsion. Introducing 1-methoxymethoxyethyl groups into the

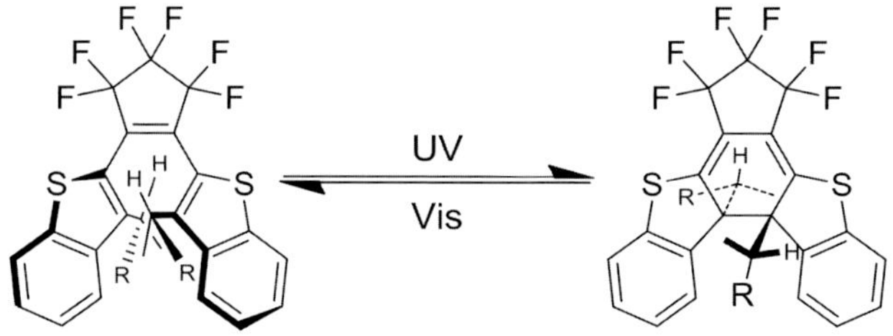

20o **20c**

Figure 3.19 Diastereoselective photochromic reaction between **20o** and **20c**.

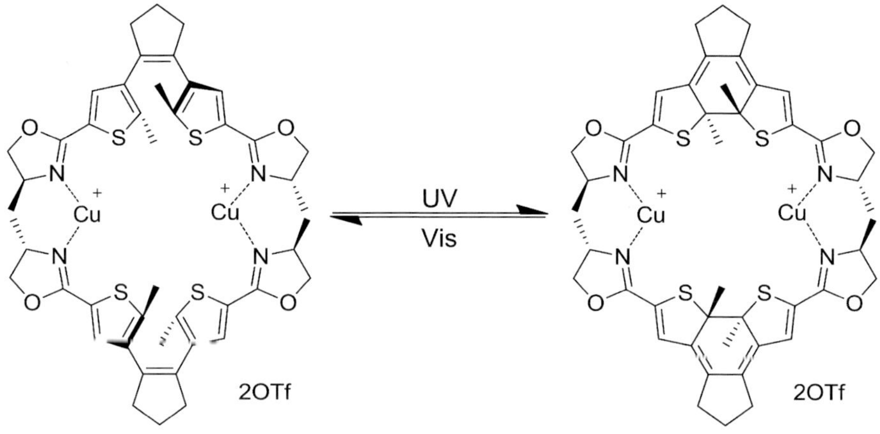

21o **21c**

Figure 3.20 Diastereoselective photochromic reaction between **21o** and **21c**.

reactive carbons can drive the cyclization progress with almost 100% diastereoselectivity at low temperature (Figure 3.19).

Generally, adding chiral auxiliaries on the other parts of the photochromic hexatriene would not achieve such satisfying selectivity. However, employing additional effects such as coordination can fix the ring-open isomer to a rigid helical structure. Then the photocyclization can selectively generate one isomer. Figure 3.20 shows the exact example (complex **21**), and the DAE–Cu complex shows large and reversible optical rotation upon irradiation with UV and visible light for nondestructive readout [49].

In a more direct way, linking the DAEs with chiral chains to form macrocycle structure (Figure 3.21, diarylethene **22**), avoiding the flip of the aryl groups, can finally lead to 100% diastereoselectivity at room temperature. The enantiomers show excellent thermal stability, thus enabling the potential for molecular photomemory [50].

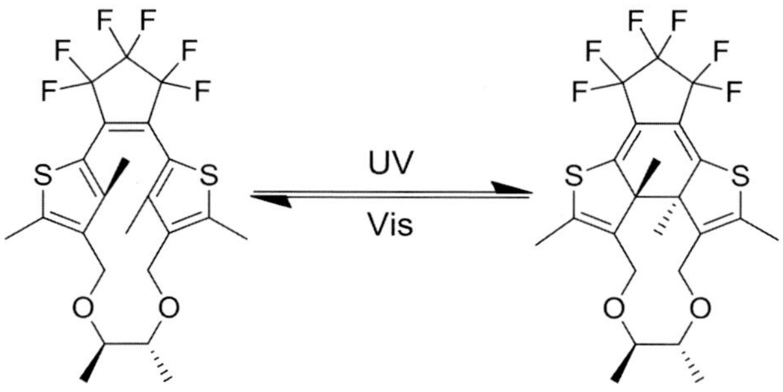

Figure 3.21 Diastereoselective photochromic reaction between **22o** and **22c**.

3.3 Gated Photochromores

Although much progress has been made in diarylethene systems, diarylethenes with gated properties, which serve as an important development in nondestructive readout, have been less reported in the literature. In gated photochromism, light does not trigger the photochromic reaction (from open isomer to closed isomer) unless another external stimulus is also present. Up to now, there are two principles to achieve gated photochromism. The first one is the conformation-restricted photochromism. As is well known, diarylethenes with five-membered heterocyclic rings possess two conformations in solution, with the two rings in a mirror symmetry parallel conformation or in a C_2 symmetry antiparallel conformation. However, only the C_2 symmetry is photoactive, and the photochromic cyclization can proceed only from this conformation [51–53]. So the first way to get gated property is to fix the heterocyclic rings in the parallel conformation. Actually, several novel-gated photochromores have already been constructed based on this strategy by introducing some external stimulus such as intramolecular hydrogen bonding, coordination, and specific reaction. Unlike the first method to induce gated photochromores, the second method to prohibit the photochromic activity is to quench the excited state of the open isomer. Indeed, the excited state can be suppressed or more easily quenched by other particular processes rather than the photocyclization reaction, although some special conditions such as quite low temperature must be presented.

3.3.1 Hydrogen Bonding

The first example of gated photochrome in the diarylethene system was reported by Irie *et al.* in 1992 [54]. Carboxyalkyl groups that are considered to make intramolecular hydrogen bonds in nonpolar solution were introduced into the side chains of the molecule as shown in Figure 3.22. The photochromic reactivity

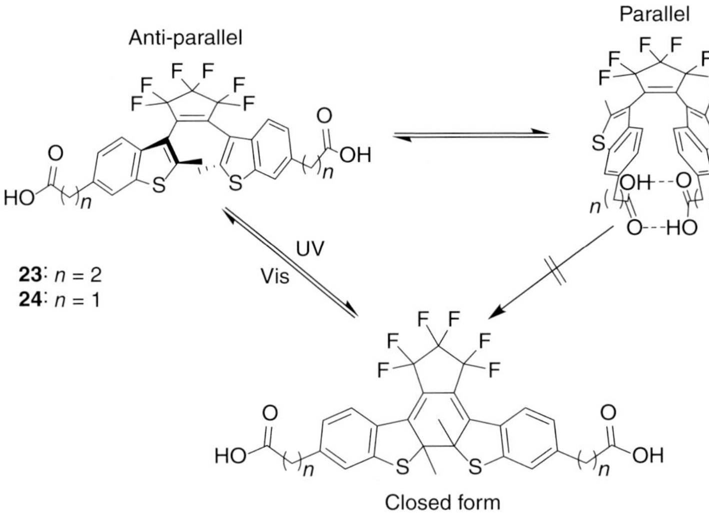

Figure 3.22 Antiparallel and parallel conformations of compounds **23** and **24** and the corresponding photochemical reaction from the antiparallel conformation.

of compound **23** with carboxyethyl group at 6-position of the benzothiophene ring was completely blocked in cyclohexane. The strong gated properties were attributed to the stable intramolecular hydrogen bonds between the carboxyalkyl groups, which were expected to fasten in all photochemically inactive parallel conformation. The photochromic performance can be recovered by the addition of a small amount of ethanol as a result of the break of the hydrogen bonds. The proposed mechanism was further conformed by the ^{1}H NMR titration. Generally, in a diarylethene system, the methyl protons at the 2-position of the heterocyclic ring give signals at different fields depending on the conformation. The protons in the parallel conformation give signal in the upper field, while signal of the protons in the antiparallel conformation was located in the lower field. The integration ratio of the lower- and upper-field signals indicates the relative population of the two conformations. The ^{1}H NMR spectrum measured in cyclohexane-d_{12} gives only lower field signals, indicating that the molecule was in the parallel conformation in cyclohexane. Consequently, no photochromism was observed in cyclohexane. However, upper-field signals appeared upon the addition of ethanol, which suggest that the parallel conformation was converted to the antiparallel conformation in the mixture solution of cyclohexane and ethanol. It is worthwhile to mention that derivative compound **24**, which contains carboxymethyl groups, can undergo normal photochromic reaction even in nonpolar solution. Owing to the short chain length of carboxymethyl groups, only weak intramolecular hydrogen bonds were formed. Consequently, in contrast with **23**, there existed some antiparallel conformers in compound **24**, and thus no gated phenomenon was observed.

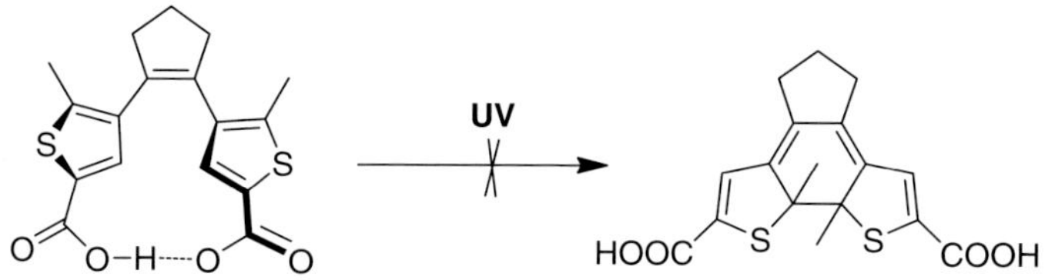

25o

25c

Figure 3.23 Intramolecular hydrogen bond between the COOH and COO^- in monoanionic form of **25o**.

A similar diacid DTE dye was also reported by Coudret *et al.* [55]. As shown in Figure 3.23, partial neutralization of the diacid open form by the addition of tetrabutylammonium hydroxide resulted in a monoanionic form (25o), whose photocyclization quantum yield was dropped from the high value of 0.9 (diacid dye) to 0.5. The remarkable gate photochromic properties could be attributed to the efficient intramolecular effects of hydrogen bonding in the monoanionic form, which kept the compound in the parallel conformation.

There is another example of using intramolecular hydrogen bonding interactions to control the photochromic reactivity [56]. As shown in Figure 3.24, the chemically gated switch system was obtained by selected oxidization/reduction of the sulfur atoms in compound **26o**. Before oxidization, the molecule exhibited good photochromic properties and could be toggled between colorless ring-open and colored ring-closed forms by alternated irradiation with UV and visible light. However, in the oxidization state, the photochromic performance was completely blocked, which is considered to be the contribution of stronger intramolecular interactions between the oxygen atoms in the *S*,*S*-dioxide moieties and the

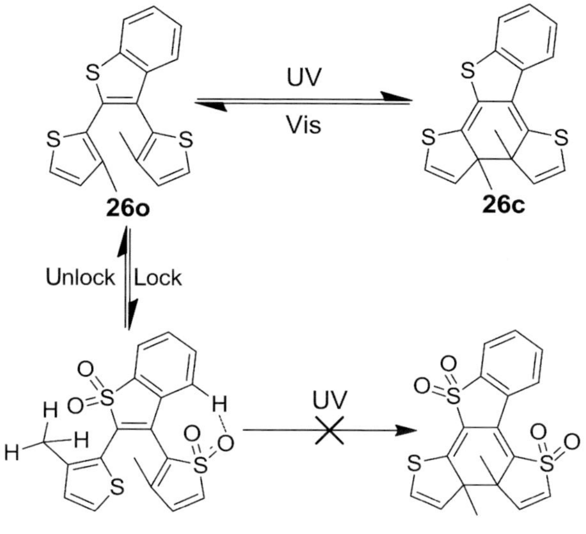

Figure 3.24 "Lock and key" process of diarylethene **26o** by selected oxidization/reduction.

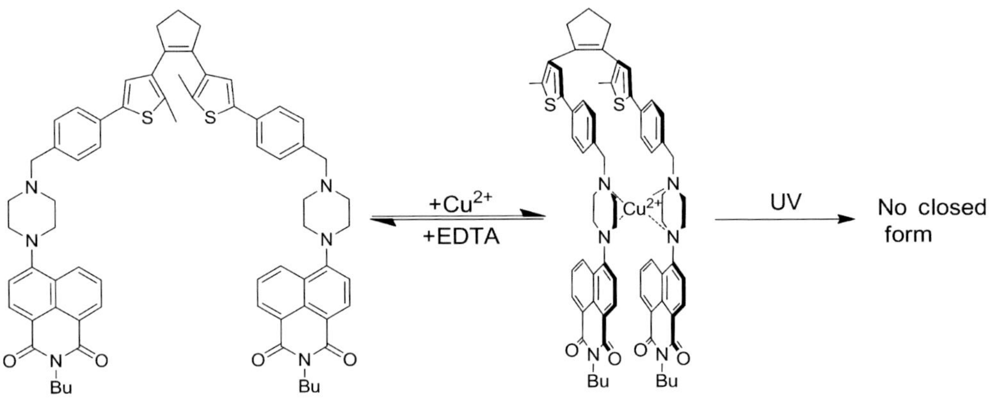

27o **27**-Cu

Figure 3.25 Gated photochromism of 1,8-naphthalimide-piperazine-tethered dithienylethene **27o** by the coordination of Cu^{2+}.

corresponding hydrogen atoms. The photochromic reactivity was obtained by reduction of the *S,S*-dioxide moieties.

3.3.2 Coordination

Except hydrogen bonding interaction, coordination effect was also developed to control photochromic reactivity. A new 1,8-naphthalimide-piperazine-tethered dithienylethene (compound **27o**) with novel gated photochromic properties was designed and synthesized by Tian *et al.* [57]. When 20 equiv. of copper(II) ion was added into the free DTE solution, the photochromic reaction was totally prohibited (Figure 3.25). Here, copper(II) ion worked as a molecular "lock" to prevent photochromism after coordination with intramolecular piperazine moieties, transforming the system into a photoinactive state. On the other hand, Ethylenediaminetetraacetic acid (EDTA) can be used as a molecular "unlock" 4 to recover the photochromic performances by extracting the copper(II) ion from DTE ligand.

More recently, a novel dithiazolethene system showing gated photochromic reactivity controlled by complexation/dissociation with BF_3 was reported by Zhu and coworkers [58]. Upon alternating UV (365 nm) and visible light (546 nm) irradiation, compound **28o** can be reversibly switched between the colorless open form and the red closed form indicating the typical photochromic properties and good fatigue resistance of the compound. Moreover, **28o** can also undergo normal photochromic reaction from colorless to a violet color in the crystal, which was obtained by the diffusion method from a mixture of ethyl acetate and octane. However, no obvious UV absorption spectral changes could be observed with an irradiation at 365 nm after the addition of $BF_3{\cdot}Et_2O$, which suggest that the photochromism was prevented to a large extent by BF_3. The photochromic performance can be recovered by adding Et_3N. Apparently, BF_3 and Et_3N acted

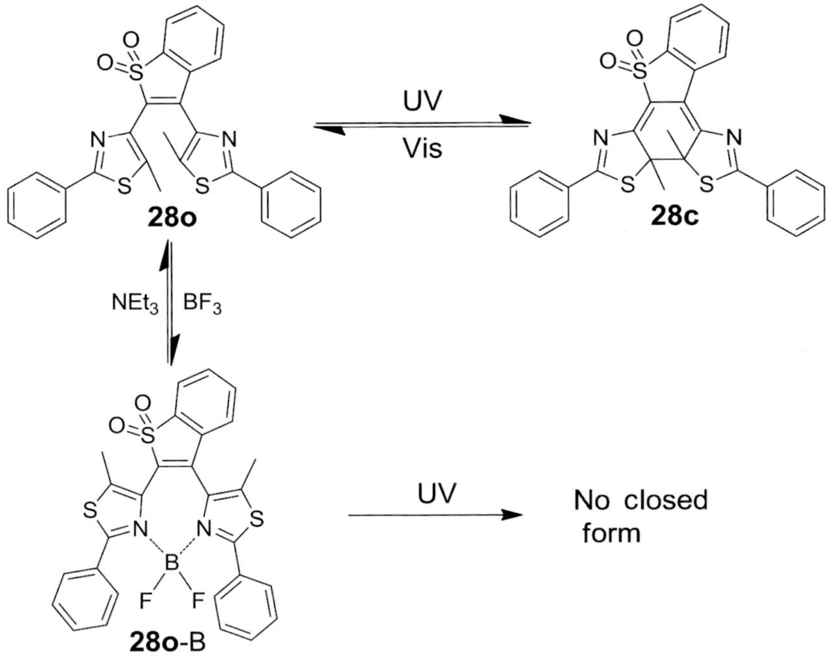

Figure 3.26 Photochromic reactions of compound **28o** and "lock and key" gated process controlled by complexation/dissociation with BF_3.

as "lock" and "key" for this special gated photochrome, respectively. A rigid seven-membered ring coordination model of **28o** with $BF_3{\cdot}Et_2O$ was proposed to explain the specific gated performance (Figure 3.26). In order to coordinate with BF_3 to form such structure complex, the two thiazole rings rotate to a parallel position, resulting in a longer distance between the two active carbon atoms and transforming **28o** into a parallel conformer. Consequently, the initial photochromic reaction was blocked. Upon adding Et_3N, the coordination effect between **28o** and BF_3 was destroyed, thus recovering the original characteristic photochromic reactivity.

3.3.3
Chemical Reaction

Gated photochromism regulated by specific chemical reactions was also reported by Branda *et al.* [59, 60]. Compound **29** possesses a butadiene backbone that cannot undergo the pericyclic photoreaction upon light irradiation. However, when the butadiene unit undergoes a Diels–Alder cycloaddition with a dienophile, the molecular structure is transformed into a typical 1,3,5-hexatriene system and then exhibited the reversible photochromism by irradiation with UV and visible light (Figure 3.27). This example of reactivity-gated photochromism has the potential application to dosimetry, along with controlling release of chemical species.

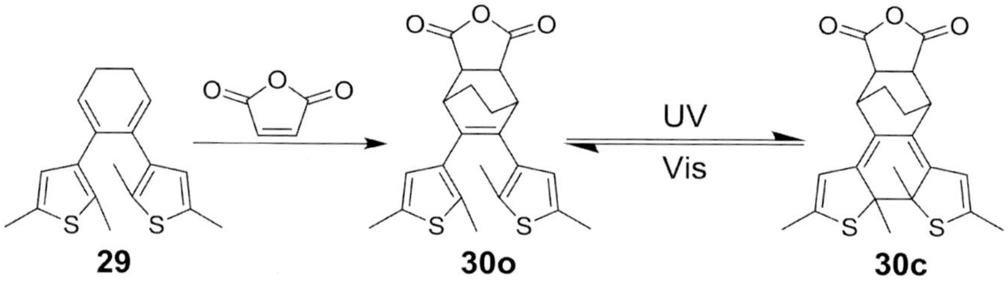

Figure 3.27 Diels–Alder reaction of diene **29** and photochromism of hexatriene **30o**.

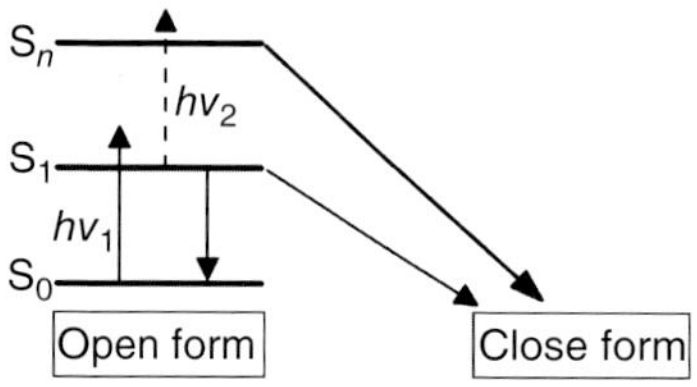

Figure 3.28 Principle of the stepwise two-photon process for the cycloreversion reaction.

In contrast with the conformation-restricted strategy, gated photochromism can also be achieved by developing some specific excited states that can undergo particular quenching processes rather than the photocyclization reaction. Irie *et al.* demonstrated the multiphoton-induced enhancement of a cycloreversion reaction in photochromic diarylethene derivatives by picosecond pulsed excitation instead of steady-state light irradiation [61, 62]. The enhancement of the cycloreversion yield (closed to 50%) is attributable to the production of a higher excited state via a successive two-photon process (Figure 3.28). The higher excited state, which is totally different from the normal excited state, is of crucial importance for the efficient cycloreversion reaction to take place. This is a new approach for one-color light control of the gated photochromic system, which can be utilized for an erasable memory system with nondestructive readout capability.

The excited states of the closed isomer can also be efficiently quenched by the intramolecular proton transfer as a result of the formation of the intramolecular hydrogen bonding, leading to a new type of chemically gated diarylethene [63]. In those diarylethene derivatives, an *N*-(2-hydroxyphenyl) group was introduced into the imide carbonyl ethene bridge. Consequently, in a nonpolar cyclohexane solution, the intramolecular hydrogen bonding was easily generated between the hydrogen atom in the phenol group and the oxygen atom in the imide carbonyl group, which induced quenching of the excited state due to the intramolecular proton transfer. Actually, no change in absorption spectrum of **31o** in cyclohexane solution was observed upon irradiation with 405 nm light. However, the pale yellow solution changed into pale red when acetic anhydride was added during irradiation with 405 nm light, showing a typical photochromism. The recovery in photochromic activity was attributed to the esterification of the hydroxyl group of **31o**, resulting a photoactive ester derivative with no proton-transfer ability (Figure 3.29).

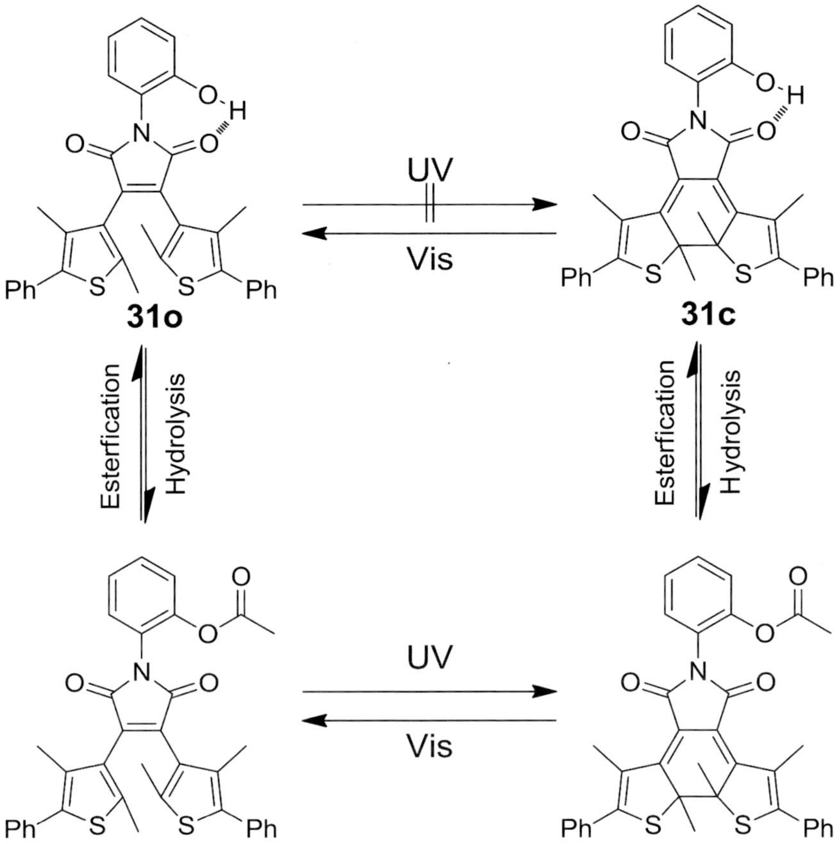

31o
31c
32o
32c

Figure 3.29 Gated photochromic process by the intramolecular proton transfer as a result of the formation of the intramolecular hydrogen bonding.

The photoreactivity of a photochromic diarylethene can be controlled by protonation. Diarylethene derivatives with conjugated (4-pyridyl)ethynyl group (**33o**, Figure 3.30) and nonconjugated (4-pyridyl)ethyl group at the reactive positions were well compared [64]. The photoreactivity of the former derivative containing conjugated (4-pyridyl)ethynyl group directly attached to the 6−π hexatriene moieties was strongly blocked by the pyridine protonation, while the photoreactivity of the latter derivative was not suppressed. The opposite photochromic behavior was due to the difference in their chemical structures. The former has an elongated π-electron system on the reactive position while the latter has no π-conjugation at the reactive position with single-bond linker. In the conjugated system, the electron-withdrawing nature of the N-methylpyridinium cation is considered to affect the 6−π electron system through π-conjugation, thus suppressing the photoreactivity.

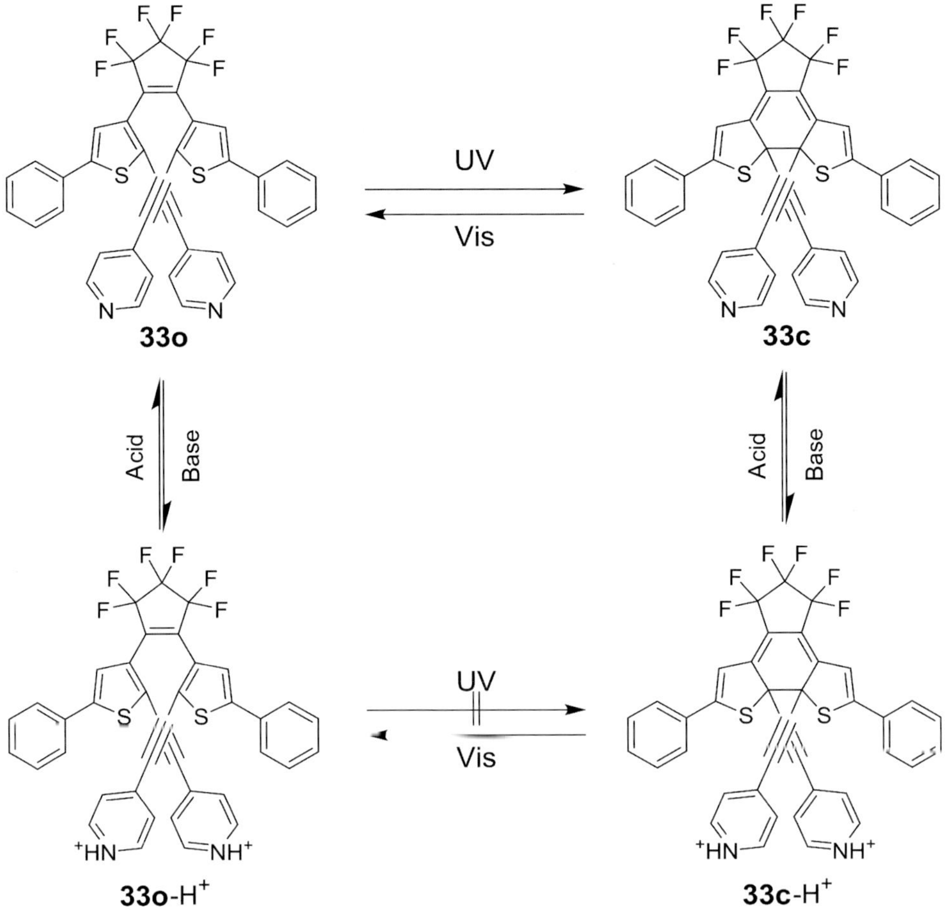

Figure 3.30 Principle of the lock and unlock process of compound **33o** containing conjugated (4-pyridyl)ethynyl group.

Belser *et al.* developed a new chemically gated DTE switch by the reversible protection and deprotection of the cyclobutene-1,2-dione skeleton [65]. Diarylethene derivative containing a cyclobutene-1,2-dione skeleton (**34o**) did not undergo any photochromic reaction with irradiation at 365 nm (Figure 3.31). Owing to the rigidity of system, the excited state of **34o** is considered to be efficiently quenched by the ring-opening reaction. The possible reason is that the resulting intermediate 1,2-bisketenes is very thermally unstable, and thermally returned into **34o** very fast, thus blocking the photochemical transformation to **34c**. Also compound **35o** with only one ketone functionality does not display any photochromic behavior due to the same reason. However, when both ketone functionalities were protected (**36o**), typical photochromic reaction took place.

Temperature could also be used for the gating of the photochromic behavior of diarylethenes. Feringa *et al.* [66] has demonstrated that the ring-opening process

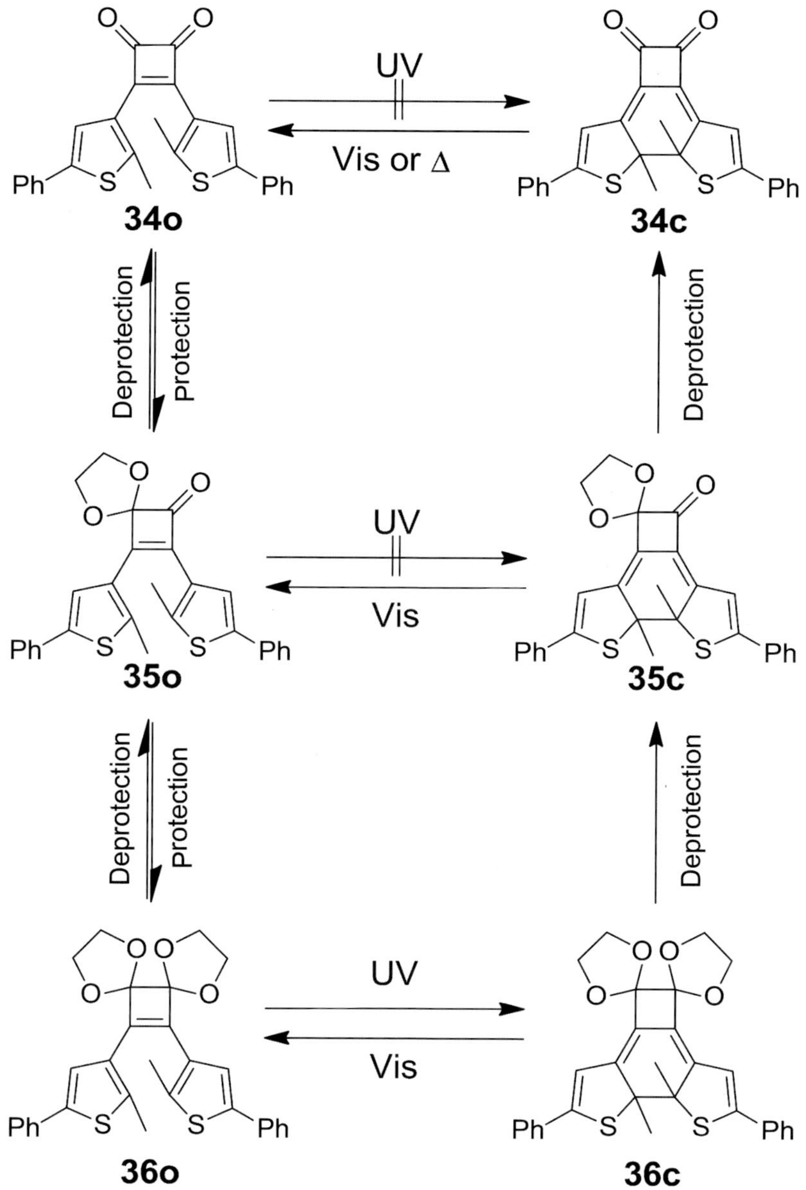

Figure 3.31 Protection reactions of **34o**, deprotection reactions of **35o**, and photochromism of **36o**.

of diarylethenes is strongly dependent on temperature, while photochemical ring-closing (photocycloreversion) processes show little temperature dependence. When the temperature was below 130 K, the photochemistry reaction can be effectively suppressed. In a typical ring-opening reaction of the closed form, the diarylethene molecule is first promoted to the thermally equilibrated excited state upon irradiation with visible light, then the excited state has to overcome a thermal barrier for further yielding the open isomer. Under the very low temperature (<130 K), the excited state cannot cross such a barrier, and the molecule is still kept in the close state. The strong temperature dependence of the ring opening provides a new way to control photochromic behavior.

Yam *et al.* provided a new approach to design gated photochromic materials by creating a highly emissive excited state with charge-transfer character [67]. In order to construct high fluorescent compound, the dimesitylboryl-2,20-bithiophene core with high fluorescence quantum yield was introduced into the framework of the photochromic DTE systems, resulting in 2 three-coordinated boron compounds **37o** and **38o**. Those two compounds display no photochromic behavior in different solution upon irradiation with UV light, while they exhibit extensive emission with high fluorescence quantum yield at 465 and 520 nm, respectively. Density functional theory (DFT) and time-dependent density functional theory (TD-DFT) calculations indicate that the high emissive state is resulted from the well conjugation between the boron pπ orbital and the bithiophene π system, leading to a decrease in the HOMO–LUMO separation. Consequently, the photochromism was suppressed by spontaneous emission derived from the excited state. A significant decrease in emission intensity of

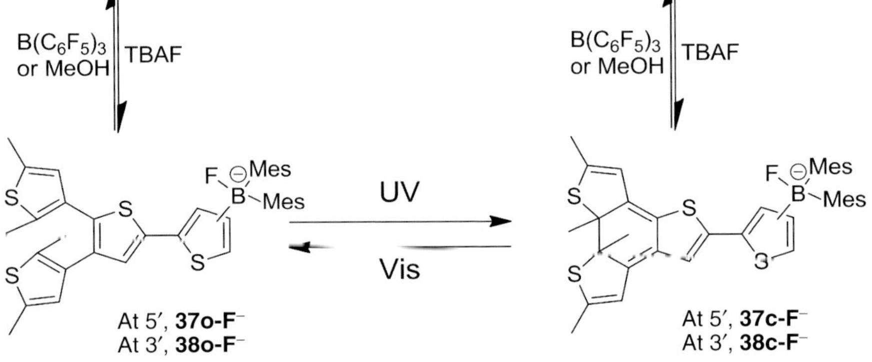

Figure 3.32 Gated photochromism of **37o** and **38o**.

37o and **38o** was observed by adding F^- as a result of the disruption of the π-conjugation extended through the boron pπ orbital by the formation of the corresponding fluoroborate (Figure 3.32). Moreover, typical photochromic reactions were observed for both compounds upon UV irradiation. The drastic photophysical changes upon fluoride-anion binding could be attributed to the disruption of $\pi-$pπ conjugation, and thus blocking the charge-transfer character in the excited state. Dissociation of F^- by adding an excess of methanol or a stronger Lewis acid such as $B(C_6F_5)_3$ can lock the photochromic system again.

References

1. Raymo, F.M. and Tomasulo, M. (2005) Electron and energy transfer modulation with photochromic switches. *Chem. Soc. Rev.*, **34**, 327–336.
2. de Silva, A.P., McClenaghan, N.D., and McCoy, C.P. (2001) *Molecular Switches*, Wiley-VCH Verlag GmbH.
3. De Silva, A.P., Gunaratne, H.Q.N., and Mccoy, C.P. (1993) A molecular photoionic AND gate based on fluorescent signalling. *Nature*, **364**, 42–44.
4. Szaciłowski, K. (2008) Digital information processing in molecular systems. *Chem. Rev.*, **108** (9), 3481–3548.
5. de Ruiter, G. and van der Boom, M.E. (2011) Surface-confined assemblies and polymers for molecular logic. *Acc. Chem. Res.*, **44** (8), 563–573.
6. Zhang, J.J., Zou, Q., and Tian, H. (2013) Photochromic materials: more than meets the eye. *Adv. Mater.*, **25** (3), 378–399.
7. Pischel, U., Andréasson, J., Gust, D., and Pais, V.F. (2013) Information processing with molecules-quo vadis? *ChemPhysChem*, **14** (1), 28–46.
8. Meng, F., Hervault, Y., Shao, Q., Hu, B., Norel, L., Rigaut, S., and Chen, X. (2014) One-way transparency of four-coloured spin-wave excitations in multiferroic materials. *Nat. Commun.*, **5**, 3203.
9. Myles, A.J., Wigglesworth, T.J., and Branda, N.R. (2003) A multi-addressable photochromic 1,2-dithienylcyclopentene-phenoxynaphthacenequinone hybrid. *Adv. Mater.*, **15** (9), 745–748.
10. Frigoli, M. and Mehl, G.H. (2005) Multiple addressing in a hybrid biphotochromic system. *Angew. Chem. Int. Ed.*, **44** (32), 5048–5052.
11. Green, K.A., Cifuentes, M.P., Corkery, T.C., Samoc, M., and Humphrey, M.G. (2009) Switching the cubic nonlinear optical properties of an electro-, halo-, and photochromic ruthenium alkynyl complex across six states. *Angew. Chem. Int. Ed.*, **48** (42), 7867–7870.
12. Tian, H. (2010) Data processing on a unimolecular platform. *Angew. Chem. Int. Ed.*, **49** (28), 4710–4712.
13. Li, B., Wang, J.-Y., Wen, H.-M., Shi, L.-X., and Chen, Z.-N. (2012) Redox-modulated stepwise photochromism in a ruthenium complex with dual dithienylethene-acetylides. *J. Am. Chem. Soc.*, **134** (88), 16059–16067.
14. Perrier, A., Maurel, F., and Jacquemin, D. (2012) Single molecule multiphotochromism with diarylethenes. *Acc. Chem. Res.*, **45** (8), 1173–1182.
15. Pu, S., Tong, Z., Liu, G., and Wang, R. (2013) Multi-addressable molecular switches based on a new diarylethene salicylal Schiff base derivative. *J. Mater. Chem. C*, **1**, 4726–4739.
16. Szalóki, G., Sevez, G., Berthet, J., Pozzo, J., and Delbaere, S. (2014) A simple molecule-based octastate switch. *J. Am. Chem. Soc.*, **136** (39), 13510–13513.
17. Wu, Y., Xie, Y., Zhang, Q., Tian, H., Zhu, W.H., and Li, A.D.Q. (2014) Quantitative photoswitching in bis(dithiazole)ethene enables modulation of light for encoding optical signals. *Angew. Chem., Int. Ed.*, **53** (8), 2090–2094.
18. Chen, S., Chen, L., Yang, H., Tian, H., and Zhu, W.H. (2012) Light-triggered reversible supramolecular transformations of multi-bisthienylethene

hexagons. *J. Am. Chem. Soc.*, **134** (33), 13596–13599.

19. Zhu, W.H., Yang, Y., Métivier, R., Zhang, Q., Guillot, R., Xie, Y., Tian, H., and Nakatani, K. (2011) Unprecedented stability of a photochromic bisthienylethene based on benzobisthiadiazole as an ethene bridge. *Angew. Chem. Int. Ed.*, **50** (46), 10986–10990.
20. Zhao, Z.J., Xing, Y.J., Wang, Z.X., and Lu, P. (2007) Dual-fluorescent donor–acceptor dyad with tercarbazole donor and switchable imide acceptor: promising structure for an integrated logic gate. *Org. Lett.*, **9** (4), 547–550.
21. Meng, X.L., Zhu, W.H., Zhang, Q., Feng, Y.L., Tan, W.J., and Tian, H. (2008) Novel bisthienylethenes containing naphthalimide as the center ethene bridge: photochromism and solvatochromism for combined NOR and INHIBIT logic gates. *J. Phys. Chem. B*, **112** (49), 15636–15645.
22. Margulies, D., Melman, G., and Shanzer, A. (2006) A molecular full-adder and full-subtractor, an additional step toward a moleculator. *J. Am. Chem. Soc.*, **128** (14), 4865–4871.
23. Bozdemir, O.A., Guliyev, R., Buyukcakir, O., Selcuk, S., Kolemen, S., Gulseren, G., Nalbantoglu, T., Boyaci, H., and Akkaya, E.U. (2010) Selective manipulation of ICT and PET processes in styryl-bodipy derivatives: applications in molecular logic and fluorescence sensing of metal ions. *J. Am. Chem. Soc.*, **132** (23), 8029–8036.
24. Chen, S.J., Yang, Y.H., Wu, Y., Tian, H., and Zhu, W.H. (2012) Multi-addressable photochromic terarylene containing benzo[*b*]thiophene-1,1-dioxide unit as ethene bridge: multifunctional molecular logic gates on unimolecular platform. *J. Mater. Chem.*, **22**, 5486–5494.
25. Andréasson, J., Pischel, U., Straight, S.D., Moore, T.A., Moore, A.L., and Gust, D. (2011) All-photonic multifunctional molecular logic device. *J. Am. Chem. Soc.*, **133** (30), 11641–11648.
26. Guo, Z., Zhu, W.H., Shen, L., and Tian, H. (2007) A fluorophore capable of crossword puzzles and logic memory. *Angew. Chem. Int. Ed.*, **46** (29), 5549–5553.
27. Strack, G., Ornatska, M., Pita, M., and Katz, E. (2008) Biocomputing security system: concatenated enzyme-based logic gates operating as a biomolecular keypad lock. *J. Am. Chem. Soc.*, **130** (13), 4234–4235.
28. Andréasson, J., Straight, S.D., Moore, T.A., Moore, A.L., and Gust, D. (2009) An all-photonic molecular keypad lock. *Chem. Eur. J.*, **15** (16), 3936–3939.
29. Zou, Q., Li, X., Zhang, J.J., Zhou, J., Sun, B.B., and Tian, H. (2012) Unsymmetrical diarylethenes as molecular keypad locks with tunable photochromism and fluorescence via $Cu^{2+}$and CN^{-}coordinations. *Chem. Commun.*, **48**, 2095–2097.
30. Chen, S.J., Guo, Z.Q., Zhu, S.Q., Shi, W.E., and Zhu, W.H. (2013) A multiaddressable photochromic bisthienylethene with sequence-dependent responses: construction of an inhibit logic gate and a keypad lock. *ACS Appl. Mater. Interfaces*, **5** (12), 5623–5629.
31. He, J.J., He, J.X., Wang, T.T., and Zeng, H.P. (2014) Ion-induced cycle opening of a diarylethene and its application on visual detection of $Cu^{2+}$and Hg^{2+} and keypad lock. *J. Mater. Chem. C*, **2** (36), 7531–7540.
32. Ceroni, P., Bergamini, G., and Balzani, V. (2009) Old molecules, new concepts: $[ru(bpy)_3]2+$as a molecular encoder–decoder. *Angew. Chem. Int. Ed.*, **48** (45), 8516–8518.
33. Balter, M., Li, S.M., Nilsson, J.R., Andréasson, J., and Pischel, U. (2013) An all-photonic molecule-based parity generator/checker for error detection in data transmission. *J. Am. Chem. Soc.*, **135** (28), 10230–10233.
34. Pischel, U. and Andréasson, J. (2010) A simplicity-guided approach toward molecular set–reset memories. *New J. Chem.*, **34**, 2701–2703.
35. Tian, H., Qin, B., Yao, R.X., Zhao, X.L., and Yang, S.J. (2003) A single photochromic molecular switch with four optical outputs probing four inputs. *Adv. Mater.*, **15** (24), 2104–2107.
36. Hirshberg, Y. (1956) Reversible formation and eradication of colors by irradiation at low temperatures. A photochemical memory model. *J. Am. Chem. Soc.*, **78** (10), 2304–2312.

37. Fukaminato, T., Sasaki, T., Kawai, T., Tamai, N., and Irie, M. (2004) Digital photoswitching of fluorescence based on the photochromism of diarylethene derivatives at a single-molecule level. *J. Am. Chem. Soc.*, **126** (45), 14843–14849.
38. Irie, M., Fukaminato, T., Sasaki, T., Tamai, N., and Kawai, T. (2002) Organic chemistry: a digital fluorescent molecular photoswitch. *Nature*, **420**, 759–760.
39. Lim, S.-J., Seo, J., and Park, S.Y. (2006) Photochromic switching of excited-state intramolecular proton-transfer (ESIPT) fluorescence: a unique route to high-contrast memory switching and nondestructive readout. *J. Am. Chem. Soc.*, **128** (45), 14542–14547.
40. Berberich, M., Krause, A.-M., Orlandi, M., Scandola, F., and Würthner, F. (2008) Toward fluorescent memories with nondestructive readout: photoswitching of fluorescence by intramolecular electron transfer in a diaryl ethene-perylene bisimide photochromic system. *Angew. Chem. Int. Ed.*, **120** (35), 6718–6721.
41. Fukaminato, T., Doi, T., Tamaoki, N., Okuno, K., Ishibashi, Y., Miyasaka, H., and Irie, M. (2001) Sing-molecule fluorescence photoswitching of a diarylethene-perylenebisimide dyad: non-destructive fluorescence readout. *J. Am. Chem. Soc.*, **133** (13), 4984–4990.
42. Nakagawa, T., Hasegawa, Y., and Kawai, T. (2009) Nondestructive luminescence intensity readout of a photochromic lanthanide(III) complex. *Chem. Commun.*, 5630–5632.
43. Uchida, K., Saito, M., Murakami, A., Nakamura, S., and Irie, M. (2003) Non-destructive readout of the photochromic reactions of diarylethene derivatives using infrared light. *Adv. Mater.*, **15** (2), 121–125.
44. Uchida, K., Saito, M., Murakami, A., Kobayashi, T., Nakamura, S., and Irie, M. (2005) Three bits eight states photochromic recording and nondestructive readout by using IR light. *Chem. Eur. J.*, **11** (2), 534–542.
45. Takeshita, M. and Yamato, T. (2002) Enantioselective photochromic reaction of a [2.2]metacyclophan-1-ene. *Angew. Chem. Int. Ed.*, **41** (12), 2156–2157.
46. Takeshita, M. and Yamato, T. (2004) Reversible optical rotation change according to the enantiospecific photochromic reaction of [2.2]metacyclophan-1-en. *Chem. Lett.*, **33** (7), 844–845.
47. Johnson, F. (1968) Allylic strain in six-membered rings. *Chem. Rev.*, **68** (4), 375–413.
48. Hoffmann, R.W. (1989) Allylic 1,3-strain as a controlling factor in stereoselective transformations. *Chem. Rev.*, **89** (8), 1841–1860.
49. Sud, D., Norsten, T.B., and Branda, N.R. (2005) Photoswitching of stereoselectivity in catalysis using a copper dithienylethene complex. *Angew. Chem. Int. Ed.*, **44** (13), 2019–2021.
50. Jin-Nouchi, H. and Takeshita, M. (2012) Syntheses and fully diastereospecific photochromic reactions of thiophenophan-1-enes with chiral bridges. *Chem. Eur. J.*, **18** (31), 9638–9644.
51. Irie, M. and Mohri, M. (1988) Thermally irreversible photochromic systems. reversible photocyclization of diarylethene derivatives. *J. Org. Chem.*, **53** (4), 803–808.
52. Irie, M. (2000) Diarylethenes for memories and switches. *Chem. Rev.*, **100** (5), 1685–1716.
53. Zhu, W.H., Meng, X., Yang, Y., Zhang, Q., Xie, Y., and Tian, H. (2010) Bisthienylethenes containing a benzothiadiazole unit as a bridge: photochromic performance dependence on substitution position. *Chem. Eur. J.*, **16** (3), 899–906.
54. Irie, M., Miyatake, O., and Uchida, K. (1992) Blocked photochromism of diarylethenes. *J. Am. Chem. Soc.*, **114** (22), 8715–8716.
55. Massaad, J., Micheau, J.C., Coudret, C., Sanchez, R., Guirado, G., and Delbaere, S. (2012) Gated photochromism and acidity photomodulation of a diacid dithienylethene dye. *Chem. Eur. J.*, **18**, 6568–6575.
56. Li, X., Ma, Y., Wang, B., and Li, G. (2008) "Lock and key control" of photochromic reactivity by controlling the oxidation/reduction state. *Org. Lett.*, **10** (16), 3639–3642.

57. Zhang, J.J., Tan, W.J., Meng, X.L., and Tian, H. (2009) Soft mimic gearshift with a multi-stimulus modified diarylethene. *J. Mater. Chem.*, **19**, 5726–5729.

58. Wu, Y., Chen, S.J., Yang, Y.H., Zhang, Q., Xie, Y.S., Tian, H., and Zhu, W.H. (2012) A novel gated photochromic reactivity controlled by complexation/dissociation with BF3. *Chem. Commun.*, **48**, 528–530.

59. Lemieux, V. and Branda, N.R. (2005) Reactivity-gated photochromism of 1,2-dithienylethenes for potential use in dosimetry applications. *Org. Lett.*, **7** (14), 2969–2972.

60. Asadirad, A.M., Boutault, S., Erno, Z., and Branda, N.R. (2014) Controlling a polymer adhesive using light and a molecular switch. *J. Am. Chem. Soc.*, **136** (8), 3024–3027.

61. Miyasaka, H., Murakami, M., Itaya, A., Guillaumont, D., Nakamura, S., and Irie, M. (2001) Multiphoton gated photochromic reaction in a diarylethene derivative. *J. Am. Chem. Soc.*, **123** (4), 753–754.

62. Murakami, M., Miyasaka, H., Okada, T., Kobatake, S., and Irie, M. (2004) Dynamics and mechanisms of the multiphoton gated photochromic reaction of diarylethene derivatives. *J. Am. Chem. Soc.*, **126** (45), 14764–14772.

63. Ohsumi, M., Fukaminato, T., and Irie, M. (2005) Chemical control of the photochromic reactivity of diarylethene derivatives. *Chem. Commun.*, 3921–3923.

64. Yumoto, K., Irie, M., and Matsuda, K. (2008) Control of the photoreactivity of diarylethene derivatives by quaternization of the pyridylethynyl group. *Org. Lett.*, **10** (10), 2051–2054.

65. Kühni, J. and Belser, P. (2007) Gated photochromism of 1,2-diarylethenes. *Org. Lett.*, **9** (10), 1915–1918.

66. Dulić, D., Kudernac, T., Pužys, A., Feringa, B.L., and van Wees B.J. (2007) Temperature gating of the ring-opening process in diarylethene molecular switches. *Adv. Mater.*, **19** (19), 2898–2902.

67. Poon, C.T., Lam, W.H., and Yam, V.W.W. (2011) Gated photochromism in triarylborane-containing dithienylethenes: a new approach to a "lock–unlock" system. *J. Am. Chem. Soc.*, **133** (49), 19622–19625.

4
Photoswitchable Supramolecular Systems

Guanglei Lv, Liang Chen, Haichuang Lan, and Tao Yi

4.1
Introduction

The spontaneous self-assembly of small molecules into complex superstructures is becoming a powerful tool toward the development of new materials and devices [1, 2]. Self-assembly systems constructed by noncovalent interactions, such as hydrogen bonding, π-stacking, metal–ligand, and van der Waals interactions, have a tunable functionality and are much more versatile than those of polymer materials because of their dynamic nature. Therefore, it is envisioned that molecularly defined smart materials can be constructed using the supramolecular concept [3–7]. In particular, such materials susceptible to the effects of external stimuli are of interest and play a crucial role as functional materials [8]. However, the regulation and control of the self-assembly process to realize a specific functionality and to obtain well-defined smart materials remain a significant challenge to scientists.

Compared with traditional stimuli, such as a change in pH values [9], temperature [10], magnetic or electric fields [11], and chemical stimulus [12], light acts as a remote stimulus with a wide range of adjustable parameters (such as wavelength, intensity, duration) and outperforms many other counterparts in terms of its broad use in molecular devices and smart materials. Photochromic systems in which each isomer of the photochromic compound can represent "0" or "1" of a digital code have garnered much interest for their potential applications in optical memory and photonic switching devices. Photochemistry is an important branch of modern science at the crossroads of chemistry, physics, and biology, with implications in the fields of engineering and medicine. Research on photoresponsive biological systems has shown that the value of the function that can be obtained from the interaction between light and matter is dependent not only on the chemical structure of the matter but also on the degree of organization and complexity of the matter. Therefore, photoresponsive supramolecular systems with an organized structure and composited functions have become increasingly significant.

Photochromic Materials: Preparation, Properties and Applications, First Edition.
Edited by He Tian and Junji Zhang.

The recent progress on photoswitchable supramolecular systems containing an amphiphilic system, a host–guest system, and metal complexes, in addition to their application as molecular devices, sensors, and bioimages, is discussed in this chapter.

4.2 Photoreversible Amphiphilic Systems

Amphiphilic compounds generally contain hydrophobic and hydrophilic moieties, possessing good solubility in organic solvents and water. Specifically, the water compatibility of the amphiphilic compounds is a very important property for their application in biological fields, such as biosensors or drug delivery systems [13]. Moreover, those amphiphilic photochromic molecules may self-assemble into supramolecular architectures whose nanostructures/microstructures can be tuned by light irradiation [14].

4.2.1 Photoreversible Diarylethene-Based Amphiphilic System

It is well known that diarylethene derivatives (DTEs) are the most promising photochromic candidates because of their notable irreversible thermal photochromic behavior, high photoisomerization quantum yields, and outstanding fatigue resistance [15]. These materials have great potential application in high-density data storage systems as a molecular switch with nondestructive readout capability and as molecular devices [16–32]. Amphiphilic diarylethene derivatives have been given great attention in recent years because they can be soluble in water and in organic solvents. Their photochromic performance can be tuned using the polarity of the solvents. By introducing a suitable fluorescent chromophore, the fluorescence of the diarylethene derivatives can be switched by light. This type of water-soluble, fluorescent switchable system can be used as a switchable mark for imaging living cells.

We recently designed and synthesized an amphiphilic diarylethene molecule with hydrophobic and hydrophilic chains at two ends of a rigid diarylethene core (**1** in Figure 4.1a) [33]. Compound **1** forms stable vesicles in water in its open form (**1O**) in a concentration greater than 1.2×10^{-6} M (critical aggregation concentration) with an enhanced green emission compared with its organic solution. The water "soluble" vesicles exhibit an obviously different fluorescence in open and closed states with alternating irradiation of UV and visible light. A scanning electron microscope (SEM) image shows a spherical morphology with a diameter of 50–200 nm for the freeze-dried sample from a low-concentration aqueous solution (1.0×10^{-5} M, Figure 4.1b). The confocal laser scanning microscopy (CLSM) images for one drop of **1O** water solution (1.0×10^{-5} M) on a glass substrate showed well-dispersed micrometer-sized fluorescent circles (Figure 4.1c), indicating that the hydrophobic interaction and the hydrophilic

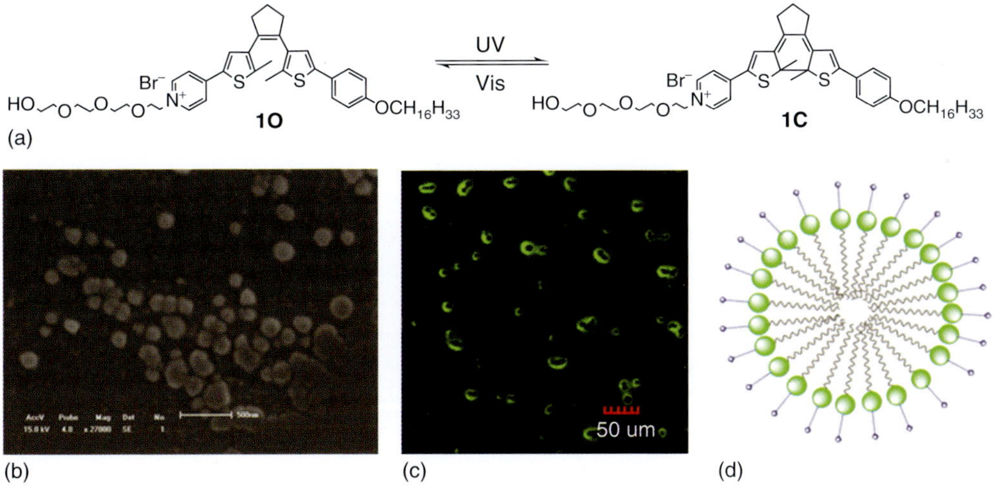

Figure 4.1 (a) Structure and photochromic process of **1**; (b) SEM and (c) CLSM images of **1O** in water (1.0×10^{-5} M, $\lambda_{ex} = 405$ nm, $\lambda_{em} = 480-580$ nm); and (d) the schematic illustration of the aggregation formed by **1O**.

repulsion between the molecules governed the arrangement of the molecules to vesicles (Figure 4.1d).

These fluorescent vesicles can enter the living cells with low cytotoxicity, achieving the presence of fluorescent switches in these cells upon alternate irradiation with 405 and 633 nm light (Figure 4.2). This process can be repeated many times without any apparent fatigue effects, revealing its potential utility as a cellular marker or fluorescent switch in living cells with high-ratio signal change and excellent resistance to fatigue.

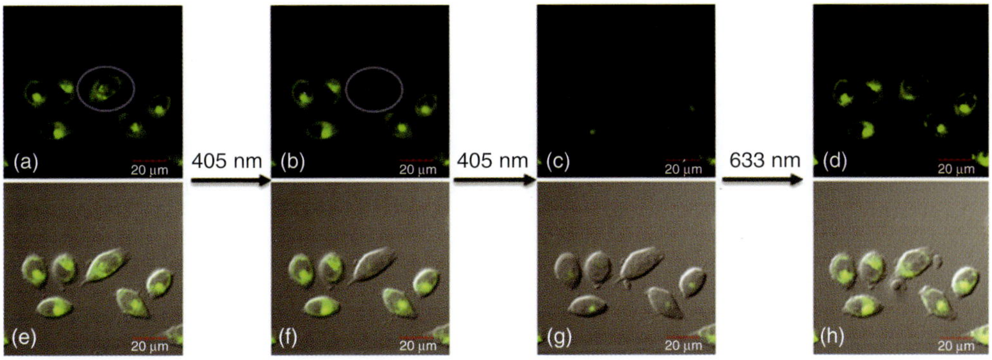

Figure 4.2 CLSM image (a–d) and the overlay image (e–h) of KB living cells incubated with **1O** for 20 min at 25 °C (a) in original state, (b) irradiated by 405 nm light (2 mW) for a single cell, (c) all cells, and (d) recovered by 633 nm light. (Adapted with permission from [33]. American Chemical Society, 2008.)

In a typical diarylethene analog, the structure of diarylethene has both parallel and antiparallel conformations in equal proportions, but only the antiparallel conformation can perform photoisomerization. The relative composition of parallel and antiparallel conformations, which may be controlled by an intramolecular hydrogen bond, electronic oxidation or reduction, solvent conditions, and so on, can be used as a trigger to tune the gated photochromic properties of diarylethene derivatives [15, 23, 26]. A biomolecule-gated photochromism might increase the applications of diarylethene derivatives to adjust a physiological process *in vivo*, including cell division and proliferation, intracellular physical movement, disease diagnosis, and therapy. Recently, we developed a DNA-gated photochromic diarylethene probe (**2**) modified with a thiazole orange (TO) group (Figure 4.3a) [34]. Compound **2** showed a difference of photophysical behavior in organic and aqueous solutions. In an organic solvent, such as acetonitrile, the TO groups in **2** are in a monomer state; however, the spectroscopic property of **2** gives a

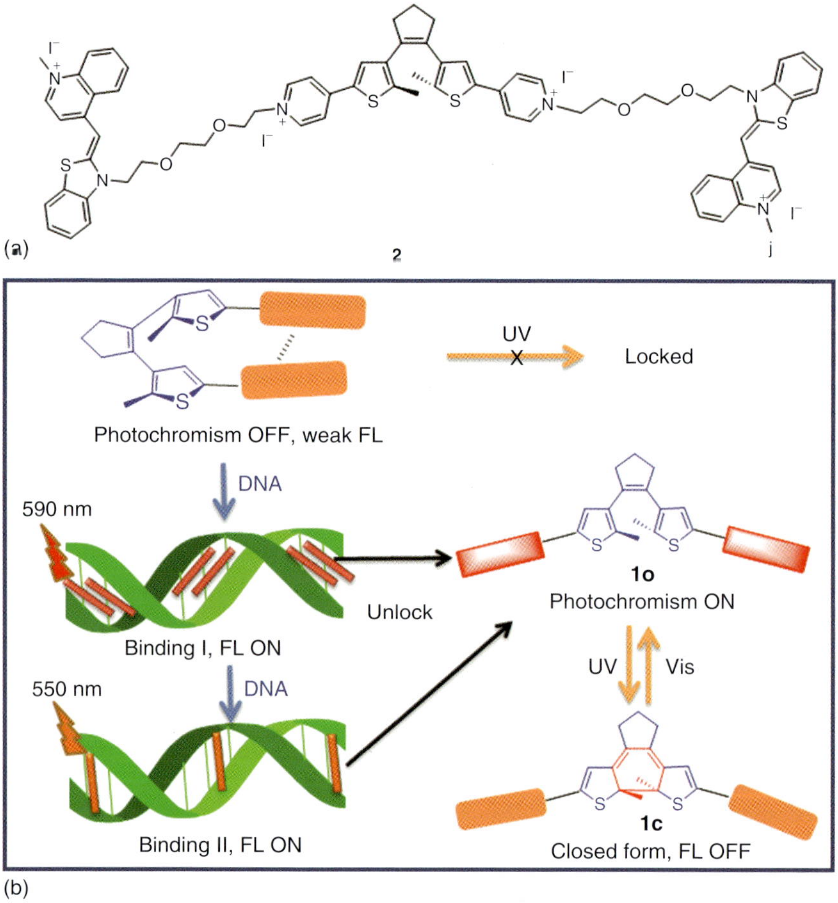

Figure 4.3 (a) The structure and (b) schematic diagram of **2** response to DNA with gated photochromism.

dimeric characteristic in water with a blue-shifted absorption and a red-shifted fluorescence of low-fluorescence quantum yield ($Q_F = 0.25\%$). However, the photocyclization reaction of **2** was inactive in aqueous medium. Due to the strong intramolecular interaction of TO in aqueous solution, **2** presumably exists in a folded parallel conformation.

When binding with DNA, the fluorescence of **2** is drastically enhanced, and the photochromic reactivity is unlocked. Compound **2** exhibited different interaction modes toward A·T and C·G base pairs with a great diversity of photophysical properties. In all of the binding modes, the reversible photoisomerization reaction could be induced by alternate irradiation with UV and visible light. A new broad absorption band centered at ~700 nm was generated, and the absorption at ~400 nm decreased after irradiation with UV light (wavelength 360 nm), accompanied by a marked decrease in fluorescence. The reverse process can be obtained by irradiation with visible light at 670 nm (Figure 4.3b).

The DNA-mediated photochromism can be performed in cell medium (Figure 4.4a–f) with low toxicity (Figure 4.4g). When a selected cell was

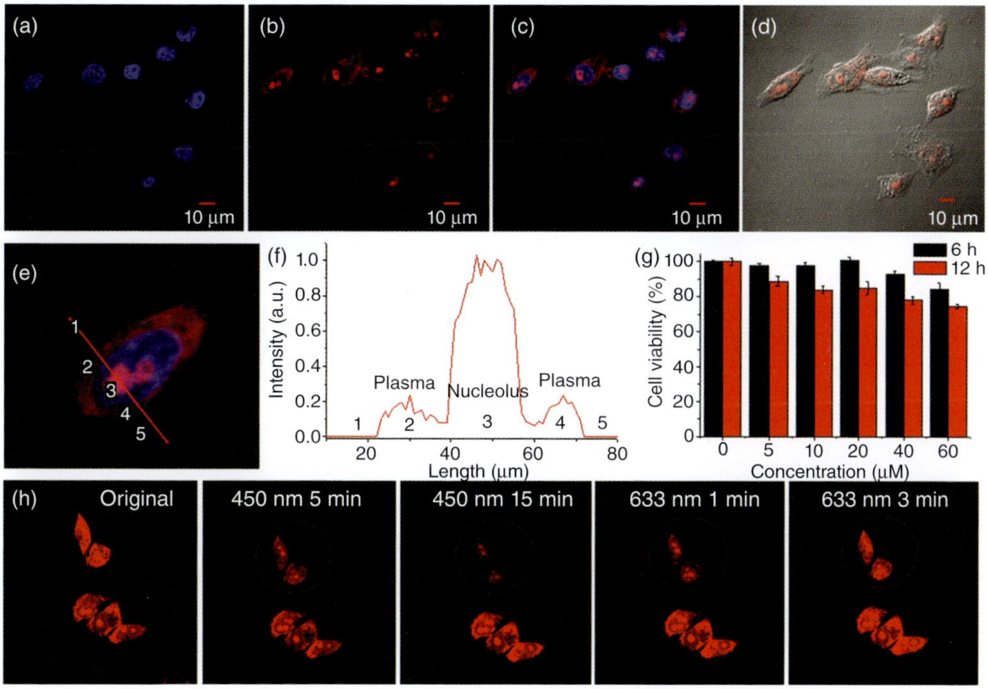

Figure 4.4 CLSM images of fixed Hela cell incubated with **2** (10 μM) for 10 min and followed with Hoechst33258 (10 μM) for 30 min at 37 °C; (a) Hoechst33258 in blue channel (420–480 nm), (b) **2** in red channel (520–620 nm), (c) overlay of Hoechst33258 and **2**, and (d) overlay image of (b) and the bright field image. (e) A selected cell and (f) cross-sectional analysis along the red line in image (e). (g) Cell viability value (%) by MTT 3-(4,5-dimethyl-2-thiazolyl)-2,5-diphenyl-2-H-tetrazolium bromide method. (h) Fluorescent images of selected cells in red circle bleached with 405 nm and recovered by 633 nm laser.

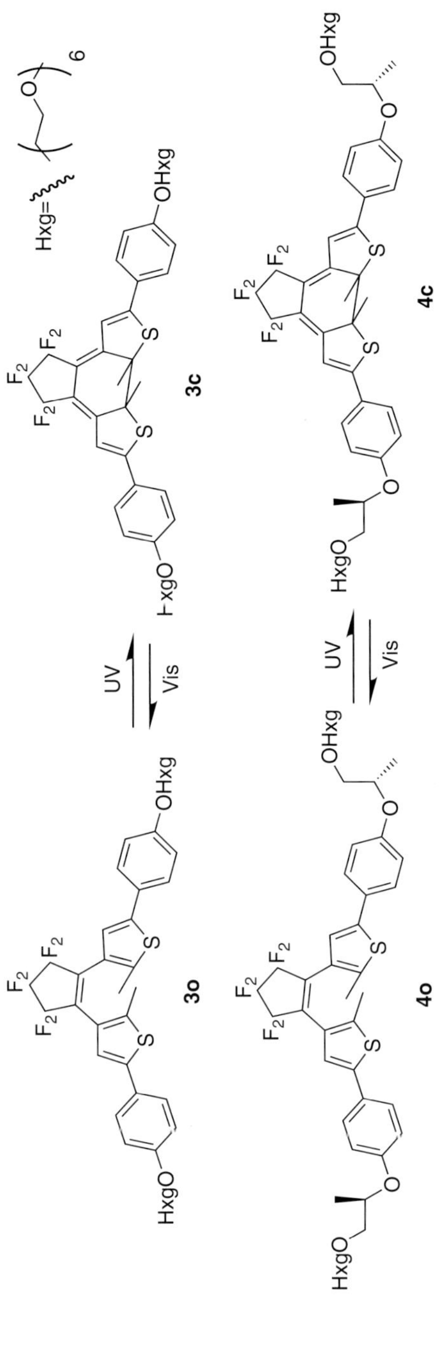

Figure 4.5 Chemical structure and photochromism of compounds **3** and **4**.

irradiated with 405 nm wavelength light for 15 min, the fluorescence was gradually bleached, which indicated that the majority of **2** was transformed from the open to the closed form. When the selected cell was treated with light at a wavelength of 633 nm, complete recovery of the fluorescence was observed in 3 min (Figure 4.4h). The optical switching process can be repeated many times without any noticeable decrease in fluorescence. The high fatigue and photobleaching resistance property of **2** is superior to other traditional fluorophores. This is a specific example of a biomolecule-controllable photoreaction, which might allow the application of light to a combination of biomolecules and stimuli-responsive materials to construct new functional molecular devices and to provide new insight into the bioapplication of photochromic materials in the future.

Irie *et al.* incorporated amphiphilic side chains into a hydrophobic diarylethene core. The self-assembling behavior and photochromic reactivity were investigated [35]. These compounds showed photochromism in organic solvents and even in water. The aqueous solution of the compounds turned turbid quickly upon heating, which revealed that the compounds self-assembled into aggregates in the aqueous solution, and the aggregates were loosened by raising the temperature. Dynamic light scattering (DLS) measurements showed that the average size of the aggregates of compounds **3** and (*S*,*S*)-**4** were approximately 100 nm in water (Figure 4.5). However, in ethyl acetate, the light-scattering signal was too weak to obtain the size distribution. The diarylethene core and the poly(ethylene glycol) (PEG) side chains are soluble in ethyl acetate; therefore, compounds **3** and (*S*,*S*)-**4** should be molecularly dispersed in ethyl acetate. The size distribution in water became narrow when the solution was heated. The transition temperature was 45–50 °C for the open-ring isomer **4a** and 42–46 °C for the photostationary-state (PSS) sample.

When asymmetric methyl groups were introduced in the amphiphilic side chains, induced circular dichroism (ICD) was observed upon irradiation with UV light in water. This ICD was explained by the difference in the self-assembling behavior between the open- and the closed-ring isomers. No cyclodextrin (CD) signal was observed before and after the photocyclization reaction in ethyl acetate, where (*S*,*S*)-**4** is molecularly dispersed. In water, the colorless open-ring isomer (*S*,*S*)-**4o** did not show any CD signal, while an intense CD signal was observed upon UV irradiation (Figure 4.6). The induced CD signal intensity depends on the twist sense bias of the transition moment and the distance between the two chromophores, which are expected to depend on the interaction between the chromophores. The closed-ring isomer has a planar structure so that the stacked structure is favorably formed by the solvophobic effect. On the other hand, the open-ring isomers have a bulky twisted conformation, which prevents the formation of the stacked structure. This difference is considered to cause the different CD spectra between the open- and closed-ring isomers.

Subsequently, the photochromic property and self-assembling behavior of diarylethene derivatives linking different lengths and the number of PEG side

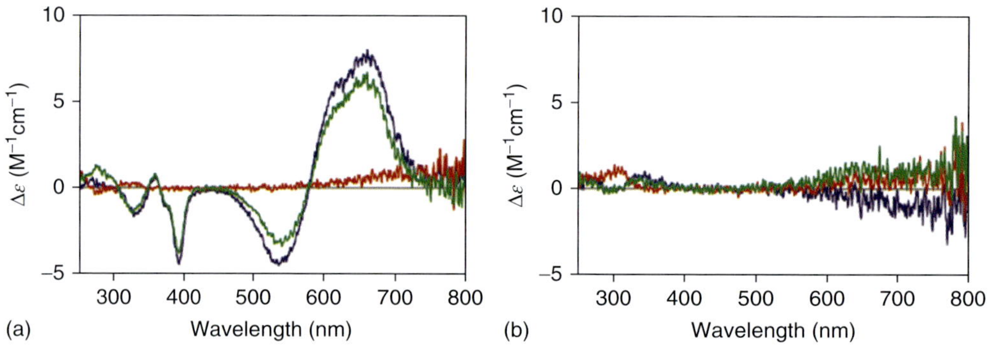

Figure 4.6 CD spectra of (*S*,*S*)-**4** (a) in water (2.6×10^{-5}) and (b) in ethyl acetate (1.6×10^{-5}). Red line, the open-ring isomer; blue line, the closed-ring isomer; and green line, the sample in the photostationary state under irradiation with 313-nm light. (Adapted with permission from [35]. American Chemical Society, 2006.)

chains have been investigated (**5**–**8** in Figure 4.7) [36]. The size of the aggregates and the self-assembling behavior depend on the proportion of the amphiphilic side chains to the hydrophobic core moiety. As the proportion of PEG side chains increased, the size distribution of aggregates became smaller and the compound showed better solubility in water. The compound that has two hexaethylene glycol side chains showed the most favorable photoswitching of the induced CD signal.

Zhou *et al.* synthesized an amphiphilic diarylethene derivative with histamines as side chains and investigated its photochromic properties and light-triggered self-assembling behavior (**9** in Figure 4.8) [37]. When the solution was irradiated with 254-nm light, the morphology changed from nanofiber to nanosphere. The conversion between nanofiber and nanosphere in the solution is reversed when the irradiation light is tuned after sufficient time for assembling. The nanostructure is stable upon shaking on the silicon chip.

Moreover, the average diameter of the nanosphere could be tuned with the pH value of the solution, transitioning from the range of 600–700 nm to the range of 200–300 nm as the pH value changed from 6.4 to 7.9. The conversion of the self-assembled nanostructure primarily originated from the structural conformational conversion between the open-ring isomer and closed-ring isomer, which influences the change of some weak force, consequently driving the reassembly.

4.2.2
Photoreversible Azobenzene-Based Amphiphilic System

Azobenzene is a widely used chromophore for the preparation of light-responsive polymers because it can undergo *trans*–*cis* photoisomerization in response to UV and visible light [38, 39]. After UV light irradiation, the distance between the two end *para* carbon atoms of azobenzene changes from 9 to 5.5 Å and the dipole

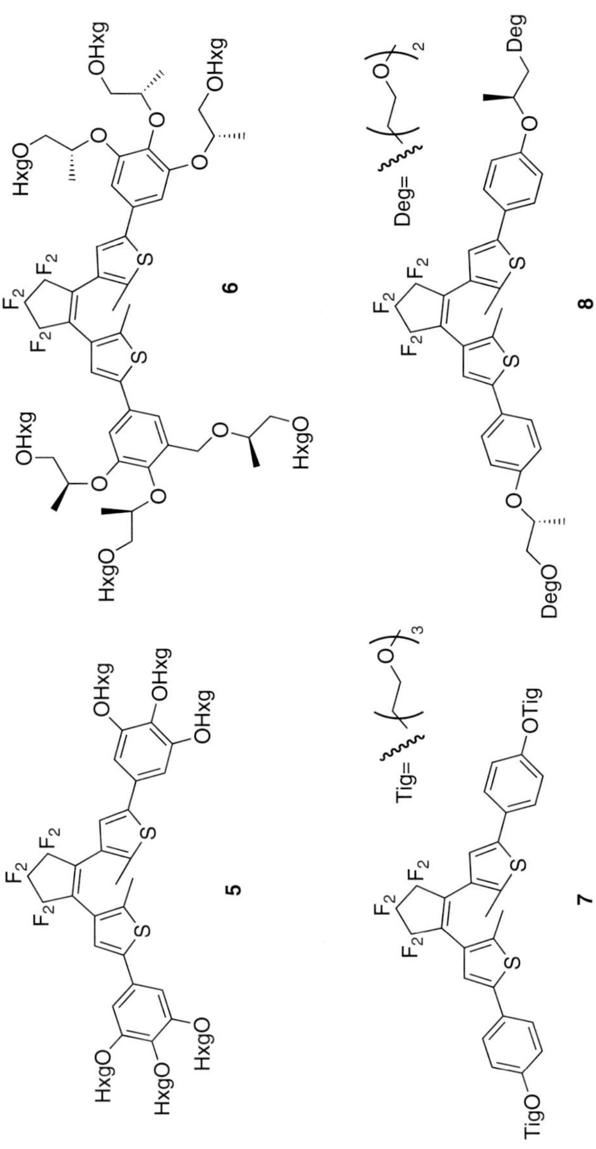

Figure 4.7 Chemical structure of compounds **5–8**.

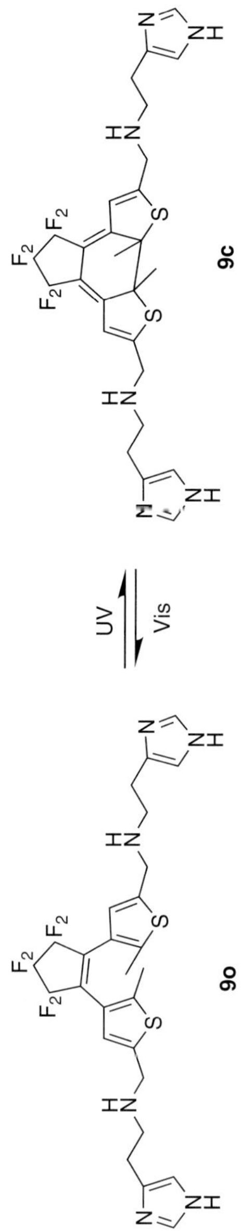

Figure 4.8 Chemical structure and photochromism of compound **9**.

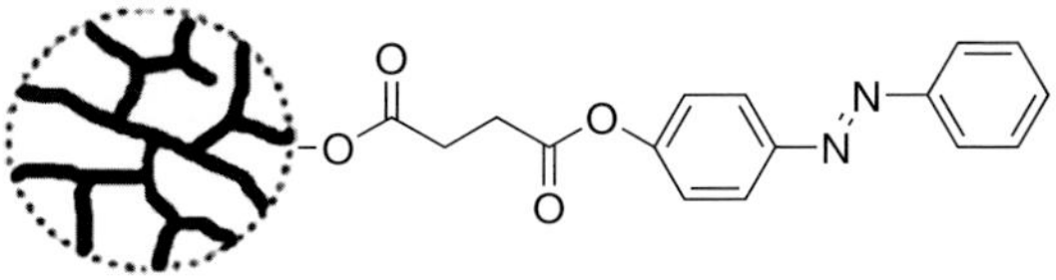

Figure 4.9 Chemical structure of compound HPHEEP-Azo.

moment differs from 0.52 to 3.08 D. This great difference between the structure of *trans-* and *cis*-isomers should drastically change the aggregation properties of those molecules, especially for the amphiphilic system.

Ji *et al.* designed and synthesized an amphiphilic hyperbranched polymer HPHEEP-Azo, which is composed of a water-soluble backbone of polyphosphate and photoresponsive hydrophobic azobenzene tails (Figure 4.9) [40]. The polymer could self-assemble into a micellar structure in aqueous solution with an average diameter of 169.9 nm and a critical micelle concentration of 0.013 mg ml^{-1}. After the addition of β-CD to the micellar solution, the micelles dissociated gradually upon increasing the addition amount of β-CD. Upon UV light irradiation, spherical micelles were formed again due to the mismatch between β-CD and *cis*-azobenzene. When the micelles were further irradiated with visible light, spherical micelles dissociated again because of the photoreverse of the *cis*-azobenzene to *trans*-azobenzene and the subsequent inclusion complexation of *trans*-azobenzene groups and β-CDs.

Zhang *et al.* demonstrated for the first time a block ionomer complex obtained by electrostatic complexation between a block copolymer and a photoresponsive azobenzene-containing surfactant (AzoC10 in Figure 4.10) [41]. More importantly, this block ionomer complex can form UV-responsive vesicles, which can be explored as supramolecular polymer nanocontainers for controlled loading and release of solutes. The photoisomerization of azobenzene moieties in the block ionomer complex can reversibly tune the amphiphilicity of the surfactants, inducing the disassembly of the vesicles. These block ionomer complex vesicles are further evaluated as nanocontainers capable of encapsulating and releasing guest solutes on demand, which is controlled by light irradiation. For example, the vesicles encapsulating the fluorescein sodium display clear spherical images, as observed by fluorescence microscopy. However, these fluorescence-marked images disappear after releasing the solute from the vesicles triggered by the UV light. This model system combines photochemistry and host–guest chemistry for a stimulus-responsive vesicle, which may have a potential application in drug delivery [42].

4.2.3 Photoreversible Spiropyran-Based Amphiphilic System

Spiropyran derivatives can adopt one of two stable states: the open-ring state, known as the merocyanine (MC) form and the closed-ring state, known as the

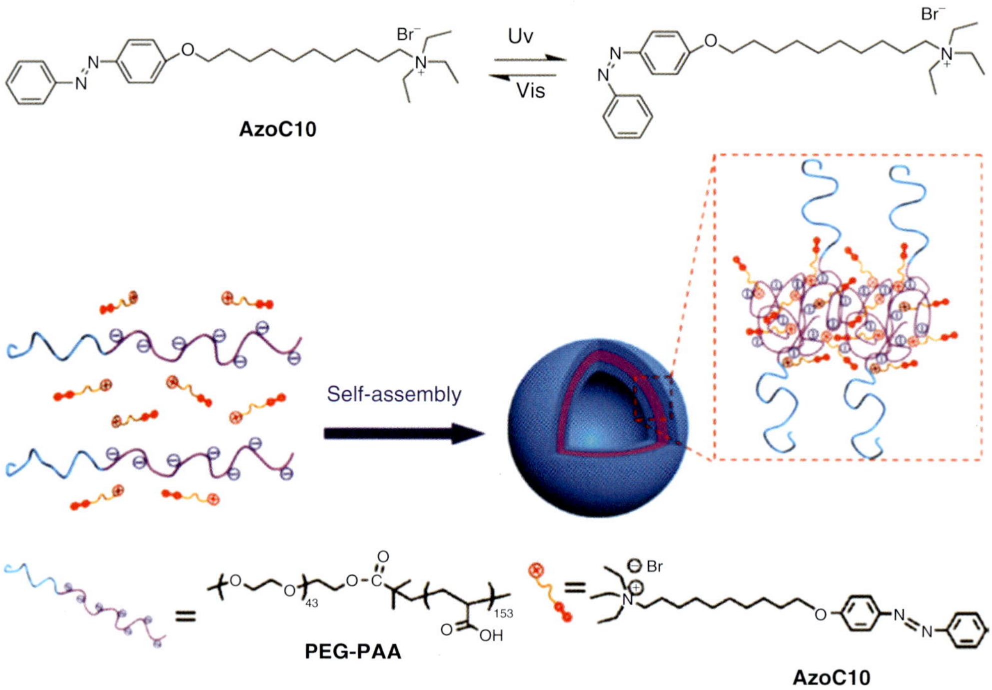

Figure 4.10 Photoisomerization of AzoC10 in the block ionomer complex vesicles ($Z = 1$). [PEG43-PAA153] = 0.06 mg ml^{-1}; [AzoC10] = 8×10^{-5} M. (Adapted with permission from [41]. American Chemical Society, 2010.)

spiro (SP) form. Irradiation with UV light in spiropyrans involves photocleavage of the C–O bond, which allows reversible switching between a colorless closed form (SP) and a strongly colored open MC form, while visible light reverses to the SP form [43, 44]. Amphiphilic polymers containing photochromic spiropyran derivatives have been developed by several research groups. For example, Landfester *et al.* designed and synthesized amphiphilic photoswitchable thiophene-based fluorescent polymer dots (Pdots) fabricated by coprecipitating a soluble and highly fluorescent polythiophene precursor with another photochromic spiropyran-linked amphiphilic copolymer into water under continuous ultrasound (Figure 4.11) [45]. The designed spiropyran-linked amphiphilic copolymer not only provided the fluorescent Pdots with unprecedented long-term photostability (>30 days) in aqueous media but also allowed the reversible modulation of the fluorescence emission of the thiophene-based polymer via alternating irradiation with UV and visible light. These fluorescent Pdots not only show a uniform small particle size (about 16 nm in diameter), excellent long-term stability, and high brightness of single particles but also reveal distinct dual-color fluorescence, fast photoresponsiveness, as well as favorable photoreversibility with a fatigue resistance over at least 10 cycles of photoconversion.

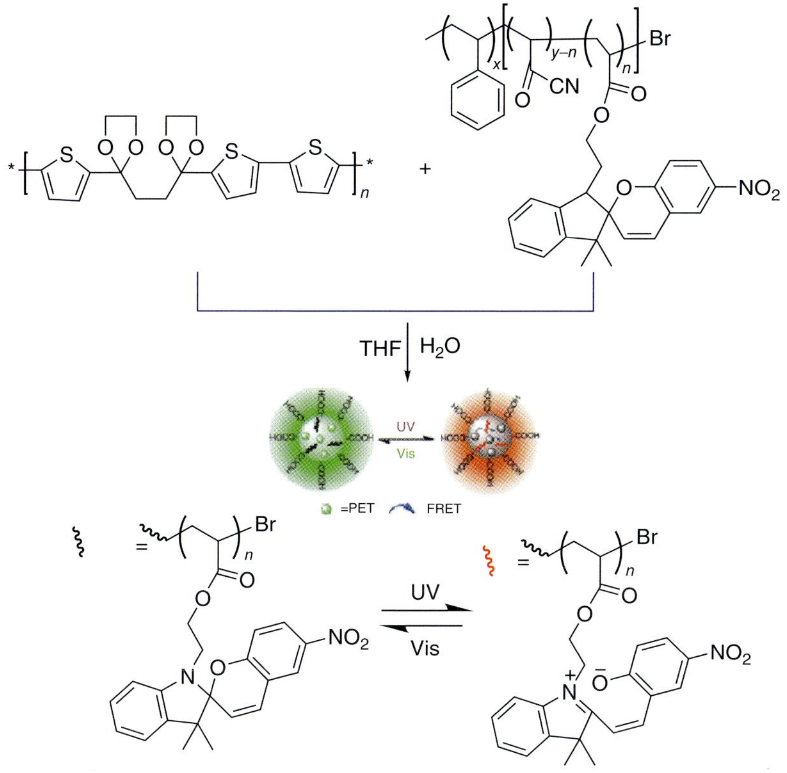

Figure 4.11 Preparation scheme for the reversibly photoswitchable amphiphilic Pdots and their FRET-mediated photoswitching property. (Adapted with permission from [45]. Royal Society of Chemistry, 2013.)

Subsequently, Chen *et al.* designed and synthesized polymer nanoparticles using a facile one-pot miniemulsion polymerization using a polymerizable nonionic surfactant, ω-methoxy poly(ethylene oxide) undecyl α-methacrylate (PEO-R-MA-40), forming optically addressable nanoparticle-based two-colored fluorescent systems in aqueous media [46]. These nanoparticles contain two energy levels of well-matched fluorophores, a fluorescein-based crosslinking monomer, fluorescein-*O,O*-bispropene (FBP), and spiropyran-linked methacrylate (SPMA) (Figure 4.12). There is no dye leakage for nanoparticles because of the covalent incorporation of dye molecules. The fluorescence emission of FBP dye in nanoparticles can be reversibly modulated by the transformation of the spiropyran moiety structures upon irradiation with UV and visible light. Overall, the resulting novel amphiphilic reversible photoswitchable fluorescent nanoparticles not only show a controllable amount and ratio of the two dyes, high fluorescence intensity, and tunable fluorescence resonance energy transfer (FRET) efficiency but also exhibit excellent photostability, relatively fast photoresponsibility, and better photoreversibility. This class of novel FRET-mediated

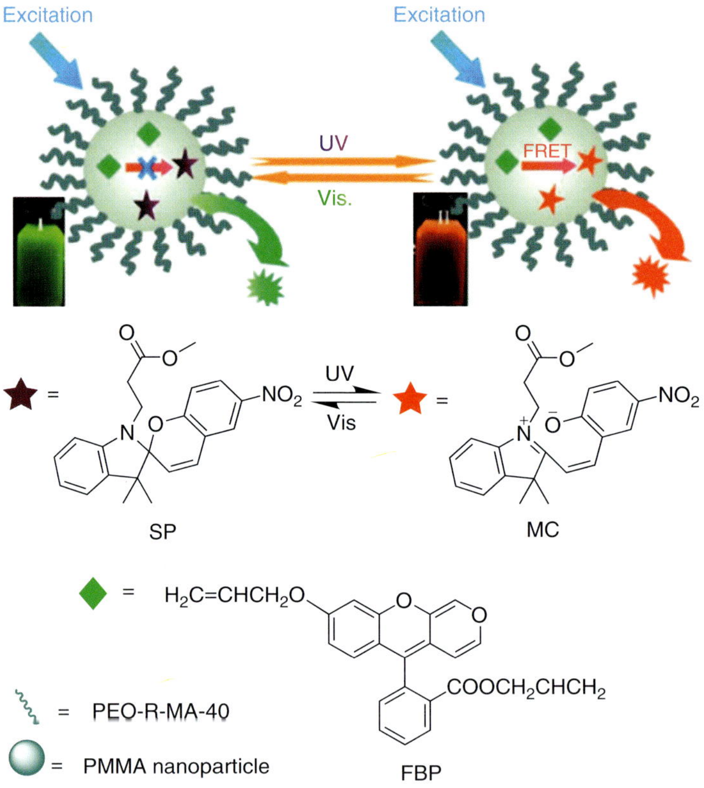

Figure 4.12 Schematic illustration of novel amphiphilic reversible photoswitchable fluorescent nanoparticles via covalently incorporating fluorescent dye (FBP) and photochromic derivative (SPMA). (Adapted with permission from [46]. Royal Society of Chemistry, 2012.)

photoresponsive nanoparticles may have a variety of interesting applications in biological labeling and imaging.

4.3 Photoswitchable Host–Guest Systems

In supramolecular chemistry, a host–guest system describes complexes that are composed of two or more molecules or ions that are held together in unique structural relationships by noncovalent bonding [47]. Host–guest chemistry encompasses the idea of molecular recognition and interactions through noncovalent bonding, which is critical in maintaining the 3D structure of large molecules, such as proteins, and is involved in many biological processes in which large molecules bind specifically to one another [48]. In the photoswitchable host–guest system, azobenzene is one of the most studied light-responsive units due to its robustness and its rapid and reversible isomerization; thus, it is commonly used to construct photoswitchable host–guest systems for practical applications [49]. In addition, fluorescent photoswitchable host–guest systems were fabricated by diarylethene derivatives due to their good photophysical properties [23, 26]. In this chapter, the advances in exploring photoresponsive building blocks, especially the azobenzene and diarylethene units, and their applications in the field of molecular self-assembly and the photoresponse capture and release of guest molecules, molecular sensors, and molecular devices were explored.

4.3.1 Photocontrolled Supramolecular Self-Assembly

Azobenzene and its derivatives can undergo an efficient and reversible *trans–cis* isomerization reaction upon irradiation with UV and visible light, which alters the distance between the two end *para* carbon atoms from 9 to 5.5 Å [50]. Therefore, the self-assembly behavior of the *trans-* and *cis-*isomers of azobenzene is different and can be controlled by light irradiation [51–53]. The first case of host–guest interactions between azobenzene derivatives was reported by Shinkai *et al.*, as shown in Figure 4.13. Compound **10** was connected via a ω-ammonium alkyl and a crown ether (Figure 4.13a). The *trans-*isomers of **10** favor intermolecular oligomerization. Under irradiation of UV light, the *cis-*isomers undergo an intramolecular interaction between the two functional groups. Thus, the

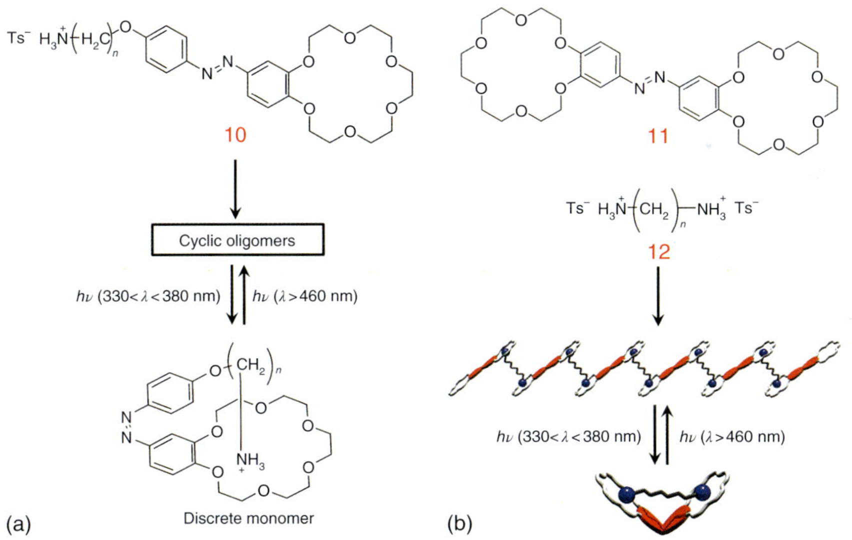

Figure 4.13 Photoresponsive self-assemblies established by Shinkai *et al.* (a) One-component system achieved by self-complementary azobenzene **10** functionalized on one end by an ω-ammonium alkyl and on the other by a crown ether. (b) Two-component system achieved by the mixture of symmetrical azocrown **11** and α,ω-diammoniumalkane **12**. (Adapted with permission from [49]. Royal Society of Chemistry, 2008.)

alternated irradiation of UV and visible light makes the molecules self-aggregate differently [55]. Furthermore, by mixing equivalent symmetrically crown-functionalized azobenzene **11** and oligomethylene-α,ω-diammonium cations, two different self-assembly products were produced (Figure 4.13b) [49, 54]. In both of the cases, the key for the formation of different self-assemblies is the conformation of azobenzene isomers. *Cis*-azobenzene acts as a nonaggregative building block by forming intramolecular supramolecular species, while *trans*-azobenzene favors intermolecular aggregation.

Yagai and Kitamura have studied a series of photoresponsive hydrogen bonded macrocycles (rosettes) formed from azobenzene-appended melamines and barbiturates/cyanurates through complementary triple hydrogen bonding interactions (Figure 4.14a) [49]. The photoresponsive process depends on the substituent of the building blocks. Rosette I was obtained by mixing AzoMel1 and BAR, both lacking sterically hindered substituents [56]. Rosette I prepared in chloroform was kinetically stable, giving rise to an irreversible precipitation due to transformation into an insoluble extended tape, such as supramolecular polymers, on the timescale of

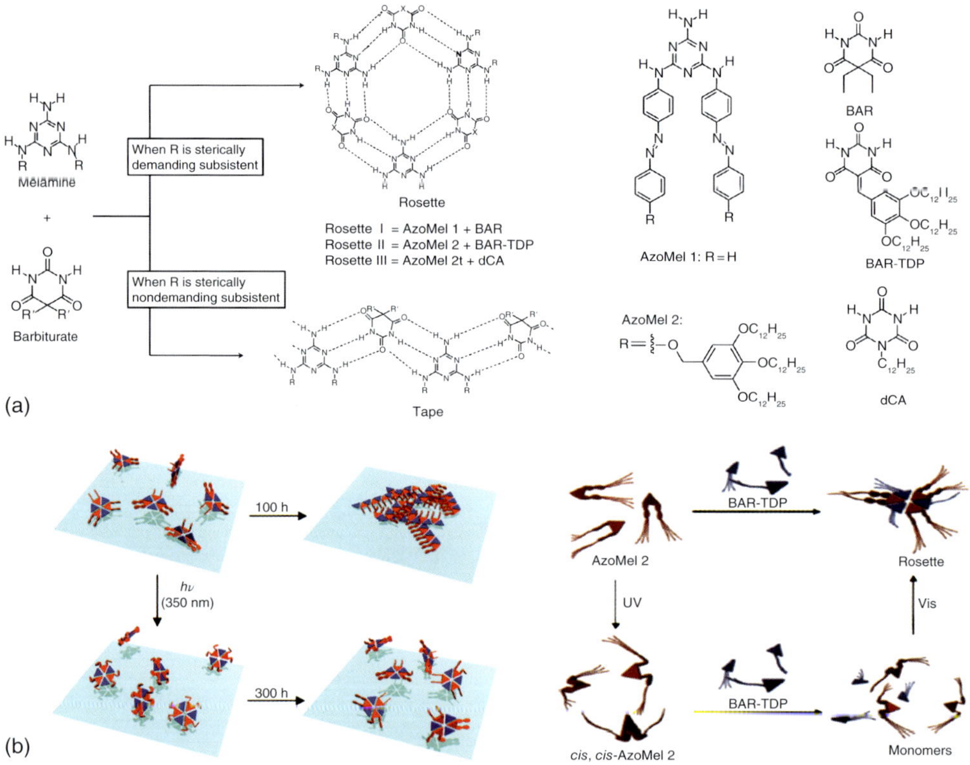

Figure 4.14 (a) Supramolecular rosettes I–III formed by mixing azobenzene-appended melamines AzoMel1 or AzoMel2 and barbiturates BAR or BAR-TDP or dodecyl cyanurate dCA. (b,c) The corresponding schematic representation of the phototriggered formation of rosettes. (Adapted with permission from [49], Royal Society of Chemistry, 2008; and [56], Wiley-VCH, 2005.)

days (Figure 4.14b). The azobenzene moieties of rosette I can be efficiently photoisomerized by UV light irradiation while maintaining the rosette architectures, as evidenced by ^{1}H NMR. The *trans:cis* ratio was 12 : 88 in a PSS, indicating that the isomerization of azobenzene moieties has no obvious effect on the self-assembly of rosette I. However, the UV-irradiated solution does not produce any precipitates upon aging over 300 h, which is different from the solution without UV irradiation. This observation clearly demonstrates that the *cis*-azobenzene moieties suppress the irreversible denaturation of rosette into tape-like assemblies.

In a further study, sterically demanding substituents, such as tridodecyloxyphenyl (TDP), were induced into azobenzene-appended melamines and barbiturates. Thus, AzoMel2 and BAR-TDP were prepared, and their mixture was shown to form thermodynamically stable rosette II in polar solvents [57]. As a compensation for the increased thermodynamic stability, however, the azobenzene moieties of rosette II displayed low photoisomerization efficiency due to the steric crowding within the rosette. The *trans:cis* ratio in a PSS achieved under the best conditions is only 75 : 25, indicating that AzoMel2 possessing two *cis*-azobenzene moieties (*cis,cis*-AzoMel2) no longer complexes with BAR-TDP to construct the rosette architecture. Thus, the *in situ* generation of aggregative *trans,trans*-AzoMel2 in the pool of monomers upon irradiation with visible light triggered the generation of rosettes (Figure 4.14c), and the amount of rosette II could be quantitatively regulated by visible (increasing the concentration of the rosette) and UV light (maintaining the concentration of the rosette). To hierarchically organize the disk-shaped rosettes into higher-order columnar structures, rosette III consisting of AzoMel2 and dodecyl cyanurate (dCA) was prepared, affording columnar assemblies stable in polar solvents [58]. In a high-concentration regime (410 mM), the bundled rosette columns formed fibrous networks, yielding organogels. The *trans–cis* photoisomerization of the azobenzene moieties occurred, even in the hierarchically organized state, showing a *cis*-content of about 45% under the best conditions. The resulting solution could be reconverted into the gel state upon irradiation with visible light and subsequent aging. Thus, by varying the number of the bulky wedges, different types of photoresponses, that is, photoinduced stabilization, phototriggered formation, and eventually photoregulatable stacking, were achieved for the hydrogen-bonded macrocycles.

Because of a characteristic structure with a hydrophilic exterior surface and a hydrophobic interior cavity that can accommodate a wide range of molecules as guests, CDs have been widely used as hosts in supramolecular chemistry for fabricating various nanostructures [59–61]. Using α-cyclodextrins (α-CDs) as a host molecule, a facile photocontrollable supramolecular route to realize the disassembly and reassembly of the pseudopolyrotaxane (PPR) hydrogels was reported by Jiang and coworkers [62]. The self-assembly of PEG as the axis and α-CDs as threaded rings can form linear PPR hydrogels [63–65]. For the photoresponsive purpose, a water-soluble competitive guest 1-[*p*-(phenylazo)benzyl]pyridinium bromide (Azo-C1-N^+), which contained a quaternary pyridine group, was mixed with the above PPR hydrogel of PEG/α-CD (Figure 4.15). After sonication, the gel

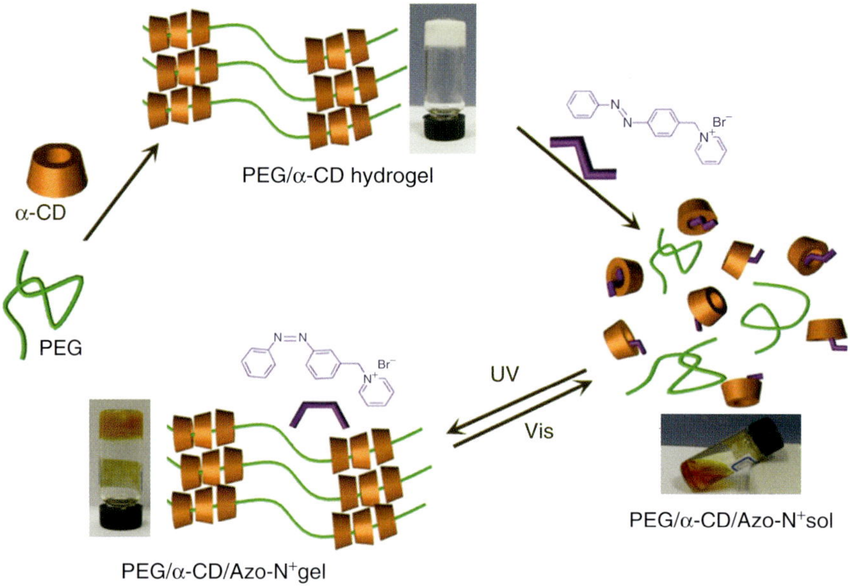

Figure 4.15 Cartoon representation of photoresponsive PPR gel–sol–gel transitions driven by competitive inclusion complexation. (Adapted with permission from [62]. Royal Society of Chemistry, 2012.)

turned into a transparent sol in a few minutes. Studies showed that the host–guest interactions between *trans*-Azo-N^+ and α-CD were stronger compared with those between the PEG chain and α-CD. Therefore, after the addition of *trans*-Azo-N^+ to the hydrogel system, the threaded CD molecules were gradually pulled off, leaving the PEG molecules as free chains and leading to the gel-to-sol transition. However, when the UV light irradiation was applied to the ternary solution, the PPR hydrogel was regenerated, implying that the interaction between α-CD and Azo-C1-N^+ becomes much weaker when the latter is converted from the *trans* to the *cis* form. Subsequent irradiation by visible light converted the hydrogel to a sol again, as expected. The photoresponsive gel to sol and sol-to-gel transitions can be repeated for several cycles without any disturbance because of the light-induced reversible *trans*–*cis* isomerization of the azobenzene moiety.

Harada and coworkers reported the first example of photoresponsive supramolecular hydrogels based on a CD host polymer and an Azo guest polymer [66]. They mixed curdlan (*b*-1,3-glucan, CUR) functionalized with CDs (CD-CUR) (4.0 wt%) and an Azo-modified poly(acrylic acid) (pAC12Azo, 8.6 wt%) in 1 : 1 ratio of the monomer units in water, resulting in the formation of a hydrogel (Figure 4.16a). The viscosity of the resultant hydrogel was measured as 54 Pa s. After photoirradiation with UV light (365 nm), the hydrogel transformed into a sol due to the isomerization of the *trans*-azo moieties in pAC12Azo to the *cis*-azo groups (*trans*:*cis* = 12 : 88), causing the average viscosity to be only 9 Pa s. Treatment with visible light or heating at 60 °C can lead to the isomerization

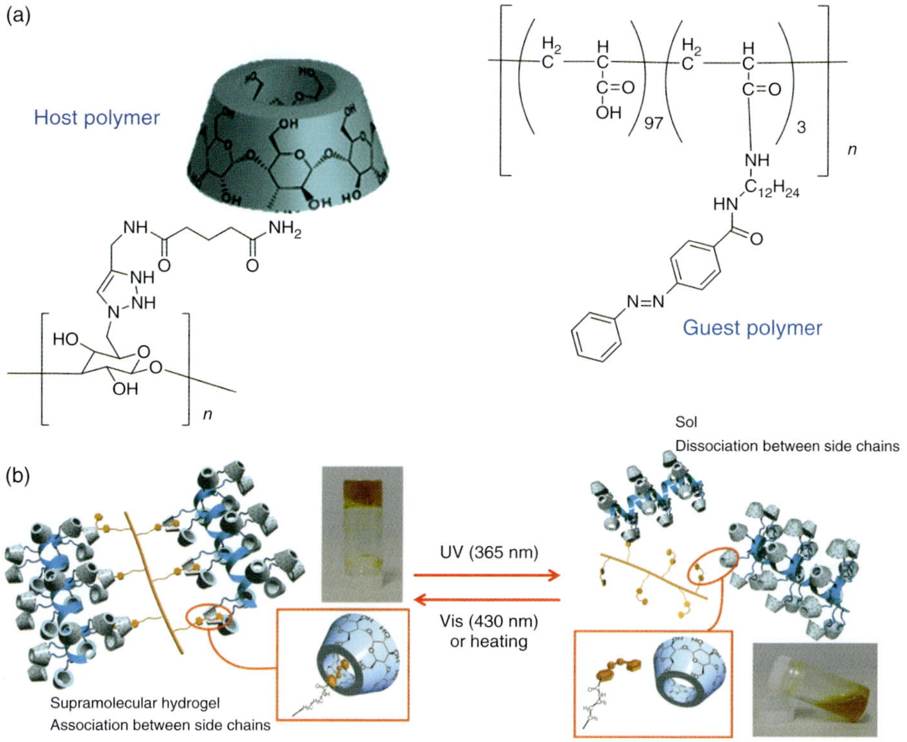

Figure 4.16 (a) Chemical structures of the host polymer and guest polymer. (b) Schematic representation of the interactions of the α-CD unit (cylindrical shapes) with azobenzene moieties (yellow shapes) upon irradiation with UV (365 nm) and visible light (430 nm) or heating at 60 °C. (Adapted with permission from [66]. Wiley-VCH, 2010.)

of the azo moieties from *cis* to *trans* (*trans*:*cis* = 75 : 25), and the homogeneous hydrogel can be obtained again within a few minutes (Figure 4.16b). The inclusion complex of azo and α-CD acts as a supramolecular fastener between the side chains; thus, the connection or disconnection between the two polymers can be controlled by light.

Although many supramolecular architectures have been well reported [67, 68], the self-assembly of macroscopic materials through molecular recognition has rarely been reported. For this purpose, Harada and coworkers reported a photoregulated gel system developed using polyacrylamide-based hydrogels functionalized with azobenzene (guest) or cyclodextrin (host) moieties [69]. Through this ingenious design, a switch between molecular self-assembly and disassembly at the macroscopic scale has been successfully realized for the first time by photoirradiation. First, the CD (α-CD or β-CD) and Azo derivatives were separately introduced to a polyacrylamide-based hydrogel prepared by radical copolymerization under mild conditions (Figure 4.17a). The host gels were stained with dye, which did not influence the association between the CDs

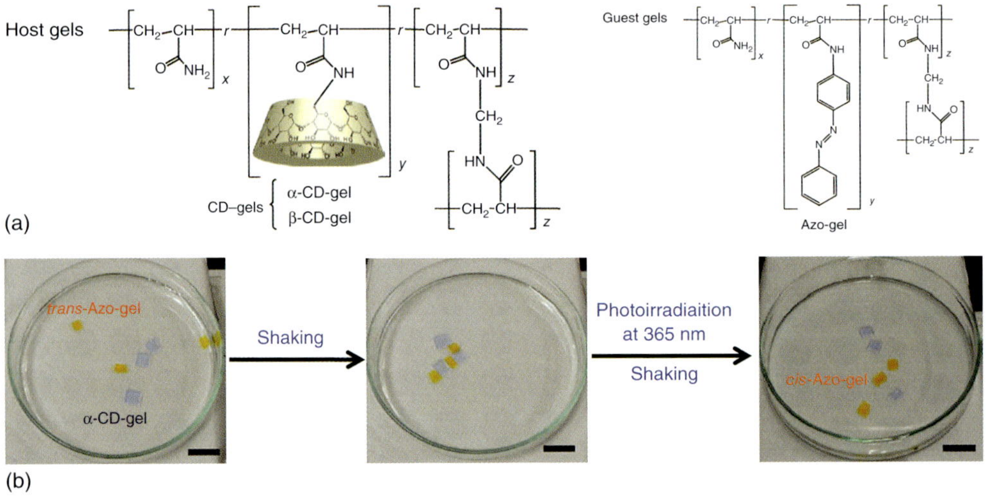

Figure 4.17 (a) The chemical structures of the host gel (α- and β-CD-gels) and the guest gel (Azo-gel). (b) Gel assembly of α-CD-gel (blue) with the *trans*-Azo-gel (orange) and gel dissociation with irradiation of UV light. The scale bar corresponds to 1.0 cm. (Adapted with permission from [69]. Nature, 2012.)

and the guest moieties in the gels. When mixing the α-CD-gel and the Azo gel in an appropriate proportion, followed by shaking these gels in water, the α-CD-gel and the Azo-gel can form a combined gel through host–guest interactions visible to the naked eye (Figure 4.17b). A comparative experiment showed that only shaking the nonfunctionalized blank gel did not induced the self-assembly, and no homogeneous assembly of α-CD-gel/α-CD-gel or Azo-gel/Azo-gel could be obtained, which further confirmed the molecular recognition between these two gels. As expected, the assembled gels dissociated upon light irradiation at 365 nm because most of the *trans*-Azo turned to *cis*-Azo, and the estimated association constant (K_a) for the α-CD/*cis*-Azo pair was too low to assemble the gels. Similarly, the separated α-CD-gels and Azo-gels were found to reassemble upon visible light irradiation (wavelength at 430 nm) for a few minutes. These results indicate that the self-assembly of the macroscopic α-CD-gels and Azo-gels can be controlled by photoirradiation.

4.3.2 Photocontrolled Capture and Release of Guest Molecules

In the past several years, great efforts have been devoted to the study of photoresponsive smart materials because of their potential use in drug delivery [70, 71]. For example, light-induced coumarin release was reported in polymeric micelles consisting of photochromic spiropyran units [72]. Recently, Zhou's group reported an azobenzene-functionalized metal–organic polyhedral (MOP) for the optically responsive capture and release of the methylene blue (MB) molecule

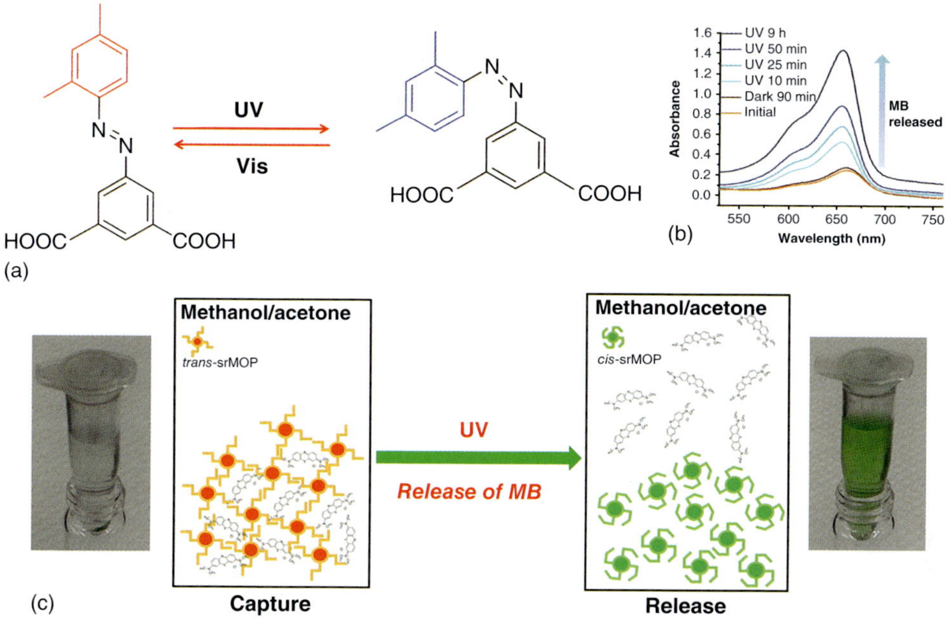

Figure 4.18 (a) Light-induced *trans–cis* isomerization of DPDP. (b) UV–vis spectra of MB in solution in 10% methanol/acetone. As MB was released from *cis*-srMOP-1, the absorbance of MB in the solution increased. (c) Schematic illustration of the capture of MB by *trans*-srMOP-1 and its release from *cis*-srMOP-1. (Adapted with permission from [73]. Wiley-VCH, 2014.)

[73]. In this study, a chloroform-soluble srMOP-1 consisting of $Cu(OAc)_2$ and the acidic form of the ligand (2,4-dimethylphenyldiazenylisophthalate, DPDP) was synthesized (Figure 4.18a). This srMOP-1 performs a *trans–cis* isomerization with irradiation of UV or blue light, and this isomerization can be utilized for light-induced guest release. As demonstrated in this study, MB, which is too large to be trapped in the core of the MOP cage but can be captured among the cages of *trans*-srMOP-1, acts as a guest molecule (Figure 4.18b). Initially, the MB in methanol solution was mixed with sr-MOP-1 in acetone and was stirred in the dark to load the MB into the *trans*-srMOP-1 cages. With irradiation of UV light, the MB molecule enclosed in the space between *trans*-srMOP-1 cages was released because the interaction between the *cis*-srMOP-1 structures was too weak to generate pockets to retain the guest molecules (Figure 4.18c). This capture and release process can be traced by UV–vis spectroscopy and can be repeated for several cycles. This was a fascinating study, but this system is far from being applicable to the controlled capture and release of guest molecules, especially drug molecules *in vivo*.

It is well known that mesoporous materials, especially mesoporous silica (MS)-based materials, are excellent candidates for drug delivery systems and have attracted extensive attention in recent years due to their stable mesostructure, lack of cytotoxicity, large surface area, and tailorable pore sizes [74, 75]. For

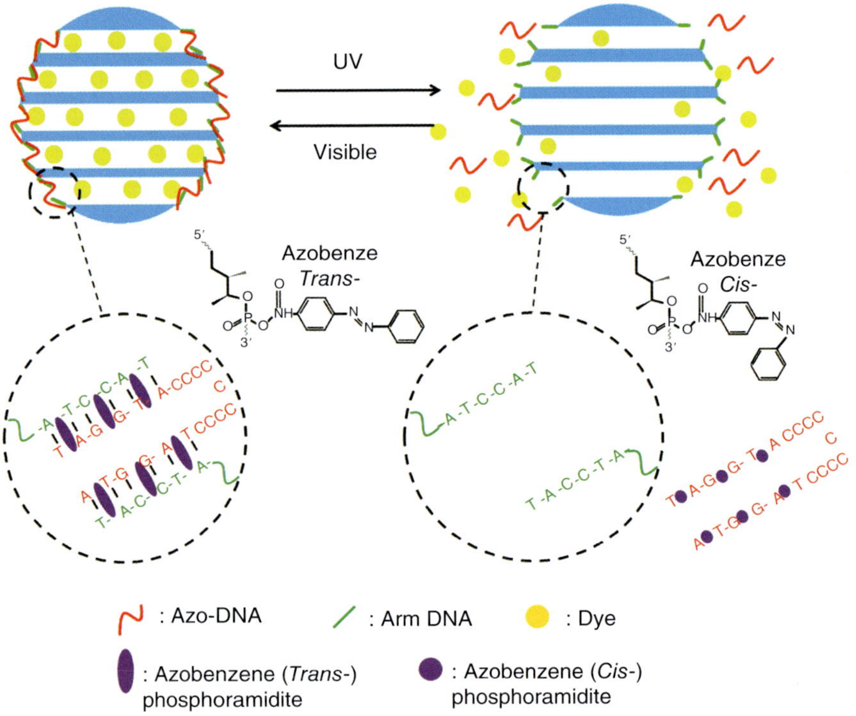

Figure 4.19 Schematic of azobenzene-modified DNA-controlled reversible release system. Visible irradiation at 450 nm (azobenzene *trans*) leads to hybridization of the linker and the complementary DNA arm. Irradiation with UV (365 nm) converts azobenzene to the *cis* form, leading to dehybridization and pore opening. (Adapted with permission from [76]. American Chemical Society, 2012.)

practical applications, zero release or controllable release before the targeted cells or tissues is required, especially for applications involving highly toxic antitumor drugs. Recently, Tan's group reported a photon-manipulated mesoporous release system based on azobenzene-modified nucleic acids (Figure 4.19) [76]. To achieve hybridization with azobenzene-modified DNA (denoted as azo-DNA) to form a cage structure, a single-stranded DNA arm (arm-DNA) was synthesized that has a complementary sequence to azo-DNA. The linker DNA strand (azo-DNA) can hybridize to the arm-DNA to form a cap over the pore mouth. Upon exposure to UV light ($\lambda = 365$ nm), the pore opens as the linker DNA is released into solution. Rhodamine 6G (Rh6G) and DOX Doxorubicin were used as model guest molecules to demonstrate the loading and releasing behavior of this system. After loading the pores with a Rh6G or DOX Doxorubicin molecule, the system was irradiated with UV light, and the guest molecules rapidly escaped into the solution because the azobenzene-incorporated DNA linker was dehybridized from the arm-DNA on the silica surface and unblocked the pore opening. The release of the entrapped guest molecules was restricted by changing to visible light irradiation at 450 nm due to the pore re-blocking. The rapid capping/uncapping response to light provided by this system allows exact point-to-point drug release, enabling this release system to have potential applications in cancer therapy in the future.

In addition, Zink *et al.* demonstrated the controlled release of the cargo accomplished using surface-mounted molecular switches based on MS materials [77, 78]. The release system was constructed in which the azobenzene groups packaged inside of the nanopores (NPs) acted as light-driven "nanoimpellers" and expulsed the pore contents. For example, when triethoxysilane with pendant azobenzene groups attached in and on the MS NPs was irradiated at 413 nm (the isosbestic point for azobenzene groups), the pendant azobenzene groups could undergo a continual dynamic wagging (Figure 4.20a) [78]. As a result, molecules trapped inside of the nanopores could be released from the pores. Another method for exploiting this dynamic motion was to attach larger azobenzene derivatives at the pore entrances so that the machines can modulate the pore openings in the dark. After loading with guest molecules within MS, the system illumination with 413 nm light causes a gradual increase in luminescence of the solution, indicating that guest molecules were gradually released from the MS (Figure 4.20b). It should be emphasized that the substituted azobenzene groups should be sufficiently large to prevent the leakage of guest molecules. In addition, a much more pronounced response was achieved with NPs prepared using the co-condensation method, as shown in Figure 4.20c. Although the luminescence spectroscopy shows no obvious change when the irradiation was stopped, the release behaviors could be recovered again when the light was turned on (inset in Figure 4.20c). The release system presented here was followed by the successful realization of a photoinduced drug delivery system [79] in which an anticancer drug, camptothecin, was released inside and induced the apoptosis of cancer cells.

In addition to conventional MS materials, MCM-41, a special type of mesoporous material with a high surface area and high pore volume, as well as a silanol-containing surface [80], was also used as an excellent candidate for controlled drug delivery systems. Two types of azobenzene derivatives prepared from 4-(3-triethoxysilylpropylureido)azobenzene (TSUA) and (*E*)-4-((4-(benzylcarbamoyl)phenyl)diazenyl)benzoic acid (BPDB) based on MCM-41 nanoparticles were reported by Zink *et al.* (Figure 4.21) [81]. The TSUA/BPDB-modified MCM-41 can be used to store small molecules and release them with the aid of light irradiation. It is well known that β-CD possesses a high binding affinity with *trans*-azobenzene and a low binding affinity, if any, with *cis*-azobenzene in aqueous solution [82]. Therefore, in the presence of β-CD and/or pyrene-modified β-CD rings, the β-CD and/or pyrene-modified β-CD rings will thread onto the stalks and bind to *trans*-azobenzene units to form pseudorotaxanes. As a result, the nanopores were sealed, and the cargo stopped release from the nanopores. Upon exposure to UV light ($\lambda = 351$ nm), the isomerization of *trans*-to-*cis* azobenzene units leads to the dissociation of the β-CD and/or pyrene-modified β-CD rings from the stalks on the surfaces of the MCM-41, thus opening the gates to the nanopores and releasing the guest molecules in a controllable manner. The

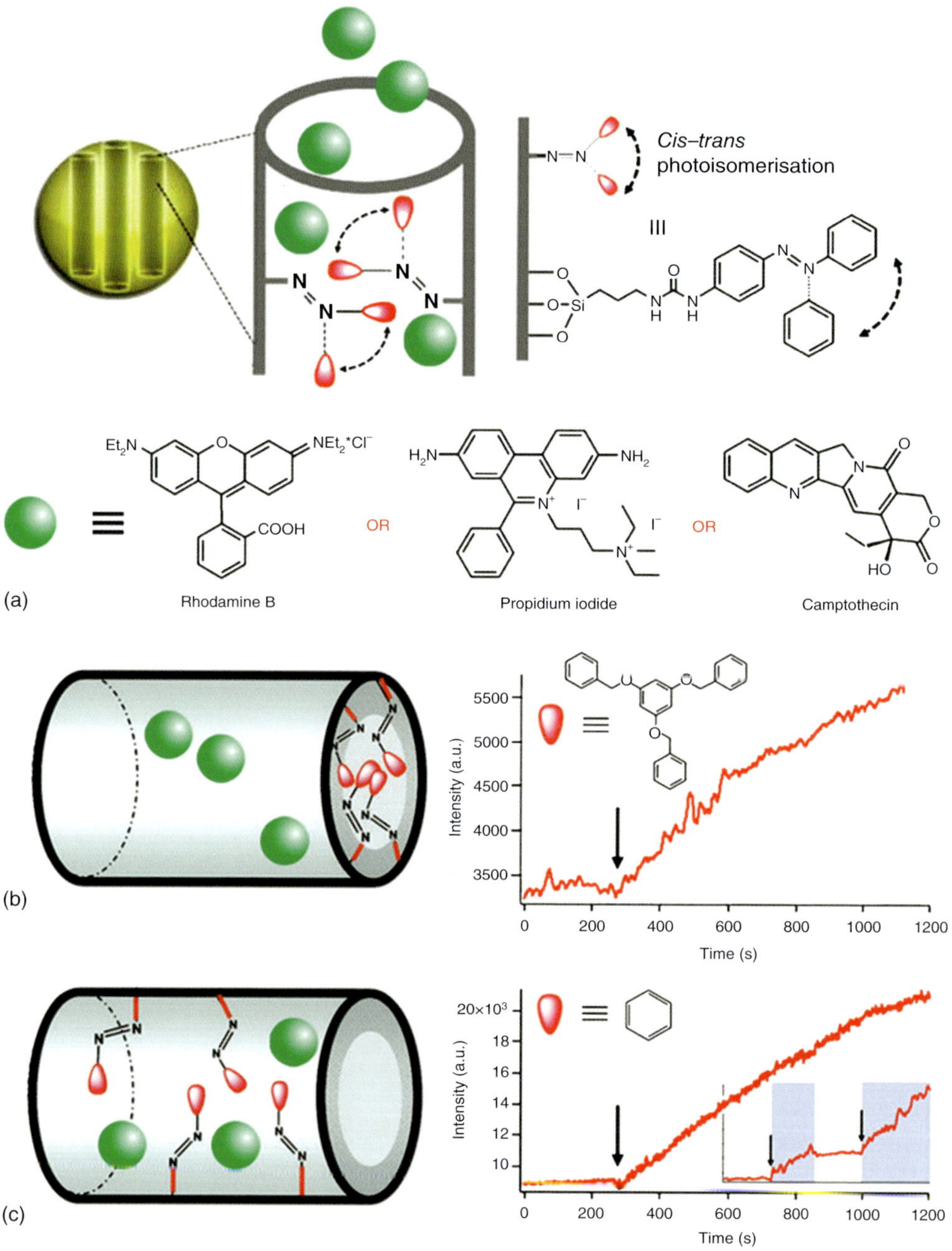

Cis–trans
photoisomerisation
OR
OR
(a)
Rhodamine B
Propidium iodide
Camptothecin
(b)
Intensity (a.u.)
Time (s)
(c)
Intensity (a.u.)
Time (s)

Figure 4.20 (a) Schematic illustration of the "nanoimpeller's" action. Irradiation of azobenzene with light of wavelength at which both isomers absorb results in a continuous "wagging" motion of the untethered terminus of the switch. (b) Release profile for MS silica nanoparticles functionalized with bulky azobenzene groups at the pore orifices. (c) Release profile for MS silica NPs functionalized with unsubstituted azobenzene groups inside pore interiors. (Adapted with permission from [78]. American Chemical Society, 2007.)

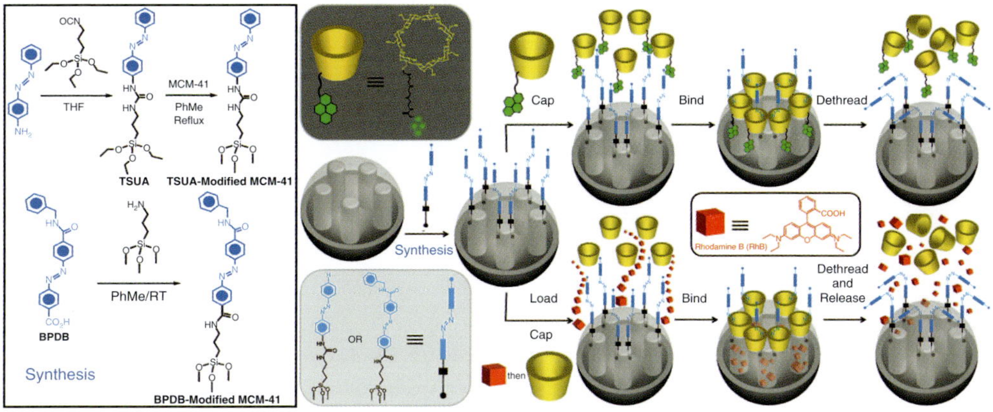

Figure 4.21 Synthesis of TSUA- and BPDB-modified MCM-41. Two approaches to the operation and function of the azobenzene-modified MCM-41 NPs carrying nanovalves.

hydrophilic nature of the mechanized MCM-41 nanoparticles and their ability to control the release of guest molecules under the stimulus of light enabled this promising release system applicable to light-driven intracellular drug delivery systems.

4.3.3 Fluorescent Switching Promoted by Host–Guest Interaction

Although a large family of diarylethene derivatives exhibits good reversibility in fluorescent intensity, most of the reported examples have relatively small fluorescent quantum yields and can only be used in solution. In our research experience, we found that the photochromism, as well as the fluorescence, could be modified by cooperation with a donor or an acceptor through self-assembly. Pyridine groups are widely used for interaction with carboxylic acid to form intermolecular hydrogen bonds, which result in novel structural generation, such as organogels, nanostructures, and optical textures in liquid crystal materials [83–88]. Therefore, we designed and fabricated switchable supramolecular self-assemblies based on the interaction between the pyridine group containing a diarylethene unit (BTEPy) and carboxylic acids (Figure 4.22) [89]. The fluorescence is obviously enhanced in the assembled systems not only in solution but also in the solid state, as well as in the nanoparticles compared with the sole

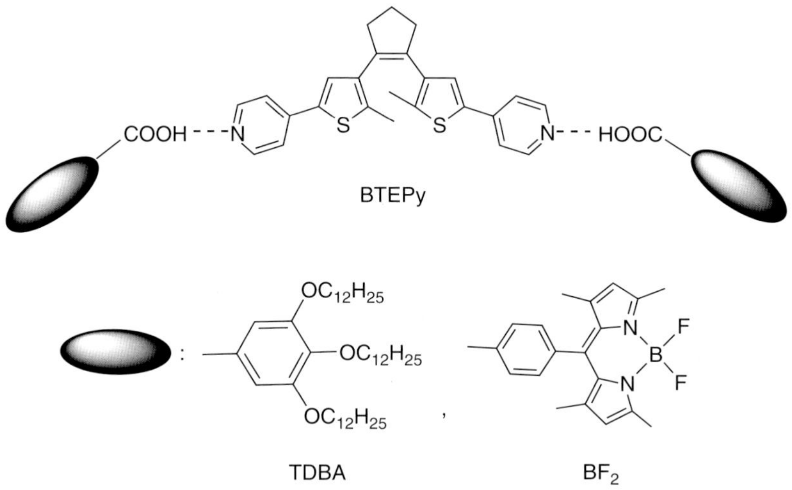

Figure 4.22 The process of the self-assemblies between **BTEPy** and carboxylic acid.

component (Figure 4.23a,b). This indicates that the formation of intermolecular hydrogen bonds may restrain the rotation of the fluorophore, leading to the fluorescence enhancement. The fluorescent photochromic organic nanoparticles were also realized in this system. In the solid state of **BTEPy·TDBA**, the fluorescent emission is red-shifted by approximately 60 nm compared with that in solution. This emission matches well with the absorption of the closed-ring isomer of BTEPy. As a result, an excellent solid fluorescent switch was explored. The ratio of the fluorescent intensity between ring-open isomer and PSS is larger than 150, whereas this value is only 1.2 in the THF Tetrahydrofuran solution. When BTEPy interacted with a proton donor containing a BODIPY Dipyrromethene boron difluoride dye, the "concentration quenching" problem of BODIPY in the solid state was alleviated due to the hydrogen bond formation and energy-transfer process. The solid-state fluorescence of all of these assemblies was effectively switched by alternating irradiation of UV and visible light. Moreover, with the separated wavelengths of writing, reading, and erasing cycles in the BODIPY system, the control of the fluorescence intensity in a reversible manner without causing destruction in the readout capability was realized (Figure 4.23c,d).

Upconversion nanophosphors (UCNPs) consisting of certain lanthanide dopants embedded in a crystalline host lattice can convert low-energy near-infrared excitation light into emission at visible wavelengths via the sequential absorption of two or more low-energy photons [90]. Different colors of visible light can be obtained from different upconversion nanophosphors when excited by the same NIR laser. In addition, UCNPs generate large anti-Stokes shifts of up to 500 nm, resulting in well-separated emission and excitation bands, making them excellent candidates for the fabrication of fluorescent switches with nondestructive readout capability. Thus, we employed a diarylethene derivative (**13**, Figure 4.24) [91] in LaF_3:Yb, Ho-loaded poly(methyl methacrylate) (PMMA)

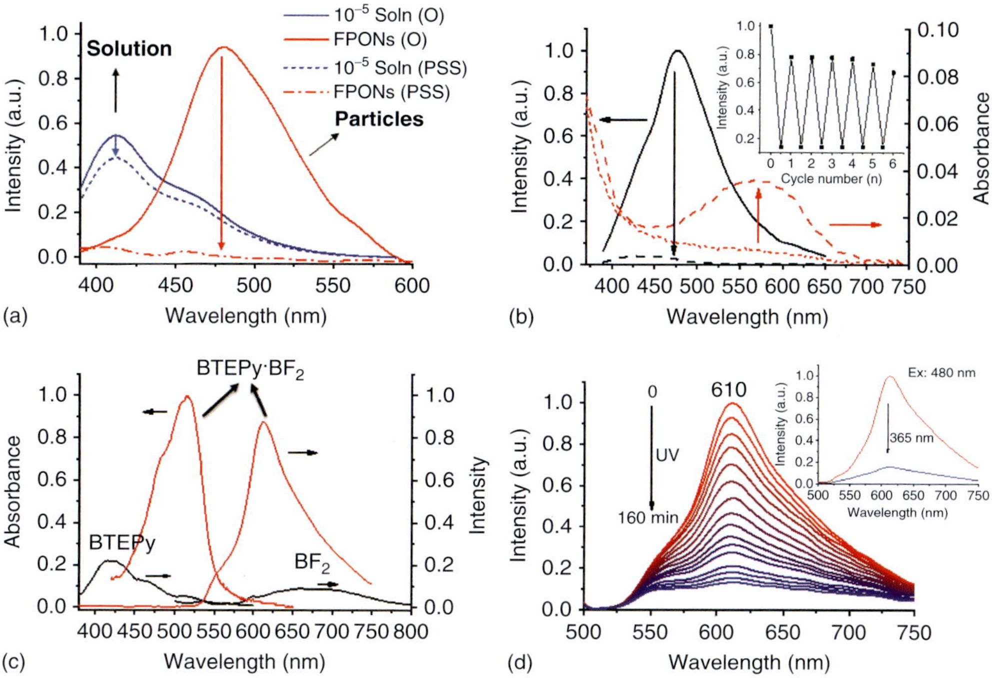

Figure 4.23 (a) Fluorescent spectra of **BTEPy·TDBA** in THF solution and the nanoparticles suspended in THF–water; (b) absorption (red line) and fluorescent (black line) spectral changes of solid film of **BTEPy·TDBA** upon irradiation of 365 nm light (inset: fluorescence switch cycles upon alternating irradiation of UV and visible light); (c) individual absorption and emission spectra of **BTEPy·BF$_2$**, **BTEPy**, and **BF$_2$** in solid film; and (d) fluorescent changes of **BTEPy·BF$_2$** during photochromism process (Ex: 365 nm); inset: excitation under 480 nm and switched under 365 nm.

film. Because **13** has negligible absorbance at 980 nm in the open form and in the PSS, and LaF_3:Yb,Ho nanophosphors can emit visible luminescence by excitation at 980 nm due to the large anti-Stokes' shift, a novel and unique route to a highly efficient nondestructive optical memory must be developed in this diarylethene/LaF_3:Yb,Ho hybrid nanosystem via an intermolecular energy transfer process.

A photochromic diarylethene moiety can act as a photoreversible switch in molecular devices. However, there is limited research concerning the morphology change of a self-assembled system based on photochromic diarylethene induced by external light stimulus. This prompted us to obtain a photochromic diarylethene self-assembly material with a high-modulating control ability of fluorescence and morphology. We designed and synthesized a new diarylethene compound with a melamine group on its side (DTE) and a naphthalimide-based gelator (**14**) that had the ability to form triple hydrogen bonds with melamine (Figure 4.25) [92]. Compound **14** can gelate some aprotic solvents, such as ethyl acetate, carbon tetrachloride, and dioxane. To obtain a photoresponsive

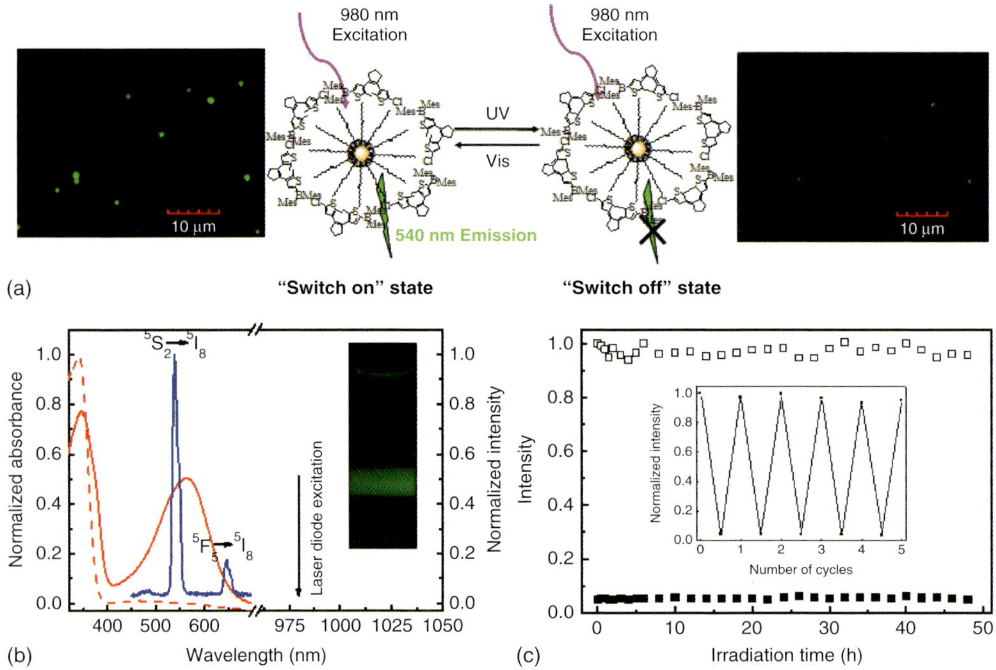

Figure 4.24 (a) Principle of the upconversion luminescent switch consisting of **13**/LaF_3:Yb,Ho-loaded PMMA film before (left), and after (right) irradiation with 365 nm light for 30 min (λ_{ex} = 980 nm). (b) UV–visible absorption spectra of loaded PMMA film before (dash line) and after (solid line) irradiation with 365 nm light for 30 min, and the normalized upconversion luminescence spectra of the prepared film (dotted line, λ_{ex} = 980 nm). Inset shows the image of the upconversion emission. (c) nondestructive readout capability of the film in the open state (□) and PSS state (■), λ_{ex} = 980 nm. Inset shows the modulated upconversion luminescence intensity at 540 nm of the film during alternating UV and visible light irradiation.

self-assembly system, DTE was added to **14** and the complex became easily soluble in a nonproton solvent because of the formation of a complementary hydrogen bond. The highly twisted structure of the dithienylethene unit in DTE destroyed the molecular packing of **14** and simultaneously improved the solubility of the complex. The absorption and fluorescence spectra of the assembly can be reversibly switched by alternating UV–vis light irradiation. Meanwhile, the morphology of the co-assembly of $\mathbf{14}_2$·DTE changed to film from the original pieces of gel **14** in ethyl acetate. When $\mathbf{14}_2$·DTE was irradiated by UV light, the film morphology was converted into aggregated flakes. Moreover, the surface wettability of the complex can also be switched by light irradiation. The photochromic diarylethene unit can modulate the fluorescence and morphology of the assembled system only by virtue of light irradiation. Therefore, these results provide further insights into the control of fluorescence and morphology, especially for application in upscale smart responsive materials.

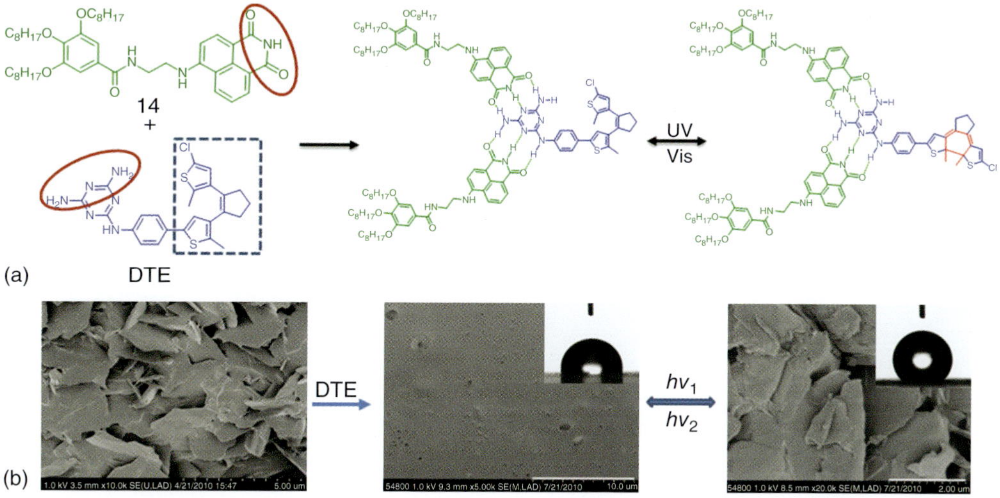

Figure 4.25 (a) Chemical structure and photochromic process of **14** + DTE and (b) SEM images of gels of **14** and **14** + DTE from ethyl acetate before and after 365 nm light irradiation (scale bar: 5, 10, and 2 μm, respectively, from left to right); insets in (b) are water contact angle of **14** + DTE before and after 365 nm light irradiation.

4.3.4 Photoswitchable Molecular Devices

Motivated by an interest in fundamental science and practical applications, researchers in molecular electronics have instigated rapid growth in this field over the past decade [93, 94]. Therefore, smart advanced materials have been used to construct various molecule devices, such as carbon nanomaterials (carbon nanotubes (CNT), graphene) and polymeric materials, as well as inorganic nonmetallic materials [95]. To our knowledge, most of the molecular devices, especially flexible electronics, were constructed based on polymeric-based materials due to their excellent stability, ductility, and processibility. In this regard, photodeformable polymeric materials outperformed many other materials because of their potential applications in various fields, such as artificial muscle, photomobile soft actuator, and micro-optomechanical systems [96, 97]. Among them, crosslinked liquid-crystalline polymers (CLCPs) containing photochromic moieties have been the most studied over the past decades [98]. Conventional CLCP-based materials suffer from broken debris and structural damage during fabrication and the accumulation of electrostatic charge on the surface during use, which hampered their application to some extent. To overcome these disadvantages, a tremendous number of studies have been developed. As an example, a photodeformable CLCP/CNT Carbon nanotube nanocomposite film was fabricated by Peng and coworkers in which an azobenzene polymer was used as a photoswitch unit [99]. To prepare the CLCP/CNT nanocomposite film, two monomers, A11AB6 and A9Bz9, a crosslinker C9A, and aligned CNT sheets

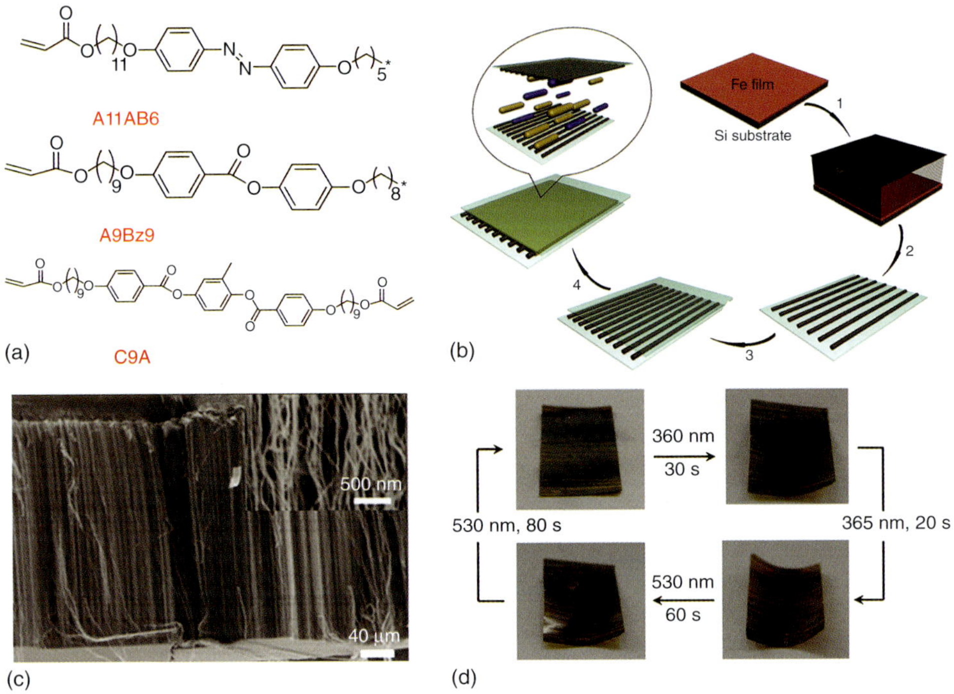

Figure 4.26 (a) Chemical structures and properties of the two monomers (**A11AB6** and **A9Bz9**) and crosslinker (**C9A**). (b) Preparation of an oriented CLCP/CNT nanocomposite film in four steps: (1) growth of a CNT array by chemical vapor deposition; (2) formation and stabilization of the CNT sheet on a glass substrate; (3) preparation of the LC (Liquid-Crystalline) cell by using two CNT-sheet-covered glass slides; and (4) injection of the molten mixture including the monomers, crosslinker, and photoinitiator into the LC cell. (c) SEM images of a CNT array (inset, high magnification). (d) Photographs of a CLCP/CNT composite film during one bending and unbending cycle after alternate irradiation by UV and visible light at room temperature. The intensities of the UV light at 365 nm and visible light at 530 nm were 100 and 35 mW cm^{-2}, respectively. (Adapted with permission from [99]. Wiley-VCH, 2012.)

were synthesized (Figure 4.26a). Following by a four-step method, a freestanding composite film can be obtained (Figure 4.26b). It was found that the aligned nanostructure of the CNT sheet (Figure 4.26c) could effectively orient the CLCP mesogens along the length of the CNTs without using any other aligning layer. Therefore, the composite film exhibited a rapid and reversible deformation under alternate irradiation by UV and visible light, as shown in Figure 4.26d. Upon exposure to UV light, the composite film bents toward the incident light along the CNT-aligned direction, and when the light was turned to visible light, the bent film completely recovered its initial morphology due to the reversible *cis*-to-*trans* isomerization of the azobenzene moieties. This photoinduced bending and unbending could be repeated over 100 cycles without obvious fatigue. In addition, the CLCP/CNT composite film also possesses enhanced mechanical properties and high conductivity compared with previously reported CLCP materials, enabling this film to be used as a new high-performance actuation material to drive actuators and microrobots.

With the rapid development of information technology, information security has attracted great attention in the past several years. Self-erasing materials based on the use of photochromic molecules play an important role in secure communications. When exposed to light with an appropriate wavelength, these photochromic molecules could isomerize and change their colors [100]. However, materials with this performance often suffered from limited colors and relatively low extinction coefficients. Recently, Klajn *et al.* reported a conceptually different self-erasing material in which Au/Ag nanoparticles acted as a smart and self-erasable "ink" whose color changed rapidly when the nanoparticles assembled and disassembled upon irradiation [101]. In their design, Au/Ag nanoparticles were coated with mixed self-assembled monolayers (mSAMs) of dodecylamine (DDA) and photoswitchable azobenzene terminated thiol 4-(11-mercaptoundecanoxy)azobenzene (MUA), as shown in Figure 4.27a. The as-prepared Au/Ag nanoparticles were dispersed in semipermeable syndiotactic PMMA paper. Despite relatively low concentrations of these NPs, the films were colored brightly: red for Au nanoparticles and yellow for Ag nanoparticles. However, upon UV light irradiation, the nanoparticles in the irradiated regions formed metastable aggregates due to the rapid *trans*-to-*cis* isomerization of the azobenzene groups of MUA (Figure 4.27b), and the aggregation of nanoparticles translated into significant color changes because of the electrodynamic coupling (Figure 4.27c). Specifically, the red Au nanoparticle films turned purple with a short irradiation of UV light and gradually turned blue when irradiated longer (Figure 4.27d, left). Similarly, the films of Ag nanoparticles changed their color from yellow to red to purple with a prolonged irradiation time (Figure 4.27d, right). The photopatterned images gradually self-erased when the nanoparticle aggregates reverted to free nanoparticles due to the reversibility of the nanoparticle aggregation. Interestingly, the erasure time can be controlled from seconds to minutes by adjusting the number of azobenzene switches per nanoparticles. In addition, the erasure times could be accelerated by increasing the intensity of visible light (Figure 4.27e). The write–erase cycles can be repeated hundreds of times without photobleaching or any noticeable damage, indicating good stability of the materials. This type of self-erasing material can be useful for storing sensitive or temporary information in the future.

In summary, a light responsive host–guest system is one type of important area regarding supramolecular systems and responsive materials. Supramolecular chemistry itself benefits from research on host–guest systems because different properties and interaction mechanisms enrich its scope, enabling the systems to be used in various fields, including chemistry, biology, and material science.

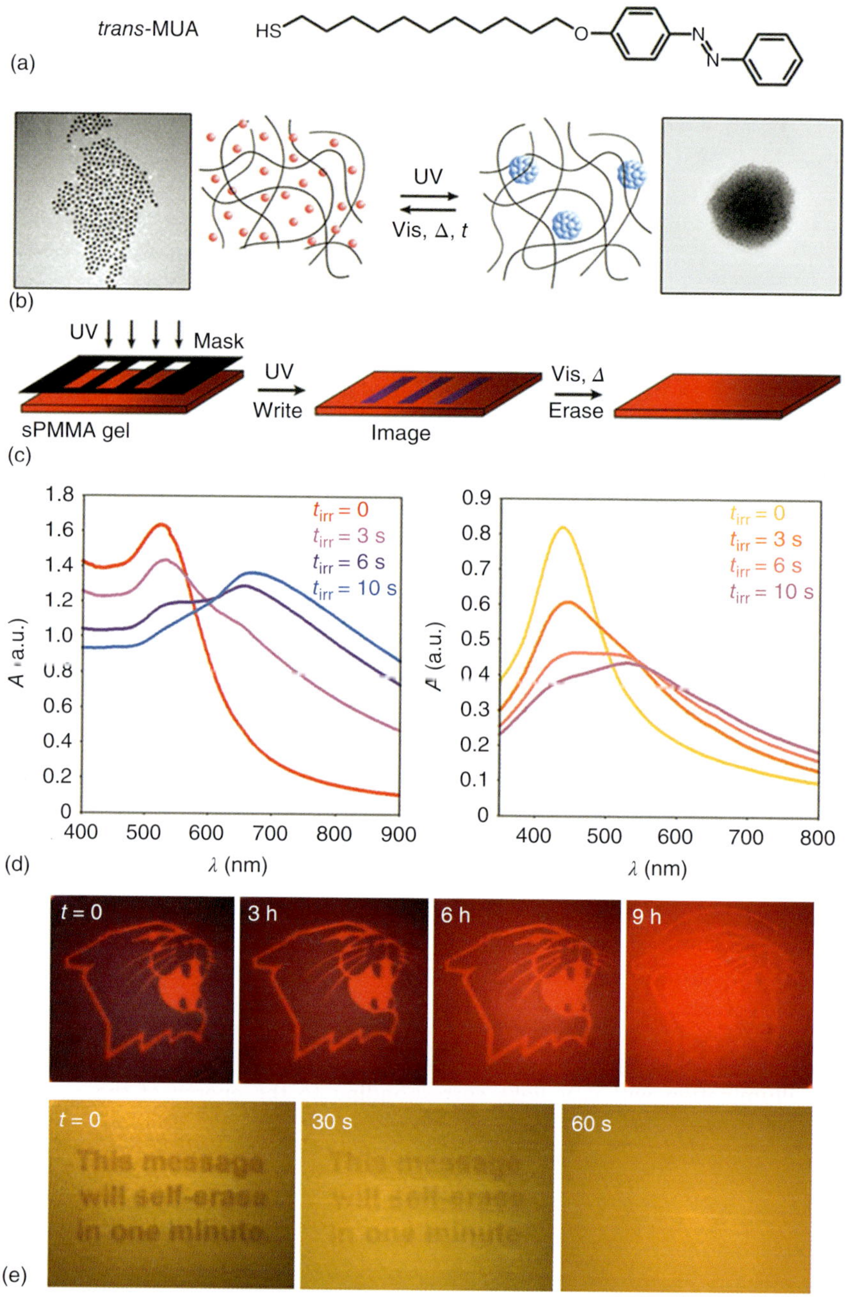

trans-MUA
HS
O
N
N
(a)
UV
Vis, Δ, t
(b)
UV
Mask
UV
Write
Image
Vis, Δ
Erase
sPMMA gel
(c)
t_irr = 0
t_irr = 3 s
t_irr = 6 s
t_irr = 10 s
A (a.u.)
λ (nm)
(d)
t = 0
3 h
6 h
9 h
t = 0
30 s
60 s
This message will self-erase in one minute
(e)

Figure 4.27 (a) Structural formula of *trans*-azobenzene-terminated thiol (*trans*-MUA). (b) Schematic illustration of the light-induced NP self-assembly in a polymer gel. TEM (transmission electron microscope) images show dispersed NPs before (left) and aggregated NPs after (right) UV irradiation. (c) Local irradiation of the sample (here, through a transparency mask) can be used to record and store graphical information. (d) Aggregate size (and color of the film) depends on the duration of UV irradiation. (e) Examples of images/messages written in Au (top) and Ag (bottom) NP-based films and their self-erasure (slow in the dark, top; rapid upon intense visible-light irradiation, bottom). (Adapted with permission from [101]. Wiley-VCH, 2009.)

4.4 Photochromic Metal Complexes and Sensors

Photochromic molecules involving metal ion coordination exhibit multiple responses to photo, metal ion, biomolecule, and other chemical inputs, which could be used in potential applications in the field of biosensors and images, storage materials, and memories [102, 103]. Metal ions are generally introduced into photochromic molecules by a coordination interaction through electron-donated atoms, such as oxygen, nitrogen, sulfur, and so on. Photochromic organometallic compounds were also developed by a metal–carbon reaction [104]. Metal ions in photochromic metal complexes are not only endowed with the capability of multiple responses but also the ability to stabilize one of the photoisomers in some cases to afford gated properties. In this section, metal complexes containing photochromic compounds, such as azobenzenes, diarylethenes, spiropyranes/spirooxazine, and rhodamine, are introduced. Interactions between the photochromic unit and metal centers and their different physicochemical properties are discussed in detail.

4.4.1 Metal Complexes with Azobenzene Groups

Azobenzene, as an organic photochromic molecule, has been previously described in detail. In this chapter, metal complexes with an azobenzene ligand will be introduced. Similar to azobenzene, azobenzene metal complexes can be photoisomerized from the more stable *trans* form to the less stable *cis* form by the illumination with UV light corresponding to the energy of the $\pi-\pi^*$ transition, while the reverse isomerization can be switched with visible light corresponding to the $n-\pi^*$ transition [105]. The *cis*-to-*trans* isomerization also occurs spontaneously by thermal back-conversion. The coordination capability of the ligands to metal ions is affected by the *trans*-to-*cis* isomerization or the type of the metal ions. The *cis* or *trans* isomers of the ligands can adjust the position of the metal centers, while introduction of the metal center may influence the conversion capability and the rate of *cis*–*trans* isomerization of the azobenzene unit, as illustrated in the following examples.

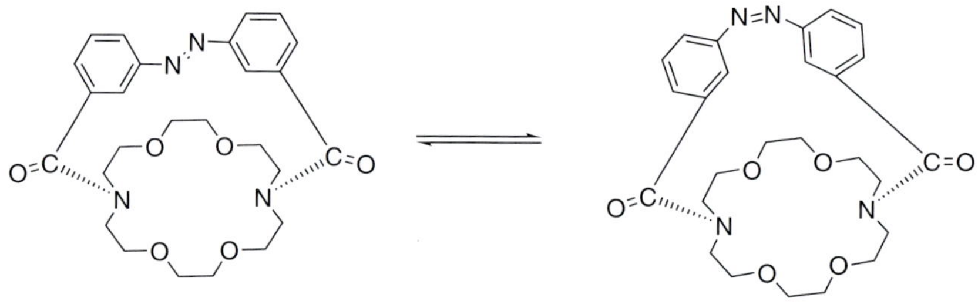

trans-**15** *cis*-**15**

Figure 4.28 Chemical structure and photochromic process of **15**.

In 1980, Shinkai and coworkers reported an azobenzene-bridged crown ether (**15**) for the first time (Figure 4.28) [106]. Compound **15** bridged with *trans*-azobenzene (*trans*-**15**) is preferably bound with an ammonium cation or methyl orange salts of Li^+/Na^+, whereas **15** bridged with photoisomerized *cis*-azobenzene (*cis*-**15**) is preferably bound with methyl orange salts of K^+/Rb^+. The steric difference between *trans*-**15** and *cis*-**15** reflected the size of the crown ether, which is sensitive to the size of the metal ions. Thus, the binding selectivity of the crown ethers in **15** toward alkali-metal ions can be reversibly controlled by photoirradiation.

Ag(I) complex (**16**) [107] based on (4,4′)azobenzeno(1,5)naphthalenophane was reported by Tamaoki and coworkers (Figure 4.29). The coordination of Ag(I) was reversibly controlled by photoisomerization of the azobenzene unit, and simultaneously, Ag(I) accelerated the *Z*–*E* thermal isomerization of the azobenzene unit. The *E*-**16** isomer only exists in solution at room temperature when stored in the dark. Irradiating *E*-**16** with a light of 365 nm converts the *E* to the *Z* isomer. The newly observed acceleration of the *Z*–*E* thermal isomerization of the azobenzene unit by Ag(I) must be related to the strong perturbations of the electronic state and the structure of the azobenzene unit in *E*-**16** and *Z*-**16** due to the direct coordination.

Photoactive dinuclear complexes (**17** and **18**) with an azobenzene linker were reported by Hirota and coworkers (Figure 4.30) [108, 109]. *Trans*-**17**-Cu

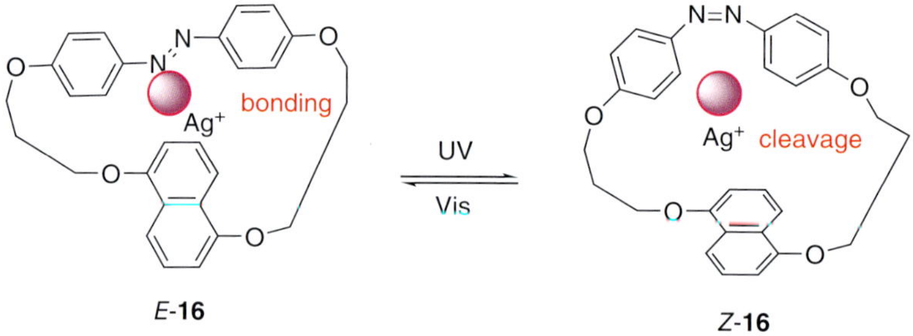

E-**16** *Z*-**16**

Figure 4.29 Chemical structure and photochromic process of **16** with existing of Ag ion.

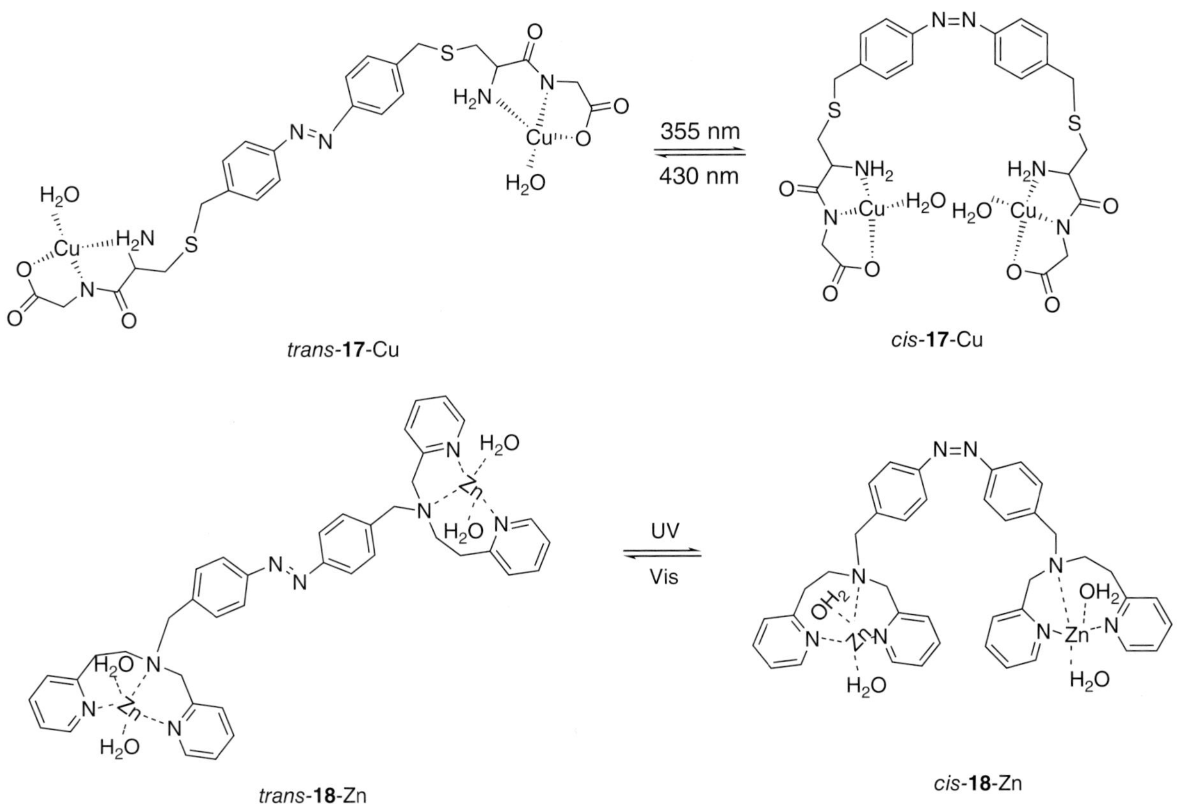

Figure 4.30 Chemical structure and photochromic process of **17** and **18** with existing of Cu and Zn ions, respectively.

is a photoresponsive molecule with two Cu(II)-bound dipeptides linked by an azobenzene chromophore in which the copper(II) ion centers are distant from each other. Under the irradiation of 355 nm light, *trans*-**17**-Cu is switched to *cis*-**17**-Cu in which the copper(II) ion centers are close to each other. These results demonstrate that the photoirradiation can control the spatial orientation of the copper(II)-bound dipeptides linked by an azobenzene derivative. Similarly, *trans*-**18**-Zn is a photoactive dinuclear zinc(II) complex with two zinc centers linked by a ligand containing an azobenzene chromophore in which the spatial disposition of the two metal centers was externally controlled by photoisomerization of the azobenzene chromophore.

4.4.2
Metal Complexes with Diarylethene Groups

With the suitable chemical modification, photoswitchable diarylethene derivatives can coordinate with metal ions to form complexes with variety properties. Along with the photoinduced reaction of diarylethene metal complexes, the photochemical, photophysical, and electrochemical properties of the metal complexes present remarkable diversification. In a typical diarylethene analog, the structure of diarylethene has parallel and antiparallel conformations in equal proportions, but only the antiparallel conformation can perform photoisomerization. The relative composition of parallel and antiparallel conformations in the complexes can be controlled by a coordination bond with the metal center. Therefore, metal ions can be used as a trigger to tune the gated photochromic properties of diarylethene derivatives. On the other hand, the open to closed isomerization of the ligand can affect the binding ability with metal ions.

Takeshita and Irie reported photochromic molecular tweezers [110] composed of two 18-crown-6 moieties and a photochromic dithienylethene (**19**, Figure 4.31). The affinity of **19** toward the cesium ion could be changed by photoirradiation. Dithienylethene undergoes a thermally irreversible and fatigue-resistant photochromic reaction. The open-ring form of dithienylethene has two conformations, parallel and antiparallel. Two crown ether moieties of the open-ring form in the parallel conformation of **19** can cooperatively bind with a large metal ion, such as the cesium ion, like tweezers, while in the photogenerated closed-ring form, the crown ether moieties are separated from each other and cannot capture the metal ion.

A multistate 1,8-naphthalimide-piperazine-tethered dithienylethene (**20**) was prepared by Tian *et al.* [111]. Copper ion (II) was introduced as a molecular "lock" to prevent photochromism (Figure 4.32). After Cu(II) coordinated with intramolecular piperazine moieties of **20**, the complex was transformed into a photo-inactive state (O-**20**-Cu). The bridging complexation of the Cu ion to the piperazine moieties of O-**20** restricts the ligand in its photo-inactive form with a parallel conformation of the bisthienylcyclopentene group, inhibiting the photocyclization process. By removing the ion from the complex, the photochromic performance of **20** could be recovered. Moreover, the addition of other metal

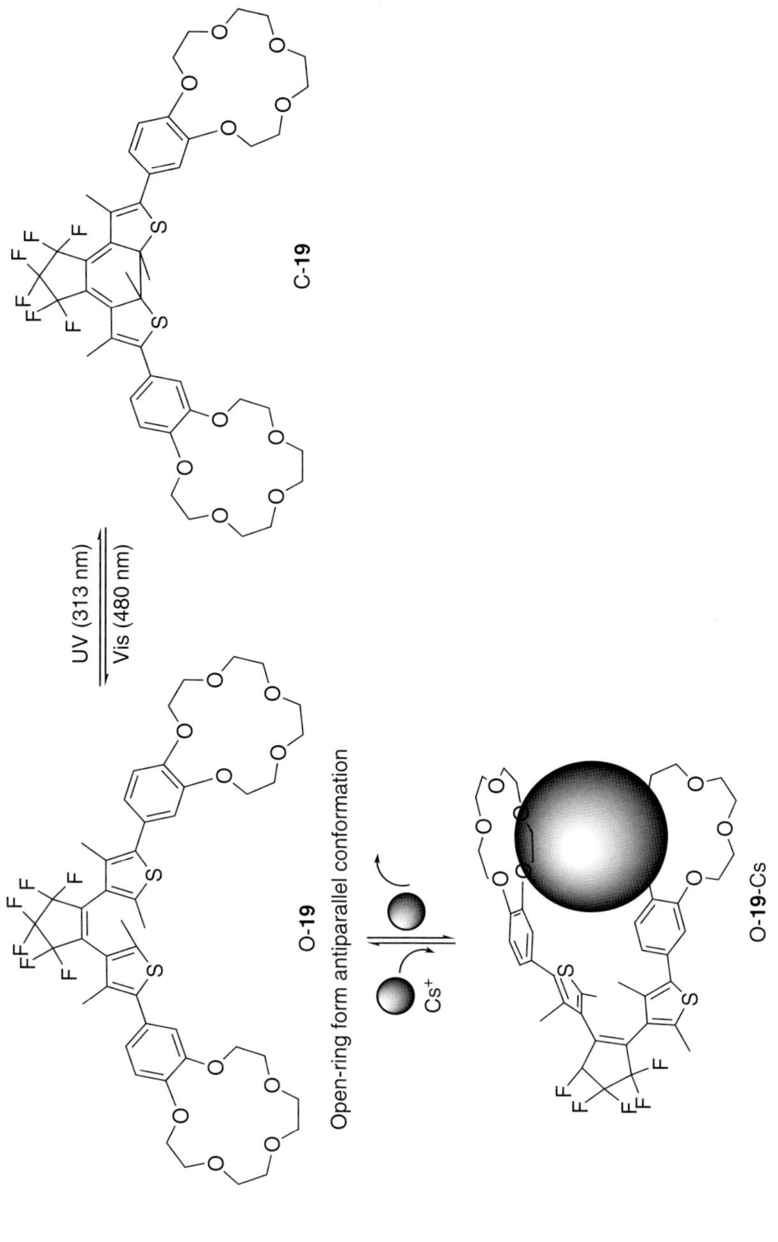

Figure 4.31 Chemical structure and photochromic process of **19** with existing of Cs ion.

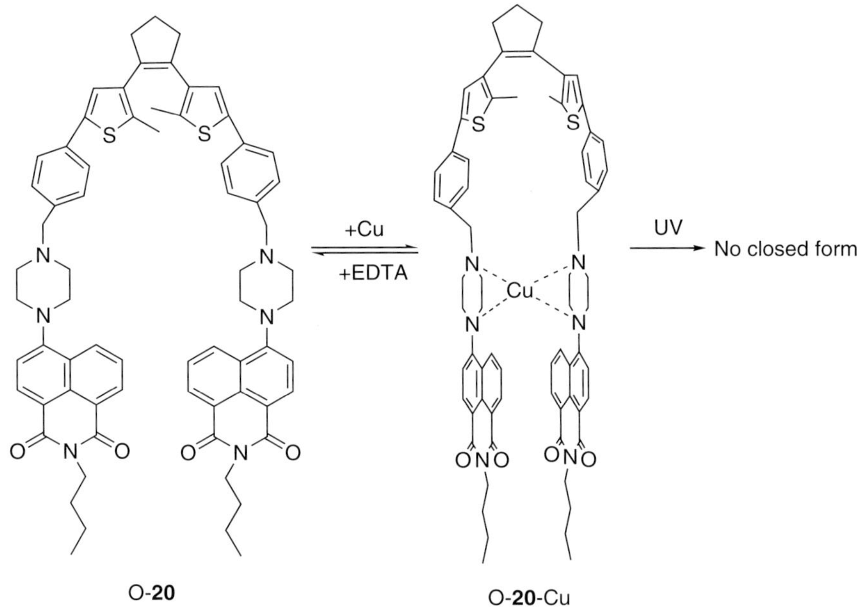

Figure 4.32 Chemical structure and photochromic process of **20** with existing of Cu ion.

ions, such as Zn^{2+}, Hg^{2+}, Fe^{3+}, and Co^{3+}, had no effect on the photochromism of **20**, indicating a selective recognition of **20** toward the copper ion.

A multiresponsive fluorescent switch based on diarylethene and terpyridine units (**21**) was developed by our group (Figure 4.33) [112]. The open-form fluorescence switch combining diarylethene and terpyridine functional units (**21o**) exhibits several clearly different and reversible fluorescence states that can be controlled by varying the light frequency and metal ion concentration. The fluorescence of **21o** could be gradually quenched by the addition of an aqueous solution of $Zn(NO_3)_2$, which was attributed to the formation of a **21o**–Zn complex, changing the charge density of the terpyridine unit. When ethylene diamine tetraacetic acid (EDTA) solution was added, the fluorescence intensity was restored, which can be explained by the difference in the association constant between EDTA–Zn and **21o**–Zn. The fluorescence intensity of **21o** can be reversibly controlled by UV–vis light or Zn/EDTA, and **21o** acts as a double-controlled molecular fluorescence switch reacting to light and chemical stimuli.

The multiresponsive properties of **21** make it an effective candidate for application as a fluorescent probe in living cells. A blue luminescence in the cytoplasm of KB human nasopharyngeal epidermal carcinoma cells was observed after incubation with a PBS (phosphate buffer saline)/DMSO (Dimethyl Sulphoxide) (100 : 2, v/v) solution of **21o** (1×10^{-5} M) for 20 min at 25 °C Using CLSM (Figure 4.34). The luminescence of **21o** could be controlled as readily in living cells as in solution using UV–vis light as the switching trigger. After irradiation

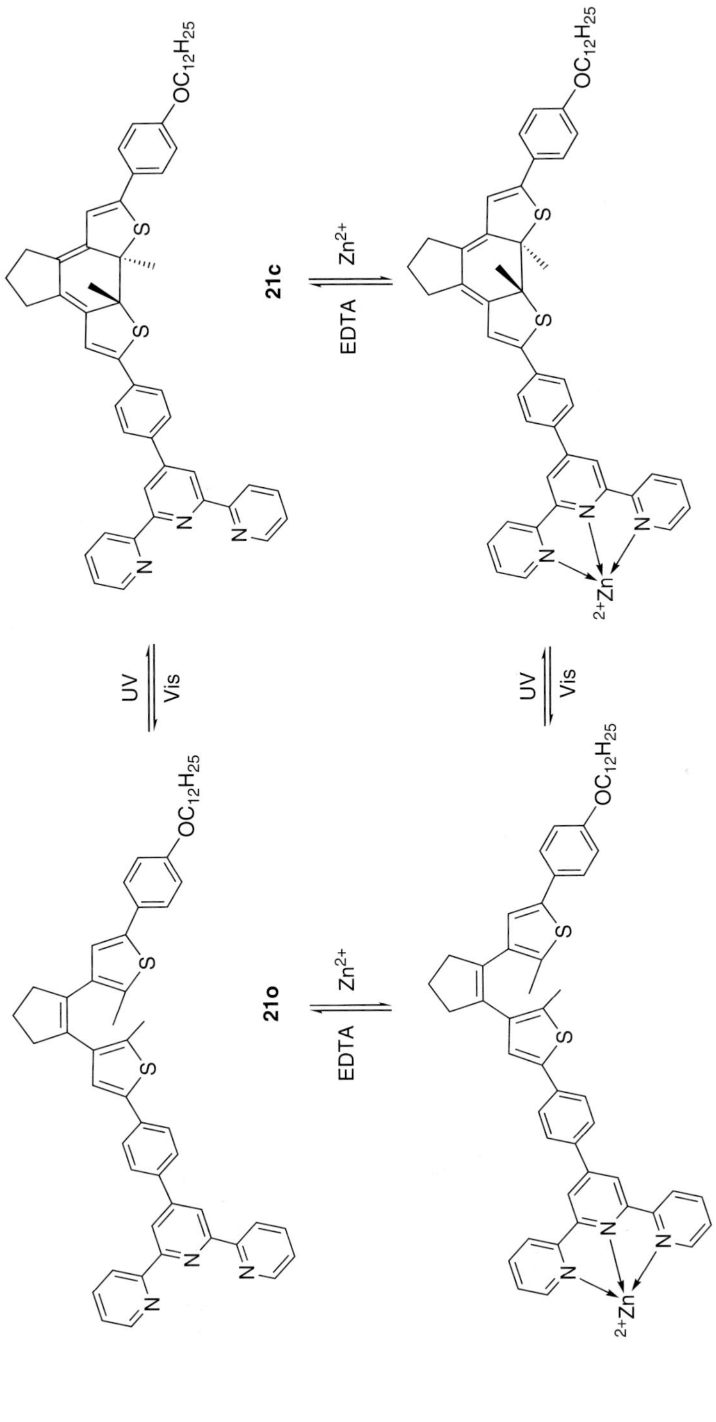

Figure 4.33 Chemical structure changes of compound **21** in different states.

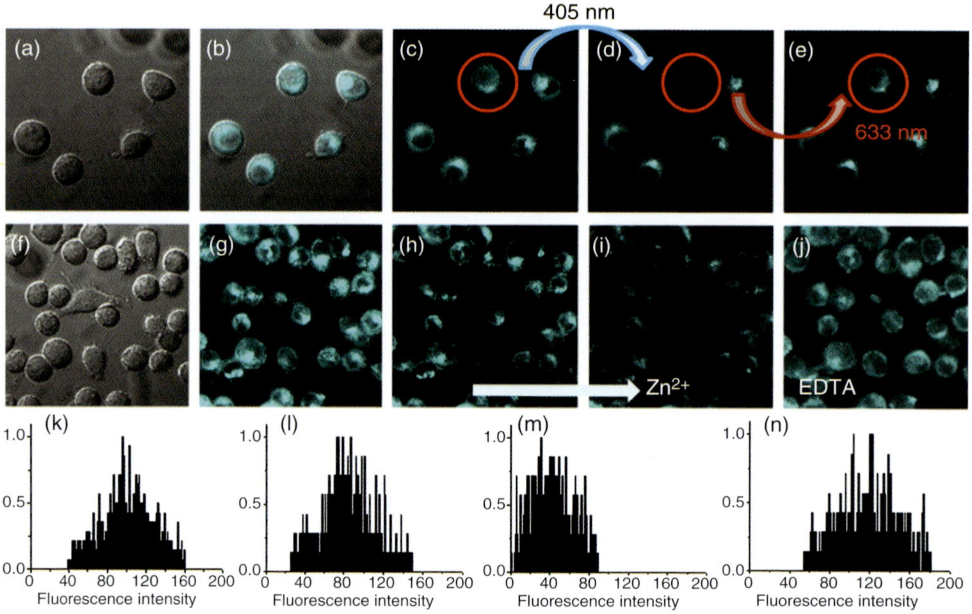

Figure 4.34 CLSM images of KB cells incubated with **21o** for 20 min at 25 °C (1×10^{-5} M in PBS/DMSO, 100 : 2, v/v). (a,f) Brightfield transmission image of KB cells. (b) Overlay image of A and C. Confocal fluorescence image of (c) original state, (d) irradiated by 405 nm light (2 mW, 3 min) for one selected cell, and (e) recovered by 633 nm light (0.7 mW, 40 min). Confocal fluorescence image of (g) original state of F, and incubation by Zn^{2+} solution with the concentrations of (h) 5×10^{-5} M and (i) 1×10^{-4} M, (j) Recovered by 5×10^{-4} M EDTA solution. Panels (k–n) were the distribution of fluorescence intensity of (g–j), respectively. (Adapted with permission from [112]. American Chemical Society, 2009.)

with 405 nm light (2 mW) for 3 min, the brightness of the selected cell noticeably decreased compared with the brightness of the surrounding cells, which remained virtually unchanged. Upon irradiation with 633 nm light (0.7 mW), the brightness of the selected cell was recovered within 40 min. In contrast, when the **21o**-labeled cells were treated with different concentrations of Zn^{2+}, an obvious change in fluorescence was observed. The luminescence intensity decreased with increasing Zn^{2+} concentration (Figure 4.34k–m). This effect provides a good method for visualizing the process of M^{2+} (M = Zn, Cu) uptake from the outside into the cell. As expected, the fluorescence was nearly restored to the original state upon the addition of EDTA solution.

The iridium complex is one of the most promising heavy-metal complexes due to its advantageous photophysical properties, such as long luminescent lifetime and distinctive wavelengths of the excitation and emission due to a metal-to-ligand charge-transfer transition (MLCT) [113–117]. The use of these complexes as photoswitchable probes may extend the excitation wavelength for photoisomerization of diarylethene to the visible range, thus affording a more biofriendly condition and avoiding autofluorescence in bioimaging. We

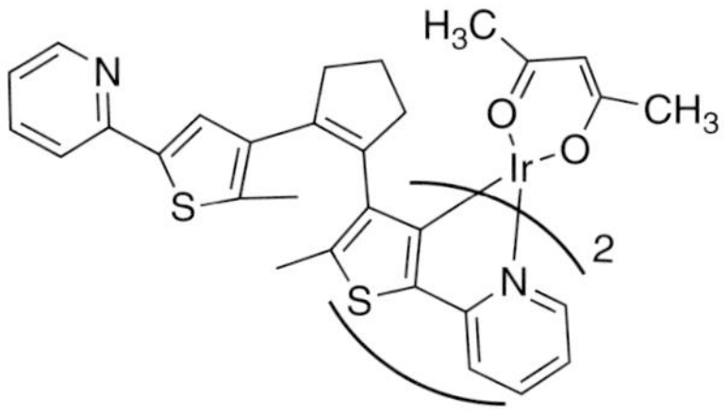

22

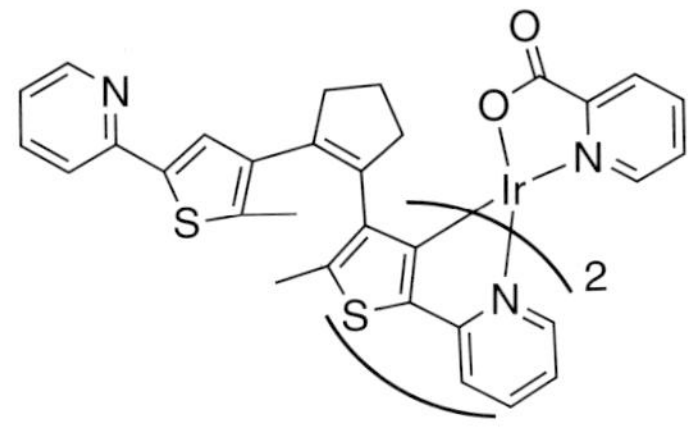

23

Figure 4.35 Chemical structures of **22** and **23**.

reported visible-light-modulated phosphorescence bioimaging probes based on photochromic iridium complexes containing a diarylethene unit (**22** and **23**) (Figure 4.35) [118, 119]. The open-form isomer of **22** shows an obvious absorption band at 450 nm in the visible region with an MLCT characteristic of the complex, red-shifted approximately 100 nm relative to the common organic diarylethenes. This characteristic is beneficial for further development of photoswitchable phosphorescent probes controlled by alternating two visible-light irradiations. To obtain better water solubility and to meet the cell-permeable demand, **23** was synthesized by replacing the ligand acac of **22** with 2-picolinic acid. The two compounds exhibit significantly reversible changes in luminescent emission by varying the visible light irradiation and can be used as a switchable luminescence probe for monitoring living cells. After 440 nm light irradiation, the emission intensity of **23** at 568 nm was decreased and was finally quenched. When the closed form of **23** was exposed to visible light with a wavelength longer than 600 nm, the emission at 568 nm could be recovered, and the cycle could be repeated many times. The complexes also showed excellent photochromic properties within the living cells. In particular, the luminescence of the living cells incubated with these complexes exhibited switchable activity between the open and closed states in response to different visible-light irradiations, revealing potential applications as a photoactive marker in biosystems.

Diarylethene-containing 1-aryl-substituted 2-(2-pyridyl)imidazole ligand derivatives and their rhenium(I) complexes (**24**) were reported by Yam and coworkers (Figure 4.36) [120]. The open forms of **L24-R** dissolve in chloroform to give colorless solutions, with an intense absorption band at about 320 nm, corresponding to the $\pi-\pi^*$ and $n-\pi^*$ transitions of the 1-aryl-2-(2-pyridyl)imidazole moiety, with mixing of $\pi-\pi^*$ and $n-\pi^*$ transitions of the thiophene moieties. Upon coordination to the Re complex, this IL (intra ligand) absorption band was shifted slightly to the red at about 352 nm. An absorption shoulder was observed at about 425 nm in the absorption spectra of complexes **24-R**, ascribed to an MLCT [dπ(Re)-π^*(**L**)] transition, with some mixing of a metal-perturbed IL ($\pi-\pi^*$) transition. Upon UV excitation at $\lambda \leq 350$ nm, the ligands showed two additional absorption bands at about 410–425 and 576–586 nm. Upon excitation of complexes **24-R** at $\lambda \leq 450$ nm into either the IL or MLCT bands, three absorption bands were generated at about 290, 475, and 712 nm and 288, 480,

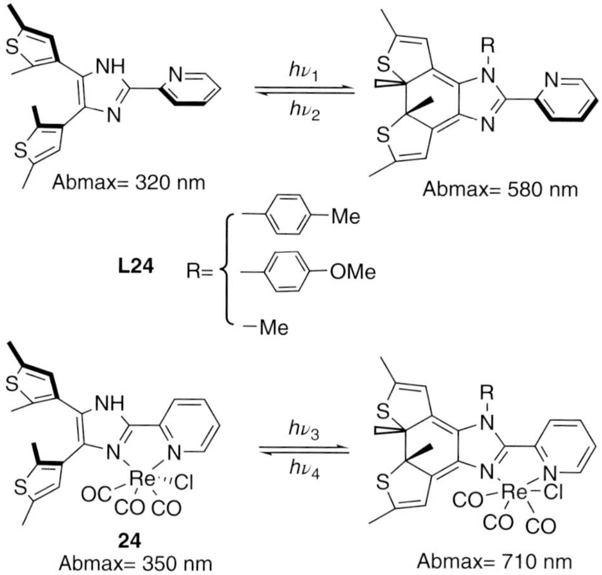

Figure 4.36 Chemical structure and photochromic process of **L24** and **24**.

and 713 nm, respectively. This large shift of the absorption bands of the closed forms of the complex to the NIR region could be attributed to the planarization of the four heterocyclic rings relative to the open forms.

4.4.3
Metal Complexes with Spirocyclic Groups

Spiropyrans and spirooxazines belong to one of the most fascinating families of photochromic compounds that undergo reversible structural isomerization between a colorless spiro form and a colored merocyanine upon either light, heat, or chemical stimulus [121]. The colored open form (MC) with extended π-conjugation is thermally unstable and readily reverts to the closed colorless SP form by thermal reaction, by being left in the dark, or by being irradiated with visible light. The open MC form has a larger dipole moment than the closed SP form; thus, the stability of the MC form strongly depends on the electronic effects of the substituents of the indolium and phenyl moieties, the solvent polarity, and on metal complexation. The conversion of the closed colorless SP form into the colored open MC form can be induced by the complexation of a metal ion with cooperative ligation of another chelating functionality attached at the key position.

A series of crown spirobenzopyrans were designed and synthesized by Inouye and coworkers [122, 123]. Here, **25** and **26** are described as examples (Figure 4.37). The isomerization of crown ether-linked spirobenzopyrans **25** to the open-chain colored MCs **25**-M^+ was induced by recognition of alkali-metal cations, as well as

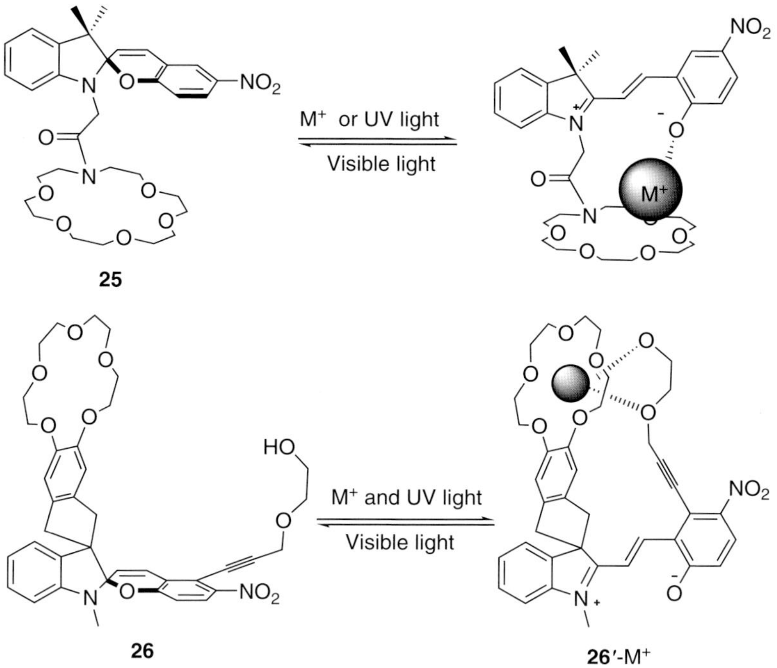

Figure 4.37 Chemical structure and multi responses of **25** and **26**.

by UV irradiation, performing "OR" gate-type signal transduction. In contrast, **26** could only be responsive to the combination of ionic and photonic stimuli in which a thermally irreversible photochromism was observed. Thus, the dual-mode signal transducer molecules **26** performed "AND" gate-type signal transduction in which only the combination of two simultaneous "chemical and physical inputs" (ion and UV light) can cause the receptors to be visible-light active. The absorption spectra of **26** were only slightly affected by the addition of any alkali-metal iodides in CH_3CN, even after several days in the dark. NMR and mass spectra suggested that the metal cations were bound to the macrocycle of **26**, and the colorless form was attributed to the lack of isomerization of **26**-M^+ to **26′**-M^+. Subsequently, irradiation (360 nm) of the alkali-metal iodide containing CH_3CN solutions of **26** caused changes in their spectra, and new absorption bands belonging to **26′**-M^+ appeared. Only photoirradiation (salt free) of **26** resulted in a slight change in its spectrum, suggesting the suppression of its photochromic property in the absence of the cations. The colored solution was stable and did not undergo thermal bleaching under dark conditions at room temperature, even after 30 days. The partial isomerization of **26′**-M^+ to the colorless spiropyran form **26**-M^+ was observed upon irradiation with >480 nm light.

The spiropyran–metal receptor (**27**) for amino acids, which might present new opportunities for the application of spiropyrans and spirooxazines, was first reported by Yang and coworkers [102]. The interaction of the free spiropyran

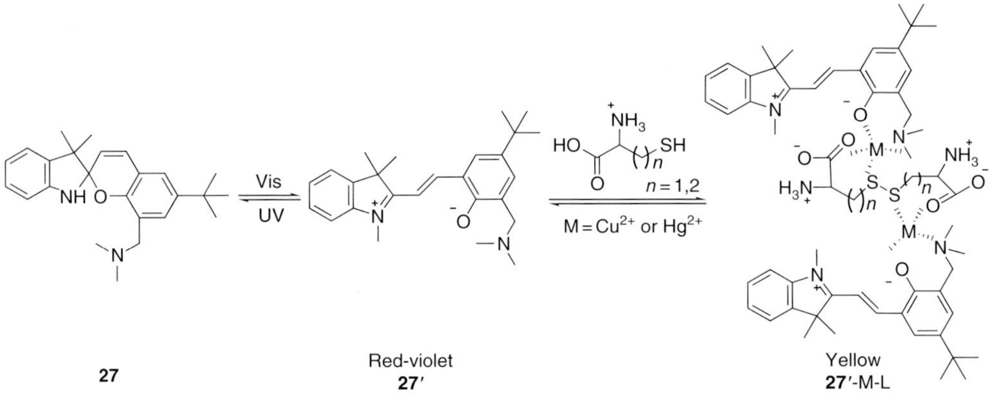

Figure 4.38 Chemical structure and multi responses of **27**.

with an amino acid is comparatively weak, but the spiropyran molecule can bind a metal ion, which can interact with an amino acid ligand (Figure 4.38). In the presence of Cu^{2+} or Hg^{2+} ions, the interaction of the spiropyran with Cys or Hcy is remarkably selective and sensitive, resulting in appreciable changes in color and absorption properties. The UV–vis spectrum of **27** only slightly changes upon the addition of 10 equiv. of Cys, indicating a weak interaction between **27** and the amino acid. In contrast, when the same amount of Cys was added to the ethanol/water solution containing **27** and Cu^{2+} or **27** and Hg^{2+} the absorbance at 532 nm greatly decreased, while the 378-nm absorption band shifted to 405 nm concomitant with an increase in intensity. In contrast, in the presence of Zn^{2+} ions and Cys, the absorption spectrum of **27** changes only slightly. Cys is deprotonated at the sulfhydryl group to form a bridged dimer (cystine) through the redox-active Cu^{2+} or Hg^{2+} ions, and then the metal center complexes with one cystine molecule and two **27** molecules to form a 2 : 1 : 2 (M^+/cystine/**27**) ternary assembly.

4.4.4 Metal Complexes with Rhodamine

Due to the long-wavelength absorption and emission, good photostability, and high absorption coefficient and quantum efficiency, rhodamine derivatives are widely available and used in industrial coloration, biomarkers, and fluorometric probes [124–126]. However, the use of rhodamine derivatives as stable photochromic molecules remains rare. Because of the short lifetime and low quantum efficiency of the photoinduced reaction, the photochromic properties of rhodamine amide were virtually disregarded. Recently, a new photochromic system based on a rhodamine B salicylaldehyde hydrazone metal complex (**28**) was reported by Tong and coworkers (Figure 4.39) [127]. Rhodamine B salicylaldehyde hydrazone, which contains a rhodamine amide moiety and a photosensitive salicylaldehyde Schiff base, was found to exhibit photochromic properties with a long lifetime when forming a metal complex. UV light promoted the isomerization

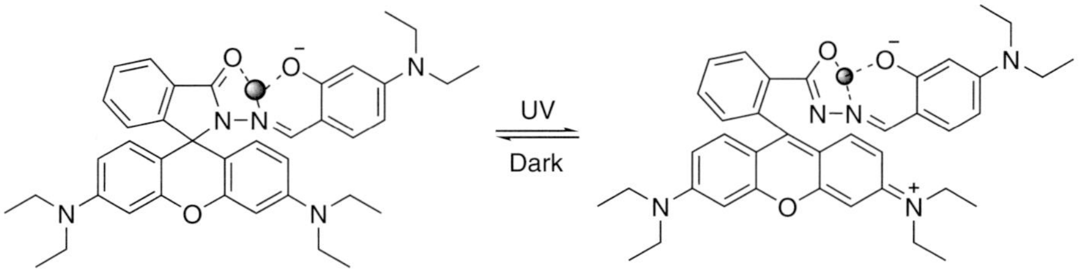

28 28′

Figure 4.39 Chemical structure and photochromic process of **28**.

of the salicylaldehyde hydrazone moiety from the enol-form to the keto-form, subsequently induced the spirolactam ring opening in the rhodamine B portion, and caused the photochromic reaction. This photochromic system exhibited good fatigue resistance, and the thermal bleaching rate was tunable using different metal ions to form the complex. Moreover, the photochromic system was successfully applied to photoprinting and UV strength measurement in the solid state.

In summary, photochemical switches based on metal complexes were described in this section. Examples of metal complexes with azobenzene, diarylethene, spirocyclic, and rhodamine ligands were analyzed. The switches exhibiting multiresponses can be extended to the design of molecular devices that perform complex roles, such as biosensors and images, storage materials, and memories.

4.5 Other Light-Modulated Supramolecular Interactions

Self-assembly triggered by light irradiation has garnered great interest in the past several years because it provides a promising strategy to obtain various types of supramolecular structures with potential applications in sensing, optics, and drug delivery systems [128, 129].

Stoddart *et al.* reported two degenerate [2]rotaxanes containing an azobenzene unit to gate degenerated donor–acceptor molecular shuttles (Figure 4.40) [130]. The authors used the modified 4,4′-azobiphenyloxy unit (ABP) as a light-operated gate by introducing (i) four methyl groups (ABP-Me_4) and (ii) four fluorine atoms (ABP-F_4) at the 3,5,3′,5′-positions of the ABP units to curtail binding by the $CBPQT^{4+}$ ring. The first approach led to a gate (ABP-Me_4) that remains closed all of the time, whereas the second approach affords a gate (ABP-F_4) that can be closed with UV light and opened with visible light. Therefore, the light can be used, in conjunction with thermal energy, to raise and lower the free energy barrier at will and to impart STOP and GO instructions upon the operation of a molecular shuttle.

Credi *et al.* also reported a self-assembling system that can be reversibly interconverted between thermodynamically stable (pseudorotaxane) and kinetically inert (rotaxane) forms by light irradiation (Figure 4.41) [131]. The system is

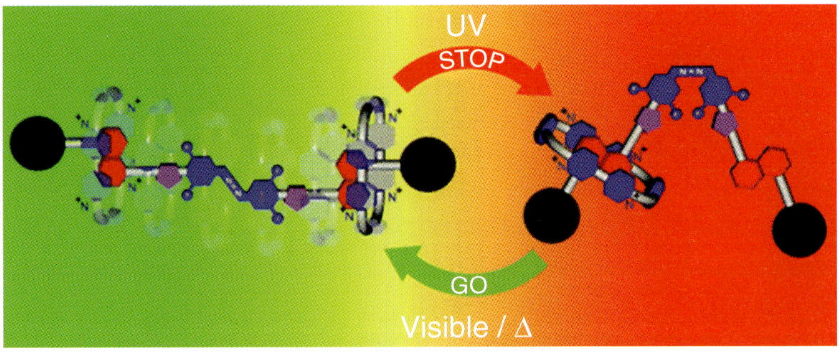

Figure 4.40 The graphical representation of the degenerate [2]rotaxane molecular shuttle gated by light irradiation and thermal energy. (Adapted with permission from [130]. American Chemical Society, 2009.)

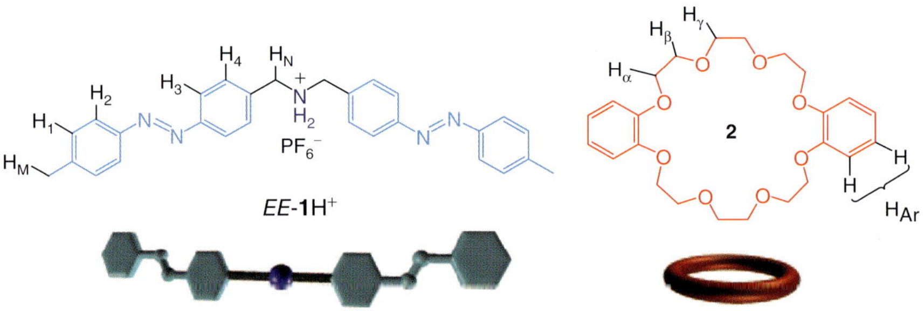

Figure 4.41 Structure of the ring and axle components, showing the atom numbering system. (Adapted with permission from [131]. Wiley-VCH, 2010.)

composed of a dibenzo[24]crown-8 ring and an axle composed of a dibenzylammonium recognition site and two azobenzene end groups. The isomeric form of the azobenzene units of the axle has little influence on the stability constants of the respective pseudorotaxanes, but greatly affects the threading–dethreading rate constants. In acetonitrile at room temperature, the EE axle is complexed by the ring in a pseudorotaxane fashion, according to a self-assembly equilibrium that is established in seconds. $E-Z$ photoisomerization of the end groups of the axle decelerates the threading–dethreading of the ring by at least four orders of magnitude, practically turning the pseudorotaxane into a kinetically inert rotaxane.

Although great progress has been realized in implementing light-induced self-assembly (LISA) at the colloidal and molecular scales [38], the assembly and disassembly of nanoscopic nanoparticles into various types and sizes "on demand" using LISA remain a great challenge. Recently, Klajn *et al.* addressed this idea by designing a LISA system based on Au nanoparticles in which photoactive *trans*-azobenzene dithiol ligands [4,4-bis (11-mercaptoundecanoxy)azobenzene] (ADT) were bound onto the surfaces of Au nanoparticles with the aid of a capping agent

(DDA) (Figure 4.42a) [132]. The modified Au nanoparticles formed a stable phase in toluene when an excess of didodecyldimethylammonium bromide (DDAB) surfactant was added. Upon exposure to UV light, a rapid *trans*-to-*cis* isomerization occurred for azobenzene groups and resulted in a significant increase in the dipole moments of the azobenzene groups (0–4.4 D). Therefore, the Au nanoparticles approached each other (Figure 4.42b, center) and underwent covalent crosslinking by the free ends of the ADT dithiols (Figure 4.42b, right). The types of resulting structures and the degree of their reversibility depended on the amount of ADT crosslinkers per NP and the solvent used for LISA (Figure 4.42c). The self-assembly process cannot occur when the number of ADT groups is less than 20 or more than 300 ADTs per nanoparticles. Between these two values, the ADT groups photoisomerized and the modified Au nanoparticles assembled into a particular morphology. This study confirmed that light can be used to modulate and control the self-assembly of nanoscopic components into larger architectures of various internal orderings and overall dimensions. In the future, more complex nanostructures with different mechanical properties and applications might be realized using the LISA strategy.

Stimuli-responsive physical gels offer researchers promising opportunities for designing and constructing new functional materials, such as sensors and actuators. Apart from thermal responsiveness, various physical gels also respond to other stimuli, such as light irradiation, which can also trigger the gel–sol (solution) transitions.

Azobenzene as a photoisomerization unit was incorporated into low-molar-weight gels (LMWGs) to generate photoresponsive gels. Shinkai and coworkers [133] initially designed LMWG **29** (Figure 4.43) bearing an azobenzene segment, and the gel–sol transitions for the resulting gels could be tuned by light irradiation. Later, Koumura and coworkers [134] reported an azobenzene compound **29** with two urethane moieties linked to two cholesteryl ester units. They also found that the gel–sol transition for gels of **30** could occur upon photoirradiation because of *trans–cis* isomerization of the azobenzene unit. Further studies manifested that the intermolecular H-bonding, which was partly responsible for stabilizing the gels, was broken or reformed during the gel–sol transition induced by light irradiation.

Yagai and coworkers [135] described gelator **31** derived from *N*,*N*′-disubstituted-4,6-diaminopyrimidin-2(1*H*)-one with two azobenzene units (Figure 4.44). ^{1}H NMR, absorption, and DLS studies clearly indicated that the self-assembly structures of **31** are strongly affected by the photoisomerization of the azobenzene units. This was ascribed to the variation of intermolecular H-bonding after *trans–cis* isomerization of azobenzene units. Exposure of the gel of **31** from heptane resulted in the collapse of the gel within a few hours. Further studies showed that the *trans–cis* isomerization of the azobenzene moieties induced the collapse of the lamellar structures into soluble supramolecular tapes.

We reported a photoresponsive gel system based on azobenzene groups in 2006 [136]. The gel was constructed with two components, photoresponsive C_3-symmetrical trisurea self-assembling building blocks containing three

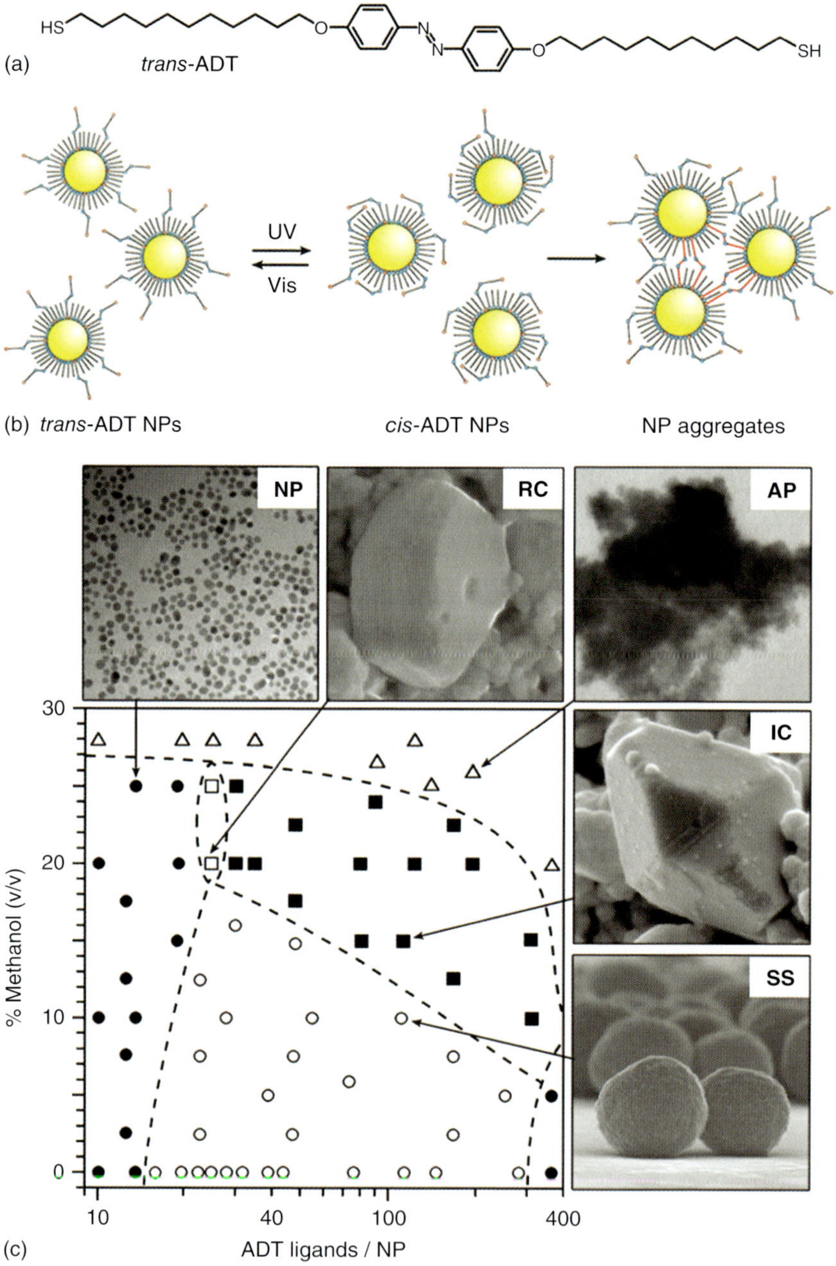
HS
O
N
N
O
SH
(a)
trans-ADT
UV
Vis
(b) trans-ADT NPs
cis-ADT NPs
NP aggregates
NP
RC
AP
IC
SS
30
20
10
0
% Methanol (v/v)
10
40
100
400
ADT ligands / NP
(c)

Figure 4.42 (a) Structural formula of *trans*-(4, 40-bis (11-mercaptoundecanoxy)azobenzene) (*trans*-ADT). (b) Light-induced self-assembly of NPs. UV irradiation of a stable solution of *trans*-ADT-coated NPs induces AP isomerization and attractive interactions between NPs. When in close proximity, covalent crosslinks between the NPs form. (c) Phase diagram of NP suprastructures. Morphology of the resulting aggregate is dictated by the ADT surface concentration and the polarity of the solvent. (Adapted with permission from [132]. Copyright (2007) National Academy of Sciences, USA.)

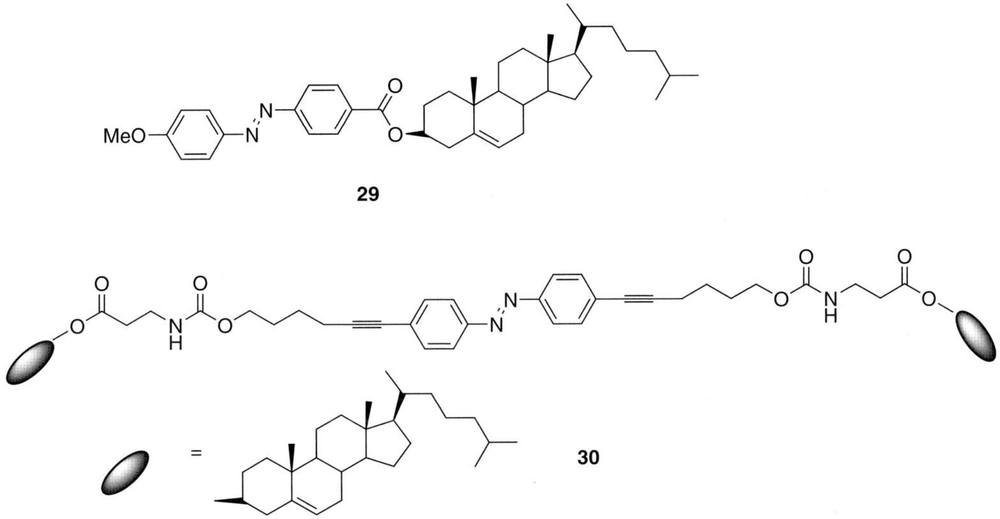

Figure 4.43 The structures of compounds **29** and **30**.

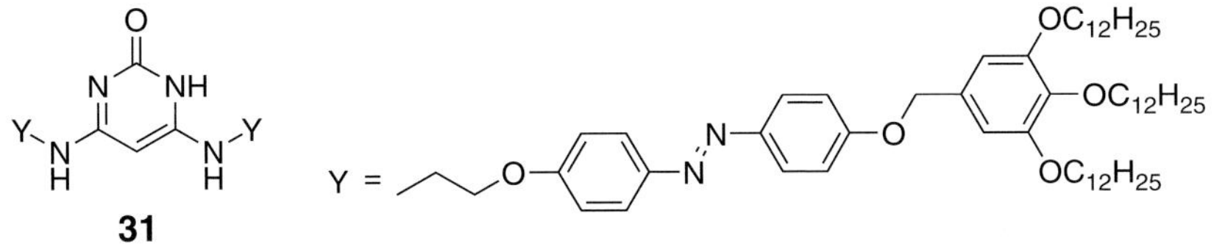

Figure 4.44 Chemical formula of **31**

azobenzene groups (**LC10** and **LC4**) and a trisamide gelator (**G18**), which can self-aggregate through hydrogen bonds of acylamino moieties to form a fibrous network (Figure 4.45). Mixing of **LC10** (or **LC4**) with **G18** forms an organogel with a coral-like supermolecular structure from 1,4-dioxane. In addition, the morphology of the gel can be tuned by varying the ratio of the two-component self-assembly building blocks or by changing the solvents within the gel, as shown in Figure 4.45e–j. In our previous study, we also observed solvent-tunable morphology in an azobenzene gelator with several competitive or cooperative

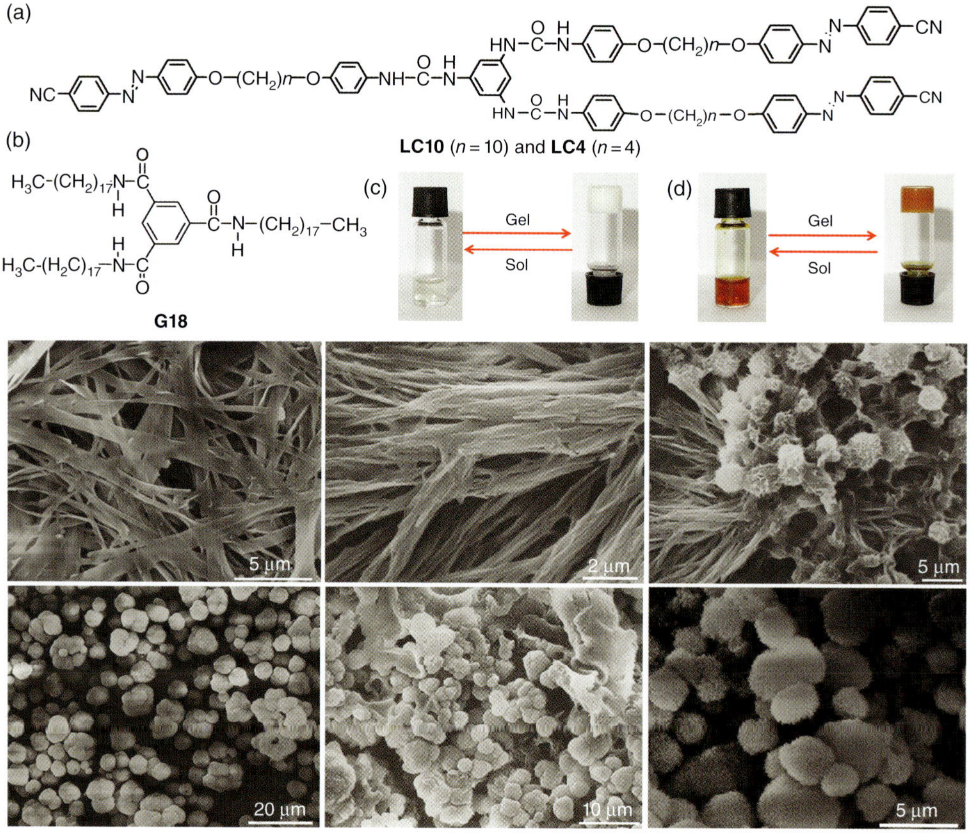

Figure 4.45 Chemical structures of **LC10** (a) and **G18** (b). Images of the sol–gel transition for the sole G18 (c) and the mixture of **LC10/G18** (1 : 19 w/w) (d) with a total concentration of 40 mg ml^{-1}. SEM images of the corresponding xerogels of **G18** from ethanol (e); 1,4-dioxane (f, main body of the fiber and g, end of fiber); 1 : 19 w/w **LC10/G18** (h). 3 : 17 w/w **LC10/G18** (i) and 1 : 9 w/w **LC4/G18** (j). The total concentration of all of the gels is 40 mg ml^{-1}. (Adapted with permission from [136]. American Chemical Society, 2007.)

interactions [137]. Moreover, the gel also exhibit photoresponsive properties, depending on the length of middle alkyl chains linked with the benzene urea groups and the photoisomerization building block of azobenzene. The *trans*-isomer of azobenzene in the gel of **LC10/G18** can be easily changed within 10 min to the *cis*-isomer by irradiation of 365 nm light (*trans*:*cis* = 29 : 71). In sharp contrast, photoisomerization between *trans* and *cis* in the **LC4/G18** gel was obviously restrained, implying the different spatial freedoms of the azobenzene moieties in the packing of these molecules. This work provides a strategy for the rational design of photoswitchable self-assembly materials for practical applications.

4.6 Conclusions and Outlook

This chapter presents a brief overview of photoswitched supramolecular systems, especially photocontrolled supramolecular self-assemblies, molecular sensors, and the capture and release of guest molecules as well as molecular devices. Azobenzene is the most used photochromic group in the self-assembly process due to its large change in the structure and dipole triggered by light. This characteristic makes it widely applied in host–guest systems, molecular devices, and delivery systems. Photoresponsive fluorescent switches and sensors have been realized in the photochromic diarylethene and spirocyclic derivatives, which may be used as biomarkers to trace a specific physiological process and to realize high-resolution mapping of a specific biomolecule. However, the stability and nondestructive capability of those materials still remain to be improved for application of the materials in real life.

The combination of photochemistry with supramolecular chemistry and, in a more context, with nanostructured assemblies has led to outstanding research achievements that may enable innovative solutions for current problems related to energy, environment, and health. Despite the great efforts, the photoswitchable supramolecular system is still in its infancy, and challenges remain, such as designing biologically compatible structures, finding new applications, obtaining efficient and optimized working materials, and applying these smart materials to real life. Overall, photoswitchable supramolecular systems play a vital role in advancing supramolecular chemistry and stimuli-responsive self-assembly research.

References

1. Chu, Z., Dreiss, C.A., and Feng, Y. (2013) Smart wormlike micelles. *Chem. Soc. Rev.*, **42** (17), 7174–7203.
2. Raeburn, J., Zamith Cardoso, A., and Adams, D.J. (2013) The importance of the self-assembly process to control mechanical properties of low molecular weight hydrogels. *Chem. Soc. Rev.*, **42** (12), 5143–5156.
3. Sijbesma, R.P., Beijer, F.H., Brunsveld, L., Folmer, B.J.B., Hirschberg, J.H.K.K., Lange, R.F.M., Lowe, J.K.L., and Meijer, E.W. (1997) Reversible polymers formed from self-complementary monomers using quadruple hydrogen bonding. *Science*, **278** (5343), 1601–1604.
4. Zhou, S.-L., Matsumoto, S., Tian, H.-D., Yamane, H., Ojida, A., Kiyonaka, S., and Hamachi, I. (2005) pH-responsive shrinkage/swelling of a supramolecular hydrogel composed of two small amphiphilic molecules. *Chem. Eur. J.*, **11** (4), 1130–1136.
5. Cordier, P., Tournilhac, F., Soulie-Ziakovic, C., and Leibler, L. (2008) Self-healing and thermoreversible rubber from supramolecular assembly. *Nature*, **451** (7181), 977–980.
6. Estroff, L.A. and Hamilton, A.D. (2004) Water gelation by small organic molecules. *Chem. Rev.*, **104** (3), 1201–1217.
7. van Bommel, K.J.C., Friggeri, A., and Shinkai, S. (2003) Organic templates for the generation of inorganic materials. *Angew. Chem. Int. Ed.*, **42** (9), 980–999.
8. Lehn, J.-M. (2007) From supramolecular chemistry towards constitutional dynamic chemistry and adaptive chemistry. *Chem. Soc. Rev.*, **36** (2), 151–160.

9. Aggeli, A., Bell, M., Boden, N., Keen, J.N., Knowles, P.F., McLeish, T.C., Pitkeathly, M., and Radford, S.E. (1997) Responsive gels formed by the spontaneous self-assembly of peptides into polymeric beta-sheet tapes. *Nature*, **386** (6622), 259–262.
10. Feng, A., Zhan, C., Yan, Q., Liu, B., and Yuan, J. (2014) A CO_2- and temperature-switchable "schizophrenic" block copolymer: from vesicles to micelles. *Chem. Commun.*, **50** (64), 8958–8961.
11. Liang, C.-Y., Keller, S.M., Sepulveda, A.E., Sun, W.-Y., Cui, J., Lynch, C.S., and Carman, G.P. (2014) Electrical control of a single magnetoelastic domain structure on a clamped piezoelectric thin film-analysis. *J. Appl. Phys.*, **116** (12), 123909.
12. Raghupathi, K.R., Guo, J., Munkhbat, O., Rangadurai, P., and Thayumanavan, S. (2014) Supramolecular disassembly of facially amphiphilic dendrimer assemblies in response to physical, chemical, and biological stimuli. *Acc. Chem. Res.*, **47** (7), 2200–2211.
13. Motulsky, A., Lafleur, M., Couffin-Hoarau, A.-C., Hoarau, D., Boury, F., Benoit, J.-P., and Leroux, J.-C. (2005) Characterization and biocompatibility of organogels based on L-alanine for parenteral drug delivery implants. *Biomaterials*, **26** (31), 6242–6253.
14. Dohner, E.R., Hoke, E.T., and Karunadasa, H.I. (2014) Self-assembly of broadband white-light emitters. *J. Am. Chem. Soc.*, **136** (5), 1718–1721.
15. Irie, M. (2000) Diarylethenes for memories and switches. *Chem. Rev.*, **100** (5), 1685–1716.
16. Xiao, S., Yi, T., Li, F., and Huang, C. (2005) A multi-photo responsive photochromic dithienylethene containing coumarin derivative. *Tetrahedron Lett.*, **46** (52), 9009–9012.
17. Xiao, S., Yi, T., Zhou, Y., Zhao, Q., Li, F., and Huang, C. (2006) Multistate molecular switches based on dithienylperfluorocyclopentene and imidazo [4,5-f] [1,10] phenanthroline. *Tetrahedron*, **62** (43), 10072–10078.
18. Zou, Y., Xiao, S., Yi, T., Zhang, H., Li, F., and Huang, C. (2007) Nanohybrid material of bilateral switch based on diarylethene. *J. Phys. Org Chem.*, **20** (11), 975–980.
19. Xiao, S., Zou, Y., Yu, M., Yi, T., Zhou, Y., Li, F., and Huang, C. (2007) A photochromic fluorescent switch in an organogel system with non-destructive readout ability. *Chem Commun.*, (45), 4758–4760.
20. Wang, S., Shen, W., Feng, Y., and Tian, H. (2006) A multiple switching bisthienylethene and its photochromic fluorescent organogelator. *Chem. Commun.*, (14), 1497–1499.
21. Tian, H. and Tu, H.-Y. (2000) Synthesis and photochromic properties of new bisthienylethene derivatives and a copolymer. *Adv. Mater.*, **12** (21), 1597–1600.
22. Chen, B., Wang, M., Li, C., Xia, H., and Tian, H. (2003) New photochromic copolymers with bisthienylethene pendant groups. *Synth. Met.*, **135**, 491–492.
23. Tian, H. and Yang, S. (2004) Recent progresses on diarylethene based photochromic switches. *Chem. Soc. Rev.*, **33** (2), 85–97.
24. Yao, X., Li, T., Wang, S., Ma, X., and Tian, H. (2014) A photochromic supramolecular polymer based on bis-p-sulfonatocalix[4]arene recognition in aqueous solution. *Chem. Commun.*, **50** (54), 7166–7168.
25. Zhang, J., Jin, J., Zou, L., and Tian, H. (2013) Reversible photocontrollable gels based on bisthienylethene-doped lecithin micelles. *Chem. Commun.*, **49** (85), 9926–9928.
26. Zhang, J., Zou, Q., and Tian, H. (2013) Photochromic materials: more than meets the eye. *Adv. Mater.*, **25** (3), 378–399.
27. Luo, Q., Cheng, H., and Tian, H. (2011) Recent progress on photochromic diarylethene polymers. *Polym. Chem.*, **2** (11), 2435–2443.
28. Zou, Q., Li, X., Zhang, J., Zhou, J., Sun, B., and Tian, H. (2012) Unsymmetrical diarylethenes as molecular keypad locks with tunable photochromism and fluorescence via Cu2+ and CN- coordinations. *Chem. Commun.*, **48** (15), 2095–2097.

29. Morimoto, M., Miyasaka, H., Yamashita, M., and Irie, M. (2009) Coordination assemblies of [Mn4] single-molecule magnets linked by photochromic ligands: photochemical control of the magnetic properties. *J. Am. Chem. Soc.*, **131** (28), 9823–9835.
30. Morimoto, M. and Irie, M. (2010) A diarylethene cocrystal that converts light into mechanical work. *J. Am. Chem. Soc.*, **132** (40), 14172–14178.
31. Fukaminato, T., Doi, T., Tamaoki, N., Okuno, K., Ishibashi, Y., Miyasaka, H., and Irie, M. (2011) Single-molecule fluorescence photoswitching of a diarylethene-perylenebisimide dyad: non-destructive fluorescence readout. *J. Am. Chem. Soc.*, **133** (13), 4984–4990.
32. Mori, K., Ishibashi, Y., Matsuda, H., Ito, S., Nagasawa, Y., Nakagawa, H., Uchida, K., Yokojima, S., Nakamura, S.-I., Irie, M., and Miyasaka, H. (2011) One-color reversible control of photochromic reactions in a diarylethene derivative: three-photon cyclization and two-photon cycloreversion by a near-infrared femtosecond laser pulse at 1.28 μm. *J. Am. Chem. Soc.*, **133** (8), 2621–2625.
33. Zou, Y., Yi, T., Xiao, S., Li, F., Li, C., Gao, X., Wu, J., Yu, M., and Huang, C. (2008) Amphiphilic diarylethene as a photoswitchable probe for imaging living cells. *J. Am. Chem. Soc.*, **130** (47), 15750–15751.
34. Liu, K., Wen, Y., Shi, T., Li, Y., Li, F., Zhao, Y.L., Huang, C., and Yi, T. (2014) DNA gated photochromism and fluorescent switch in a thiazole orange modified diarylethene. *Chem. Commun.*, **50** (65), 9141–9144.
35. Hirose, T., Matsuda, K., and Irie, M. (2006) Self-assembly of photochromic diarylethenes with amphiphilic side chains: reversible thermal and photochemical control. *J. Org. Chem.*, **71** (20), 7499–7508.
36. Hirose, T., Irie, M., and Matsuda, K. (2009) Self-assembly of photochromic diarylethenes with amphiphilic side chains: core-chain ratio dependence on supramolecular structures. *Chem. Asian J.*, **4** (1), 58–66.
37. Zhou, X., Duan, Y., Yan, S., Liu, Z., Zhang, C., Yao, L., and Cui, G. (2011) Optical modulation of supramolecular assembly of amphiphilic photochromic diarylethene: from nanofiber to nanosphere. *Chem. Commun.*, **47** (24), 6876–6878.
38. Liu, X. and Jiang, M. (2006) Optical switching of self-assembly: micellization and micelle-hollow-sphere transition of hydrogen-bonded polymers. *Angew. Chem. Int. Ed.*, **45** (23), 3846–3850.
39. Jochum, F.D. and Theato, P. (2010) Thermo- and light responsive micellation of azobenzene containing block copolymers. *Chem. Commun.*, **46** (36), 6717–6719.
40. Chen, C.-J., Liu, G.-Y., Liu, X.-S., Li, D.-D., and Ji, J. (2012) Construction of photo-responsive micelles from azobenzene-modified hyperbranched polyphosphates and study of their reversible self-assembly and disassembly behaviours. *New J. Chem.*, **36** (3), 694–701.
41. Wang, Y., Han, P., Xu, H., Wang, Z., Zhang, X., and Kabanov, A.V. (2010) Photocontrolled self-assembly and disassembly of block ionomer complex vesicles: a facile approach toward supramolecular polymer nanocontainers. *Langmuir*, **26** (2), 709–715.
42. Wang, Y., Ma, N., Wang, Z., and Zhang, X. (2007) Photocontrolled reversible supramolecular assemblies of an azobenzene-containing surfactant with α-cyclodextrin. *Angew. Chem.*, **119** (16), 2881–2884.
43. Zhu, L., Wu, W., Zhu, M.-Q., Han, J.J., Hurst, J.K., and Li, A.D.Q. (2007) Reversibly photoswitchable dual-color fluorescent nanoparticles as new tools for live-cell imaging. *J. Am. Chem. Soc.*, **129** (12), 3524–3526.
44. Chen, J., Zeng, F., Wu, S., Zhao, J., Chen, Q., and Tong, Z. (2008) Reversible fluorescence modulation through energy transfer with ABC triblock copolymer micelles as scaffolds. *Chem. Commun.*, (43), 5580–5582.
45. Chen, J., Wang, D., Turshatov, A., Munoz-Espi, R., Ziener, U., Koynov, K., and Landfester, K. (2013) One-pot fabrication of amphiphilic photoswitchable

thiophene-based fluorescent polymer dots. *Polym. Chem.*, **4** (3), 773–781.

46. Chen, J., Zhang, P., Fang, G., Weng, C., Hu, J., Yi, P., Yu, X., and Li, X. (2012) One-pot synthesis of amphiphilic reversible photoswitchable fluorescent nanoparticles and their fluorescence modulation properties. *Polym. Chem.*, **3** (3), 685–693.
47. Cram, D.J. and Cram, J.M. (1974) Host-gest chemistry. *Science*, **183** (4127), 803–809.
48. Yang, H., Yuan, B., and Zhang, X. (2014) Supramolecular chemistry at interfaces: host–guest interactions for fabricating multifunctional biointerfaces. *Acc. Chem. Res.*, **47** (7), 2106–2115.
49. Yagai, S. and Kitamura, A. (2008) Recent advances in photoresponsive supramolecular self-assemblies. *Chem. Soc. Rev.*, **37** (8), 1520–1529.
50. Kumar, G.S. and Neckers, D.C. (1989) Photochemistry of azobenzene-containing polymers. *Chem. Rev.*, **89** (8), 1915–1925.
51. Dawn, S., Dewal, M.B., Sobransingh, D., Paderes, M.C., Wibowo, A.C., Smith, M.D., Krause, J.A., Pellechia, P.J., and Shimizu, L.S. (2011) Self-Assembled phenylethynylene bis-urea macrocycles facilitate the selective photodimerization of coumarin. *J. Am. Chem. Soc.*, **133** (18), 7025–7032.
52. Yu, Z. and Hecht, S. (2011) Reversible and quantitative denaturation of amphiphilic oligo(azobenzene) foldamers. *Angew. Chem. Int. Ed.*, **50** (7), 1640–1643.
53. Wu, W., Yao, L., Yang, T., Yin, R., Li, F., and Yu, Y. (2011) NIR-light-induced deformation of cross-linked liquid-crystal polymers using upconversion nanophosphors. *J. Am. Chem. Soc.*, **133** (40), 15810–15813.
54. Shinkai, S., Yoshida, T., Manabe, O., and Fuchita, Y. (1988) Photoresponsive crown ethers. Part 20. Reversible photocontrol of association-dissociation equilibria between azobis (benzo-18-crown-6) and diammonium cations. *J. Chem. Soc., Perkin Trans.*, **1** (6), 1431–1437.
55. Shinkai, S., Yoshida, T., Miyazaki, K., and Manabe, O. (1987) Photoresponsive crown ethers. 19. Photocontrol of reversible association-dissociation phenomena in "tail(ammonium)-biting" crown ethers. *Bull. Chem. Soc. Jpn.*, **60** (5), 1819–1824.
56. Yagai, S., Karatsu, T., and Kitamura, A. (2005) Photocontrollable self-assembly. *Chem. Eur. J.*, **11** (14), 4054–4063.
57. Yagai, S., Nakajima, T., Karatsu, T., Saitow, K., and Kitamura, A. (2004) Phototriggered self-assembly of hydrogen-bonded rosette. *J. Am. Chem. Soc.*, **126** (37), 11500–11508.
58. Yagai, S., Nakajima, T., Kishikawa, K., Kohmoto, S., Karatsu, T., and Kitamura, A. (2005) Hierarchical organization of photoresponsive hydrogen-bonded rosettes. *J. Am. Chem. Soc.*, **127** (31), 11134–11139.
59. Kim, K., Selvapalam, N., Ko, Y.H., Park, K.M., Kim, D., and Kim, J. (2006) Functionalized cucurbiturils and their applications. *Chem. Soc. Rev.*, **36** (2), 267–279.
60. Liu, Y. and Chen, Y. (2006) Cooperative binding and multiple recognition by bridged bis(β-cyclodextrin)s with functional linkers. *Acc. Chem. Res.*, **39** (10), 681–691.
61. Nepogodiev, S.A. and Stoddart, J.F. (1998) Cyclodextrin-based catenanes and rotaxanes. *Chem. Rev.*, **98** (5), 1959–1976.
62. Liao, X., Chen, G., Liu, X., Chen, W., Chen, F., and Jiang, M. (2012) Photoresponsive pseudopolyrotaxane hydrogels based on competition of host–guest interactions. *Angew. Chem. Int. Ed.*, **49** (26), 4409.
63. Harada, A. and Kamachi, M. (1990) Complex formation between poly(ethy1ene glycol) and α-cyclodextrin. *Macromolecules*, **23** (10), 2821–2823.
64. Harada, A., Li, J., and Kamachi, M. (1992) The molecular necklace: a rotaxane containing many threaded α-cyclodextrin. *Nature*, **356**, 325–327.
65. Harada, A., Li, J., and Kamachi, M. (1994) Double-stranded inclusion complexes of cyclodextrin threaded

on poly(ethylene glycol). *Nature*, **370**, 126–128.

66. Tamesue, S., Takashima, Y., Yamaguchi, H., Shinkai, S., and Harada, A. (2010) Photoswitchable supramolecular hydrogels formed by cyclodextrins and azobenzene polymers. *Angew. Chem. Int. Ed.*, **49** (41), 7461–7464.
67. Oshovsky, G.V., Reinhoudt, D.V., and Verboom, W. (2007) Supramolecular chemistry in water. *Angew. Chem. Int. Ed.*, **46** (14), 2366.
68. Harada, A., Hashidzume, A., Yamaguchi, H., and Takashima, Y. (2009) Polymeric rotaxanes. *Chem. Rev.*, **109** (11), 5974–6023.
69. Yamaguchi, H., Kobayashi, Y., Kobayashi, R., Takashima, Y., Hashidzume, A., and Harada, A. (2012) Photoswitchable gel assembly based on molecular recognition. *Nat. Commun.*, **3**, 603.
70. Cohen Stuart, M.A., Huck, W.T.S., Genzer, J., Muller, M., Ober, C., Stamm, M., Sukhorukov, G.B., Szleifer, I., Tsukruk, V.V., Urban, M., Winnik, F., Zauscher, S., Luzinov, I., and Minko, S. (2010) Emerging applications of stimuli-responsive polymer materials. *Nat. Mater.*, **9** (2), 101–113.
71. Croissant, J. and Zink, J.I. (2012) Nanovalve-controlled cargo release activated by plasmonic heating. *J. Am. Chem. Soc.*, **134** (18), 7628–7631.
72. Lee, H.-I., Wu, W., Oh, J.K., Mueller, L., Sherwood, G., Peteanu, L., Kowalewski, T., and Matyjaszewski, K. (2007) Light-induced reversible formation of polymeric micelles. *Angew. Chem. Int. Ed.*, **46** (14), 2453–2457.
73. Park, J., Sun, L.B., Chen, Y.P., Perry, Z., and Zhou, H.C. (2014) Azobenzene-functionalized metal–organic polyhedra for the optically responsive capture and release of guest molecules. *Angew. Chem. Int. Ed.*, **53** (23), 5842–5846.
74. Balas, F., Manzano, M., Horcajada, P., and Vallet-Regi, M. (2006) Confinement and controlled release of bisphosphonates on ordered mesoporous silica-based materials. *J. Am. Chem. Soc.*, **128** (25), 8116–8117.
75. Vallet-Regi, M., Ramila, A., Del Real, R.P., and Perez-Pariente, J. (2001) A new property of MCM-41: drug delivery system. *Chem. Mater.*, **13** (12), 308–311.
76. Yuan, Q., Zhang, Y., Chen, T., Lu, D., Zhao, Z., Zhang, X., Li, Z., Yan, C., and Tan, W. (2012) Photon-manipulated drug release from a mesoporous nanocontainer controlled by azobenzene-modified nucleic acid. *ACS Nano*, **6** (7), 6337–6344.
77. Klajn, R., Stoddart, J.F., and Grzybowski, B.A. (2010) Nanoparticles functionalised with reversible molecular and supramolecular switches. *Chem. Soc. Rev.*, **39** (6), 2203–2237.
78. Angelos, S., Choi, E., Voegtle, F., De Cola, L., and Zink, J.I. (2007) Photo-driven expulsion of molecules from mesostructured silica nanoparticles. *J. Phys. Chem. C*, **111** (18), 6589–6592.
79. Lu, J., Choi, E., Tamanoi, F., and Zink, J.I. (2008) Light-activated nanoimpeller-controlled drug release in cancer cells. *Small*, **4** (4), 421–426.
80. Kresge, C.T., Leonowicz, M.E., Roth, W.J., Vartuli, J.C., and Beck, J.S. (1992) Ordered mesoporous molecular-sieves synthesized by a liquid-crystal template mechanism. *Nature*, **359** (6397), 710–712.
81. Ferris, D.P., Zhao, Y.-L., Khashab, N.M., Khatib, H.A., Stoddart, J.F., and Zink, J.I. (2009) Light-operated mechanized nanoparticles. *J. Am. Chem. Soc.*, **131** (5), 1686–1687.
82. Tomatsu, I., Hashidzume, A., and Harada, A. (2006) Contrast viscosity changes upon photoirradiation for mixtures of poly(acrylic acid)-based alpha-cyclodextrin and azobenzene polymers. *J. Am. Chem. Soc.*, **128** (7), 2226–2227.
83. Luo, X.F., Goh, S.H., Lee, S.Y., and Tan, K.L. (1998) Interpolymer complexation between poly(N-methyl-4-piperidinyl methacrylate) and acidic polymers. *Macromolecules*, **31** (10), 3251–3254.
84. Lu, X., He, C., and Griffin, A.C. (2003) Crystallization of sequence random multicomponent supermolecular complexes. *Macromolecules*, **36** (14), 5195–5200.

85. Bazuin, C.G. and Brandys, F.A. (1992) Novel liquid-crystalline polymeric materials via noncovalent "grafting". *Chem. Mater.*, **4** (5), 970–972.
86. Bao, C.Y., Lu, R., Jin, M., Xue, P.C., Tan, C.H., Liu, G.F., and Zhao, Y.Y. (2005) L-Tartaric acid assisted binary organogel system: strongly enhanced fluorescence induced by supramolecular assembly. *Org. Biomol. Chem.*, **3**, 2508–2512.
87. Li, R., Yang, X., Li, G., Li, S., and Huang, W. (2006) Core-corona polymer composite particles by self-assembled heterocoagulation based on a hydrogen-bonding interaction. *Langmuir*, **22** (19), 8127–8133.
88. Kato, T. and Frechet, J.M.J. (1989) Stabilization of a liquid-crystalline phase through noncovalent interaction with a polymer side chain. *Macromolecules*, **22** (9), 3818–3819.
89. Xiao, S., Zou, Y., Wu, J., Zhou, Y., Yi, T., Li, F., and Huang, C. (2007) Hydrogen bonding assisted switchable fluorescence in self-assembled complexes containing diarylethene: controllable fluorescent emission in the solid state. *J. Mater. Chem.*, **17** (24), 2483–2489.
90. Zhou, J., Liu, Z., and Li, F. (2012) Upconversion nanophosphors for small-animal imaging. *Chem. Soc. Rev.*, **41** (3), 1323–1349.
91. Zhou, Z., Hu, H., Yang, H., Yi, T., Huang, K., Yu, M., Li, F., and Huang, C. (2008) Up-conversion luminescent switch based on photochromic diarylethene and rare-earth nanophosphors. *Chem. Commun.*, (39), 4786–4788.
92. Cao, X., Zhou, J., Zou, Y., Zhang, M., Yu, X., Zhang, S., Yi, T., and Huang, C. (2011) Fluorescence and morphology modulation in a photochromic diarylethene self-assembly system. *Langmuir*, **27** (8), 5090–5097.
93. Lindsay, S.M. and Ratner, M.A. (2007) Molecular transport junctions: clearing mists. *Adv. Mater.*, **19** (1), 23–31.
94. Tao, N.J. (2006) Electron transport in molecular junctions. *Nat. Nanotechnol.*, **1**, 173–181.
95. Cao, Y., Dong, S., Liu, S., Liu, Z., and Guo, X. (2013) Toward functional molecular devices based on graphene–molecule junctions. *Angew. Chem. Int. Ed.*, **52** (14), 3906–3910.
96. Zhao, Y. and Ikeda, T. (2009) *Smart Light-Responsive Materials*, John Wiley & Sons, Inc., Hoboken, NJ, pp. 95–136.
97. Ikeda, T., Mamiya, J., and Yu, Y. (2007) Photomechanics of liquid-crystalline elastomers and other polymers. *Angew. Chem. Int. Ed.*, **46** (4), 506–528.
98. Li, M.H., Keller, P., Li, B., Wang, X., and Brunet, M. (2003) Light-driven side on nematic elastomer actuators. *Adv. Mater.*, **15** (7), 569–572.
99. Wang, W., Sun, X., Wu, W., Peng, H., and Yu, Y. (2012) Photoinduced deformation of crosslinked liquid-crystalline polymer film oriented by a highly aligned carbon nanotube sheet. *Angew. Chem. Int. Ed.*, **51** (19), 4644–4647.
100. Li, F., Zhuang, J., Jiang, G., Tang, H., Xia, A., Jiang, L., Song, Y., Li, Y., and Zhu, D. (2008) A rewritable optical data storage material system by 2+2 photocycloreversion-photocycloaddition. *Chem. Mater.*, **20** (4), 1194–1196.
101. Klajn, R., Wesson, P.J., Bishop, K.J.M., and Grzybowski, B.A. (2009) Writing self-erasing images using metastable nanoparticle "Inks". *Angew. Chem. Int. Ed.*, **48** (38), 7035–7039.
102. Shao, N., Jin, J.Y., Cheung, S.M., Yang, R.H., Chan, W.H., and Mo, T. (2006) A spiropyran-based ensemble for visual recognition and quantification of cysteine and homocysteine at physiological levels. *Angew. Chem. Int. Ed.*, **45** (30), 4944–4948.
103. Tsujioka, T., Kume, M., and Irie, M. (1996) Optical density dependence of write/read characteristics in photon-mode photochromic memory. *Jpn. J. Appl. Phys.*, **35** (8), 4353–4360.
104. Chan, J., Lam, W.H., Wong, H.L., Zhu, N.Y., Wong, W.T., and Yam, V. (2011) Diarylethene-containing cyclometalated platinum (II) complexes: tunable photochromism via metal coordination and rational ligand design. *J. Am. Chem. Soc.*, **133** (32), 12690–12705.

105. Astrand, P.O., Ramanujam, P.S., Hvilsted, S., Bak, K.L., and Sauer, S. (2000) Ab initio calculation of the electronic spectrum of azobenzene dyes and its impact on the design of optical data storage materials. *J. Am. Chem. Soc.*, **122** (14), 3482–3487.
106. Shinkai, S., Nakaji, T., Nishida, Y., Ogawa, T., and Manabe, O. (1980) Photoresponsive crown ethers. 1. Cis–trans isomerism of azobenzene as a tool to enforce conformational-changes of crown ethers and polymers. *J. Am. Chem. Soc.*, **102** (18), 5860–5865.
107. Oka, Y. and Tamaoki, N. (2010) Structure of silver(I) complex prepared from azobenzenonaphthalenophane, photochemical coordination change of silver(I) and silver(I)-induced acceleration of Z-E thermal isomerization of azobenzene unit. *Inorg. Chem.*, **49** (11), 4765–4767.
108. Prakash, H., Shodai, A., Yasui, H., Sakurai, H., and Hirota, S. (2008) Photocontrol of spatial orientation and DNA cleavage activity of copper (II)-bound dipeptides linked by an azobenzene derivative. *Inorg. Chem.*, **47** (12), 5045–5047.
109. Panja, A., Matsuo, T., Nagao, S., and Hirota, S. (2011) DNA cleavage by the photocontrolled cooperation of Zn-II centers in an azobenzene-linked dizinc complex. *Inorg. Chem.*, **50** (22), 11437–11445.
110. Takeshita, M. and Irie, M. (1998) Photoresponsive cesium ion tweezers with a photochromic dithienylethene. *Tetrahedron Lett.*, **39** (7), 613–616.
111. Zhang, J.J., Tan, W.J., Meng, X.L., and Tian, H. (2009) Soft mimic gearshift with a multi-stimulus modified diarylethene. *J. Mater. Chem.*, **19** (32), 5726–5729.
112. Piao, X., Zou, Y., Wu, J., Li, C., and Yi, T. (2009) Multiresponsive switchable diarylethene and its application in bioimaging. *Org. Lett.*, **11** (17), 3818–3821.
113. Zhao, Q., Huang, C., and Li, F. (2011) Phosphorescent heavy-metal complexes for bioimaging. *Chem. Soc. Rev.*, **40** (5), 2508–2524.
114. Puckett, C.A. and Barton, J.K. (2007) Methods to explore cellular uptake of ruthenium complexes. *J. Am. Chem. Soc.*, **129** (1), 46–47.
115. Ko, C.C. and Yam, V. (2010) Transition metal complexes with photochromic ligands – photosensitization and photoswitchable properties. *J. Mater. Chem.*, **20** (11), 2063–2070.
116. Lo, K., Louie, M.W., Sze, K.S., and Lau, J. (2008) Rhenium(I) polypyridine biotin isothiocyanate complexes as the first luminescent biotinylation reagents: Synthesis, photophysical properties, biological labeling, cytotoxicity, and imaging studies. *Inorg. Chem.*, **47** (2), 602–611.
117. Yu, M., Zhao, Q., Shi, L., Li, F., Zhou, Z., Yang, H., Yi, T., and Huang, C. (2008) Cationic iridium(III) complexes for phosphorescence staining in the cytoplasm of living cells. *Chem. Commun.*, (18), 2115–2117.
118. Tan, W., Zhang, Q., Zhang, J., and Tian, H. (2009) Near-infrared photochromic diarylethene iridium (III) complex. *Org. Lett.*, **11** (1), 161–164.
119. Tan, W., Zhou, J., Li, F., Yi, T., and Tian, H. (2011) Visible light-triggered photoswitchable diarylethene-based iridium(III) complexes for imaging living cells. *Chem. Asian J.*, **6** (5), 1263–1268.
120. Lee, P.H., Ko, C., Zhu, N., and Yam, V.W. (2007) Metal coordination-assisted near-infrared photochromic behavior: a large perturbation on absorption wavelength properties of N,N-donor ligands containing diarylethene derivatives by coordination to the rhenium(I) metal center. *J. Am. Chem. Soc.*, **129** (19), 6058–6059.
121. Tanaka, M., Nakamura, M., Salhin, M., Ikeda, T., Kamada, K., Ando, H., Shibutani, Y., and Kimura, K. (2001) Synthesis and photochromism of spirobenzopyran derivatives bearing an oxymethylcrown ether moiety: metal ion-induced switching between positive and negative photochromisms. *J. Org. Chem.*, **66** (5), 1533–1537.
122. Inouye, M., Noguchi, Y., and Isagawa, K. (1994) Sensitive and selective coloration of cryptand-type crown

spirobenzopyrans for alkaline-earth metal-cations. *Angew. Chem. Int. Ed.*, **33** (11), 1163–1166.

123. Inouye, M., Akamatsu, K., and Nakazumi, H. (1997) New crown spirobenzopyrans as light- and ion-responsive dual-mode signal transducers. *J. Am. Chem. Soc.*, **119** (39), 9160–9165.

124. Yu, M., Shi, M., Chen, Z., Li, F., Li, X., Gao, Y., Xu, J., Yang, H., Zhou, Z., Yi, T., and Huang, C. (2008) Highly sensitive and fast responsive fluorescence turn-on chemodosimeter for Cu^{2+} and its application in live cell imaging. *Chem. Eur. J.*, **14** (23), 6892–6900.

125. Wang, Q., Li, C., Zou, Y., Wang, H., Yi, T., and Huang, C. (2012) A highly selective fluorescence sensor for Tin (Sn^{4+}) and its application in imaging live cells. *Org. Biomol. Chem.*, **10** (33), 6740–6746.

126. Lan, H., Wen, Y., Shi, Y., Liu, K., Mao, Y., and Yi, T. (2014) Fluorescence turn-on detection of Sn^{2+} in live eukaryotic and prokaryotic cells. *Analyst*, **139** (20), 5223–5229.

127. Li, K., Xiang, Y., Wang, X.Y., Li, J., Hu, R.R., Tong, A.J., and Tang, B.Z. (2014) Reversible photochromic system based on rhodamine B salicylaldehyde hydrazone metal complex. *J. Am. Chem. Soc.*, **136** (4), 1643–1649.

128. Oh, M. and Mirkin, C.A. (2005) Chemically tailorable colloidal particles from infinite coordination polymers. *Nature*, **438** (7068), 651–654.

129. Balzani, V., Credi, A., Silvi, S., and Venturi, M. (2006) Artificial nanomachines based on interlocked molecular species: recent advances. *Chem. Soc. Rev.*, **35** (11), 1135–1149.

130. Coskun, A., Friedman, D.C., Li, H., Patel, K., Khatib, H.A., and Stoddart, J.F. (2009) A light-gated STOP–GO molecular shuttle. *J. Am. Chem. Soc.*, **131** (7), 2493–2495.

131. Baroncini, M., Silvi, S., Venturi, M., and Credi, A. (2010) Reversible photoswitching of rotaxane character and interplay of thermodynamic stability and kinetic lability in a self-assembling ring–axle molecular system. *Chem. Eur. J.*, **16** (38), 11580–11587.

132. Klajn, R., Bishop, K.J.M., and Grzybowski, B.A. (2007) Light-controlled self-assembly of reversible and irreversible nanoparticle suprastructures. *Proc. Natl. Acad. Sci. U.S.A.*, **104** (25), 10305–10309.

133. Murata, K., Aoki, M., Suzuki, T., Harada, T., Kawabata, H., Komori, T., Ohseto, F., Ueda, K., and Shinkai, S. (1994) Thermal and light control of the sol-gel phase transition in cholesterol-based organic gels. Novel helical aggregation modes as detected by circular dichroism and electron microscopic observation. *J. Am. Chem. Soc.*, **116** (15), 6664–6676.

134. Koumura, N., Kudo, M., and Tamaoki, N. (2004) Photocontrolled gel-to-sol-to-gel phase transitioning of meta-substituted azobenzene bisurethanes through the breaking and reforming of hydrogen bonds. *Langmuir*, **20** (23), 9897–9900.

135. Yagai, S., Iwashima, T., Kishikawa, K., Nakahara, S., Karatsu, T., and Kitamura, A. (2006) Photoresponsive self-assembly and self-organization of hydrogen-bonded supramolecular tapes. *Chem. Eur. J.*, **12** (15), 3984–3994.

136. Zhou, Y., Xu, M., Yi, T., Xiao, S., Zhou, Z., Li, F., and Huang, C. (2007) Morphology-tunable and photoresponsive properties in a self-assembled two-component gel system†. *Langmuir*, **23** (1), 202–208.

137. Zhou, Y., Yi, T., Li, T., Zhou, Z., Li, F., Huang, W., and Huang, C. (2006) Morphology and wettability tunable two-dimensional superstructure assembled by hydrogen bonds and hydrophobic interactions. *Chem. Mater.*, **18** (13), 2974–2981.

5
Light-Gated Chemical Reactions and Catalytic Processes

Robert Göstl, Antti Senf, and Stefan Hecht

5.1
Introduction

Apart from its analytical and theoretical element, chemistry has always been a creative and constructive science. The synthetic chemist utilizes compounds that are readily available and converts them to desired products through chemical transformations, thereby constructing new materials, drugs, and technologically relevant fine chemicals. As society faces ever-growing environmental challenges, efforts to perform "green" chemistry have been undertaken that lead to better atom economy and energy efficiency, facilitated by the development of catalysts with boosted efficiency and selectivity as well as advanced reaction engineering and improved purification methods. Hence, the impact of chemistry on society largely depends on the success of making new molecules in an environmentally benign and sustainable manner. For this reason over the past decades, chemical research, in particular in industry, has been primarily concerned with developing new chemical transformations and applying them to generate new small molecules as well as polymers for pharmaceuticals and materials with improved properties. As the demand for complexity and finesse of these products increases with the proceeding industrial advancement of our society, it becomes imperative that the methods to control a chemical reaction must also keep pace. While so far it was enough to exert control over a chemical transformation by varying reaction temperature and time controlling *what* and *how* to make, the tailor-made molecules of the future clearly demand for stimuli featuring higher precision and orthogonality enabling control over *when* and *where* bonds are formed. The eventual goal is to perform chemistry with high spatial and temporal resolution, which would, for example, allow for the timing of reaction cascades and the localization of 2D patterns for array chip technologies. Ideally, control over time and space of the chemical transformation of choice is provided by a gate, which can act as a "remote control" upon the action of an external stimulus.

Many physical and chemical stimuli exist that could possibly be employed to incorporate a higher dimension of control into a reaction system. Regarding the

Photochromic Materials: Preparation, Properties and Applications, First Edition.
Edited by He Tian and Junji Zhang.

quest to attain the best possible spatial control, superior resolution is obviously provided by the tip of a scanning tunneling microscope (STM), which can be used to precisely induce chemical reactions on the molecular scale [1]. Yet, the impracticality of an STM lies in its slow speed, serial processing, and limited scale-up. Other stimuli such as temperature or mechanical forces suffer from poor spatial resolution as well as high inertia and offer little parallelization capability [2]. Light, however, with its superior spatial, temporal, and energetic resolution in combination with its noninvasive character offers the possibility to remote control different processes simultaneously with molecular scale resolution. While the choice of the wavelength and intensity allow for precise triggering of a specific photoreaction, the exposure can be carried out using modern optics, thereby enabling highly parallel processes. In addition to providing outstanding temporal control over illumination down to the subfemtosecond range, spatial resolution that is traditionally limited due to the law of diffraction has recently been improved significantly by super-resolution techniques [3] and multiphoton processes [4]. It is thus only consequential to design molecules that respond to light as a stimulus and that can transfer the acquired information to manipulate the outcome of a chemical reaction.

Gating a chemical reaction with light can be carried out in multiple ways [2], the most direct way of which it is to induce excited-state reactivity by absorption of a photon and subsequent transition of the respective chemical entity to a reactive excited state (Figure 5.1a). While unique chemical transformations, most notably electron-transfer-induced redox reaction and catalysis as well as cycloaddition reactions, can take place on the upper potential energy surface, the reactivity and lifetime of the involved excited states are generally hard to predict and thus this concept leaves little room for molecular structure–property engineering. Another possibility is that a dormant "cage" entity undergoes an irreversible transformation, the so-called "uncaging," upon photoexcitation (Figure 5.1b). While the reactivity of the resulting uncaged species is easier to predict and to analyze as compared to the excited-state photoreactivity, this concept resembles a *fuse* and therefore allows for a singular activation or deactivation event only. In strong contrast, the introduction of a photoreversible "gate" provides the opportunity to toggle between two states multiple times allowing to *switch* the system and hence the entire chemical process (Figure 5.1c). Therefore, only reversible photoreactions, which are typically associated with photochromes or nowadays more popularly called *photoswitches*, provide true "remote control" over the chemical process by light.

In this book chapter that is based on a recent review [5], we cover the increasing number of such photoswitchable systems, which display external control over thermal (ground-state) chemical reactions (Figure 5.1c). After describing key design criteria for merging photochromic and reactive units, we first discuss approaches that utilize the photoswitch in a stoichiometric manner, for example, to control substrate, product, or a template of a reaction, followed by a survey of photoswitchable catalytic systems [6]. Finally, we identify future challenges and opportunities of this exciting and rapidly developing field.

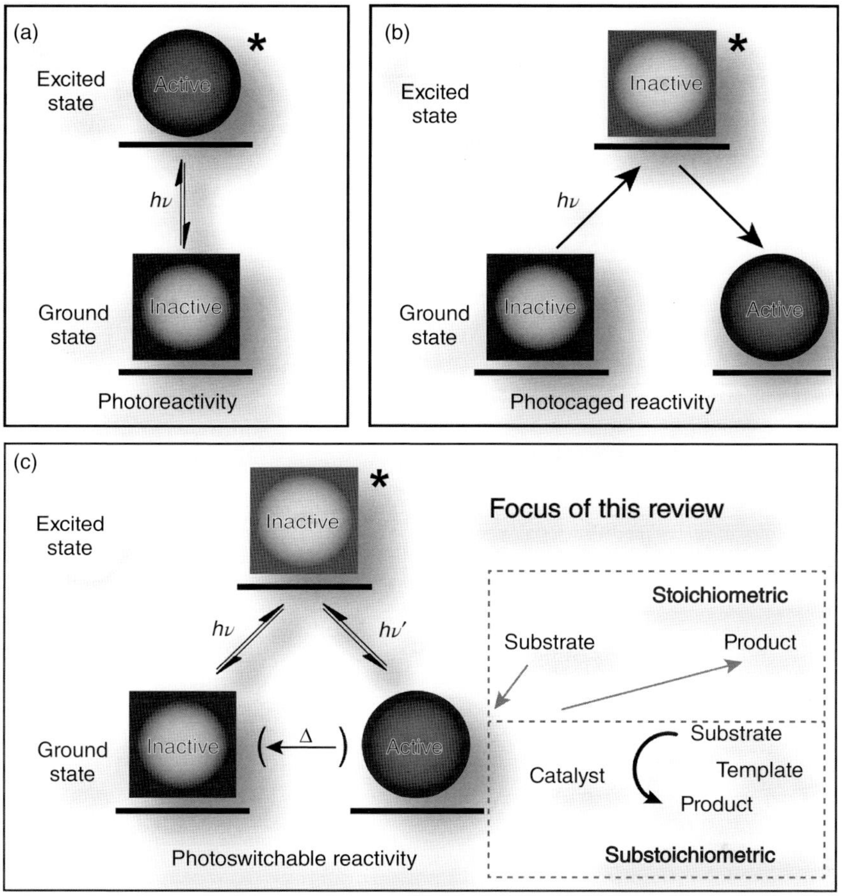

Figure 5.1 General approaches toward photocontrol of reactivity: (a) excited-state reactivity, where after photoexcitation reactions are performed in the excited state; (b) photocaged reactivity, where in the course of an irreversible photochemical transformation a reactive ground-state species is generated ("uncaged"); and (c) photoswitchable reactivity, where a reversible photochemical reaction (photochromism) allows for switching between unreactive and reactive ground-state species. In the last approach, which is the focus of this review, the photoswitchable system can be exploited either using stoichiometric processes (serving as substrate, product, or template) or substoichiometric processes (serving as the catalyst).

5.2 General Design Considerations

The design of photoswitchable systems that are able to control a ground-state chemical reaction requires the incorporation of suitable photochromic gates [7] into the system in a manner that the structural differences between their two switching states produce different chemical reactivity. Therefore, it is necessary to optimize intrinsic switching properties of the photoswitch itself as well as the way it is intersecting with the gated chemical reaction.

The performance of a photoswitch is best described by the efficiency and robustness of the light-induced forward and backward switching. Two parameters generally account for the photochemical efficiency of a photochromic system: (i) the reaction quantum yield ϕ for the desired photoprocess and (ii) the composition of the two isomers at the photostationary state (PSS). The latter expresses the degree of attainable photoconversion for forward (or backward) switching via the following equation:

$$\frac{c_A^\infty}{c_B^\infty} = \frac{\phi_{BA} \cdot \varepsilon_B}{\phi_{AB} \cdot \varepsilon_A}$$

that is governed by the molar absorptivities ε as well as the reaction quantum yields ϕ for the forward and backward photochemical reactions of both switching forms A and B at the used irradiation wavelength. Ideally, one form of the switch can be selectively excited and undergoes the respective photochemical reaction with a high quantum yield. Note that there are cases in which the quantum yield is dependent on the irradiation wavelength [8]. In addition to efficiency, the repeated operation of the switch requires uniform and highly reversible photochemistry, that is, high fatigue resistance. Specific irreversible side reactions can be generally suppressed by molecular design considerations, whereas unspecific degradation via the triplet-sensitized formation of $^1O_2$ can be inhibited by removing oxygen from the solvent.

Depending on the employed photoswitch, either both switching forms are thermally stable and can only be interconverted by light (P-type photochromic compounds, such as dithienylethenes (DTEs) and dithiazolylethenes shown in Figure 5.2b) or one of them is metastable and reverts back thermally to the more stable form (T-type photochromic compounds, such as azobenzenes or stilbenes shown in Figure 5.2a). While the first requires the use of two different irradiation wavelengths to toggle the system between its ON and OFF forms, the latter is only ON (or OFF) when being irradiated (assuming a fast dark reaction to the OFF (or ON) state).

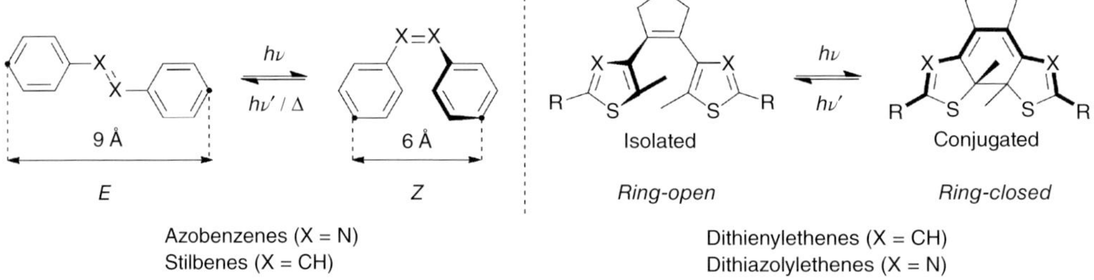

Figure 5.2 Molecular structure changes upon either *E*/*Z*-isomerization of azobenzene and stilbene or 6π-electrocyclization of dithienylethene and dithiazolylethene photoswitches, typically used as photochromic gates (see examples).

We here associate the ON state with a higher reactivity as compared to the OFF state and suggest that the switching states' reactivity difference is the handle to remote-controlling the chemical process of choice. To transform molecular alterations during the course of photoswitching into reactivity differences, the photoswitch should undergo significant geometrical or electronic changes upon the switching event (Figure 5.2). Large geometrical changes are typically achieved via *E/Z*-isomerization reactions of azobenzenes and stilbenes leading to a relatively smaller distance between the phenyl termini in the bent, nonplanar *Z*-configuration when compared to the extended, planar *E*-isomer. Complementary to the systems with large geometrical changes, the electrocyclic ring-closing/ring-opening reactions of 1,3,5-hexatriene systems, for example, in DTEs and dithiazolylethenes result in substantially modulated electronic properties of the ring-closed isomer as compared to the ring-open form, as a conjugation pathway in the ring-closed derivative is opened up. It is important to maintain the switching behavior throughout the reaction, and therefore the excitation of the photochromic moiety should be selective and local, without interfering quenching processes by energy or electron transfer [9].

The photochromic gate can be incorporated either directly by covalent connection to one of the components participating in the reaction or indirectly by noncovalent interaction with the same. This review will focus on the former as covalent constructs offer the advantage that they generally operate independent of concentration as opposed to supramolecular approaches. For an overview regarding the noncovalent interactions of switches on reactivity-controlling entities, the reader is referred to the existing literature [6]. In general, the coupling of a photochromic system to gate another thermal reaction can be realized by various concepts depending on the particular geometrical or electronic alteration of the switch and the point of interference. Such interference can occur either at the level of the substrate, product, or template, leading to stoichiometric processes with maximum overall reaction quantum yields of unity ($\phi \leq 1$), or at the level of the catalyst or effector, giving rise to an amplification of the light stimulus and overall reaction quantum yields exceeding unity ($\phi > 1$). All examples have been classified according to the general approach used and are being discussed in the next two sections.

5.3 Photoswitchable Stoichiometric Processes

The introduction of light as a gating stimulus means enabling manipulation of one reaction component in a way that it can be switched from an unreactive state to a reactive state or vice versa. This can be achieved through four different approaches. (i) Switching one of the starting materials from an inactive to an active form allows the respective compound to participate in the reaction and to form the desired product on demand (starting material control). (ii) Switching the reaction product from an active form into an inactive form removes the

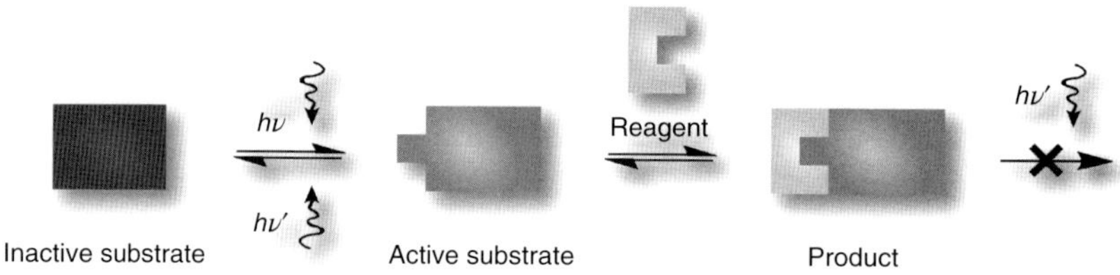

Figure 5.3 Concept of starting material control: photoswitching a substrate between inactive and active forms allows for controlled feeding or removal of a starting material to or from a dynamic covalent equilibrium. Reproduced from Ref. [5] with permission from The Royal Society of Chemistry.

compound from the initial equilibrium between the reactants and thus locks the product in an unreactive state (product control). (iii) Switching the starting materials as well as the products from their inactive to their active forms or vice versa combines approach (i) and (ii) thus enabling complete photocontrol over a reaction. (iv) Switching neither of the substrates but a "reaction mediator," namely a template, from an inactive to an active form and thereby controlling conversion by supramolecular complexation (template control).

5.3.1 Starting Material Control

The photoreversible activation of a starting material allowing it to participate in a chemical reaction is conceptually depicted in Figure 5.3. As opposed to photocaging, true reversibility can only be achieved if the photodynamic reaction is coupled to a dynamic thermal equilibrium, that is, the starting material can be reformed thermally and subsequently deactivated by light.

First, Kawashima and coworkers reported silicon fluoride derivatives bearing a covalently attached azobenzene ligand as early as 2001 [10]. Photochemical isomerization of the azobenzene from the *Z*- to the *E*-form could reversibly change the complex from pentacoordinate to hexacoordinate geometry. The group then transferred this reversible geometry change to an azobenzene-substituted allyldifluorosilane **1** [11]. Photoisomerization of *Z*-**1** to *E*-**1** followed by subsequent treatment with 18-crown-6 and potassium fluoride induces an allyl shift to the azobenzene reducing it to diphenylhydrazine **2** (Scheme 5.1). However, the original *Z*-isomer is unreactive under those reaction conditions. Further work on this system enabled photocontrol not only over the allyl shift but also over disiloxane formation (**3** in Scheme 5.1) [12] and intermolecular hydrosilylation and desilylation reactions [13]. The latter is an exceptional example, as it not only allows the ON/OFF switching of a reaction but also offers the choice between two different reaction types.

Another elaborate example is the incorporation of an aromatic imidazolium moiety into the bridge position of diarylethene **4o** reported by Kawai and

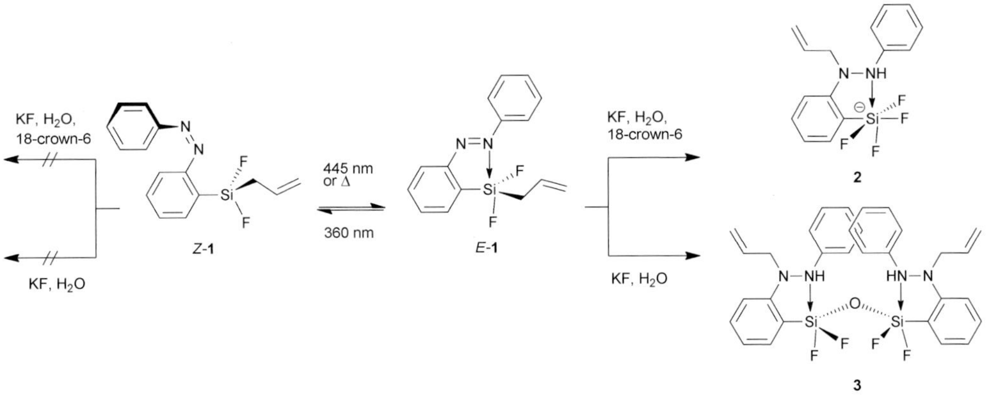

Scheme 5.1 Photocontrolled allyl transfer reaction of silyl azobenzenes with fluoride yielding hydrazine rearrangement products [11].

coworkers (Scheme 5.2) [14]. Due to its aromatic stabilization, the imidazolium ion is not prone to nucleophilic addition of methanolate anions. However, upon photocyclization to **4c**, the bridge moiety is transformed into an imidazolinium ion removing the crucial double bond as well as the aromatic stabilization. The closed-form switch can then react reversibly to yield the methanol adduct **5**.

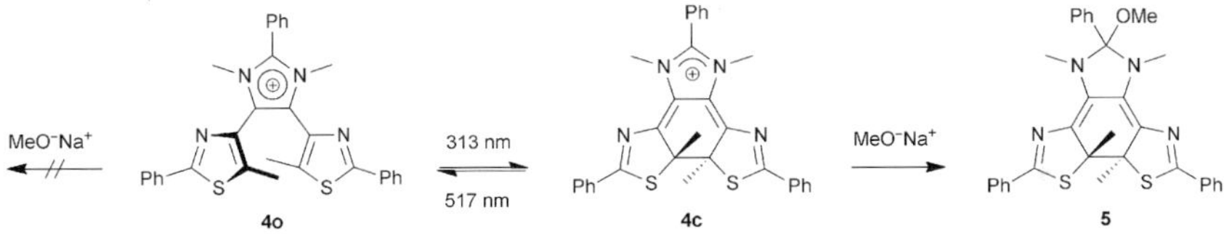

Scheme 5.2 Photocontrolled addition of methanol to imidazol(in)ium ions embedded in a diarylethene framework [14].

Kawai and coworkers further exploited the combination of photoswitching the possibility of aromatic stabilization to benzothienyl-substituted terarylenes [15]. The crucial methyl substituent in 2-position of the aryl moieties was omitted on one benzothiophene and replaced by a methoxy group on the other. UV-induced photocyclization of this switch to the dihydrophenanthrene derivative enables the elimination of methanol under acidic conditions while generating the aromatically stabilized phenanthrene derivative. This system could be tuned to the point where the authors were also able to achieve photocontrol over the elimination of ethanol and reversible nucleophilic addition/elimination cascades [16].

Photomodulation of nucleophilicity could be achieved by Branda and coworkers, who reported diarylethene **6** incorporating a pyridyl substituent on one aryl

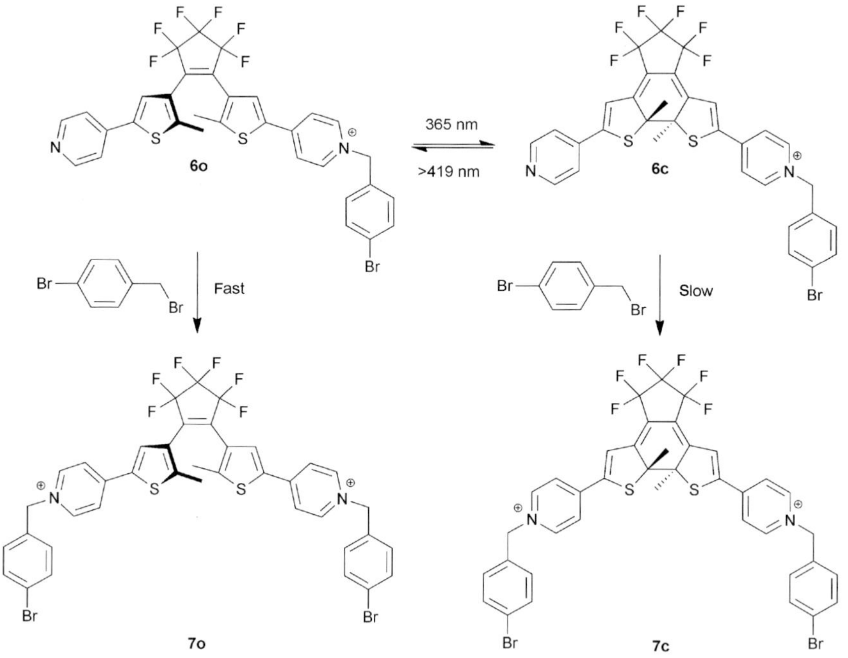

Scheme 5.3 Photomodulation of nucleophilicity by switchable coupling to an acceptor moiety through a diarylethene bridge [17].

moiety and an electron-withdrawing pyridinium group on the other (Scheme 5.3) [17]. In the ring-open form **6o,** the two aryl moieties are not in conjugation and thus the pyridyl-lone pair's nucleophilicity is unbiased in a reaction with *p*-bromobenzyl bromide to **7o**. However, upon UV-induced photocyclization of **6o** to **6c,** a conjugation pathway between the two aryl moieties is opened up and electron density is removed from the pyridyl-lone pair, lowering its nucleophilic character and hampering the reaction with *p*-bromobenzyl bromide to **7c** severely.

In the same manner, an example of switching on the starting materials of a chemical reaction was reported by Moritomo *et al.* in 2011 (Scheme 5.4) [18]. They synthesized a phenylacetylene-substituted DTE **8o** that upon photocyclization to the ring-closed form **8c** coupled an electron-donating dimethylaminophenyl residue into the conjugated π-system of the molecule rendering the alkyne moiety reactive toward the addition of tetracyanoethylene yielding product photoswitch **9c**. The ring-open form **8o** was not found to add tetracyanoethylene under these conditions. However, **9c** exhibited thermal reversibility and thus yielded the ring-open product **9o** after thermal equilibration. Even though it was found that **9o** could undergo UV-induced cyclization to **9c** again, it was not possible to exploit that behavior for further control over the reaction as the addition of tetracyanoethylene to **8c** was found to be irreversible.

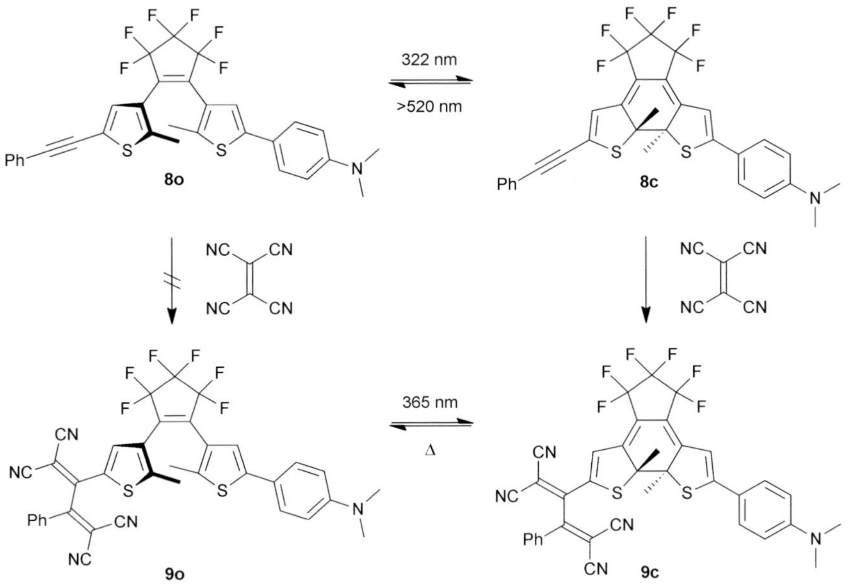

Scheme 5.4 Photoswitching of the nucleophilic properties of an acetylene moiety by coupling to an electron-donating residue through a diarylethene bridge [18].

5.3.2
Product Control

In an analogous manner to switching ON the starting materials of a covalent chemical reaction, switching OFF a reaction product crucially demands for the thermal reaction being in a dynamic equilibrium. Only in this way, reaction products can be removed from the initial reaction mixture and locked in an unreactive state (Figure 5.4).

Branda and coworkers synthesized a number of pro-diarylethenes incorporating a *cisoid* diene moiety that can undergo a Diels–Alder reaction with a corresponding dienophile. Upon adduct formation, a 6π-electron system

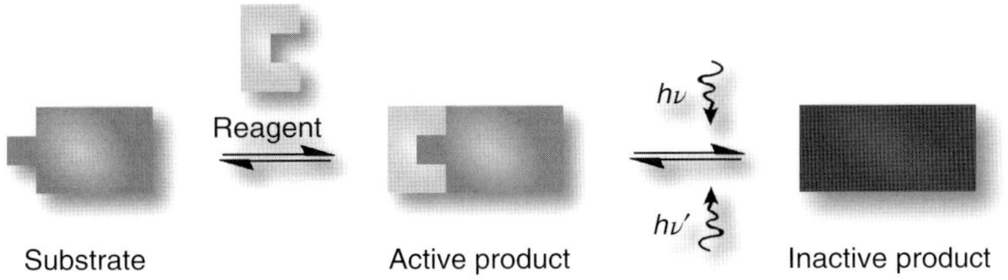

Figure 5.4 Concept of product control: photoswitching a product to an inactive "locked" form removes it from a dynamic covalent equilibrium, while switching it back into its active "unleashed" form and re-introduces it to the system. Reproduced from Ref. [5] with permission from The Royal Society of Chemistry.

is generated and photocyclization can be induced by UV-light removing the diarylethene adduct from the chemical equilibrium and locking it in an unreactive state.

The first work employing this concept relied on a hexadiene motif in the bridge position of a pro-diarylethene that could undergo a Diels–Alder reaction with maleic anhydride to yield the target photoswitch [19]. Since the equilibrium was shifted to the product side, this system proved to be especially useful for gating photochromism through chemical reactivity. However, the potential of the overall concept was quickly recognized and a publication followed that evolved the bridge motif of pro-diarylethene **10** to a photochromically more advantageous fulvenyl moiety [20]. Reaction of this pro-photoswitch with dicyanofumarate yielded the respective diarylethene **11o** that could undergo photocyclization to **11c**, locking it in its unreactive state (Scheme 5.5).

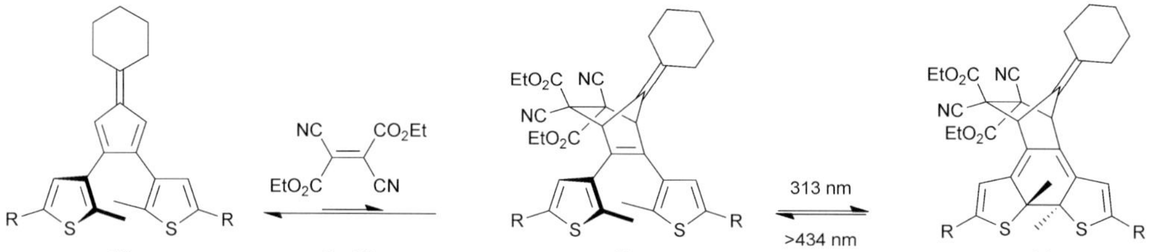

10 **11o** **11c**

Scheme 5.5 Photoswitching locks the Diels–Alder adduct (**11o** → **11c**) and removes it from the dynamic equilibrium (**10** → **11o**) [20].

However, the equilibrium between diene and dienophile was lying strongly on the starting materials side at room temperature. This allows for exploitation of the design for photorelease of the dicyanofumarate from the pure locked Diels–Alder adduct **11c**, which was isolated beforehand. Only recently, the scope was expanded by incorporating a furyl moiety in the bridge position as the diene in pro-DTE **12** enabling the use of maleimide derivatives as dienophiles (Scheme 5.6) [21] Since the furan–maleimide couple exists in a highly reversible

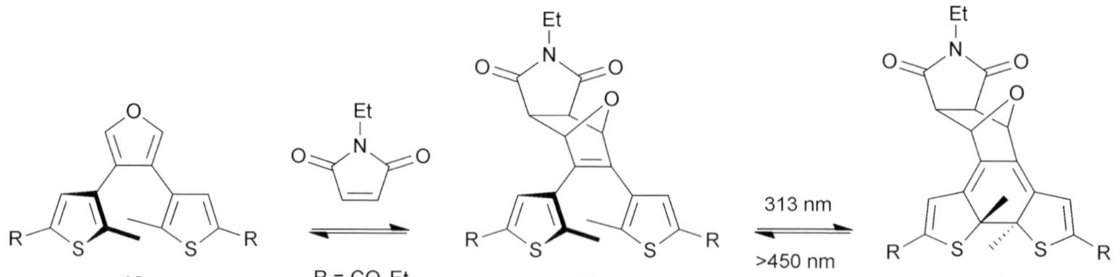

12 **13o** **13c**

Scheme 5.6 Photoswitching locks the Diels–Alder adduct (**13o** → **13c**) and removes it from the dynamic equilibrium (**12** → **13o**) [21].

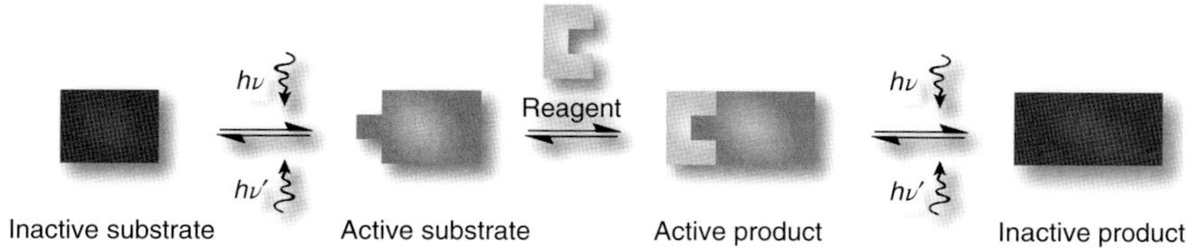

Figure 5.5 Concept of starting material and product control: photoswitching starting material as well as product to an inactive "locked" form removes them from a dynamic covalent equilibrium, while switching them back into their active "unleashed" forms re-introduces them to the system.

regime at ambient conditions, the system could be exploited far beyond the simple photorelease from adduct **13o** to **13** adding new potential applications for the light-controlled adduct locking in materials sciences and biological environments. Consequentially, the motif was exploited to design an elaborate photoswitchable adhesive relying on the switching ON of the retro Diels–Alder reaction [22].

5.3.3 Starting Material and Product Control

When starting materials and reaction products are photoreactive, it is possible to switch both components of the reaction ON or OFF enabling complete photocontrol over a reaction that exists in dynamic equilibrium (Figure 5.5). In an analogous manner to the examples presented earlier, the starting materials can be locked in an unreactive state preventing them from participation in the reaction. Transformation of the photoreactive product to an inactive state furthermore removes it from the dynamic equilibrium preventing the back-reaction.

Göstl and Hecht expanded the scope of the photoswitchable Diels–Alder reaction established by Branda and coworkers by designing a system in which the starting materials and the products were photoactive diarylethenes (Scheme 5.7) [23]. A furyl-substituted diarylethene **14o** was synthesized that could reversibly undergo the Diels–Alder reaction with maleimide to yield adduct **15o**. Either ring-open switch **14o** could be cyclized to **14c** inhibiting the conversion to the products or ring-open Diels–Alder adduct **15o** could be cyclized to **15c** inhibiting the retro Diels–Alder reaction. Furthermore, by performing irradiation with multiple wavelengths *in situ*, this dual-switching behavior could either amplify or inhibit the conversion of the Diels–Alder reaction by removing the respective starting material **14c** or product **15c** from the ground-state thermal equilibrium between **14o** and **15o**. Through this versatility, the developed system becomes practical for a broad range of applications where control over covalent connection or disconnection is demanded, such as the reversible crosslinking of polymers, the covalent functionalization of sp^2-carbon allotropes, or the photogated release of dienophiles.

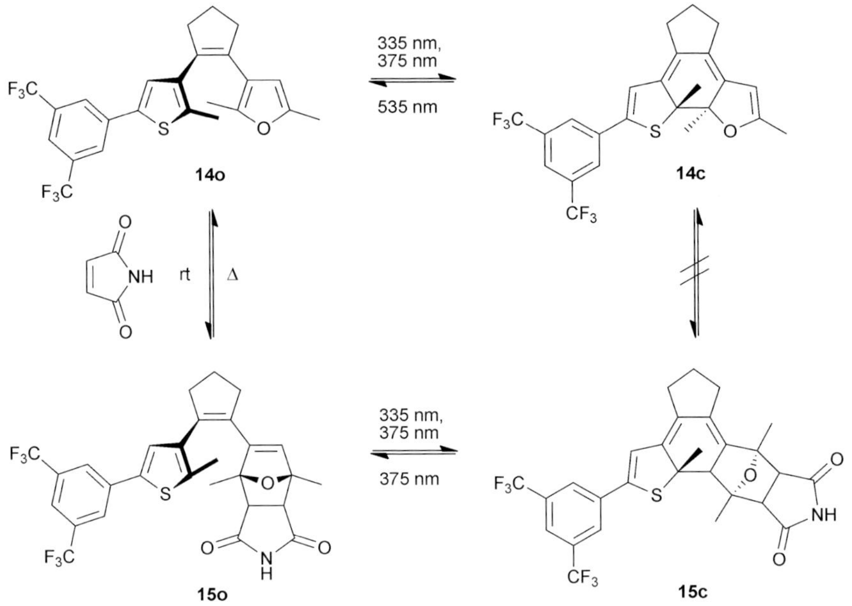

Scheme 5.7 Photoswitching locks either the unreacted switch (**14o** → **14c**) or the Diels–Alder adduct (**15o** → **15c**) and thus removes the respective component from the dynamic equilibrium (**14o** → **15o**) allowing for amplification or inhibition of the Diels–Alder reaction [23].

5.3.4
Template Control

Substrate switching offers convenient possibilities to enable or disable a chemical reaction. However, since the photoswitchable unit has to be incorporated into one of the reaction partners, its inherent disadvantage lies in the lack of flexibility in choosing the substrates. Template switching addresses this issue by outsourcing the photocontrollable unit from the substrate to an external molecule that pre-arranges the reactive centers of the starting materials so that a previously inhibited reaction can take place (Figure 5.6).

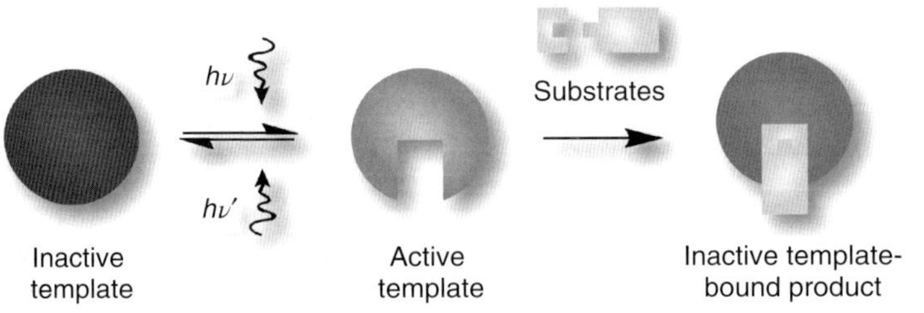

Figure 5.6 Concept of template control: photoswitching a template between inactive and active forms allows for control of conversion of starting materials to products. Reproduced from Ref. [5] with permission from The Royal Society of Chemistry.

The first photoswitchable template was synthesized as early as 1995 by Würthner and Rebek [24]. They designed a complex biscarbazolylimide terminated azobenzene **16** bearing four imide arms (Scheme 5.8). In the *E*-form, this template is inactive and shows no conversion in the coupling reaction between aminoadenosine **17** and *p*-nitrophenyl ester **18** to amide **19**, whereas upon UV-induced isomerization to the *Z*-**16**, the formed cavity acts as ideal host to promote the coupling reaction. However, the coupling product remains bound to the host upon bond formation, thereby not completely fulfilling the ambitious task of designing a photoswitchable catalyst. The authors later showed that product inhibition as well as inhibition by competitive binders are observed during the reaction, and neither photochemical nor thermal $Z \rightarrow E$ isomerization could be induced to the host while bound to the coupled product [25].

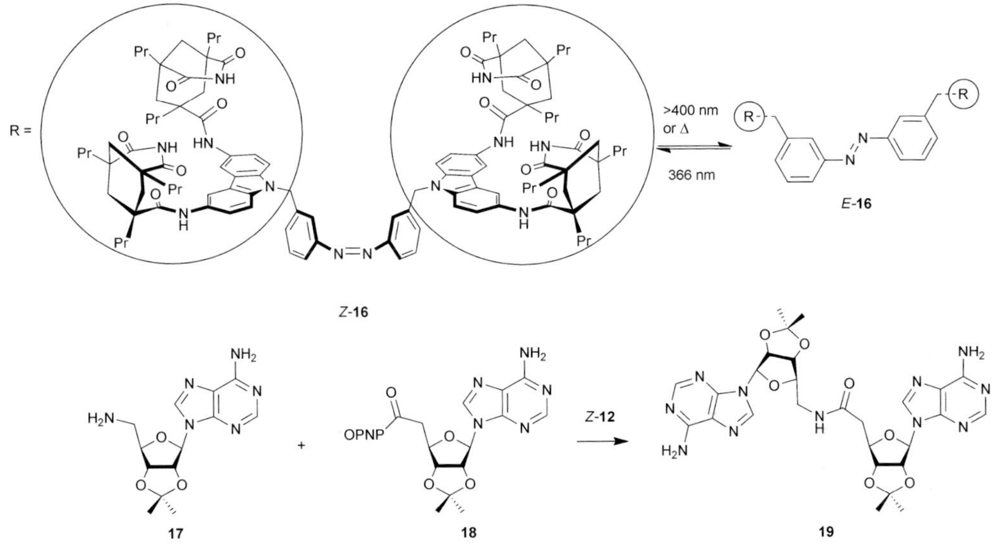

Scheme 5.8 Photoswitchable template **16**, which in its *Z*-isomer enabling the coupling of amine **17** and activated acid **18** to amide **19** [24].

Switching of the template can, in a broader sense, also be achieved by manipulating the track of a molecular walker by light [26]. Leigh and coworkers synthesized walker-track conjugate **20** offering different possibilities for hydrazone and disulfide binding (Scheme 5.9). The key step in the movement of the walker is the isomerization of the stilbene moiety incorporated in the center of the track. By UV-induced isomerization from *E*-1,2-**20** to *Z*-1,2-**20**, the equilibrium between disulfide formation on 1-position and 3-position is highly biased toward the latter position and the walker moves along the track to form *Z*-2,3-**20**. Subsequent visible-light-induced back-isomerization from *Z*-2,3-**20** to *E*-2,3-**20** shifts the equilibrium between the hydrazone formation on 2- and 4-position toward 4-position to form *E*-3,4-**20**. Overall, the walker was moved successfully along

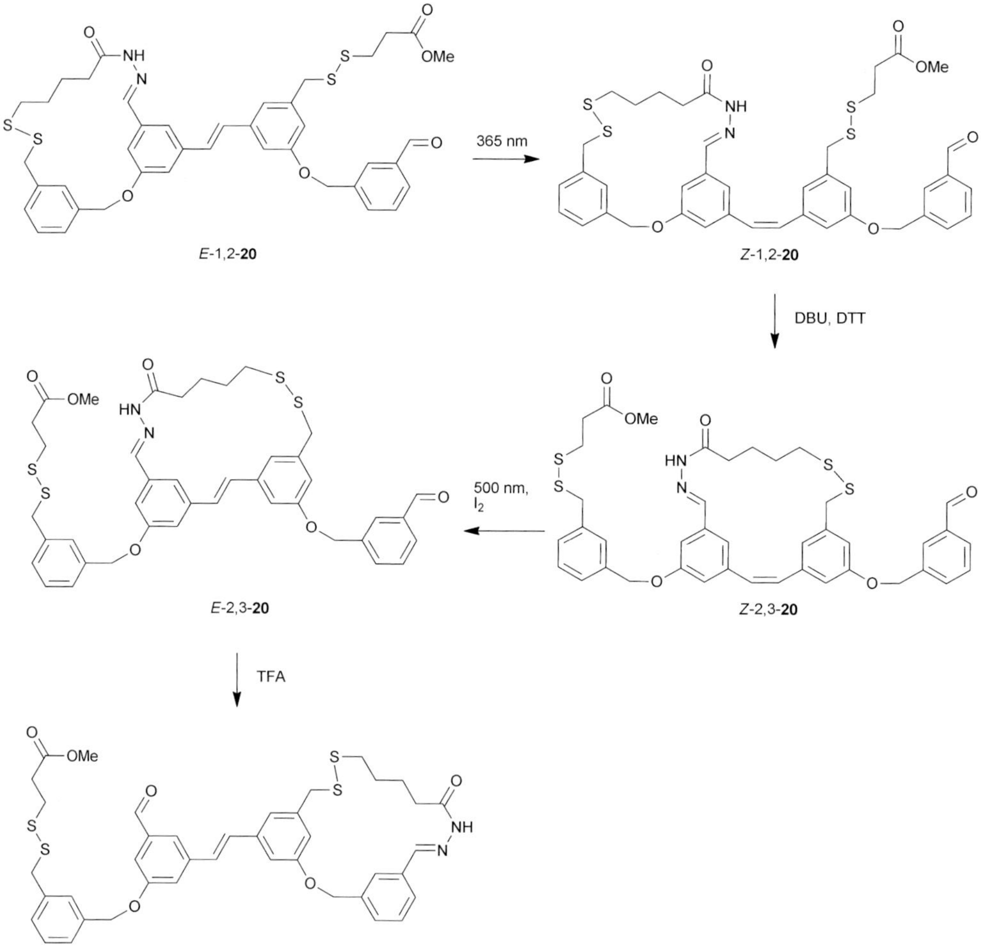

Scheme 5.9 In a molecular walker–track conjugate, a photoswitchable stilbene moiety in the track guides the walker along the track by biasing the individual dynamic covalent equilibria for disulfide and hydrazone formation [26].

the track by dynamically breaking and forming covalent bonds and biasing the associated equilibria by photoisomerization.

Apart from directly switching the template ON or OFF, the assembly of the template itself can be dynamic and controlled by light (Figure 5.7). As opposed to the switching of a stable and inert template molecule, the formation of the template itself can be rendered photoswitchable by coupling the template assembly to dynamic covalent chemistry. Typically, a dynamic constitutional library consists of a large number of equilibrated host structures, derived from a number of building blocks connected via reversible covalent bonds, and the equilibria present in the library can be biased toward one or few particular structures by stabilizing noncovalent interactions with a certain guest molecule,

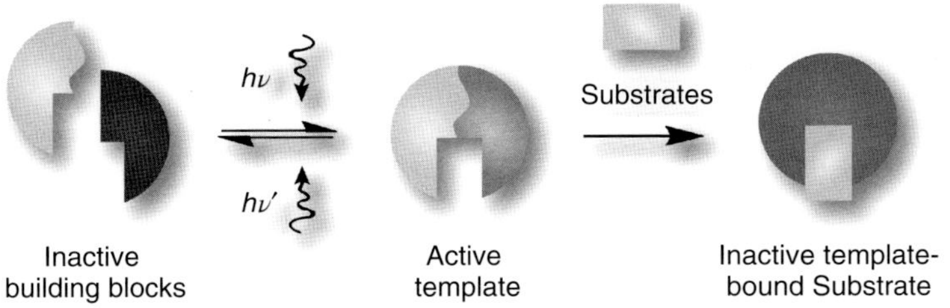

Figure 5.7 Concept of dynamic template control: photoswitching a building block of a dynamic constitutional library between reactive and nonreactive forms allows for the assembly of an active template, which facilitates the conversion of starting materials to products. Reproduced from Ref. [5] with permission from The Royal Society of Chemistry.

thereby leading to a selection and amplification of the best binder. If one of the building blocks of the library consists of a photoswitch, one should be able to modulate the initial composition of the unbiased library and also the selection and amplification process by light.

Such system has been described by Waters and Ingerman, who employed a dynamic constitutional library consisting of one azobenzene *E*-**21** and two proline (**22** and **23**) building blocks (Scheme 5.10) [27]. The library was equilibrated in the presence of an oligoproline guest and yielded a distinct distribution of host molecules. However, upon UV-induced isomerization of building block *E*-**21** to *Z*-**21**, the distribution among the possible macrocycles changed significantly, therefore amplifying macrocyclic *E*-**21** by more than 60%. This shows that in addition to switching the template ON or OFF, it is also possible to switch between different receptor states.

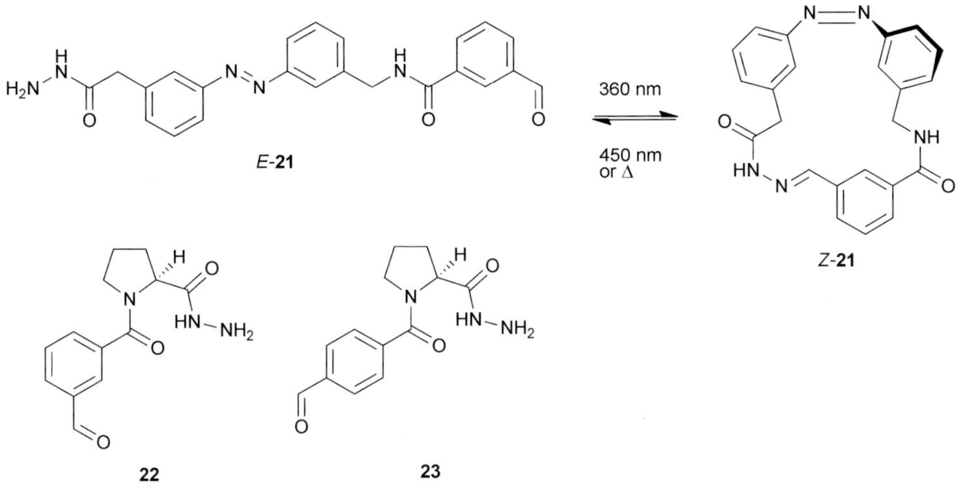

Scheme 5.10 Photoswitching the composition of a dynamic constitutional library by incorporating a photochromic building block leads to a change of the host distribution in the presence of an oligoproline guest [27].

5.4 Photoswitchable Catalytic Processes

All of the above examples suffer from the drawback that they need a stoichiometric amount of a photoswitch to achieve complete conversion. The previously mentioned system by Würthner and coworkers, for example, shows that a template can be used to reversibly control a reaction, but due to product inhibition, a stoichiometric amount of template is needed to reach full conversion. This of course has an impact on the photoefficiency with which the reaction can take place. In order to achieve full conversion, each template molecule needs to be switched, which leads to problems when trying to address the photoresponsive groups *in situ*. Most reactions take place at a reasonable rate only at concentrations, which exceed the ones typically employed in photochemistry by several orders of magnitude. This problem can be overcome to a certain extent by using a photoresponsive catalyst, which lowers the overall concentration of the absorbing photoswitch in the reaction mixture. Furthermore, with a photoresponsive catalyst, it is possible to amplify the light stimulus in the system, since the photoreaction that is needed to isomerize the catalyst molecule triggers multiple chemical conversions (turnover). Reaction control in catalysis can be divided into two approaches: one is to control the activity of the catalyst and the other is controlling the selectivity. In this review, both of these concepts are covered.

5.4.1 Activity Control

Almost all photoswitchable catalyst systems reported up-to-date focus on modulating the catalytic activity. For this, a catalytically active molecule is functionalized with a photoswitchable entity. Ideally, the difference in catalytic activity between the two forms of the photoswitch varies significantly enough, so it leads to an ON/OFF-switching of the catalyst (Figure 5.8). There are several strategies that can be employed to achieve control over the activity of a catalyst. Herein, we focus on the control of cooperative effects, steric shielding of the

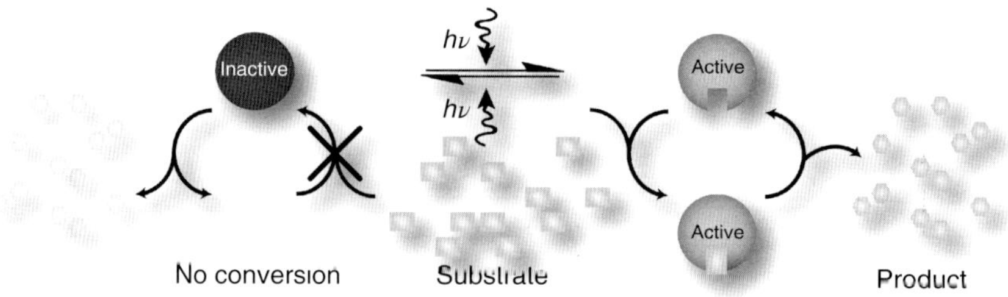

Figure 5.8 Concept of photoswitching the activity of a catalyst: photoswitching converts a catalyst from an inactive to an active form, which turns substrate over to product, while the inactive form shows no conversion (turnover). Thereby, one switching event can lead to the formation of many product molecules (amplification). Reproduced from Ref. [5] with permission from The Royal Society of Chemistry.

active site, and modulating the electronic properties of a catalyst, which are the more promising approaches. For additional approaches using photochromic effector molecules, activating or inhibiting the catalyst system, the reader is referred to previous reviews [6].

Cooperative effects play an important role in biological and synthetic catalytic systems and have therefore also been exploited to photomodulate reactivity. The basic principle relies on a large geometrical change during the isomerization process that allows the variation of the distance between to catalytically active sites. Cacciapaglia *et al.* reported the first successful use of a photoswitchable cooperative effect in catalysis [28]. The *bis*-barium complex of an azobis(benzo-18-crown-6) ether **24** was used to catalyze a basic ethanolysis of tertiary anilides (Scheme 5.11). By reversible *E*/*Z*-isomerization of the azobenzene spacer, the catalytic activity could be phototuned. The thermodynamically more stable *E*-**24** has only a low catalytic activity. Photoswitching the azobenzene moiety into the *Z*-form changes the geometry of the *bis*-barium complex into a more favorable concave conformation in which the two barium centers are in a close proximity to each other. In the catalytically active complex, one barium center serves as a binding site for a carboxylate-anchoring group on the anilide substrate while the other barium center binds a nucleophilic ethoxide ion. The close proximity of these two pre-organized starting materials gives rise to an increased catalytic activity of the *Z*-isomer.

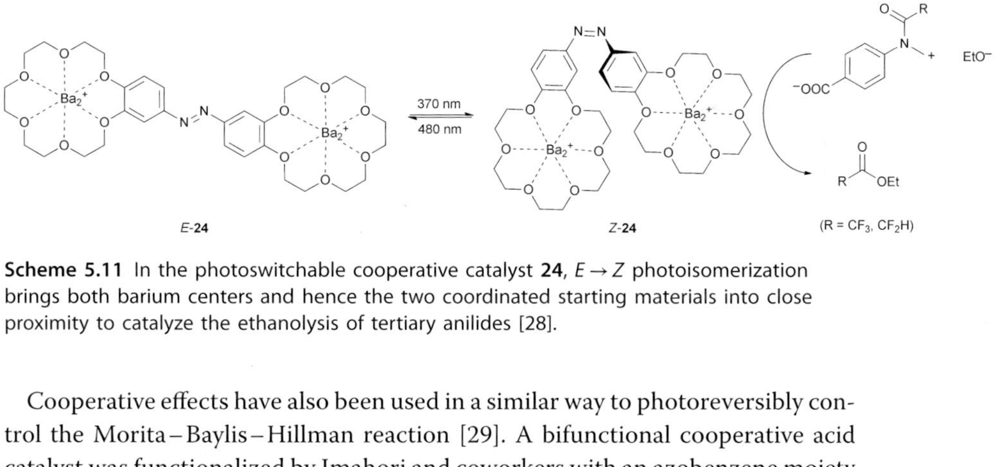

E-**24** *Z*-**24**

Scheme 5.11 In the photoswitchable cooperative catalyst **24**, $E \rightarrow Z$ photoisomerization brings both barium centers and hence the two coordinated starting materials into close proximity to catalyze the ethanolysis of tertiary anilides [28].

Cooperative effects have also been used in a similar way to photoreversibly control the Morita–Baylis–Hillman reaction [29]. A bifunctional cooperative acid catalyst was functionalized by Imahori and coworkers with an azobenzene moiety to activate or deactivate the cooperative effect, which led to a reversible control of the reaction rate.

Steric effects can influence the activity as well as selectivity of a catalyst for a given reaction. To gain control over the activity of a catalyst using this concept, a photoswitch is needed that induces a large geometrical change to shield or deshield the substrate binding site of the catalyst. An approach where this concept was successfully used has been reported previously by our group [30].

The lone pair of the *N*-alkylated piperidine base **25** can reversibly be shielded by *E*/*Z*-isomerization of a rigidly connected azobenzene "wiper" (Scheme 5.12). It was shown by titration experiments that the piperidine base has a lower basicity in the resting state *E*-**25**. Switching the azobenzene moiety into the *Z*-conformation deshields the piperidine electron lone pair, which led to an increase in basicity. The different basicities were used to photocontrol the conversion of a base-catalyzed nitroaldol (Henry) reaction. Later on, a related catalyst was immobilized on various solid surfaces, such as silica gel or silicon wafers, to prevent the loss of spatial resolution due to diffusion of the catalyst in solution [31].

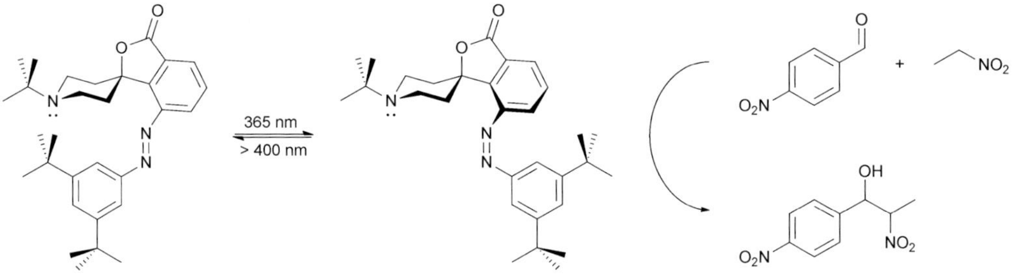

E-**25** *Z*-**25**

Scheme 5.12 *E/Z*-photoisomerization of an azobenzene moiety controls the accessibility of a piperidine base, which can be used to catalyze a nitroaldol (Henry) reaction [30].

Electronic effects can significantly influence the active site, and hence electronic fine-tuning constitutes one of the main strategies in catalyst design. Therefore, it is no surprise that efforts were put into photomodulating the electronic properties of a catalyst. The basic concept is founded on breaking or forming a conjugated system between the active site and an electronically activating group. Recently Bielawski and Neilson used this concept to control the activity of a catalyst [32]. An *N*-heterocyclic carbene (NHC) functionality was incorporated into the backbone of a DTE backbone **26** (Scheme 5.13). The length of the conjugated π-system modulates the electronic properties of the NHC functionality. In the

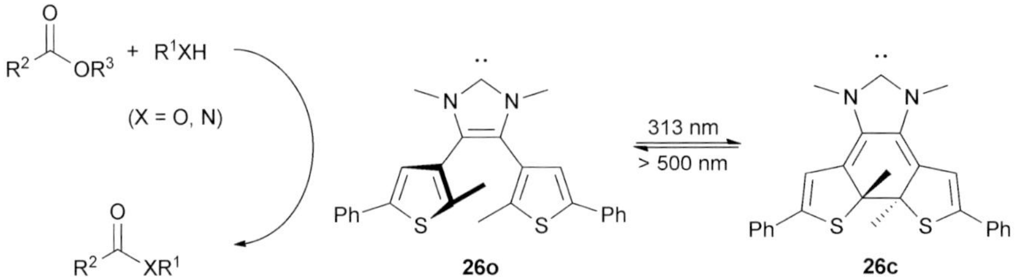

26o **26c**

Scheme 5.13 Ring-open *N*-heterocyclic carbene (NHC) **26o** catalyzes transesterification, amidation, and ring-opening polymerization reactions, yet upon irradiation its corresponding ring-closed isomer **26c** exhibits significantly reduced catalytic activity [32, 33].

presence of visual light and a base, the NHC **26o** catalyzes transesterification and amidation reactions. Upon irradiation with UV light to the ring-closed derivative **26c**, the rate of a transesterification and an amidation reaction was significantly decreased. This process was reversibly switched several times between a slow and a fast reaction rate. On the basis of NMR experiments with an isotopic label at the C2 "carbene" carbon, the authors could rationalize the observed activity differences by showing that the ring-open form **26o** exists as an imidazolium species while the ring-closed species **26c** forms the less active alcohol adduct.

This photomodulation of the nucleophilicity of the NHC center was further used to switch the activity of a Rh(I)-complex **27** (Scheme 5.14) [34]. It was shown that the Rh-metal center functionalized with the photochromic NHC ligand **27o** can catalyze the hydroboration of various alkenes, such as styrene, with modest activity differences between the two different switching forms. In this case, the rate-determining reductive elimination step was attenuated by the lower donor capability of the NHC ligand in its ring-closed form.

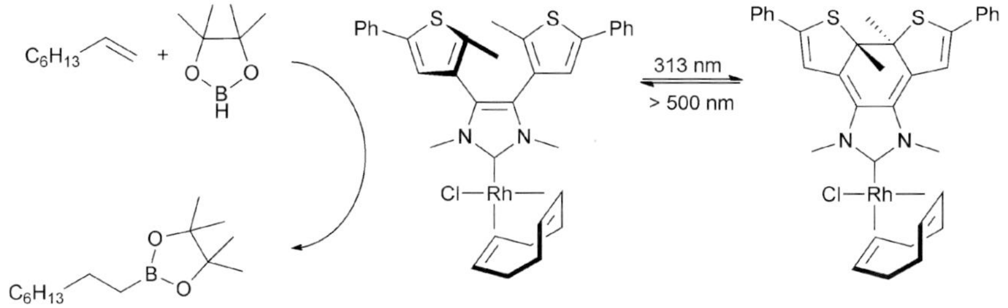

Scheme 5.14 Photoswitching of a dithienylethene-based *N*-heterocyclic carbene (NHC) ligand modulates the activity of the derived Rh(I)-complex in the hydroboration of styrene [34].

More recently, Bielawski and coworkers were also able to show the first photoswitchable attenuation for the ring-opening polymerization (ROP) of δ-valerolactone as well as ε-caprolactone using their DTE-based NHC **26** (see Scheme 5.11) [33].

5.4.2
Selectivity Control

The previous examples impressively show how much progress has been made in the field of photoinduced activity control of catalysts. Although efforts on activity control have thus far been the main focus, approaches to photoswitch the selectivity of catalysts are equally important. For this, a catalytically active entity is combined with a photoswitch that leads to a difference of chemo-, regio-, and stereo-selectivity between the two forms (Figure 5.9). Until now, efforts have been limited to reversible switching of the catalyst's chirality, thereby trying to modulate its stereoselectivity.

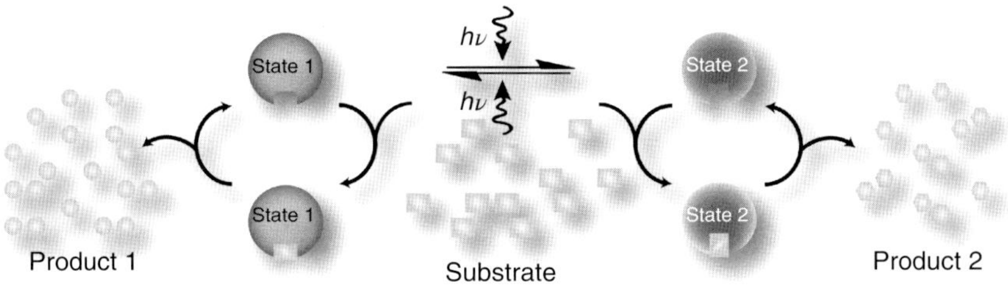

Figure 5.9 Concept of photoswitching the selectivity of a catalyst: photoswitching interconverts a catalyst between two forms exhibiting different selectivity in a given transformation. Again, one switching event can lead to the formation of many product molecules (amplification). Reproduced from Ref. [5] with permission from The Royal Society of Chemistry.

The first successful approach to photochemically switch the stereoselectivity of a catalyst was reported by Branda and coworkers [35]. Their DTE-based chiral bis(oxazoline) ligand **28** is only able to chelate to the catalytically active copper center in its more flexible ring-open form (Scheme 5.15). This complexation generates a chiral environment around the copper center, thereby allowing a cyclopropanation reaction to take place stereoselectively. Hence, in the open-form **28o,** an *ee* of 30–50% was observed, whereas the closed-form **28c**, where the rigidity of the ligand prevents chelation, only showed a very low *ee* of 5%. Irradiating the sample with visible light led to recovery of the original chiral information and gave an ee of 11–37%.

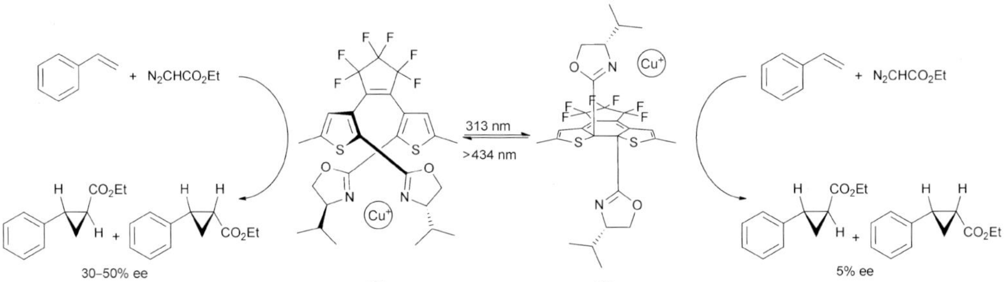

28o **28c**

Scheme 5.15 Photoswitching of a dithienylethene-based bisoxazoline ligand leads to modulation of its chelation ability and hence chirality of the corresponding copper complexes, which display different degrees of stereoselectivity for the cyclopropanation of styrene [35].

Feringa and Wang combined both of the two concepts of photoswitching activity and selectivity (Scheme 5.16) [36]. For this purpose, their rotatory molecular motor was transformed into a photoswitchable bifunctional organocatalyst by attaching a Brønsted base and a thiourea hydrogen-bonding donor group, which are known to cooperate in the catalysis of Michael additions, among other reactions. The thermodynamically stable (*P*,*P*)-*E*-**29** isomer shows a negligible

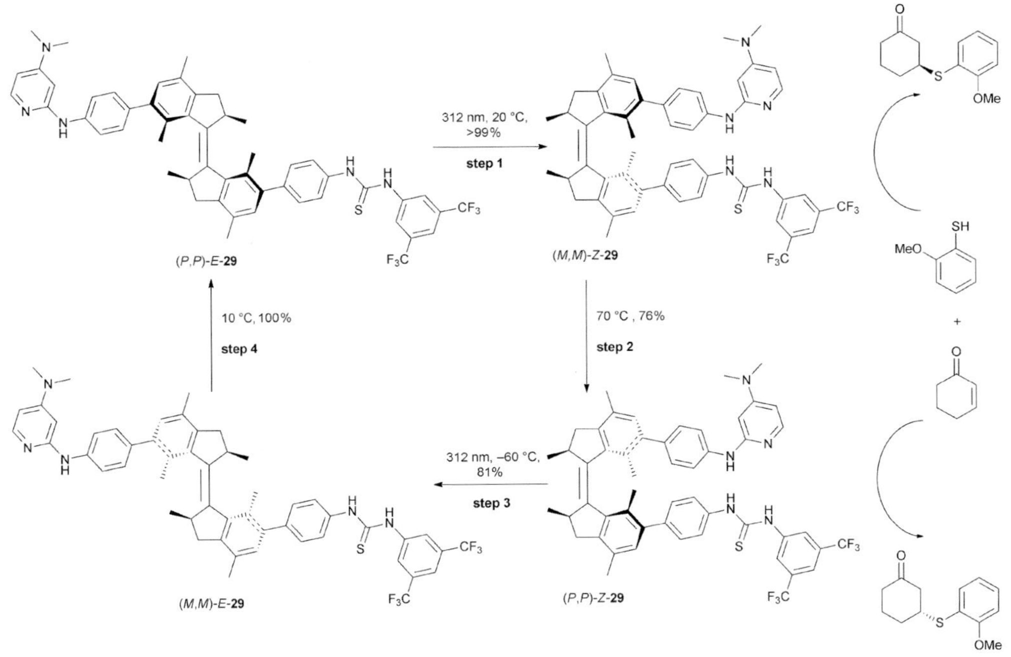

Scheme 5.16 Modulation of the relative orientation of a pyridine basic and a thiourea hydrogen-bonding site embedded in a molecular motor leads to photoswitchable bifunctional organocatalyst, which allows for control over the activity and stereoselectivity of a Michael reaction [36].

catalytic activity with no stereoselectivity (*e.r.*, S:R = 49 : 51) in the Michael addition. However, upon irradiation, helix inversion takes places and the (*M*,*M*)-*Z*-**29** isomer is formed, which shows a higher activity in the Michael addition and forms the product in considerable enantiomeric excess (*e.r.*, S:R = 75 : 25). Heating the (*M*,*M*)-*Z*-**29** isomer to 70 °C triggers a thermal isomerization step that forms the (*P*,*P*)-*Z*-**29** isomer, which also catalyzed the Michael addition but yielding the opposite stereoisomer (*e.r.*, S:R = 23 : 77). Subsequent photochemical and thermal isomerizations give rise to reformation of the original (*P*,*P*)-*E*-**29** isomer via (*M*,*M*)-*E*-**29**.

5.5 Outlook

In this chapter, we have summarized the existing body of literature in the field of photoswitchable gating of chemical reactions and arranged the examples according to the different conceptual approaches explored thus far. The various designs have been divided into two main categories depending on the stoichiometric or substoichiometric use of the photoswitch (and hence incoming photons). After discussing all of the specific examples, we want to seize this opportunity to point out some general drawbacks, emphasize remaining challenges, and highlight opportunities. Please be aware that this future vision of the field is naturally based on our personal and current perspective [5]. To realize the full potential of photoswitches to gate chemical processes, there are a number of critical issues that have to be addressed:

***In situ* switching** has to be accomplished by carefully engineering the entire reaction set-up. The reader will have noticed that most studies are typically performed using both forms of the photoswitch for reactivity studies only after they have been isolated. While such *ex situ* experiments establish the theoretically achievable ON/OFF reactivity ratios of both switching forms, the practically accessible window may be much narrower due to nonquantitative photochemical interconversion yielding an isomer mixture at the PSS. This resulting "window of opportunity" has to be used in real *in situ* switching experiments to exploit the desired advantages of "remote-controlling" chemical reactions.

Concentration effects limit the applicability of light in bulk and solution. According to the Beer–Lambert law, photochemical processes work best in diluted solution, that is, on the micromolar scale, as in this way the total absorption phenomena are avoided and all compartments of the sample are irradiated uniformly. However, regular ground-state chemical reactions usually follow bimolecular kinetics and thus become increasingly slow upon dilution of the reaction mixture beyond the millimolar scale. To perform reactions in a concentration regime suitable for bimolecular processes, several parameters have to be optimized to assure for instant and complete exposure of the entire sample volume to light. For example, both the absorbance (dictated by the concentration of the absorbing photoswitch and its molar absorptivity) at the irradiation wavelength (potentially exciting at the absorption edge, not the maximum) and the sample thickness should be minimized to allow for maximum light penetration. While the first aspect clearly shows the advantage of using catalytic approaches because the amount or loading of absorbing photoswitch can be much lower, the second aspect illustrates the attractiveness of working with monolayers at interfaces, for example, SAMs. Photoswitching bimolecular connections would enable the control over polymerization processes, for example, by the covalent connection (and disconnection) of monomers to linear chains or crosslinking of polymer chains into networks. However, thus far there has been only Bielawski's work on photoswitchable catalysts for the ROP of lactones. Another important area would be the control over covalent surface functionalization, for example, to allow for dynamic patterning of SAMs. In this context, the work by Kawai on photoswitching reversible NHC-adduct formation (compare Scheme 5.2) could prove useful [14–16]. Without doubt, such photoswitchable bimolecular connection reactions will provide ample opportunities in the field of soft matter materials science.

Increasing the overall efficiency and robustness of the light-controlled chemical process is an important ongoing effort to design photochromes with ever-improving switching characteristics. The composition at the PSS dictates the population difference of both switches associated with different reactivities. Therefore, to optimize the photoswitching behavior and achieve quantitative photoconversions, the absorption spectra of both switching forms have to be separated to assure for selective excitation, and the reaction quantum yields have to be maximized. In addition to improving the efficiency of the switching process, which also enhances the system's sensitivity in particular when coupled to a catalytic amplification mechanism, it is important to design reliable and hence robust switches. The necessary high reversibility of both photochemical reactions associated with the photochromicity requires that all other competing side reactions, in particular if irreversible (the so-called fatigue), have to be shut down effectively. Although there has been lots of progress made over the years designing fatigue-resistant switches in the context of optical data storage [7e], we want to emphasize here that there is an ongoing need to improve photochromic performance with regard to this key criterion, which is often overlooked in academic research. Last but not least, we would like to point out that T-type photochromes, which typically are not valued highly by most people in the community, are very interesting as gates. By engineering the half-lives of their metastable form, which we assume to exhibit higher reactivity (ON), to allow for the chemical transformation yet otherwise reverting quickly back to the less reactive (OFF) form, only one light source in needed to control the reaction. Such combination of light-induced ON-switching and fast thermal OFF-switching should prove increasingly important for many applications (even beyond the gating of reactivity).

Maximizing the reactivity differences of the two switching forms is equally important to optimizing their population changes accessible by illumination. This aspect is perhaps the least straightforward to engineer and is primarily tied to the creative force of the chemist. With regard to the stoichiometric use of photochromes, it is rather obvious to integrate it into a chemical reaction, which is sensitive to the structural or geometrical changes during the photochromic reaction. Due to the available portfolio of photoswitches and the myriads of known chemical reactions, there clearly is lots of room to be explored along this broad direction. More difficult to achieve and much harder to predict is the interaction of photochromes with catalysts. The most obvious approach will be based on an established catalyst structure, ideally with a mechanism known in great detail, and merging it with a known photoswitch. From our own experience and keeping in mind that a low catalyst loading is necessary to enhance the penetration of light through the sample, we have found it more promising to render a highly active catalyst less active (ON → OFF) as compared to the opposite approach (OFF → ON) [30]. However, the reader should be reminded that there is plenty of room for higher reactivity differences (ON/OFF-ratios) and qualitatively new creative designs are needed.

At this point, we want to return to our initial motivation to develop photoswitchable reactive systems as "remote-controlled synthetic tools" and critically reflect on their (potential) capability to carry out chemical processes with **high spatial and temporal resolution**. Clearly, the achievable spatial resolution has physical limitations inherent to the wavelength of the light used, and it furthermore requires "defocusing" pathways, such as delocalization of excitation as well as diffusion, to be shut down. The first point is most readily achieved by removing, or at least sufficiently separating, all possibly interfering energy and electron-transfer quenching pathways. The latter point necessitates immobilization of (one of) the reactants (and the product) either directly or via the photoswitch or by the absence of any solvent. Taking these aspects into consideration, patterning of surfaces or even structuring in 3D with spatial resolution characteristic for (nonlinear) optical techniques should be achieved. While the spatial resolution achievable with light is limited to length scales much larger than individual molecules, the temporal resolution provided by modern laser technology easily allows that individual chemical reactions can be followed in real time. Therefore, the potential of the "remote controls" described herein lies perhaps more in their superior temporal resolution, which in principle allows to *time* individual successive reaction steps. Such unprecedented control over the sequence of chemical events should have a large impact on our ability to prepare complex chemical structures, in particular macromolecules. This is perhaps most readily illustrated by considering a photoswitchable polymerization catalyst. By toggling between its ON and OFF states, one should be able to program a sequence of blocks with exact control over their individual length (activity), their monomer composition (chemoselectivity), and their tacticity (stereoselectivity), thereby creating completely new and otherwise inaccessible polymer architectures.

When considering the opportunities of this field, it is clear that we have to continue to design improved photoswitchable reactive systems and exploit their unique features to precisely and externally control chemical processes in time and space. Such high-resolution, remote-controlled chemical tools will prove extremely powerful in basic research, ranging from synthetic chemistry to cellular biology, and have tremendous impact on future materials applications.

References

1. (a) For an authorative review on single molecule chemistry conducted in the STM, see: Ho, W. (2002) Single-molecule chemistry. *J. Phys. Chem.*, **117**, 11033–11061; (b) Representative examples include: Single covalent bond formation: Hla, S.-W., Bartels, L., Meyer, G., and Rieder, K.-H. (2000) Inducing all steps of a chemical reaction with the scanning tunneling microscope tip:

towards single molecule engineering. *Phys. Rev. Lett.*, **85**, 2777–2780 as well as polymerization: Okawa, Y. and Aono, M. (2001) Materials science: nanoscale control of chain polymerization. *Nature*, **409**, 683-684; (c) Note that within the electric field in the STM junction, multiple local processes can be induced in parallel as shown in a periodic switching array: Dri, C., Peters, M.V., Schwarz, J., Hecht, S., and Grill, L. (2008) Spatial periodicity in molecular switching. *Nat. Nanotechnol.*, **3**, 649–653.

2. An in-depth analysis of the available control stimuli is given in: Stoll, R.S. and Hecht, S. (2010) Artificial light-gated catalyst systems. *Angew. Chem. Int. Ed.*, **49**, 5054–5075.
3. Hell, S.W. (2007) Far-field optical nanoscopy. *Science*, **316**, 1153–1158.
4. (a) Originally shown in: Zhou, W., Kuebler, S.M., Braun, K.L., Yu, T., Cammack, J.K., Ober, C.K., Perry, J.W., and Marder, S.R. (2002) An efficient two-photon-generated photoacid applied to positive-Tone 3D microfabrication. *Science*, **296**, 1106–1109; (b) A review about multiphoton fabrication is given by:LaFratta, C.N., Fourkas, J.T., Baldacchini, T., and Farrer, R.A. (2007) Multiphoton fabrication. *Angew. Chem. Int. Ed.*, **46**, 6238–6258.
5. Göstl, R., Senf, A., and Hecht, S. (2014) Remote-controlling chemical reactions by light: Towards chemistry with high spatio-temporal resolution. *Chem. Soc. Rev.*, **43**, 1982–1996.
6. (a) Imahori, T. and Kurihara, S. (2014) Stimuli-responsive cooperative catalysts based on dynamic conformational changes toward spatiotemporal control of chemical reactions. *Chem. Lett.*, **43**, 1524–1531; (b) Neilson, B.M. and Bielawski, C.W. (2013) Illuminating photoswitchable catalysis. *ACS Catal.*, 1874–1885; (c) Stoll, R.S. and Hecht, S. (2010) Artificial light-gated catalyst systems. *Angew. Chem. Int. Ed.*, **49**, 5054–5075.
7. (a) Feringa, B.L. and Browne, W.R. (eds) (2011) *Molecular Switches*, Wiley-VCH Verlag GmbH, Weinheim; (b) Russew, M.-M. and Hecht, S. (2010) Photoswitches: From molecules to materials. *Adv. Mater.*, **22**, 3348–3360; (c) Hecht, S. (2005) Optical switching of hierarchical self-assembly: towards "enlightened" materials. *Small*, **1**, 26–29; (d) Dürr, H. and Bouas-Laurent, H. (eds) (2003) *Photochromism: Molecules and Systems*, Elsevier Science, Amsterdam; (e) Irie, M. (2000) Diarylethenes for memories and switches. *Chem. Rev.*, **100**, 1685–1716; (f) Crano, J.C. and Guglielmetti, R.J. (eds) (1999) *Organic Photochromic and Thermochromic Compounds*, Vol. 1: Photochromic Families, Springer, New York; (g) El'tsov, A.V. (ed) (1990) *Organic Photochromes*, Springer, New York.
8. (a) Sumi, T., Takagi, Y., Yagi, A., Morimoto, M., and Irie, M. (2014) Photoirradiation wavelength dependence of cycloreversion quantum yields of diarylethenes. *Chem. Commun.*, **50**, 3928–3930; (b) Santos, A.R., Ballardini, R., Belser, P., Gandolfi, M.T., Iyer, V.M., and Moggi, L. (2009) Photochemical investigation of a photochromic diarylethene compound that can be used as a wide range actinometer. *Photochem. Photobiol. Sci.*, **8**, 1734–1742.
9. Peters, M.V., Goddard, R., and Hecht, S. (2006) Synthesis and characterization of azobenzene-confined porphyrins. *J. Org. Chem.*, **71**, 7846–7849.
10. Kano, N., Komatsu, F., and Kawashima, T. (2001) Reversible photocontrol of the coordination number of silicon in a tetrafluorosilicate bearing a 2-(phenylazo)phenyl group. *J. Am. Chem. Soc.*, **123**, 10778–10779.
11. Kano, N., Yamamura, M., and Kawashima, T. (2004) Reactivity control of an allylsilane bearing a 2-(phenylazo)phenyl group by photoswitching of the coordination number of silicon. *J. Am. Chem. Soc.*, **126**, 6250–6251.
12. Yamamura, M., Kano, N., and Kawashima, T. (2007) Intramolecular allylation of the azo group of 2-(allylsilyl)azobenzenes and its photocontrol. *J. Organomet. Chem.*, **692**, 313–325.
13. Yamamura, M., Kano, N., and Kawashima, T. (2007) Photoswitching of the reactivity involving hydrosilylation of a 1,1,3,3-tetrahydrodisiloxane bearing

two azo groups. *Tetrahedron Lett.*, **48**, 4033–4036.
14. Nakashima, T., Goto, M., Kawai, S., and Kawai, T. (2008) Photomodulation of ionic interaction and reactivity: reversible photoconversion between imidazolium and imidazolinium. *J. Am. Chem. Soc.*, **130**, 14570–14575.
15. Nakagawa, H., Kawai, S., Nakashima, T., and Kawai, T. (2009) Synthesis and photochemical reactions of photochromic terarylene having a leaving methoxy group. *Org. Lett.*, **11**, 1475–1478.
16. Nakagawa, H., Nakashima, T., and Kawai, T. (2012) Subsequent chemical reactions of photochromic 4,5-dibenzothienylthiazoles. *Eur. J. Org. Chem.*, 4493–4500.
17. (a) Samachetty, H.D. and Branda, N.R. (2006) Integrating molecular switching and chemical reactivity using photoresponsive hexatrienes. *Pure Appl. Chem.*, **78**, 2351–2359; (b) Samachetty, H.D., Lemieux, V., and Branda, N.R. (2008) Modulating chemical reactivity using a photoresponsive molecular switch. *Tetrahedron*, **64**, 8292–8300.
18. Morimoto, M., Murata, K., and Michinobu, T. (2011) Photochemical control of a highly efficient addition reaction between electron-rich alkynes and tetracyanoethylene. *Chem. Commun.*, **47**, 9819–9821.
19. Lemieux, V. and Branda, N.R. (2005) Reactivity-gated photochromism of 1,2-dithienylethenes for potential use in dosimetry applications. *Org. Lett.*, **7**, 2969–2972.
20. Lemieux, V., Gauthier, S., and Branda, N.R. (2006) Selective and sequential photorelease using molecular switches. *Angew. Chem. Int. Ed.*, **45**, 6820–6824.
21. Erno, Z., Asadirad, A.M., Lemieux, V., and Branda, N.R. (2012) Using light and a molecular switch to 'lock' and 'unlock' the Diels–Alder reaction. *Org. Biomol. Chem.*, **10**, 2787–2792.
22. Asadirad, A.M., Boutault, S., Erno, Z., and Branda, N.R. (2014) Controlling a polymer adhesive using light and a molecular switch. *J. Am. Chem. Soc.*, **136**, 3024–3027.
23. Göstl, R. and Hecht, S. (2014) Controlling covalent connection and disconnection with light. *Angew. Chem. Int. Ed.*, **53**, 8784–8787.
24. Würthner, F. and Rebek, J. (1995) Light-switchable catalysis in synthetic receptors. *Angew. Chem. Int. Ed.*, **34**, 446–448.
25. Würthner, F. and Rebek, J. (1995) Photoresponsive synthetic receptors: binding properties and photocontrol of catalytic activity. *J. Chem. Soc. Perkin Trans.*, **2**, 1727–1734.
26. Barrell, M.J., Campaña, A.G., von Delius, M., Geertsema, E.M., and Leigh, D.A. (2011) Light-driven transport of a molecular walker in either direction along a molecular track. *Angew. Chem. Int. Ed.*, **50**, 285–290.
27. Ingerman, L.A. and Waters, M.L. (2009) Photoswitchable dynamic combinatorial libraries: coupling azobenzene photoisomerization with hydrazone exchange. *J. Org. Chem.*, **74**, 111–117.
28. Cacciapaglia, R., Di Stefano, S., and Mandolini, L. (2003) The bis-barium complex of a butterfly crown ether as a phototunable supramolecular catalyst. *J. Am. Chem. Soc.*, **125**, 2224–2227.
29. Imahori, T., Yamaguchi, R., and Kurihara, S. (2012) Azobenzene-tethered bis(trityl alcohol) as a photoswitchable cooperative acid catalyst for Morita–Baylis–Hillman reactions. *Chem. Eur. J.*, **18**, 10802–10807.
30. (a) Peters, M.V., Stoll, R.S., Kühn, A., and Hecht, S. (2008) Photoswitching of basicity. *Angew. Chem. Int. Ed.*, **47**, 5968–5972; (b) Stoll, R.S., Peters, M.V., Kuhn, A., Heiles, S., Goddard, R., Bühl, M., Thiele, C.M., and Hecht, S. (2008) Photoswitchable catalysts: correlating structure and conformational dynamics with reactivity by a combined experimental and computational approach. *J. Am. Chem. Soc.*, **131**, 357–367.
31. Stoll, R.S. and Hecht, S. (2009) Immobilization of a photoswitchable piperidine base. *Org. Lett.*, **11**, 4790–4793.
32. Neilson, B.M. and Bielawski, C.W. (2012) Photoswitchable organocatalysis: using light to modulate the catalytic activities of N-heterocyclic carbenes. *J. Am. Chem. Soc.*, **134**, 12693–12699.
33. Neilson, B.M. and Bielawski, C.W. (2013) Photoswitchable NHC-promoted

ring-opening polymerizations. *Chem. Commun.*, **49**, 5453–5455.
34. Neilson, B.M. and Bielawski, C.W. (2013) Photoswitchable metal-mediated catalysis: remotely tuned alkene and alkyne hydroborations. *Organometallics*, **32**, 3121–3128.
35. Sud, D., Norsten, T.B., and Branda, N.R. (2005) Photoswitching of stereoselectivity in catalysis using a copper dithienylethene complex. *Angew. Chem. Int. Ed.*, **44**, 2019–2021.
36. Wang, J. and Feringa, B.L. (2011) Dynamic control of chiral space in a catalytic asymmetric reaction using a molecular motor. *Science*, **331**, 1429–1432.

6
Surface and Interfacial Photoswitches

Junji Zhang and He Tian

Photochromic molecular switches, which change their chemical and physical properties (e.g., absorption/fluorescence, electronic conductance, anti/paramagnetism, catalysis) as well as geometrical conformations under irradiation of light, have found themselves of great potential in the fields of molecular logic/memory, sensors, catalysis, and supramolecular smart materials as mentioned in previous chapters. This great photoswitchable performance endows them with ultimate limit of miniaturization for constructing functional molecular devices [1]. Up until now, a lot of research on photochromic molecular switches has been carried out in solution and is distributed and oriented randomly without certain placement, individual addressability, as well as cooperation as those smart functional materials on surfaces or chips [2]. This prompts us to develop photoresponsive "smart surfaces," which can take orders from light signals to carry on specific tasks. Smart surface materials, considered as the bridge between photochromic molecules and optoelectronic devices, are of burgeoning interest in research areas as organic electronics [3], functional coatings [4], microfluidics [5], ion channels [6], and biological imaging together with medicinal diagnosis and therapeutics [7]. Light or photons, as an energy-efficient, ultrafast, and environment-friendly power source, would provide these smart surface with "clean" orders without causing unwanted wastes. Therefore, more and more attentions have been turned to design and prepare light-responsive surficial materials with photochromic family members in the past decade [2a, 8].

Different from their "free" counterparts in the solution, photochromic molecules on the surface sometimes function differently, either positive (enhanced functionality) or negative (photochemical quantum yields impairment or even loss of photoactivities). Thus, it is important to characterize the surface morphology [2b] and effects of surface assembly on molecular structures, arrangements, and functions. In this chapter, we will focus our view on the surface assembly of photochromic compounds on traditional metal electrodes, nanoparticles, and single molecular junctions together with their novel functionalities and potential applications.

Photochromic Materials: Preparation, Properties and Applications, First Edition.
Edited by He Tian and Junji Zhang.

6.1 Photochromic SAMs

Photochromic self-assembled monolayers (SAMs) are defined as photochromic molecules assembled as Langmuir–Blodgett monolayers deposited either in vacuum or from solution. With the advantage of ordered, dense structures and directed coupling between surface and molecules, herein we mainly focus on monolayers self-assembled from solution.

Generally, conducting/metallic surfaces (e.g., gold, platinum, ITO) are among the best choice for electrode materials due to their easy modification (e.g., through thiol–gold affinity and pyridyl absorption) and fast, sensitive electric signals. Recently, planting photoswitchable molecules on semiconductor substrates such as silicon is becoming an emerging force, providing a prospect of adding photoswitching element into the application of organic electronics and photovoltaics, for example, solar cells, molecular storage devices, and field-effect transistors (Chapter 7).

6.1.1 Photochromic Electrode SAMs

One distinct drawback of photochromic SAMs is the limited amount (approximately $10^{-11}-10^{-9}$ mol·cm^{-2} or mol/cm^2) of compounds assembled on the surface. To study the photoswitching behaviors of these functional surfaces, methods with high sensitivity are required, that is why electrochemistry plays an important role in the research of photochromic SAM materials [9]. Endowed with advantages such as rapid addressability of entire sample, precise control of the surface modified with molecules in different redox states, as well as high sensitivity, electrochemistry has established a "dominant" place in the research field of photochromic SAM materials. In contrast, the limited amount of material requires highly sensitive spectroscopic detection, including Localized Surface Plasmon Resonance (SPR), Raman Spectroscopy, UV–Vis Spectroscopy, and *In Situ* Sum Frequency Generation Spectroscopy, and so on, has become generally available only in recent years.

In 1994, Willner group have reported a simple yet very efficient spiropyran-modified gold electrode [10]. With this photoswitchable SAMs, a bioelectrochemical hybrid device model was constructed. As shown in Figure 6.1a, under alternate irradiation of UV and visible light, a remarkable polarity change occurred between spiropyran and its positively charged photoisomer merocyanine. This simple molecular change can achieve, among other applications, the inhibition or enhancement of the rate of electron transfer between the surface and electro-active substrate in solution via "capture and repel" behavior. In this work, the electro-active substrate, cytochrome *C* herein, could absorb to the spiropyran-immobilized gold electrode and thus facilitates an efficient electrode transfer from electrode to the substrate in the solution, resulting in a redox current flow. After irradiation of the electrode, the merocyanine-rich electrode repelled the cytochrome *C* (which is also rich of positive charges), resulting

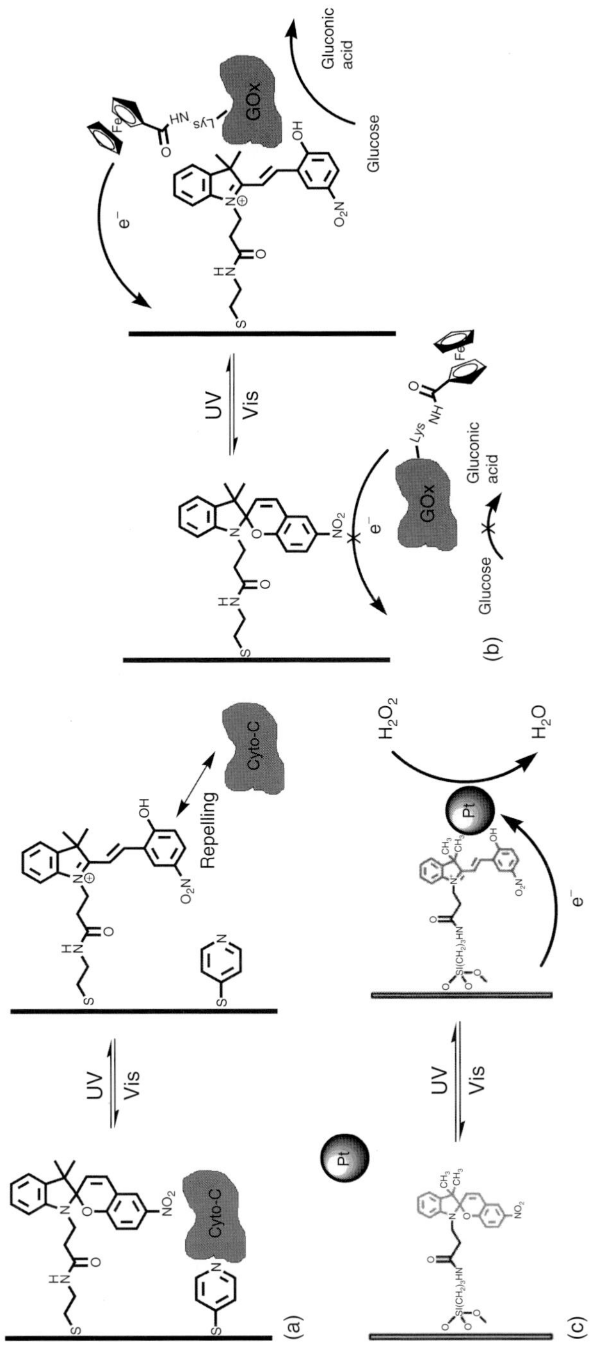

Figure 6.1 Spiropyran-modified electrode with photoreversible interactions with (a) cytochrome *C*, (b) glucose oxidase, and (c) Pt nanoparticles.

in the decrease of electron transfer from the electrode to the electro-active cytochrome *C* substrate and the consequential redox current. Related approaches were implemented to photoswitch the bioelectrocatalytic oxidation of glucose by glucose oxidase, GO_x (Figure 6.1b) [11]. Through the reversible photochemical isomerization of the monolayer between the spiropyran and merocyanine states, the bioelectrocatalytic oxidation of glucose was cycled between "OFF" and "ON" states, respectively.

Later, in 2007, Willner group advanced their research on spiropyran electrode-controlled electrochemical device [12]. This time, catalytic platinum nanoparticles (Pt-NPs) were chosen as the electro-active substrate. Unlike the positively charged cytochrome *C*, the Pt-NPs are negatively charged that intends to interact with the merocyanine dominant, positively charged electrode (Figure 6.1c). Hence, the electrocatalysis and the chemiluminescence were turned on after UV irradiation while turned off under visible light due to the generation of neutral spiropyran-confined surface.

Apart from spiropyran-based phototunable tunable electrostatic interactions, photo-/electro-active surfaces using dithienylethene (DTE) photochromic molecules, have also been constituted. In contrast to spiropyran, DTE undergoes photoisomerization to generate two thermally stable forms: open form (usually colorless) and closed (usually colored) form. These two photoisomers exhibit different electrochemical activities attributed to the conjugation changes after photoinduced electrocyclic reaction. Willner and Yehezkeli reconstituted apo-GO_x with its FAD (flavin adenine dinucleotide) co-factor tethered DTE monolayer on a gold electrode, which leads to the photonic wiring of the enzyme with the electrode and to the optical/electrochemical switching of the bioelectrocatalytic functions of the enzyme (Figure 6.2) [13]. Different from photoisomerization-induced electrostatic changes of spiropyran, DTE here acts as a photoswitchable electric mediator. As the catalytic active site, FAD is surrounded with insulating protein structures of GO_x, DTE molecule here works as the photoswitchable mediator for the transfer of electrons from the active site to the electrode. In this work, the DTE open isomer performed as a better electron mediator while the closed isomer, generated either by electrochemical or photochemical cyclization, had a much lower electrochemical activity. Therefore, a molecular wire that switched between a charge-transporting state and an insulating configuration by means of external photonic or electrical signals was designed.

6.1.2
Photoreversible Functional Surfaces

6.1.2.1 Photoswitchable Surface Wettability

Wettability is one of the important properties for a surface since it determines the hydrophobicity of a material. Surface hydrophobicity is generally characterized by measuring the contact angle (CA) of a liquid (usually water) droplet on the surface, with $CA > 90°$ as hydrophobic and $CA < 90°$ as hydrophilic. Currently, more and more superhydrophobic surface ($CA > 150°$) and superhydrophilic

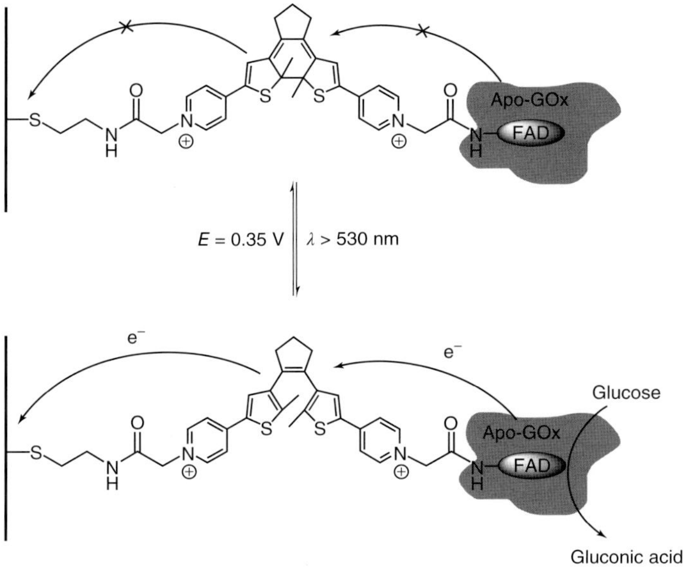

Figure 6.2 Dithienylethene-based electrochemical/photoelectrochemical switchable wired glucose oxidase electrode and the structure of amino-FAD.

surface (CA < 5°) have been developed because of their significant applications in novel functional coatings, biomimetic materials, novel drug delivery systems, and other practical and fundamental researches [14].

To control surface wettability, photochromic smart surfaces have attracted more and more attentions nowadays because these wettability modulations are reversible and can be remotely controlled at high time and space resolution. The surface wettability starts from the easy modification of photochromophores with hydrophobic (e.g., $-CF_3$)/hydrophilic (e.g., –COOH, pyridinum) groups or metal ion/water soluble molecules chelating groups or even simply photoinduced polarity and electrostatic property changes of the molecule itself (e.g., spiropyran). Azobenzene is the most investigated and popular molecule in the photocontrolled surface wettability area. Apart from their *trans–cis* dipole change induces surface energy differences, the geometric transformation also leads to the "shield" or "expose" the hydrophobic/hydrophilic groups modified either with azobenzene or on the surface, resulting in the surface wettability variations.

The unique wetting property of trifluoromethyl makes it a perfect group to decrease the surface free energy in organic molecules [15]. When fluorine-containing surface-confined azobenzene molecules are in the *trans* form, the $-CF_3$ groups are exposed to the monolayer/water droplet interface, which makes the surface exhibit hydrophobicity. In contrast, $-CF_3$ group will be "hidden" from the interface when the azobenzene is in its *cis* conformation due to *trans*-to-*cis* isomerization upon UV irradiation, changing the surface to a more hydrophilic state (Figure 6.3a). Not only azobenzene, $-CF_3$ group has also been attached

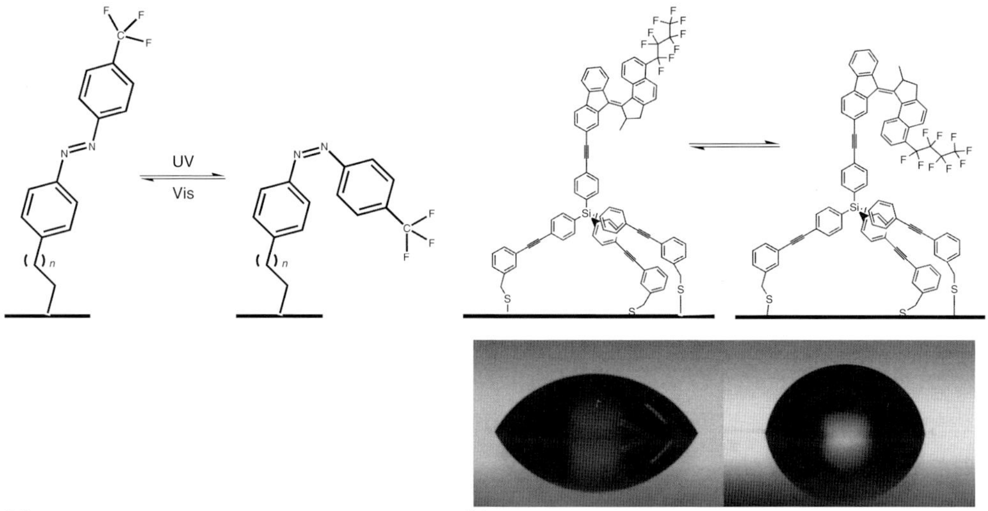

(a) (b)

Figure 6.3 Photoinduced wettability changes of perfluoroalkyl attached (a) azobenzene and (b) molecular rotor modified functional surfaces. (Adapted with permission [15d]. Copyright 2014, American Chemical Society.)

to other photoswitchable species to perform the photoreversible wettability. Overcrowded alkene molecular rotor, synthesized by Feringa group, has been covalently attached to the gold surface by either 1,3-dipolar cycloaddition (or Click reaction) [16] or rigid phenyl-acetylene-based tripod as a stator [15d]. The tripod stator used here are necessary for (i) a fixed and altitudinal orientation with respect to the surface; (ii) sufficiently isolating the motor from the surface in order to prevent the unwanted quenching of the excited state of the rotor, allowing efficient photoisomerization; and (iii) preventing the high-density packing of the rotor on the surface, ensuring its rotation speed. As shown in Figure 6.3b, the $-CF_3$ group would face outside or to the gold surface with the photo-driven facile rotation of the stated rotor, displaying an efficient surface wettability change between $60° \pm 1°$ and $82° \pm 1°$.

Similarly, introduction of hydrophilic group, for example, carboxylic group to the surface-confined azobenzene, could also build a wettability switchable functional surface. A carboxylated azobenzene-modified silicon surface was developed by Demirel *et al.* [17a], and the wettability of this silicon surface could be well modulated by alternate UV–Vis light, attributed to the "hidden" and "expose" of hydrophilic carboxylic group by *trans–cis* isomerization of the azobenzene photochromophore. This photocontrolled silicon surface can be further functionalized by covalently linking single-stranded DNA (ssDNA) with carboxylated azobenzene for photoregulated hybridization of double-stranded DNA (dsDNA) as well as enhanced wettability modulation [17b].

Spiropyran, which will undergo a zwitterionic conformation change under UV–Vis irradiation, is another type of photochromic organic molecule possessing

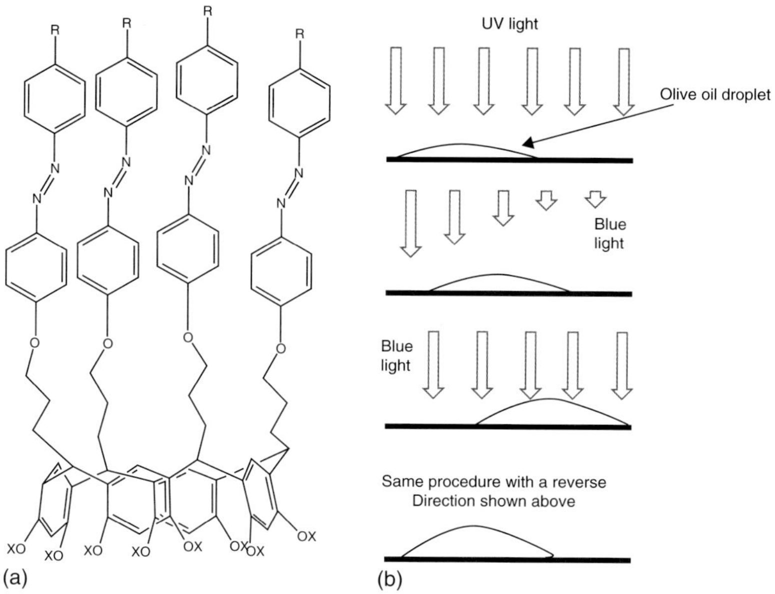

Figure 6.4 (a) Molecular structure of per-azobenzene-modified macrocyclic amphiphile. (b) Lateral photographs of light-driven motion of an olive oil droplet on a silica plate modified with macrocyclic amphiphile.

a light-switchable wetting property [18]. Its closed neutral form with hydrophobicity and its charge-separated zwitterionic open isomer with hydrophilicity could be reversibly tuned upon alternate light stimulation. Obviously, the surface wettability of surfaces modified by spiropyrans can be precisely tuned. Recently, DTE bulk crystalline with reversible surface wettability has been reported by Kobatake *et al.*, which will be discussed in Chapter 8.

One of the most interesting application of these wettability phototunable surface materials is the mechanical macroscopic motion driven by microscopic components, or *a que fuerza*. In 2000, Ichimura *et al.* reported a photo-driven macroscopic motion of liquids on a flat solid surface covered by the photoisomerization of azobenzene-tethered calix[4]resorcinarene (Figure 6.4) [19]. Asymmetrical photoirradiation caused a gradient in surface free energy due to the photoisomerization of the azobenzene moieties, leading to the directional motion of the droplet. The direction and velocity of the motion were tunable by varying the direction and steepness of the gradient in light intensity. In 2005, further work on the photo-driven liquid droplet was reported by Leigh *et al.*, which is based on a photoresponsive rotaxane shuttle with a diphenylethene photochromophore (Figure 6.5) [20]. Liquid transportation using photoresponsive surfaces may prove useful for delivering analytes in lab-on-a-chip environments or for performing chemical reactions on a tiny scale without reaction vessels by bringing individual drops containing different reactants together.

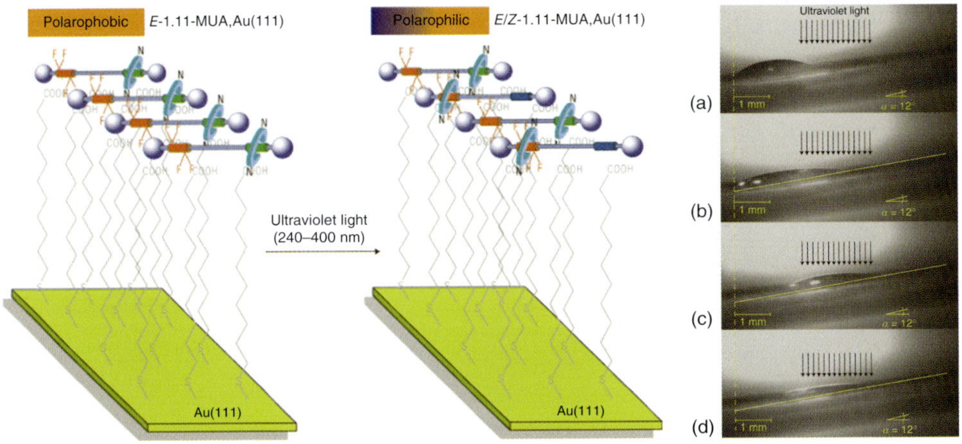

Figure 6.5 (left) A photoswitchable fluorinated molecular shuttles modified surface. (right) Lateral photographs of light-driven transport of a 1.25 μl diiodomethane drop on an *E*-1,11-MUA (11-mercaptoundecanoic acid). Au(111) substrate on mica up a 12° incline. (Adapted with permission [20]. Copyright 2005, Nature Publishing Group.)

6.1.2.2 Photocontrolled Capture-and-Release System

In some photopatterning, delivery or disposable surface systems, only "one-way" optic operation is needed. Therefore, photoirreversible measures such as photocleavage/caging, photodimerization, or even photodegradation are generally used. Yet, in many photoresponsive surface systems, secondary operation and recycling are demanded, thus promoting photochromic molecules as promising candidates for surfaces requiring reversibility.

The most common photoreversible surface systems are those with the function of capture-and-release. Surfaces with metal-ion chelating group modified azobenzenes could perform good metal-ion capture-and-release performance with high selectivity and efficiency by photochemically induced geometric changes to shield or expose their binding sites [21]. Azobenzene surfaces that capture and release charged small molecules via electrostatic interactions were also reported [22]. Compared to azobenzene, spiropyran has its own advantage of significant polarity changes after photostimulations to form charge-separated merocyanine isomer. This bestows them with the modification-free "gift," that is, capturing metal ions with their charged merocyanine isomer and then releasing them after isomerizing back to spiropyran form [23]. The whole process can be carried out only by themselves. The capture-and-release surface based on DTEs are relatively rare [24] since they neither possess significant geometric changes nor polarity changes compared to their azobenzene and spiropyran brothers. However, it does not mean that they are out of the game. The marvelous electronic property differences due to the π-electron conjugation change after photoisomerization grants them with the ability to play the game in another way. An optic/electrochemical molecular imprinting electrode based on dithienylethene pyridinum derivative (DTE-Py) and gold nanoparticles (AuNPs) matrix was reported by Willner and Tian as a molecular "sponge" (Figure 6.6) [25]. On accounts of the lengthened

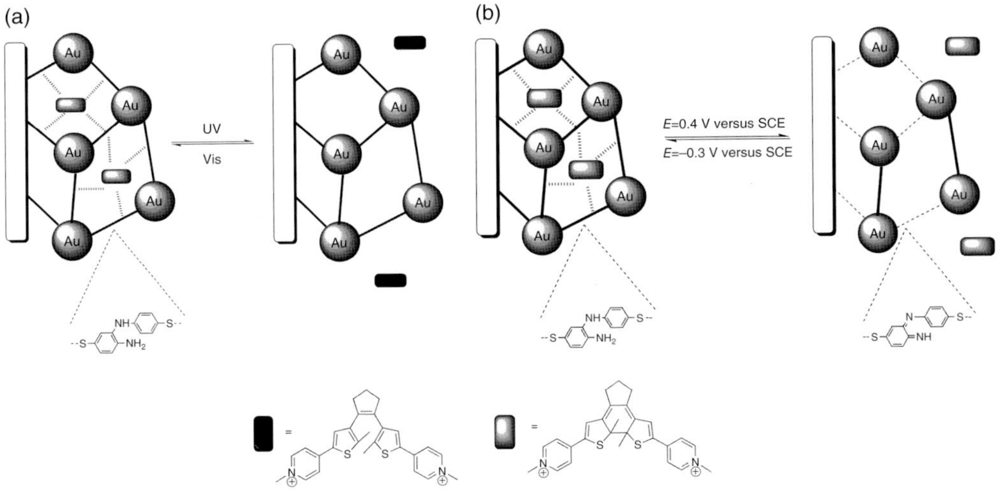

Figure 6.6 Photochemical (a) and electrochemical (b) uptake and release of DTE into and from the imprinted AuNPs matrix.

conjugation structure of the closed isomer of DTE-Py, it exhibits the property of an electron acceptor. This facilitates the interaction between the closed isomer and AuNPs matrix, which contains bisaniline donor groups as linking bridges via π-donor/acceptor interactions. The open isomer of DTE-Py, on the other hand, lacks electron acceptor feature, thus it is destabilized and released from the donor-bridged AuNPs matrix. This leads to the photocontrolled release of substrate from the polymer matrix. For the electrochemically controlled process, DTE substrates stay in its closed acceptor isomer form, while the initial donor bridges are varied between their oxidized acceptor (quinone) and reduced donor forms (bisaniline), controlling the uptake and release performances of the DTE-Py substrate. The π-donor/acceptor interactions between the DTE-Py photoisomers and AuNPs matrix are disturbed reversibly on both photoirradiation and redox operation, providing the DTE-based electrode with opto/electrochemically switchable capture-and-release molecular "sponge" behavior.

6.1.2.3 Smart Photochromic Surface Based on Supramolecular Systems

As an important field of supramolecular chemistry, spontaneous host–guest interactions between small molecules and macrocyclic hosts (e.g., cyclodextrins, crown ethers, cucurbiturils, and calixarenes) into complex superstructures by various noncovalent interactions has attracted much interest as it is a powerful approach toward the development of new materials and devices. Photocontrolled supramolecular systems become more attractive since they provide a very broad range of tunable parameters, such as wavelength, duration, and intensity (see Chapter 4). Some successful examples of tuning surface performances with photoregulated supramolecular host–guest systems have been presented in this chapter.

The most typical photoactive host–guest system is the azobenzene-cyclodextrin (Azo-CD) molecular shuttle [26]. Zhang group has systematically

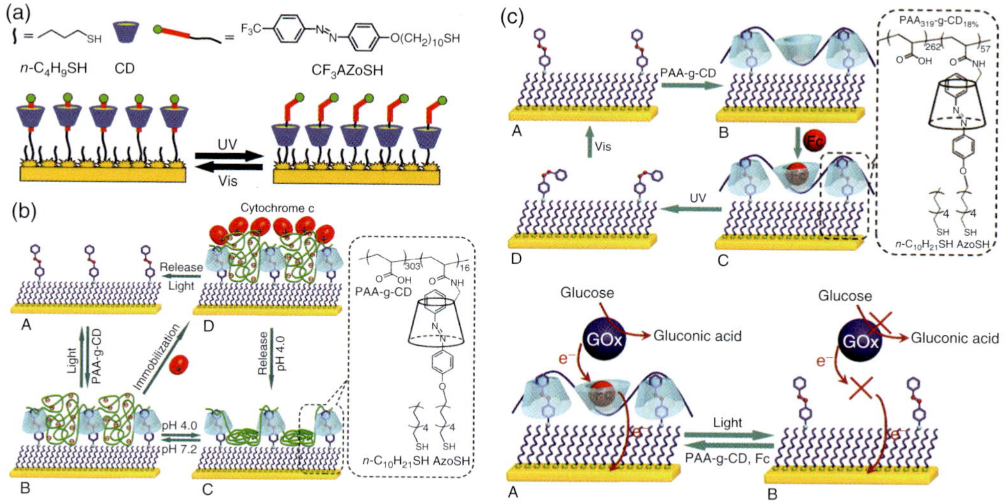

Figure 6.7 (a) Photoreversible azobenzene shuttles on the surface. (Adapted with permission [27]. Copyright 2008, Royal Society of Chemistry.) (b) Light/pH dual-responsive capture and release surface system based on azobenzene-CD shuttles. (Adapted with permission [28]. Copyright 2009, Wiley-VCH Verlag GmbH & Co. KGaA, Weinheim.) (c) Light-switchable catalysis on photocontrolled host–guest functionalized surface. (Adapted with permission [29]. Copyright 2011, Royal Society of Chemistry.)

studied the photoswitchable behaviors of Azo-CD-modified surfaces. They first fabricated this kind of molecular shuttles onto the gold electrode to form a SAM, whose wettability is responsive to the photostimuli (Figure 6.7a) [27]. As the cyclodextrin bears water solubility, the movement of cyclodextrin can induce the hydrophobicity change of the surface. Before UV-light irradiation, CD host stayed on the top of the surface and the SAM displayed more hydrophilic properties. After UV-light irradiation, CD host slid down onto the alkyl chain and the SAMs became hydrophobic. As an indicator, a change of nearly 50° for the surface CA with a water droplet was found before and after UV-light irradiation. This great change in wettability could be cycled several times by alternating between UV- and visible-light irradiation. Based on this, a dual-way responsive biointerface for capture-and-release of cytochrome *C* was fabricated (Figure 6.7b) [28]. First, azobenzene was immobilized on the gold electrode. The pH-responsive poly(acrylic acid) polymer grafted β-cyclodextrin (PAA-*g*-CD) was sequentially self-assembled on the surface-confined *trans*-azobenzene. At low pH ($pH < 4.0$), the poly(acrylic acid) (PAA) is in its neutral form that is reluctant to bind with positively charged cytochrome *C*. The surface, then, could be activated by increasing the pH to above 7.2 when PAA was deprotonated and fully negatively charged, resulting in a strong binding with cytochrome *C* in the solution. This cytochrome *C* affinitive biointerface could be reversibly activated and reactivated by manipulating the solution pH and could be further “cleaned” or “reset” by photoisomerization of azobenzene. The *cis*-azobenzene could exclude the PAA-*g*-CD binding site due to the crowded steric hindrance, thus resetting the biointerface to its originate state. Further work on photoswitchable bioelectrocatalysis has been carried out on this photo-/pH dual-responsive functional surfaces (Figure 6.7c) [29]. The integration of the photochromism and host–guest supramolecular system on surfaces to form multiresponsive biointerfaces for reversible sorption and release of functional guest molecules meets the modern requirements of developing bioscience and biotechnology, and it is anticipated to provide an excellent platform for potentially wide-ranging applications in biomimetics, biomembranes, controlled bioseparation, stimuli-responsive biomedical technologies, biosensing, and optobioelectronic devices.

Azo-CD photoreversible host–guest surface materials have been recently applied on repeatable analysis of cardiac biomarker, photoresponsive molecular imprinting system, and so on [30], exhibiting their great value of potential applications in both industrial and scientific researches.

6.1.2.4 Photochromic Surface for Molecular Data Processing

Since the multi-addressable photochromic materials have been well discussed in Chapter 3, we will here briefly introduce some representative work on photochromic data processing surfaces.

In 2006, Willner and Tian together constructed an electro-active and photoisomerizable DTE monolayer associated with an Au electrode, which acted as a write–read–erase information processing system and as a flip-flop set/reset memory element (Figure 6.8a) [31]. In this DTE associated electrode system, the open form could act as an information recording interface. The information was encoded electrochemically by either electrochemically or photochemically induced isomerization to closed form. The encoded information could be read out by cyclic voltammetry (CV) spectra, and the stored information was erased by a photochemical cyclic reversion step (from closed to open). The analysis of the electrochemical/photochemical properties of the monolayer-functionalized electrode revealed a unique system, where electronic or optical signals may lead to write–read–erase functions. In the same year, Feringa group reported a DTE-based read–write–erase information storage system on ITO surface (Figure 6.8b) [32]. In this study, the reversible conversion between open and closed isomers of DTE can be realized by both photochemical and electrochemical ways. The open form of DTE-ITO performed as an information recording interface. Either UV light or oxidation potential (1.2 V) could generate the closed form of DTE-ITO, defining the "written" process. Similarly, the "stored" information could be "erased" by either visible light or reduction potential (0.6 V) induced cyclic reversion to the open DTE-ITO. Meanwhile, to "read" the information, the reversible first oxidation potential (0.0–0.5 V) of closed form could be used to monitor the process. It should be noted that this "read" potential interval does not interfere with the write–erase potentials, demonstrating a very interesting nondestructive "read out," which is very essential for the practical design of data storage systems.

Willner group highlighted the formation of spiropyran/Co-NP photopatterning surface to encode the optical information [33]. As shown in Figure 6.9,

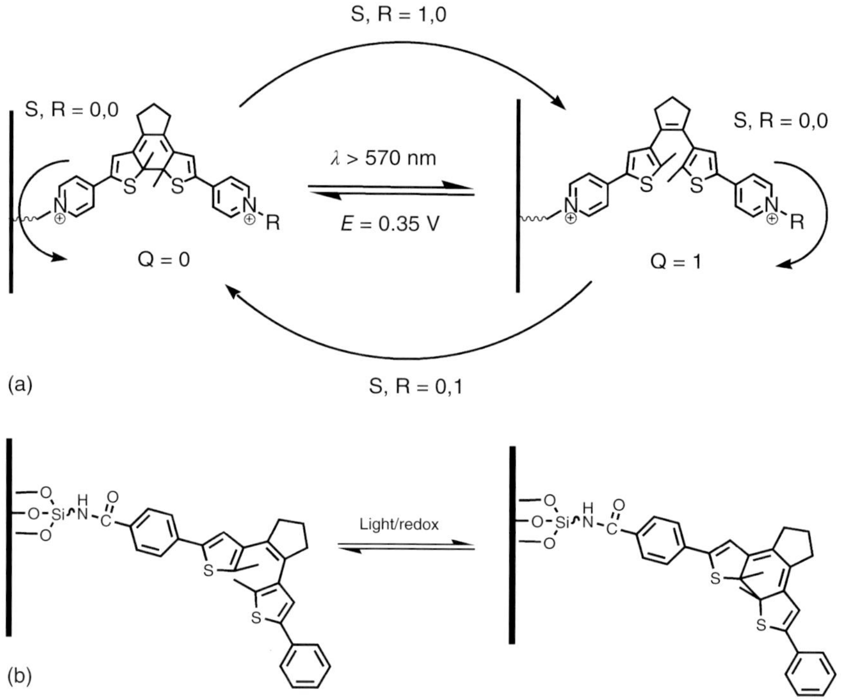

Figure 6.8 (a) The flip flop set/reset memory device based on DTE-modified gold electrode. (b) Photo "write/erase," electronic "read" system based on DTE-modified ITO electrode.

the electrode was modified with spiropyran and the surface was patterned by photoirradiation. The photoencoded/generated merocyanine isomer acted as a template for binding Co(II) ions. To read out the light-encoded information, a redox potential was applied to reduce the captured Co(II) ions to magnetic Co-NP functionalized patterns. The main part of this work lies in that the magnetic information could be erased by electrochemical breakdown of the particles while leaving the Co(II) ion associated with the surface. In other words, the encoded information could be retained in this system, making this merocyanine/Co(II) surface as a "secret ink."

Other smart data processing systems based on spiropyran and azobenzene-modified surfaces have also been reported recently [34], regretfully all of them would not be discussed in detail due to the space limitations.

6.2 Photoregulated Nanoparticles

In the past decades, the relation between organic and inorganic chemistry led to an explosive uprise of novel hybrid materials, combing beneficial characteristics and properties of both components [35]. Nanoparticles, defined as particles

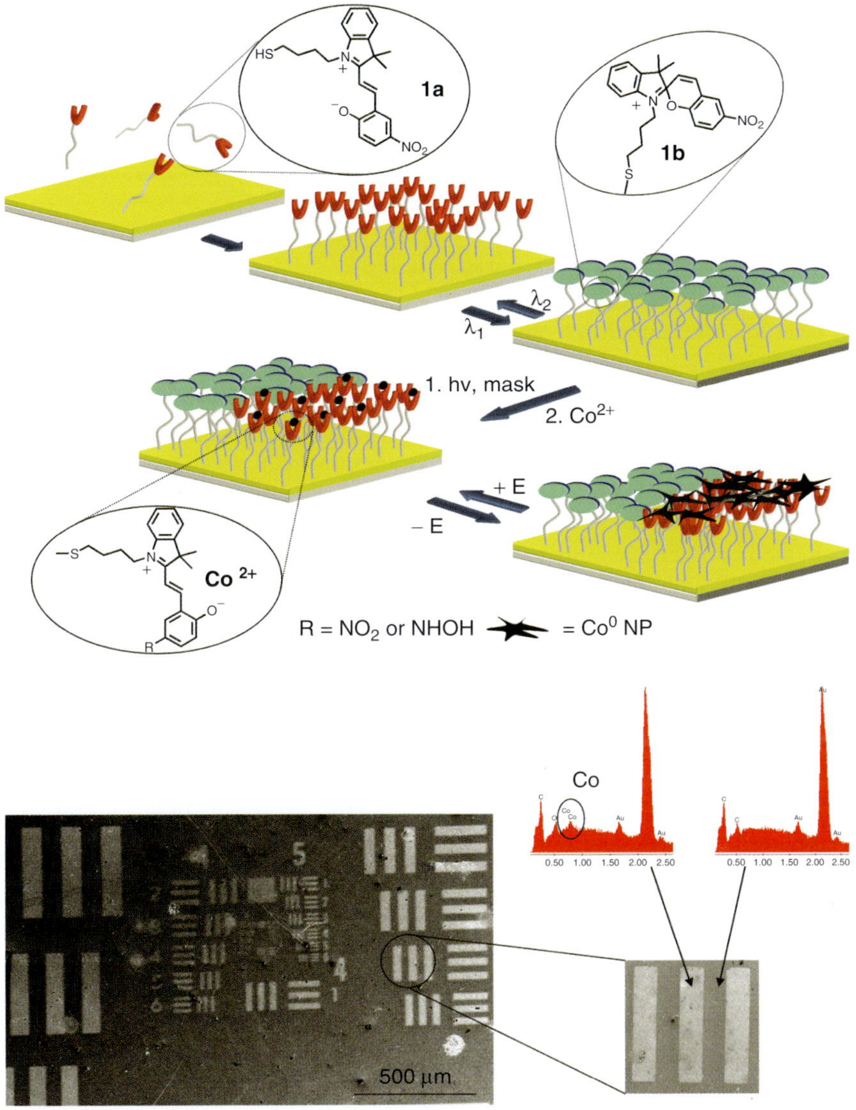

Figure 6.9 Spiropyran/Co-functionalized erasable magnetic patterning surfaces for photo/electrochemical encoding system. (Adapted with permission [33]. Copyright 2008, Wiley-VCH Verlag GmbH & Co. KGaA, Weinheim.)

sized between 1 and 100 nm, are one of the shining stars in the scientific study of functional hybrid materials. Nanoparticles possess unique chemical/physical properties such as size-dependent absorption/emission of quantum dots (QDs), quantum confinement in semiconductor particles, SPR in some noble metal particles (e.g., AuNPs and Ag-NPs), superparamagnetism in magnetic nanoparticles (MNPs), and optic upconverting effects in upconversion nanoparticles (UCNPs). Recently, nanoparticles directed biosensors/imaging, catalysis, data storage, solar cells, molecular imprinting polymers, and optoelectronic devices have attracted many attentions in various scientific areas [36–38]. In this section, we will go further and review the photochromic hybrid nanoparticles which acquire the ability of photoswitching.

6.2.1 Photochromic Switches on Traditional Metal Nanoparticles

6.2.1.1 Photoswitching on the Metal Nanoparticles

As mentioned at the beginning of this chapter, either enhancement or decrease or even loss of photoswitch ability would occur when photochromic molecules are anchored on the surface, especially, noble metal nanoparticles surfaces. Steric and electronic effects are the two most significant challenges to influence the photochromic activities of the anchored molecules. For the steric effect, the tightly packing of molecules assembled on the monolayers might affect the photoisomerization efficiency; while for the electronic effect, the excited states of anchored molecules for photoisomerization could be quenched by the NPs' SPR. Therefore, meticulous surface assembly strategies should be implemented.

Steric induced loss or decrease in photoisomerization activity by overcrowded density is common to SAMs. Small nanoparticles (~2 nm), fortunately, can avoid such problem by virtue of their curvature effect, which makes additional free volume for the photoisomerization [39]. For larger nanoparticles (>5 nm), however, are not so lucky to make it right [40]. Thus, additional methods have to be taken to optimize the photoisomerization on the nanoparticle surfaces. One generally used method is to modify other "assistant" ligands (usually not stimuli responsive) and create a "mixed" monolayer, thus "diluting" the surface density of the photochromophores to recover their photoactivities [41, 42]. The problem with this regular method is that it is difficult to control the ratio of photochromophore ligands and assistant ligands [43], hampering the repeatability of the experiments. Two alternative strategies thus have been brought out to further tackle this "squeeze" problem. One effective method to make more space for switching is to use the photochromophore ligands with "multi-legged" anchor group, for example, disulfides or even more complex tripod. An elegant example has been demonstrated by Feringa group, who attached the crowded alkene molecular motor onto 2 nm Au NPs via two alkane thiol "legs" (similar example was used in constructing motor on electrode SAMs, see Section 6.1) [44]. Two anchor points successfully created volume for the light-triggered rotation of the motor. In addition, long alkyl chains were used here to limit electronic interactions between the gold core and excited states of the photochromophore (the effect of electronic interactions will be discussed later in this chapter). With these precautions, the motor's performance on NPs was virtually identical to that in solution (Figure 6.10a). Another method is to utilize bulky spacers to make sufficient "room" for isomerization. The typical example for this is the introduction of cyclodextrin with the azobenzene ligand to form a host–guest photoisomerizable molecular shuttle on the nanoparticles (Figure 6.10b) [45]. Because of their bulky volume, the distance between

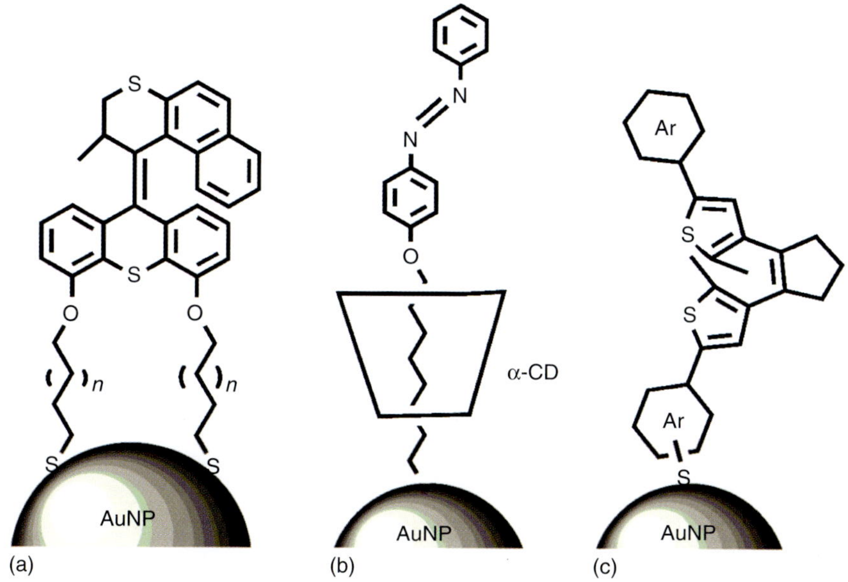

Figure 6.10 (a) Molecular rotor immobilized on AuNPs via two thiol anchors. (b) α-CD "makes room" for immobilized azobenzene to perform photoswitching on AuNPs. (c) Immobilized DTE on the AuNPs via arylthiol linkers (Ar = benzene, thiophene).

azobenzene ligands has been extended, facilitating their effective *trans–cis* photoisomerization.

SPR is the collective oscillation of electrons in a solid or liquid stimulated by incident light. The resonance condition is established when the frequency of the light photons matches the natural frequency of surface electrons oscillating against the restoring force of positive nuclei. The mechanism of photochromic excited-state quenching by SPR is attributed to the resonance energy transfer, which is determined by two factors: (i) overlap between the photochromophore absorption and NP SPR band and (ii) distance between photochromophore and NPs [46]. Since the absorption bands of conventional photochromophores more or less inevitably overlap with the NP SPR band, the most efficient solution for tackling the quenching effects is the separation of photochromophore from NP surfaces by long spacers [47].

Feringa and coworkers investigated the photochromic behaviors of DTEs directly linked to the AuNP surface via different aromatic spacers (*para-/meta-*phenylene and thiophene) (Figure 6.10c) [48]. As expected, both photocyclization and cycloreversion reactions of DTE were affected. The quantum yields of ring closure and open process decreased to certain extent according to different spacers ($\Phi_{O-C} \sim 0.07$ for *meta-/para-*phenylene bridged DTE, while completely quenched excited state for thiophene-bridged DTE; $\Phi_{C-O} < 0.01$ for all three DTEs on AuNPs), respectively. These differences in quenching phenomena may lie in the conjugation extent between DTE and AuNP surfaces. Later, further studies were carried out in Feringa group. They discovered that the efficient energy transfer from the excited open-ring isomer of thiophene-bridged DTE to AuNP surface was attributed to a greater overlap between the highest occupied molecular orbital of DTE (open) and the metal density band of states near the Au Fermi level [49].

To further investigate the quenching process between photochromophore and AuNPs, Kobatake *et al.* synthesized different sizes of AuNPs covered with different lengths of low polydispersity and thiol-terminated DTE polymers (poly(DTE)) [50]. It is found that in poly(DTE)–AuNP systems, the photocyclization reactivity of the DTE polymer decreased with the decreasing distance between the DTE and AuNP (96% conversion of 20-nm long-chain poly(DTE) compared to 81% of poly(DTE) with chain length of 11 nm). In another DTE/styrene block copolymer (poly(St)-*b*-poly(DTE)) AuNP system, in which the polystyrene segments (12 nm) were closer to the AuNP surface and poly-DTE segments were further away from the surface (6.3 nm), the photocyclization conversion (94%) was not obviously affected. This further confirms the distance effect that influences the photoconversion of photochromophore on the AuNP surface. Interestingly, in recent researches, photocycloreversion reaction of DTEs was found to be enhanced near the AuNP surfaces [51]. The enhancement factor and enhanced area were determined to be two- to fivefold and the area is of 9–12 nm from the gold surface. The dependence of particle size, distance from the particle surface, and irradiation wavelength have been further studied [52].

In conclusion, demonstration of spacer-dependent photoswitching behaviors emphasizes the importance of the spacer connecting the switching units to the nanoparticle.

6.2.1.2 Photoinduced Reversible Aggregation of Nanoparticles and Their Versatile Applications

Traditionally, nanoparticles are dispersed and stabilized in solution by modifying them with ligands carrying the same charge (e.g., sodium citrate, thiol carboxylic acid). Disrupting the stability with counterions or linker ligands (dithiols for instance) would cause the aggregation of the nanoparticles, changing their spectroscopic as well as other unique chemical and physical properties. In this section, photoinduced aggregation of nanoparticles and their unique properties and regulations are discussed.

The most studied photoinduced nanoparticle aggregation system is based on the azobenzene family [53], which displays a significant dipole change after photoisomerization (~4–5 D for *cis* and ~0 D for *trans*). Grzybowski group first thoroughly investigated the light-controlled reversible and irreversible 3D supra-structures and aggregations of AuNPs by azobenzene ligands based on the different surface concentrations of the photochromic ligand [40].

Some metal nanoparticles are known to have interesting capabilities in chemical and biological catalyses. Based on the photoswitchable nanoparticle system mentioned earlier, a photoswitchable catalysis of hydrosilylation reaction was consequentially introduced [54]. As shown in Figure 6.11, the *trans*-azobenzene-modified AuNPs are highly catalytic and are able to catalyze the hydrosilylation of 4-methoxybenzaldehyde in organic solution at 39 °C and under argon. When

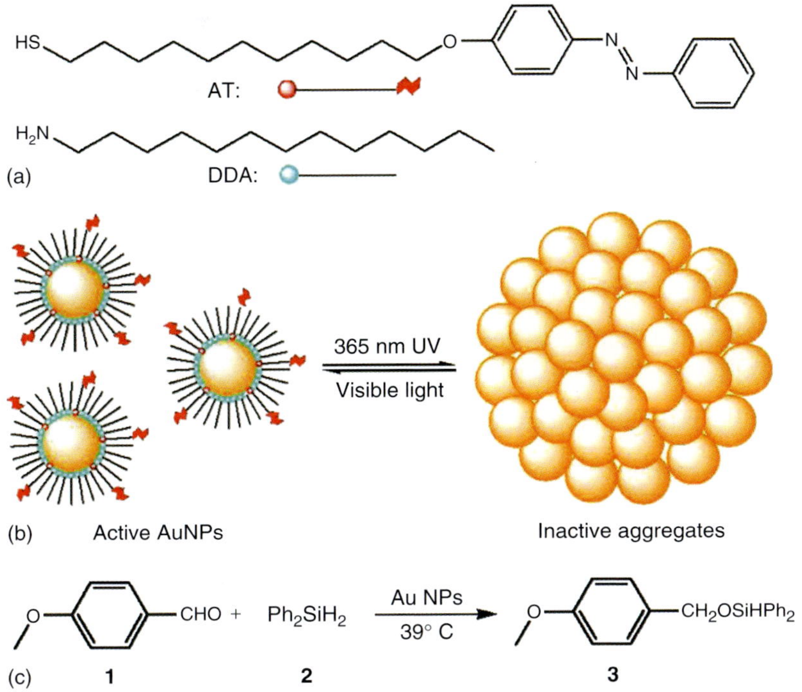

Figure 6.11 (a) Molecular structures of the photoresponsive azobenzene–thiol ligand. (b) Photoswitching of AuNPs aggregation. Dispersed NPs are catalytically active; aggregated NPs are catalytically inactive. (c) Hydrosilylation of 4-methoxybenzaldehyde catalyzed by AuNPs. (Adapted with permission [54]. Copyright 2010, American Chemical Society.)

exposed to UV light, the AuNPs covered with azobenzene ligands aggregate and the catalysis is effectively switched "off." After exposed to the visible light, the particles re-dispersed and the catalysis could proceed. Though the catalytic efficiency and the number of cycles should be further improved, this interesting work has provided a novel concept for future light-controlled catalysis in both chemistry and biology.

Another derivative application of this photoregulated AuNPs system is the development of a class of photocontrolled self-erasable and rewritable materials in which information was written into metastable nanoparticle "inks" [55]. In a typical procedure, the nanoparticles were dispersed in syndiotactic poly(methyl methacrylate) (sPMMA) organogel laminated between two poly(vinyl chloride)-coated poly(ethylene terephthalate) sheets. The writing procedure is to apply the UV irradiation that triggered the aggregation of the azobenzene toggled nanoparticles. In the absence of UV irradiation or in the presence of visible light, the written information/images would be gradually self-erased, with the erasure times controlled by the composition of the mixed self-assembled monolayers (mSAMs) coating the NP inks (Figure 6.12). Apart from the "numbers of cycles" cliché, maybe the most needed improvement for practical application of this

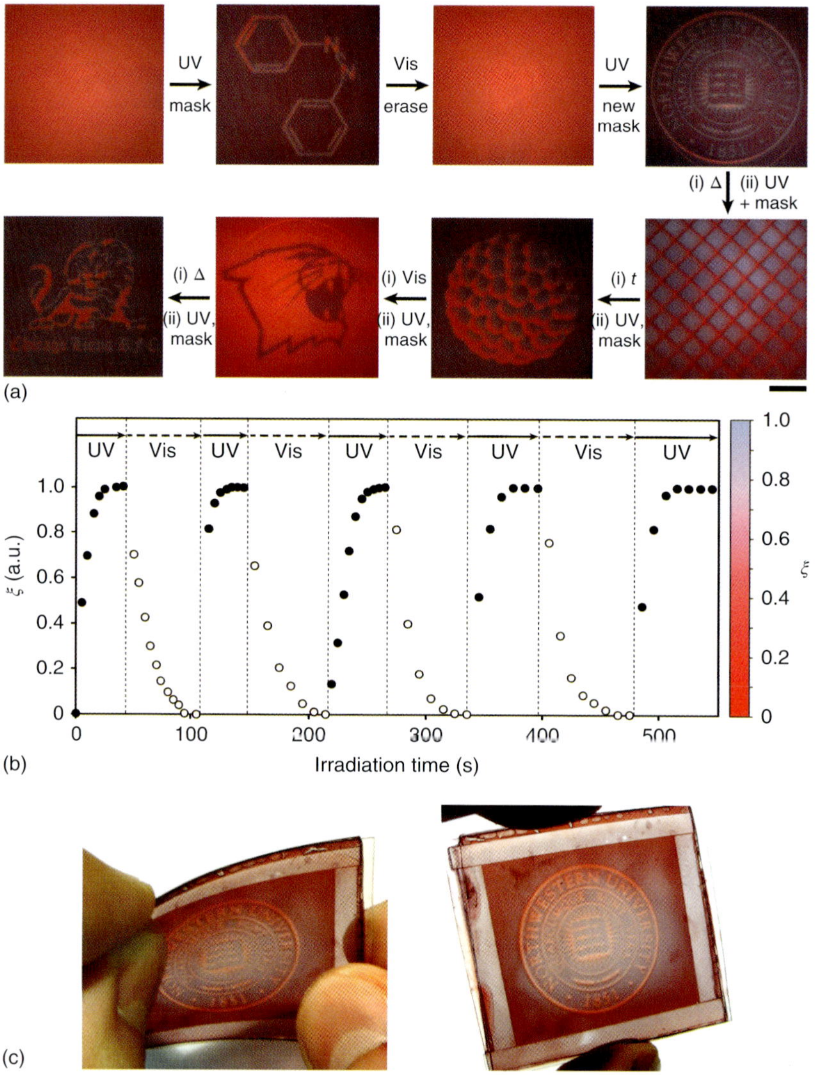

Figure 6.12 Rewritable and flexible films. (a) Sequential writing into and erasing from the same AuNP film. (b) Reversible spectral changes of an AuNP film upon alternating exposures to UV and visible light. (c) Patterned films can be mechanically distorted without disrupting the imprinted image. (Adapted with permission [55]. Copyright 2009, Wiley-VCH Verlag GmbH & Co. KGaA, Weinheim.)

nanoparticle ink is the organogels used as the supporting medium, which is not the best desired for environment. Since the azobenzene and long-chain alkyl thiol-modified AuNPs are lipophilic, hydrophilic functional AuNPs should be developed corresponding to the use of alternative water-soluble supporting materials.

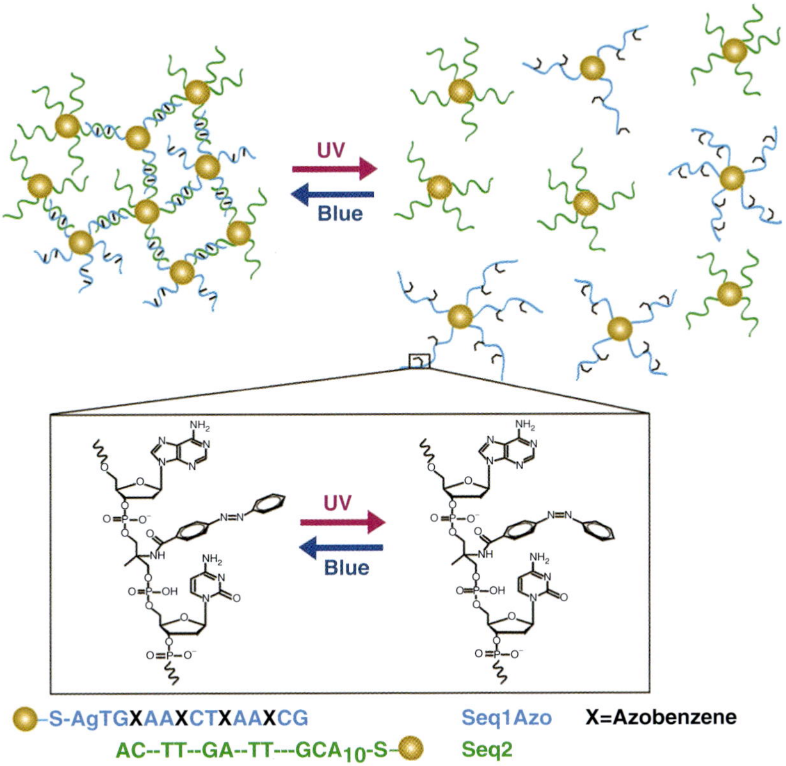

Figure 6.13 Photoswitchable DNA-functionalized gold nanoparticle conjugates. (Adapted with permission [57]. Copyright 2012, American Chemical Society.)

The photoregulated DNA hybridization by linking azobenzene moieties onto the oligonucleotide backbone to perform biomimetic operations [56] has been demonstrated by Asanuma and Tan, respectively. Ginger *et al.* recently further attached this azobenzene-linked photocontrollable oligonucleotides onto the AuNPs (Figure 6.13) [57]. It could be predicted that the photoisomerization-induced hybridization and dissociation between matched ssDNAs could trigger the reversible aggregation and dissociation of AuNPs. More meaningfully, the perfect complementary and partially mismatched DNA strands exhibit clearly distinguishable photoinduced melting properties. Thus, this photocontrolled AuNPs systems bear the ability to discriminate single-base mismatches, demonstrating that photon dose holds the potential to be used to replace the temperature and ionic strength controlled hybridization. Using photostringency strategy has several advantages as follows: (i) light intensity can be controlled more readily than temperature, pH, or ionic strength; (ii) photomelting can be accelerated at higher intensity, so the stringency wash could potentially be faster using more intense illumination (indeed we achieved full dissociation in minutes in the optical microscope geometry); (iii) photostringency could reduce the complexity

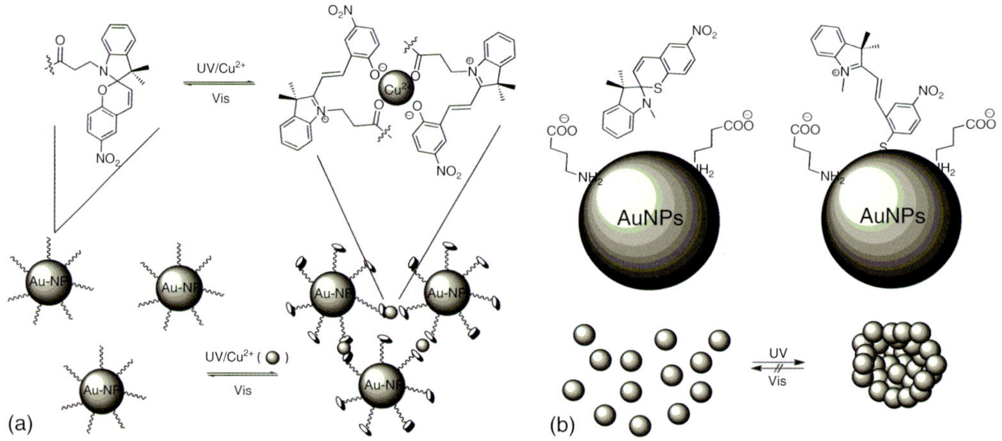

Figure 6.14 (a) Spiropyran-AuNPs based logic gates. (b) Photoinduced aggregation of AuNPs with spirothiopyran.

of microfluidic systems such as heaters/mixers and valves in lab-on-a-chip hybridization applications; (iv) photostringency enables remote manipulation without contacting the sample; and (v) reversibility could provide a chance to recover DNA after the stringency "wash."

Different from dipole interactions of azobenzene, spiropyran exerts its photoinduced charge separation "trick" to become another efficient molecular "glue" for AuNPs aggregation [58].

Based on the reversible aggregation of spiropyran-AuNPs, Wang and Jiang presented a resettable and multi-readout logic system that includes AND, OR, and INHIBIT logic operations (Figure 6.14a) [59]. As mentioned earlier, spiropyran can undergo reversible photoisomerization between neutral spiropyran and zwitterionic merocyanine isomers under alternate irradiations of UV–Vis light. The merocyanine isomer has been reported to have a coordination ability with several metal ions such as copper(II) and ferric(III) ions [60]. Therefore, the complexation of metal ions between merocyanine photoisomer ligands triggered the aggregation of AuNPs. In this logic system, UV/Cu^{2+} (AND), Fe^{3+}/Cu^{2+} (OR), and Cu^{2+}/EDTA (INHIBIT) were used as input signals, respectively. On the other hand, the ratio between the absorption at $\lambda = 520$ and 670 nm, which is characteristic for the aggregation form, was defined as the output signal. The distinctive advantage of this system is that molecular events in aqueous solution could be translated into a color change of the solution, which can be monitored by several readouts, such as UV–Vis spectroscopy, zeta potential, dynamic light scattering (DLS), and even with the naked eye.

In conventional designed photocontrollable NPs, the photochromic ligands are all planted on the surface of NPs [61]. This requires further modification of photochromophores with anchoring groups (e.g., thiols) and consequential synthesis of functionalized NPs. Although technologically it is not a difficult task to fulfill, the photochromophore modified NPs still bear the disadvantages as it is unable to precisely control the ratio between photochromophore ligands and assistant ligands. Recently, a smart molecule design has been reported by Shiraishi *et al.* In this work, spirothiopyran was synthesized, in which the oxygen atom in normal spiropyran was replaced by the sulfur atom (Figure 6.14b) [62]. After the UV irradiation, the open form of spirothiopyran bears a thiolate "protruding" outside which facilitates their covalent binding with the AuNPs surface. Sequentially, positively charged spirothiopyran molecules neutralize the negative charges on the AuNPs surface, leading to their aggregation. This design demands no pre-synthesis of functional AuNPs, thus simplifying the whole operation process using totally free photochromic molecules in solution. Smart it is, yet the irreversibility of this system cannot be neglected. The visible light-induced back-reaction of covalently bound spirothiopyran on the AuNPs could not dissociate the aggregated AuNPs to their dispersed state, which appears as the Achilles heel for this smart design. Future work may consider the utilization of noncovalent interactions, such as donor–acceptor interaction, to design reversibly controlled NPs aggregation process with free photochromic molecules in bulk solution.

6.2.2 Photochromic Switches on Other Novel Functional Nanoparticles

6.2.2.1 Photoswitchable Magnetic Nanoparticles

MNPs are a class of nanoparticles that can be manipulated using magnetic field. Such particles commonly consist of magnetic elements such as iron, nickel, and cobalt and their oxides. The MNPs have been the focus of a lot of research recently because they possess attractive properties that could see potential use in nanomaterial-based catalysts [63], biomedicine [64], magnetic resonance/particle imaging [65], data storage [66], defect sensors [67a], and cation sensors [67b].

Photoswitchable MNPs have also been presented in the past decade. The magnetic properties and performances of MNPs can be reversibly controlled by azobenzene (through perturbing the electrostatic field around the NPs) [68] and spiropyran (reversible aggregation via photoinduced electrostatic interaction changes between spiropyran ligands [69]. For DTEs, a fluorescent sulfur-oxidized DTE derivative was encapsulated onto the MNPs, and their fluorescence and flocculation/dispersion could be reversibly switched by light and magnetic field, respectively [70]. This hybridized photochromic MNPs could provide useful tools for imaging, tracking, enrichment, extraction, and purification of complicated biological samples.

6.2.2.2 Photomanipulated Quantum Dots

QDs have unique photophysical properties that offer significant advantages as optical labels [37c]. Typical for QDs are high-fluorescence quantum yields, stability against photobleaching [71], and size-controlled luminescence properties [72]. The efficient fluorescence and stability of QDs improve sensitivity and prolong lifetime in their use as optical labels. The size-controlled luminescence functions of QDs illustrate the major advantages of QDs as optical labels, as particles of the same material with variable sizes may be used as fluorescent labels for the parallel multiplexed analysis of different analytes. Another advantageous property of broad absorption bands allows the excitation of different QDs at a common wavelength manifold while emitting various fluorescence, respectively. In recent years, effective methods to synthesize QDs with functional capping monolayers or thin films were developed [73]. These functional QDs allow the secondary tethering of ligands or receptor units to the surface of the QDs, thus yielding QD–ligand conjugates in which the NP acts as an optical transducer for recognition of sensing events occurring at the surfaces of the NPs [74]. Herein, we introduce novel QDs functionalized with photochromic ligands.

Modulation of QDs fluorescence by photochromism was first reported by Li *et al.*, with spiropyrans as photochromic ligands [75]. The fluorescence of QDs was switched off via fluorescence resonance energy transfer (FRET) to photogenerated merocyanine acceptors after UV irradiation, while switched on after isomerizing back to spiropyran form.

Since the development of nanoparticle fluorescent labels is considered revolutionary in bioimaging, the photochromic modified QDs are also keen to exert their power in future biological and medicinal applications. Although promising, a lot of limitations still remain to be pushed. The first challenge is the thermostability of the photochromic ligand, in which azobenzene and spiropyran suffer a lot. DTE, enjoying its advantage of unique thermo-irreversibility, is considered a good candidate for bioimaging. Yet, there is no paragon, a reduced efficiency and stability of photochromism were observed when DTE is in aqueous media, in which most biological processes take place. While the water solubility of QDs can be settled by attaching hydrophilic ligands, the modification of lipophilic DTE ligands to the QDs surface not only leave their problems unsolved but also decrease the water solubility of QDs to some extent. To tackle this problem, in 2011, Jares-Erijman and coworkers developed a photochromic DTE-tethered amphiphilic polymer capping for QDs [76]. Since the amphiphilic polymer capping had a hydrophobic region with DTE between the QD surface and a hydrophilic exterior, this enabled the phase transfer of this QD from an organic to an aqueous medium without notable decrease in desirable properties of the original QD. More importantly, as protected by the interior lipophilic polymer chains, the photochromic behavior and the FRET efficiency were successfully retained (over 15 times of cycling and ~40% quenching efficiency). Therefore, an efficient photomodulated fluorescent QDs system was successfully invented in aqueous solution, pushing forward the photoswitchable QDs in promising bioimaging applications (Figure 6.15a).

The above-mentioned QDs-DTE system represents a "Turn-OFF" model for labeling, which would fall into a trouble of background interference. To optimize this switchable labeling system and reject the background, an improvement was further made as additional fluorophores were attached to the amphiphilic polymer chain to build a dual-color photoswitchable QDs [77]. In subsequent studies (Figure 6.15b), different fluorophores (Alexa 647 and Lucifer Yellow) were comodified with DTE into the polymer coatings on different sizes of QDs (diameter = 5.5 (green) and 10 nm (red)), respectively. Upon alternate UV–Vis irradiation, the dual-color/photoswitchable ratiometrically modulated QD systems based on FRET (with high tunable ratio over 100%) were realized with

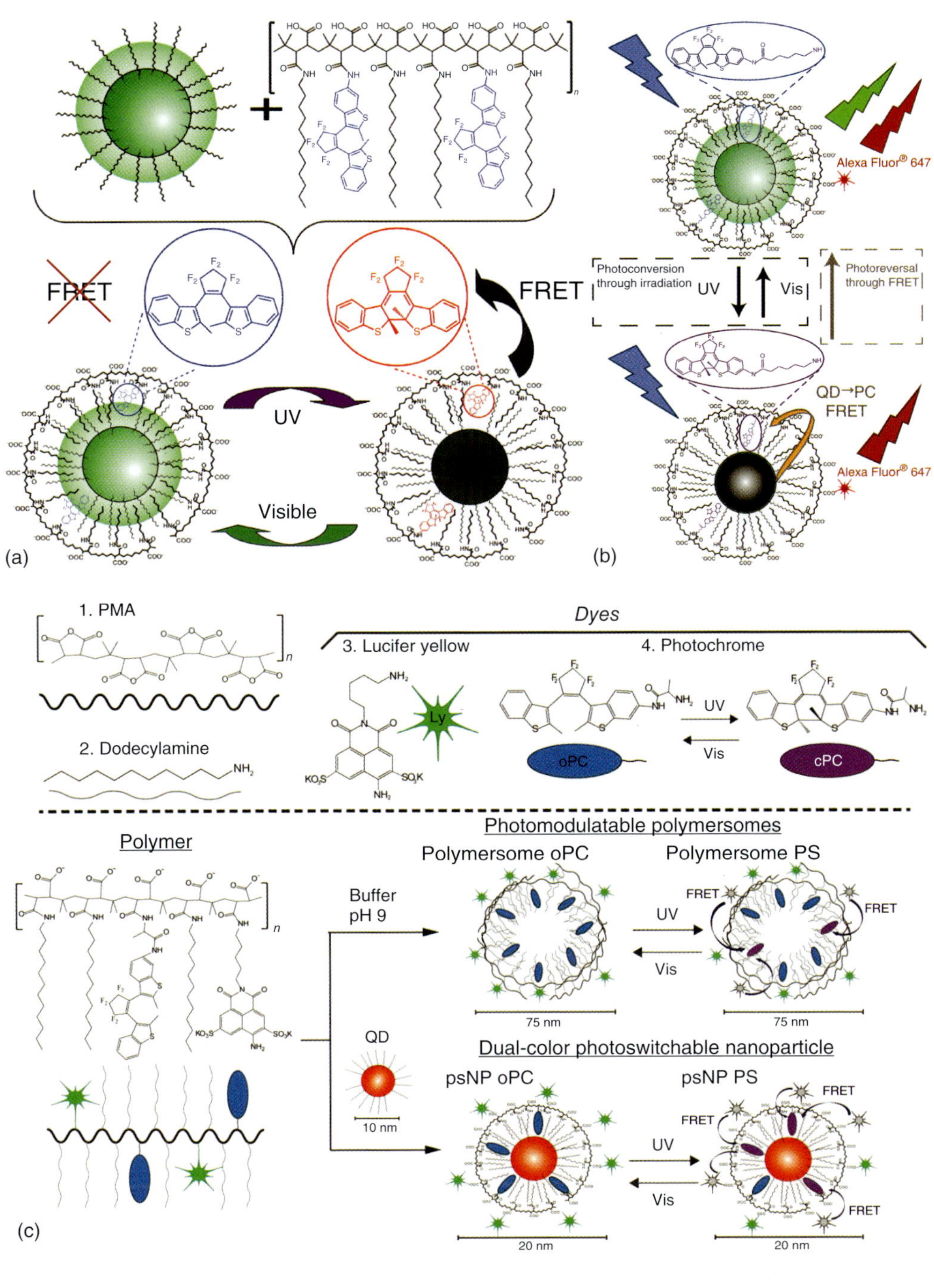
FRET
FRET
UV
Visible
(a)
Alexa Fluor® 647
Photoconversion through irradiation
UV
Vis
Photoreversal through FRET
QD→PC FRET
Alexa Fluor® 647
(b)
1. PMA
2. Dodecylamine
Dyes
3. Lucifer yellow
4. Photochrome
Ly
UV
Vis
oPC
cPC
Polymer
Photomodulatable polymersomes
Polymersome oPC
Polymersome PS
Buffer pH 9
FRET
FRET
UV
Vis
75 nm
75 nm
QD
10 nm
Dual-color photoswitchable nanoparticle
psNP oPC
psNP PS
FRET
FRET
FRET
UV
Vis
20 nm
20 nm
(c)

Figure 6.15 (a) Photoswitching QDs coated with an amphiphilic photochromic polymer. (Adapted with permission [76]. Copyright 2011, American Chemical Society.) (b) Dual-color photoswitching QDs coated with an amphiphilic photochromic polymer and Alexa 647. (Adapted with permission [77]. Copyright 2012, American Chemical Society.) (c) Dual-color photoswitching QDs coated with an amphiphilic photochromic polymer and Lucifer Yellow. (Adapted with permission [78]. Copyright 2013, American Chemical Society.)

red fluorescence (either from Alexa 647 dye or from 10 nm QDs, respectively) as "reference" signal (Figure 6.15c). This design could lead to manifold applications in cellular imaging due to its greater sensitivity, selectivity, and background rejection, facilitating the achievement of modulation required sensitive lock-in detection [78].

In the last decades, QDs have been constantly putting on their glamor in biosensor labeling. Yet, many still concern about their biotoxicity because poisonous metals, such as Cd, are used in QDs. As mostly biolabeling and imaging are carried out *in vivo*, toxicity, indeed, should be the primacy for consideration. Fluorescent carbon nanoparticles (CNPs), whose photoluminescence behavior is similar to QDs, are elected as a candidate for biolabeling because of its low cytotoxicity and excellent environmental and biological compatibility. Recently, photoswitchable CNPs with spiropyrans as photochromic ligands was reported [79]. Although the overlap between emission spectra of merocyanine acceptor and CNPs is not that large, there is still a considerable FRET efficiency for this photochromic CNPs system.

So far, solutions for challenges of thermostability, biotoxicity, and water solubility have all been presented, yet that does not mean that photochromic fluorescent nanoparticles are all set for practical biolabeling and sensing. One should note that the lights to excite the photochromic reactions are located in the UV region, which will do harm to our tissues as well as suffering from low tissue penetration, background fluorescence, and photobleaching. This indeed hampers the practical applications of photochromic fluorescent nanoparticle labels (actually not only for NPs but also for the whole photochromic materials family) in biological and medicinal sensing, diagnosis, and therapeutics, which is discussed in detail in the following section.

6.2.2.3 Photochromic with Upconversion Nanoparticles

UCNPs have the ability to generate visible or near-infrared (NIR) emissions under continuous-wave NIR excitation. Utilizing this special photoluminescent properties, UCNPs can be used as key components in complex nanocomposites for a wide range of applications [38b,d, 80].

The first example of combining photochromism with UCNPs is reported by Yi and Li [81], with DTE as photochromophore to reversibly switch the upconversion luminescence of UCNPs. In this work, UV light was still used to excite the photochromic reaction of DTE. In 2010, Branda group successfully prepared UCNPs functionalized with DTE molecules to achieve a mono-wavelength, reversible remote-control fashion of photochromism by the NIR light at 980 nm

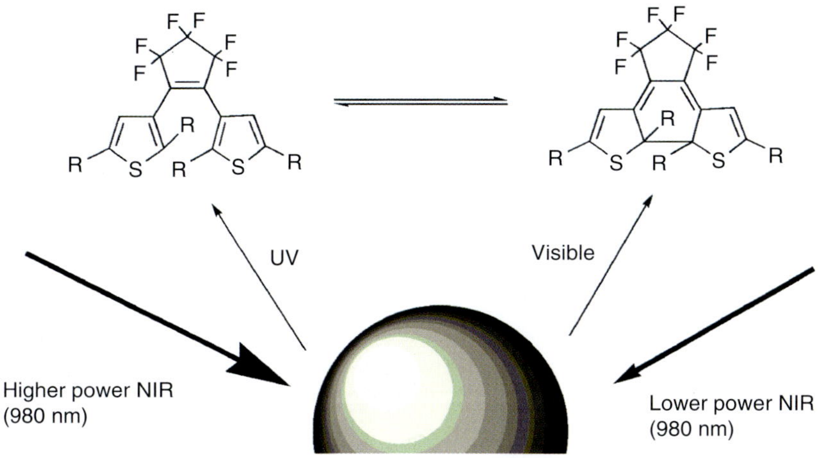

Figure 6.16 Single wavelength near-infrared-induced photochromism based on upconversion nanoparticles by changing the light intensity.

[82]. The highlight of this work is the nonlinearity of the upconversion mechanisms, which realizes the selective generation of UV and visible light under different power intensities. The high-power infrared irradiation of UCNPs generated UV light at $\lambda = 365$ nm, which triggered the ring-closing reactions of DTEs. Conversely, the ring-opening reactions of DTEs were triggered simply by reducing the power density of the excitation infrared light to the UCNPs when visible light was generated (Figure 6.16). This encouraging discovery, indeed, offers promising strategy for *in vivo* photochromic nanoparticles labeling with not only the NIR light-triggered photochromism but also a highly convenient and versatile method to spatially and temporally regulate photochromic reactions using a single light source by changing only its power or focal point. Further work on covalently linking DTE with UCNPs by click chemistry and encapsulating DTE with UCNPs with amphiphilic polymer coatings as "plug-and-play" to modulate upconversion luminescence were sequentially reported in following years [83].

Apart from their potential application in biolabeling/imaging, photochromic UCNPs also find their utilities in nondestructive molecular storage systems. Photochromic data storage systems usually suffer from unavoidable destructive readout since most reading process would somehow trigger further unwanted photochromic reactions and erase recorded data. Several methods have been developed to tackle this problem (see Chapter 3). Yan group demonstrated a DTE-based UCNPs data recording system with nondestructive readout [84]. In this device, the original data (unpatterned UCNPs luminescence) was erased by UV light (luminescence quenching by closed DTE). The new recorded data (patterned UCNPs luminescence) was written by visible light to block the FRET via opening the closed DTE to its open form. Different from other photorecording/readout system, the readout of the patterned data of UCNPs herein was achieved by NIR-induced upconversion luminescence where both open and closed DTE have no absorption, preventing the unwanted data loss by using conventional visible light excitation. One of the disadvantage of this design is that the multi-wavelength light sources are demanded, yet this upconversion-based photoluminescence provides a readily readable window with the advantages of low background and high sensitivity, which are important for super-resolution techniques. Based on this 2D recording media, we can foresee the fabrication of 3D assemblies of photochromic UCNPs for ultra-high-density 3D optical data storage in the near future.

6.2.3 Photocontrolled Mesoporous Silica Nanoparticles

Different from nanoparticles mentioned earlier, mesoporous silica nanoparticles (MSNs) are much larger (~100–800 nm) while containing ordered arrays of nanopores. These unique nanopore arrays grants the MSN with promising cargo carrier ability for potential biocompatible vehicles such as drug delivery [85]. Photochromic functionalized MSNs are commonly divided into two categories: (i) photocontrolled "Nanovalves" and (ii) photoregulated "Nanoimpellers." In the case of photo-nanovalves, the photoswitching is achieved by irradiation at the maximum wavelength of only one isomer, ensuring the increase of the population of the wanted photoisomer to the maximum extent. While in the case of "nanoimpellers," the system is irradiated at a wavelength at which both isomers have the same extinction coefficients, the isosbestic point, for instance. As a consequence, the modified photochromic compounds undergo a continual dynamic wagging, thus imparting motion upon the molecules trapped inside the nanopores, and eventually forcing their expulsion from the pores.

6.2.3.1 Photo-nanovalves

The stimuli that trigger the open/close behaviors of nanovalves varied from electro-/chemically induced redox reactions [86] to acid-/base-induced pH changes [87] and, even recently, biological enzymes/DNAs caused specific cutting of peptide/nucleotide sequences [88, 89]. Although more specific, biological triggers suffer from the lack of reversibility comparing to their chemical counterparts. As mentioned earlier, among the chemical stimuli, light enjoys the benefit of quick responsibility, remote control ability, time/space resolution, and environmental friendliness with no waste produced during the whole operation.

With the fast development of photocontrolled MSNs, more and more azobenzene-based MSN cargo carrier systems have sprung up [90] and, more recently, have found their promising applications in drug delivery and cancer therapy [91]. In 2009, Zink *et al.* associated the azobenzene photochromophore on the nanopore rim of the MSN with β-cyclodextrin as the "cap" and Rhodamine B as cargo fluorophore molecule. Thus, the first light-operated mechanized MSN was constructed [92]. As shown in Figure 6.17a, the external light stimulus could precisely control the release of the fluorescent cargo by the dissociation of β-cyclodextrin cap from the MSN due to the geometric *trans–cis* isomerization of azobenzene. In the same year, the rapid emergence of multi-stimulated photochromic MSN was witnessed as photocontrolled operation assisted synergistically with other stimuli. A multi-stimuli triggered MSN carrier was reported

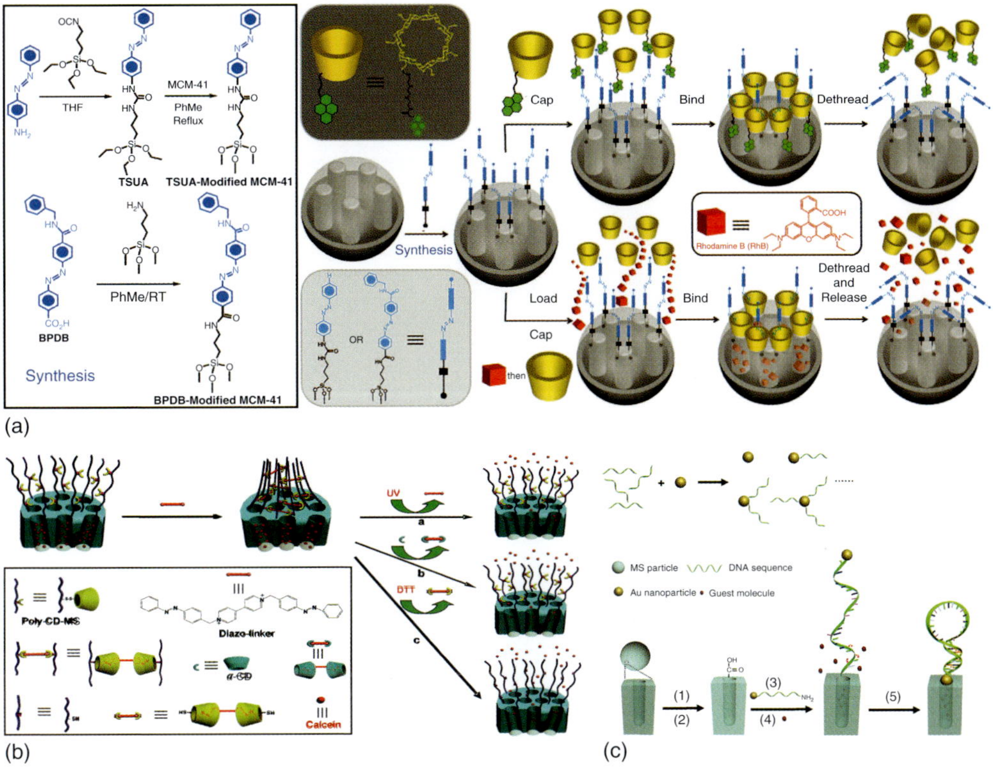

Figure 6.17 (a) Mesoporous silica nanoparticle (MSN) carriers with azobenzene as the photovalve. (b) Multistimulated MSN with light/α-CD/DTT as triggers. (Adapted with permission [92, 93]. Copyright 2009, American Chemical Society.) (c) MSN carriers with azobenzene-tethered DNA hairpin as gate keeper. (Adapted with permission [94]. Copyright 2012, Royal Society of Chemistry.)

by Feng *et al.* [93], in which light (azobenzene), host–guest (α-CD), and redox Dithiothreitol (DTT) were used as triggers to release the cargo (Figure 6.17b). An interesting work using photoswitchable DNA hairpin as cap molecule was demonstrated recently [94]. The *trans*-azobenzene-tethered DNA could form hairpin structure to cap the MSN, preventing the release of trapped cargo. After UV irradiation, the formed *cis*-azobenzene dehybridzed the DNA hairpin structure to ssDNA, which uncovered the MSN and released the cargo (Figure 6.17c).

Spiropyran, as another branch of photochromic materials, also plays an active role in photoswitchable functionalized mesoporous silica materials. Martínez-Máñez and coworkers modified the mesoporous MCM-41 framework with photochromic spiropyran derivatives for the photocontrolled release of substrates [95]. Spiropyran underwent a reversible photoisomerization to its zwitterionic merocyanine isomer under the irradiation of UV light (Figure 6.18a). In this system, $Ru(bipy)_3^{2+}$ was used as the substrate stuffed in the mesoporous MCM-41 framework and the negatively charged generation 1.5 poly(amidoamine) (G1.5 PAMAM) dendrimers were designed as nanoscopic molecular stoppers,

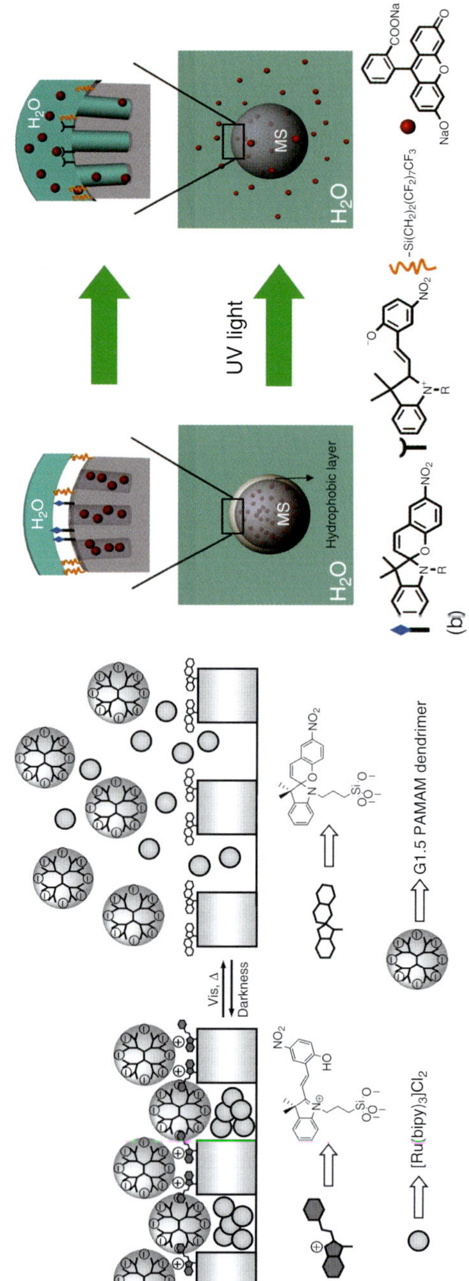

Figure 6.18 (a) Phototriggered release of cargo from MCM-41 modified with spiropyran. (Adapted with permission [95]. Copyright 2007, Wiley-VCH Verlag GmbH & Co. KGaA, Weinheim.) (b) Releasing cargo from MSN by photoswitchable wettability. (Adapted with permission [96]. Copyright 2014, American Chemical Society.)

which synergistically worked with spiropyran to perform as photoreversible caps. When the anchored spiropyran derivatives were in their zwitterionic merocyanine form, they interacted with negative-charged G1.5 PAMAM dendrimers via an electrostatic interaction and blocked the $Ru(bipy)_3^{2+}$ from releasing from the mesoporous MCM-41 framework. Irradiation of the system with the visible light turned the merocyanine isomers back to neutral spiropyran form, resulting in the dissociation of the blocking dendrimers from the MCM-41 and a consequent release of $Ru(bipy)_3^{2+}$ to the outside solution. One should note that this spiropyran-controlled release system requires a cooperation with another capping molecule, increasing the complexity of the materials design. This deficiency has been improved after a spiropyran/MSN carrier system was introduced by applying photoinduced wettability [96]. In this light-responsive MSN system, Jiang and coworkers functionalized MSN with optimal ratio of spiropyran and perfluorodecyltriethoxysilane (0.249 : 1). Under this ratio, the surface could be prevented from being wetted by blocking the loaded cargos inside the MSN. After the UV irradiation, the originally hydrophobic spiropyran transformed into its relatively hydrophilic merocyanine isomer, which enhanced the hydrophilicity of the surface and led to the wetted surface by outside water. In this situation, the carried water-soluble fluorescein disodium molecules were able to diffuse from the nanopores as shown in Figure 6.18b. Further applications in drug delivery and cancer therapy were then carried out. Here, anticancer drug camptothecin (CPT) was used as the loaded cargo and UV light that stimulated the enhanced cytotoxicity for EA.hy926 and HeLa cells was observed.

6.2.3.2 **Photo-nanoimpellers**

By modifying both the outside and inside of the MSN with azobenzene, Zink *et al.* carried out the proof-of-concept experiment on the MSN photo-nanoimpeller performance [90a,b]. It was discovered that (i) only under continual light irradiation of 413 nm (isosbestic point of azobenzene) could the loaded cargo be released; (ii) no increase in luminescence of the released cargo was observed when the irradiation was halted; and (iii) the release could be resumed when the 413 nm light was irradiating again. These phenomena confirmed the dynamic nature of this photo-nanoimpeller.

Based on their pioneering work, Zink *et al.* further constructed a photo-/pH dual-responsive MSN, which exhibited an AND logic [97]. As shown in Figure 6.19, apart from the modification of MSN with azobenzene inside, a pH-sensitive cucurbit[6]uril (CB[6])-based molecular shuttle was planted at the rim of the nanopore as a cap. Only when both light (phototriggered azobenzene wagging) and base (deprotonation of the shuttle to release the CB[6] cap) were applied could the trapped cargo be released from the MSN carrier, which formed an AND logical releasing of trapped cargos. These fundamental studies were followed by the successful realization of a phototriggered drug delivery system, which carried and released anticancer drug CPT inside the cancer cells, causing the apoptosis process [91a].

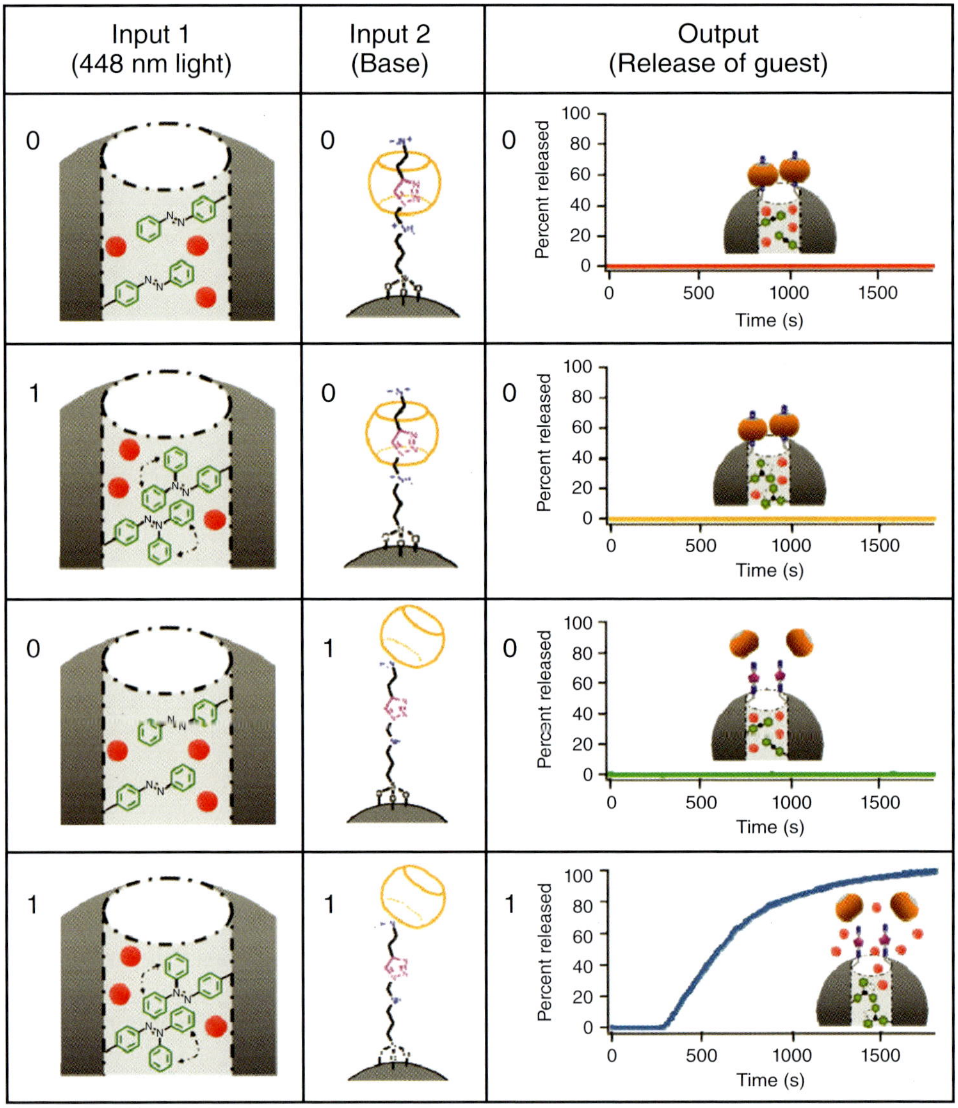

Figure 6.19 Truth table for an AND logic gate based on azobenzene-modified MSN. (Adapted with permission [97]. Copyright 2009, American Chemical Society.)

6.2.3.3 NIR Light-Triggered MSN Drug Delivery and Therapeutic Systems

The above-mentioned photochromic MSN-based drug delivery and cancer therapy experiment were mostly carried out *in vitro*. Many existing problems and limitations for practical applications still need to be improved and solved. One of most concerned limitations of phototriggered systems in biological and medicinal areas is the UV/short-waved visible light as the stimulation power source because of their poor tissue penetration ability and unwanted damage to the tissues. To solve

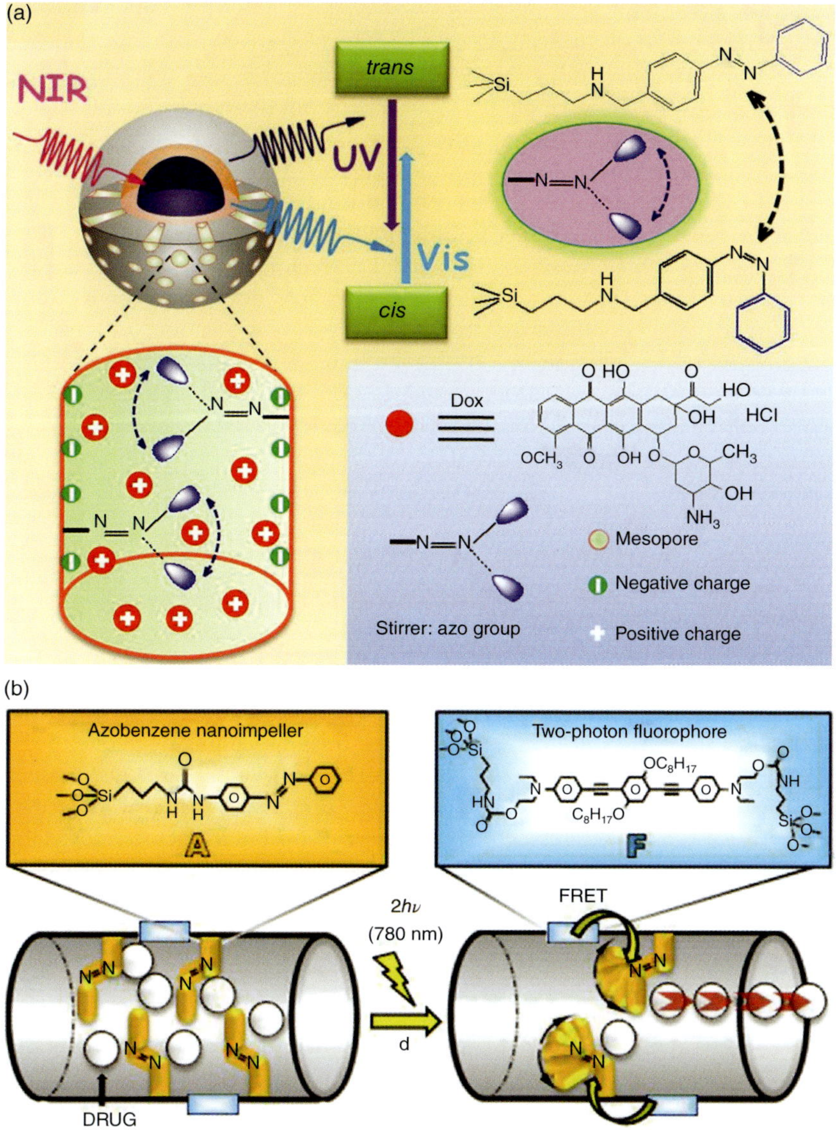

Figure 6.20 (a) Upconversion and (b) two-photon-based photorelease of drugs loaded in MSN. (Adapted with permission [91]. Copyright 2013, Wiley-VCH Verlag GmbH & Co. KGaA, Weinheim.)

this problem, two-photon excitation (TPE) and UCNPs technologies, providing long-waved visible and NIR light irradiation, are considered the best candidates.

In 2013, Shi group synthesized an azobenzene-modified, NaYF4:TmYb UCNPs coated mesoporous silica (Figure 6.20a). In this drug delivery system, the controlled release of cancer drug doxorubicin (Dox) was realized by using NIR light (980 nm) instead of UV light [91b].

In the same year, Zink *et al.* modified azobenzene along with a two-photon fluorophore into the MSN to realize a two-photon-triggered MSN nano-impeller for cancer drug delivery and release [91a]. The highlight of this interesting work is that the photo-driven wagging of azobenzene is not directly from the outside light source but from the efficient intermolecular FRET from the two-photon-stimulated fluorescence of TPE fluorophore (Figure 6.20b). This ingenious design ensures both the light-triggered cancer cell death and cell imaging, and meanwhile avoids damages to other normal cells under TPE irradiation.

Using biocompatible UCNP coatings and TPE technology has several advantages than traditional UV irradiation: (i) remarkable deep tissue penetration depth; (ii) avoiding photodamage to tissues by UV irradiation; (iii) precise control of the release amount by varying the intensity or time duration of NIR light irradiation; and (iv) no autofluorescence from biological samples and high photostability. Although only in its infancy, we could foresee the bright future of photocontrolled MSN drug carriers in clinical applications for battling vicious malignant tumors.

6.3 Photocontrolled Surface Conductance

As the precursor of developing optoelectronic device, photochromic functional surfaces are among the interests of many material scientists and chemists. Nonetheless, there are still many challenges in this burgeoning field, one of the key challenges is the observation of photoregulated switching of the conductance, particularly under the working condition similar to those real molecular devices. Important issues at present are charge transport in molecules; stochastic conformational changes versus conductance changes; stochastic switching versus controlled switching; stability of the monolayer or single-molecule bridge under working potentials, as well as the methods to obtain data of the photoswitching behavior from the surface. Most commonly used techniques available for measuring these well-defined SAMs or single-molecule nanoelectrodes nowadays are STM (Scanning Tunneling Microscopy), (Atomic Force Microscopy), mechanically controlled break junction (MCBJ), and so on. In this section, we will go on our journey on the conductance photoswitching from electrode/nanoparticle surface to the single-molecule level.

6.3.1 Photochromic Conductance Switching Based on SAMs

Since the first photochromic conductance study carried out by Shigekawa *et al.* on azobenzene by STM [98], a series of work have been devoted to this *cis*–*trans* induced conductance change system using either STM or c-AFM, or even Hg-drop electrode [99]. Generally speaking, the *cis*-isomer enjoys higher conductance due to their lower tunnel barrier length (decrease in the length of the molecule) and a lowering of the energy position of the LUMO (Lowest Unoccupied

Molecular Orbital) with respect to the electrode Fermi energy. To further increase the ON/OFF ratio of photoconductance between *cis–trans* isomers, different detection methods, anchoring/linking groups, and azobenzene molecular structures have been optimized. Lenfant and Vuilaume designed an azobenzene–thiophene derivative SAMs and measured the *trans–cis* induced conductance ON/OFF by c-AFM [99c]. This system achieved a precedent high ON/OFF conductance ratio up to average 1.5×10^3 (7×10^3 maximum). With these results, a fast solid-state molecular switch and memory device with a high ON/OFF conductance ratio could be envisioned using azobenzene molecules as building blocks (theoretically, with $\sigma \approx 10^{-18}$ cm^2 switching time about 1–10 μs could be achieved with a high-intensity light source).

One may notice that the photoconductance switching of azobenzene derivatives rely on their height/molecular length switching. As a result, devices that are sensitive to geometric changes (e.g., STM, AFM) are required during I/V measuring. In contrast, DTEs, with their molecular conjugation and electrochemical redox properties totally modified after photocyclization, are chosen as another promising family for photochromic conductance devices. In this molecule, the light-triggered conductance variation is mainly from a direct switching of intrinsic molecular resistivity (the geometrical structure of DTE is not altered significantly). The first demonstrated DTE-based conductance study was reported by Feringa *et al.* with thiolated DTE chemisorbed on MCBJ electrodes [48b]. These experiments demonstrated that the switching of a molecule from the closed to the open form results in a significant resistance increase in three orders of magnitude. This is another indication that closed and open forms exhibit intrinsically different charge-transport properties. However, once the switch is connected to gold via the Au–S bond, it can only be switched from the closed to the open form. This is because of the quenching of the excited state of the open form by the gold electrode, as discussed in Section 6.2.1. Leading by the systematic research of DTE photochromic behavior on gold with different linker by Feringa group, which demonstrated that the quenching effects were much lower when the linker group was *meta*-thiolated phenyl group, a new reversible DTE-based MCBJ was constructed [100] (Figure 6.21a). A photoreversible macroscopic alteration of the current in this solid-state molecular electronic device was achieved with a decent ON/OFF ratio of 16, under alternate irradiation of UV–Vis lights. Apart from the flat electrode, photoconductance was also performed on AuNPs, demonstrated independently by Matsuda, Irie, and van der Molen, Feringa, respectively [101, 102]. These devices were fabricated with DTE-bridged interdigitated nanoparticles (Figure 6.21b). Upon irradiation of UV light, the generated closed DTE exhibited a higher conductance while the visible light reversed open isomer presented a relatively lower conductance. Through careful manipulations, ON/OFF ratios ranged from 1.2 to 3.8 were achieved.

Other DTE-based conductance work also sprung up in the past decades either on STM [103] and AFM [104]. Nishihara and coworkers used an ethylene anchor to attach DTE on a hydrogen-terminated silicon (111) surface [104]. This system underwent photochemically reversible current changes upon alternate irradiation with UV–Vis light. c-AFM was used to measure the current changes on the

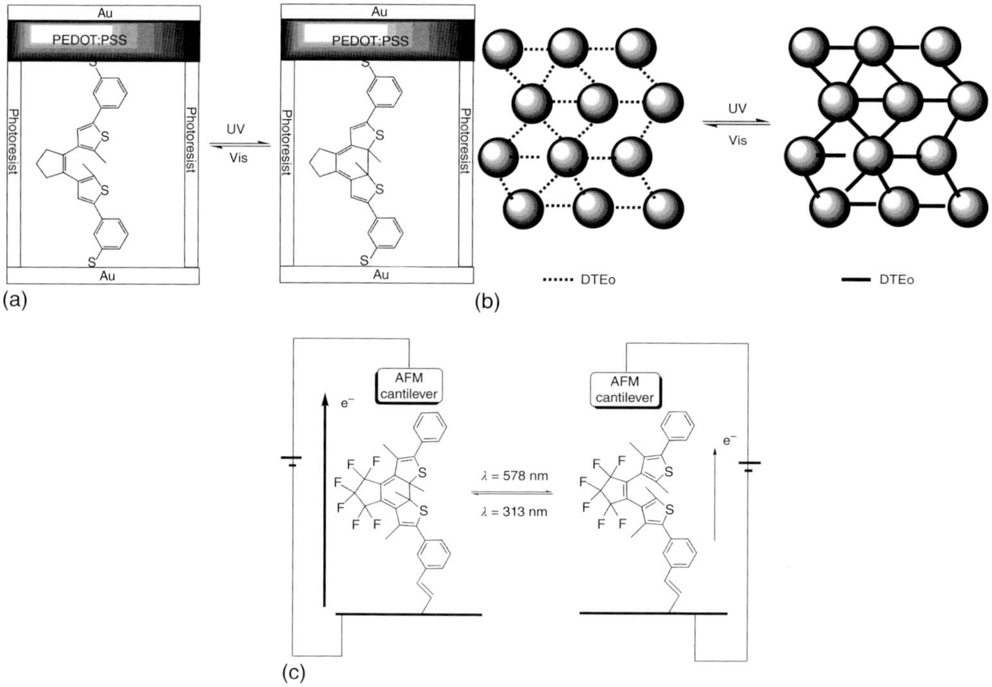

Figure 6.21 (a) Schematic cross section of the device layout of a large-area molecular junction in which the diarylethene is sandwiched between Au and poly(3,4-ethylenedioxythiophene):poly(4-styrenesulfonic acid) (PEDOT:PSS)/Au. (b) Schematic picture of a photoswitchable nanoparticle network modified with diarylethenes. (c) Current changes of the open and closed DTE junctions by UV and visible light.

modified silicon electrodes (Figure 6.21c). The closed isomer showed a relatively higher current value than that of the open isomer, which indicated a better conductance of closed isomer compared to the open counterpart. The reversibility of current switching upon irradiation was directly related to the π-conjugation change between the open and closed DTE isomers. Understanding the interplay between the molecular structures and the switching properties can provide firm design principles for the creation of optoelectronic devices.

6.3.2 Photochromic Conductance on Single-Molecule Level

To realize the functional electronic devices such as transistors and storage devices, further investigation of charge transport to single-molecule level is necessary and has aroused more and more interest. Since the commencement of electrical studies of photochromic molecules, measurements of the charge transport properties of photoreversible ensembles by using SAMs [100], nanoparticle networks [102], AFM [104], STM [105], MCBJs [48b, 106], and carbon nanomaterials [107] have been performed successfully and started to be down to single-molecule level. Containing a core of aromatic ring that can be switched open and closed by irradiation, diarylethenes (mainly DTEs) are one of the most popular example of photochromic molecules studied in single-molecule level. Upon their ring-opening/ring-closure reaction, the conjugation of the electronic π-system and therefore the conductance are supposed to be strongly affected as well. This ring-opening/ring-closure reaction is accompanied by only a small geometrical change, which makes diarylethene molecules the promising building blocks for optoelectronic applications [108]. Related theoretical studies have also been done to explain the mechanism and assist the design of smart diarylethene switches to perform the optically controlled single-molecule conductance alternation [109].

Early work mainly used the thiolated DTE, which has been argued that the photoswitching behaviors may be severely affected or even blocked through strong electronic coupling by direct covalent linking of DTE on the electrode [46–49]. One feasible solution to circumvent this problem is to use the physical adsorption instead of covalent-linked chemisorption. Ralph *et al.* investigated the photoswitchable conductance of break-junctions with pyridine-terminated DTEs. In this study, the forming of the DTE junctions was based on the adsorption of pyridine group on the gold surfaces, avoiding the direct interaction of the surface and the molecule. The well-defined photoswitchable conductance changes were observed with an ON/OFF ratio of at least 30 ($(3.3 \pm 0.5) \times 10^{-5}$ G_0 for closed form and $(1.5 \pm 0.5) \times 10^{-6}$ G_0 for open form) [110]. Sheer group further synthesized sulfur-free difurylethene derivatives (using furan moieties instead of traditional thiophene moieties) with different terminating non-thiol groups, namely pyridine (4Py), cyanide (MN), thiourea (TSC), conjugated acrylthiol (YnPhT), and nonconjugated acrylthiol (ThM) (see Figure 6.22 [111]). The conductance switching differences on ending groups, molecular length, and conjugation were systematically studied (Table 6.1). It is found that difurylethene-4Py exhibited the best switching performance, while nonconjugated difurylethene-ThM behave differently from its conjugated YnPhT counterpart and displayed a lowest ON/OFF ratio. This reveals that both the conductance properties and the switching ratios are significantly influenced by the end groups and by the side chains of the molecules. By analyzing the I/V curves within the framework of the single-level transport model, it is discovered that the alignment of the dominant transport level and the level broadening vary unambiguously between open and closed forms, and that the delocalization of the π-electron system significantly amplifies the coupling. The higher conductance values of the closed forms of the difurylethene molecules as compared to the open forms are explained by the strong increase in this broadening, which overcompensates even a more off-resonant transport situation found in some cases for the closed forms. This work reveals the importance of molecular design in construction of highly efficient molecular devices. However, these experiments were taken under low temperatures. While in the practical applications, the thermostability should be taken into consideration since the furyl group is less thermostable than its thienyl brother [112]. The heat generated in the electric

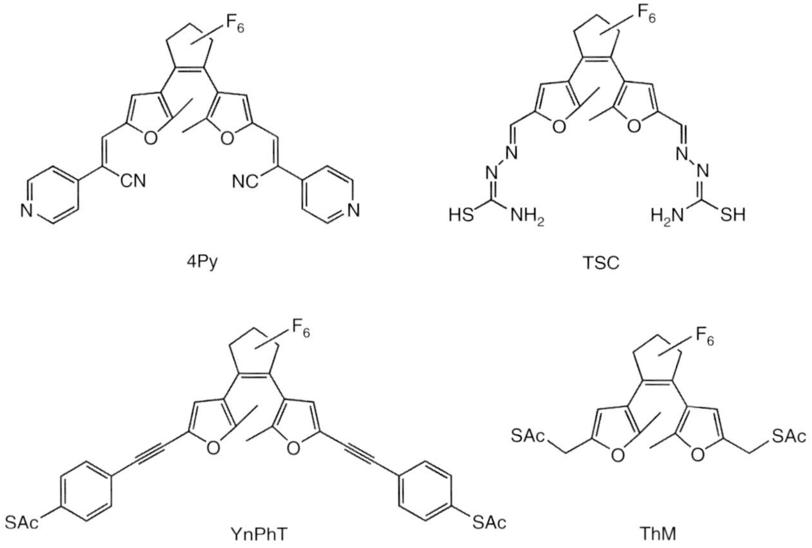

Figure 6.22 Structures of four molecules used in the difurylethene single-molecule conductance study.

Table 6.1 Length of the molecules in their open and closed forms, conductance values, and conductance switching ratios [111a].

	Length of molecule (Å)		Conductance (G_0)		Conductance switching ratio
	Open	Closed	Open	Closed	
4Py	19.05	16.66	$(1.2 \pm 0.5) \times 10^{-7}$	$(4.6 \pm 0.9) \times 10^{-6}$	38.3 ± 17.6
TSC	16.45	16.04	$(7.2 \pm 3.2) \times 10^{-8}$	$(7.5 \pm 1.9) \times 10^{-7}$	10.4 ± 5.3
YnPhT	22.80	19.77	$(1.1 \pm 0.2) \times 10^{-7}$	$(1.3 \pm 0.4) \times 10^{-6}$	11.8 ± 4.2
ThM	9.46	8.64	$(1.4 \pm 1.0) \times 10^{-7}$	$(8.3 \pm 4.5) \times 10^{-}7$	5.9 ± 5.3

circuit at ambient environment would lead to the thermo-induced opening of the closed diarylethene switches, which impairs the fatigue and sustainability of the device and should be avoided in practical applications.

Due to their high electronic conductivity, carbon nanomaterials including single-wall nanotubes (SWNT) and graphenes have recently joined the molecular junction family. Early in 2007, Guo and Nuckolls have discovered the reversible switching of diarylethenes in carbon nanotube-based junctions with semi-conjugated amide linkers [107a]. Recently, optically switchable graphene molecular junctions with azobenzenes and diarylethenes have been demonstrated by Guo group [107b,c]. For the diarylethene junction, high and gradually increasing ON/OFF ratios from 60 to 300 have been observed for three diarylethene derivatives investigated. The tendency of ON/OFF ratio change highly depends on the side groups of the DTE switching core. Thus, the conductance switching is dominated by the changes in the molecular energy levels as a result of either conformation transformation or the side group substitution.

Diarylethene-based single-molecule junctions can also be constructed through organometallic side chains [113] and rotaxane wirings [114]. Other photochromic single-molecule junctions based on azobenzenes and dimethyldihydropyrenes (DHPs) have also been reported [99a, 107b, 115], which we would not go further here.

References

1. (a) Balzani, V., Gomez-Lopez, M., and Stoddart, J.F. (1998) Molecular machines. *Acc. Chem. Res.*, **31** (7), 405–414; (b) Balzani, V., Credi, A., Raymo, F.M., and Stoddart, J.F. (2000) Artificial molecular machines. *Angew. Chem. Int. Ed.*, **39** (19), 3349–3391; (c) Orgiu, E. and Samori, P. (2014) 25th anniversary article: organic electronics marries photochromism: generation of multifunctional interfaces, materials, and devices. *Adv. Mater.*, **26** (12), 1827–1845; (d) Balzani, V., Credi, A., and Venturi, M. (2008) *Molecular Devices and Machines: Concepts and Perspectives for the Nanoworld*, Wiley-VCH Verlag GmbH, Weinheim.
2. (a) Browne, W.R. and Feringa, B.L. (2009) Light switching of molecules on surfaces. *Annu. Rev. Phys. Chem.*, **60**, 407–428; (b) Pathem, B.K., Claridge, S.A., Zheng, Y.B., and Weiss, P.S. (2013) Molecular switches and motors on surfaces. *Annu. Rev. Phys. Chem.*, **64**, 605–630; (c) Collin, J.P., Dietrich-Buchecker, C., Gavina, P., Jimenez-Molero, M.C., and Sauvage, J.P. (2001) Shuttles and muscles: linear molecular machines based on transition metals. *Acc. Chem. Res.*, **34** (6), 477–487; (d) Feringa, B.L. (2001) *Molecular Switches*, Wiley-VCH Verlag GmbH, Weinheim.
3. Rogers, J.A., Bao, Z., Baldwin, K., Dodabalapur, A., Crone, B., Raju, V.R., Kuck, V., Katz, H., Amundson, K., Ewing, J., and Drzaic, P. (2001) Paper-like electronic displays: large-area rubber-stamped plastic sheets of electronics and microencapsulated electrophoretic inks. *Proc. Natl. Acad. Sci. U.S.A.*, **98** (9), 4835–4840.
4. Russell, T.P. (2002) Surface-responsive materials. *Science*, **297** (5583), 964–967.
5. (a) Abbott, N.L., Gorman, C.B., and Whitesides, G.M. (1995) Active control of wetting using applied electrical potentials and self-assembled monolayers. *Langmuir*, **11** (1), 16–18; (b) Mu, L., Liu, Y., Cai, S.Y., and Kong, J.L. (2007) A smart surface in a microfluidic chip for controlled protein separation. *Chem. Eur. J.*, **13** (18), 5113–5120.
6. (a) Wen, L., Hou, X., Tian, Y., Nie, F.Q., Song, Y., Zhai, J., and Jiang, L. (2010) Bioinspired smart gating of nanochannels toward photoelectric-conversion systems. *Adv. Mater.*, **22** (9), 1021–1024; (b) Hou, X., Zhang, H., and Jiang, L. (2012) Building bio-inspired artificial functional nanochannels: from symmetric to asymmetric modification. *Angew. Chem. Int. Ed.*, **51** (22), 5296–5307; (c) Banghart, M.R., Mourot, A., Fortin, D.L., Yao, J.Z., Kramer, R.H., and Trauner, D. (2009) Photochromic blockers of voltage-gated potassium channels. *Angew. Chem. Int. Ed.*, **48** (48), 9097–9101; (d) Tochitsky, I., Banghart, M.R., Mourot, A., Yao, J.Z., Gaub, B., Kramer, R.H., and Trauner, D. (2012) Optochemical control of genetically engineered neuronal nicotinic acetylcholine receptors. *Nat. Chem.*, **4** (2), 105–111; (e) Stein, M., Middendorp, S.J., Carta, V., Pejo, E., Raines, D.E., Forman, S.A., Sigel, E., and Trauner, D. (2012) Azo-propofols: photochromic potentiators

of GABAA receptors. *Angew. Chem. Int. Ed.*, **51** (42), 10500–10504; (f) Ying, Y.L., Zhang, J.J., Meng, F.N., Cao, C., Yao, X., Willner, I., Tian, H., and Long, Y.T. (2013) A stimuli-responsive nanopore based on a photoresponsive host-guest system. *Sci. Rep.*, **3**, 01662; (g) Zhang, X., Zhang, J.J., Ying, Y.L., Tian, H., and Long, Y.T. (2014) Single molecule analysis of light-regulated RNA: spiropyran interactions. *Chem. Sci.*, **5** (7), 2642–2646.

7. (a) Velema, W.A., Szymanski, W., and Feringa, B.L. (2014) Photopharmacology: beyond proof of principle. *J. Am. Chem. Soc.*, **136** (6), 2178–2191; (b) Zhang, J.J., Wang, J., and Tian, H. (2014) Taking orders from light: progress in photochromic bio-materials. *Mater. Horiz.*, **1** (2), 169–184.
8. (a) Pace, G., Ferri, V., Grave, C., Elbing, M., von Hanisch, C., Zharnikov, M., Mayor, M., Rampi, M.A., and Samorì, P. (2007) Cooperative light-induced molecular movements of highly ordered azobenzene self-assembled monolayers. *Proc. Natl. Acad. Sci. U.S.A.*, **104** (24), 9937–9942; (b) Dri, C., Peters, M.V., Schwarz, J., Hecht, S., and Grill, L. (2008) Spatial periodicity in molecular switching. *Nat. Nanotechnol.*, **3** (11), 649–653; (c) London, G., Carroll, G.T., Landaluce, T.F., Pollard, M.M., Rudolf, P., and Feringa, B.L. (2009) Light-driven altitudinal molecular motors on surfaces. *Chem. Commun.*, **13**, 1712–1714; (d) Davis, J.J., Orlowski, G.A., Rahman, H., and Beer, P.D. (2010) Mechanically interlocked and switchable molecules at surfaces. *Chem. Commun.*, **46** (1), 54–63.
9. Bard, A.J. and Faulkner, L.R. (2001) *Electrochemical Methods: Fundamentals and Applications*, John Wiley & Sons, Inc., New York.
10. Lion-Dagan, M., Katz, E., and Willner, I. (1994) Amperometric transduction of optical signals recorded by organized monolayers of photoisomerizable biomaterials on Au electrodes. *J. Am. Chem. Soc.*, **116** (17), 7913–7914.
11. (a) Willner, I., Doron, A., Katz, E., Levi, S., and Frank, A.J. (1996) Reversible associative and dissociative interactions of glucose oxidase with nitrospiropyran monolayers assembled onto gold electrodes: amperometric transduction of recorded optical signals. *Langmuir*, **12** (4), 946–954; (b) Blonder, R., Willner, I., and Bückmann, A. (1998) Reconstitution of apo-glucose oxidase on nitrospiropyran and FAD mixed monolayers on gold electrodes: photostimulation of bioelectrocatalytic features of the biocatalyst. *J. Am. Chem. Soc.*, **120** (36), 9335–9341.
12. Niazov, T., Shlyahovsky, B., and Willner, I. (2007) Photoswitchable electrocatalysis and catalyzed chemiluminescence using photoisomerizable monolayer-functionalized surfaces and Pt nanoparticles. *J. Am. Chem. Soc.*, **129** (20), 6374–6375.
13. Yehezkeli, O., Moshe, M., Tel-Vered, R., Feng, Y., Li, Y., Tian, H., and Willner, I. (2010) Switchable photochemical/electrochemical wiring of glucose oxidase with electrodes. *Analyst*, **135** (3), 474–476.
14. (a) Sun, T., Feng, L., Gao, X., and Jiang, L. (2005) Bioinspired surfaces with special wettability. *Acc. Chem. Res.*, **38** (8), 644–652; (b) Shi, F., Song, Y., Niu, J., Xia, X., Wang, Z., and Zhang, X. (2006) Facile method to fabricate a large-scale superhydrophobic surface by galvanic cell reaction. *Chem. Mater.*, **18** (5), 1365–1368; (c) Liu, Y., Mu, L., Liu, B., and Kong, J. (2005) Controlled switchable surface. *Chem. Eur. J.*, **11** (9), 2622–2631; (d) Öner, D. and McCarthy, T.J. (2000) Ultrahydrophobic surfaces. Effects of topography length scales on wettability. *Langmuir*, **16** (20), 7777–7782; (e) Wang, R., Hashimoto, K., Fujishima, A., Chikuni, M., Kojima, E., Kitamura, A., Shimohigoshi, M., and Watanabe, T. (1998) Photogeneration of highly amphiphilic TiO_2 surfaces. *Adv. Mater.*, **10** (2), 135–138.
15. (a) Lim, H.S., Han, J.T., Kwak, D., Jin, M., and Cho, K. (2006) Photoreversibly switchable superhydrophobic surface with erasable and rewritable pattern. *J. Am. Chem. Soc.*, **128** (45), 14458–14459; (b) Min, M., Bang, G.S., Lee, H., and Yu, B.C. (2010) A photoswitchable methylene-spaced

fluorinated aryl azobenzene monolayer grafted on silicon. *Chem. Commun.*, **46** (29), 5232–5234; (c) Feng, C., Zhang, Y., Jin, J., Song, Y., Xie, L., Qu, G., Jiang, L., and Zhu, D. (2001) Reversible wettability of photoresponsive fluorine-containing azobenzene polymer in Langmuir-Blodgett films. *Langmuir*, **17** (15), 4593–4597; (d) Chen, K.Y., Ivashenko, O., Carroll, G.T., Robertus, J., Kistemaker, J.C.M., London, G., Browne, W.R., Rudolf, P., and Feringa, B.L. (2014) Control of surface wettability using tripodal light-activated molecular motors. *J. Am. Chem. Soc.*, **136** (8), 3219–3224.

16. Carroll, G.T., London, G., Landaluce, T.F., Rudolf, P., and Feringa, B.L. (2011) Adhesion of photon-driven molecular motors to surfaces via 1, 3-dipolar cycloadditions: effect of interfacial interactions on molecular motion. *ACS Nano*, **5** (1), 622–630.
17. (a) Demirel, G.B., Dilsiz, N., Çakmak, M., and Çaykara, T. (2011) Molecular design of photoswitchable surfaces with controllable wettability. *J. Mater. Chem.*, **21** (9), 3189–3196; (b) Demirel, G.B., Dilsiz, N., Ali Ergün, M., Çakmak, M., and Çaykara, T. (2011) Photocontrollable DNA hybridization on reversibly photoresponsive surfaces. *J. Mater. Chem.*, **21** (28), 10415–10420.
18. (a) Rosario, R., Gust, D., Garcia, A.A., Hayes, M., Taraci, J.L., Clement, T., Dailey, J.W., and Picraux, S.T. (2004) Lotus effect amplifies light-induced contact angle switching. *J. Phys. Chem. B*, **108** (34), 12640–12642; (b) Rosario, R., Gust, D., Hayes, M., Jahnke, F., Springer, J., and Garcia, A.A. (2002) Photon-modulated wettability changes on spiropyran-coated surfaces. *Langmuir*, **18** (21), 8062–8069.
19. Ichimura, K., Oh, S.K., and Nakagawa, M. (2000) Light-driven motion of liquids on a photoresponsive surface. *Science*, **288** (5471), 1624–1626.
20. Berná, J., Leigh, D.A., Lubomska, M., Mendoza, S.M., Pérez, E.M., Rudolf, P., Teobaldi, G., and Zerbetto, F. (2005) Macroscopic transport by synthetic molecular machines. *Nat. Mater.*, **4** (9), 704–710.
21. Takahashi, I., Honda, Y., and Hirota, S. (2009) Regulating copper-binding affinity with photoisomerizable azobenzene ligand by construction of a self-assembled monolayer. *Angew. Chem. Int. Ed.*, **48** (33), 6065–6068.
22. Callari, F.L. and Sortino, S. (2007) "Catch-and-release" of porphyrins by photoswitchable self-assembled monolayers. *J. Mater. Chem.*, **17** (39), 4184–4188.
23. (a) Wen, G., Yan, J., Zhou, Y., Zhang, D., Mao, L., and Zhu, D. (2006) Photomodulation of the electrode potential of a photochromic spiropyran-modified Au electrode in the presence of Zn^{2+}: a new molecular switch based on the electronic transduction of the optical signals. *Chem. Commun.*, **28**, 3016–3018; (b) Fries, K., Samanta, S., Orski, S., and Locklin, J. (2008) Reversible colorimetric ion sensors based on surface initiated polymerization of photochromic polymers. *Chem. Commun.*, **47**, 6288–6290.
24. Zhang, J.J., Riskin, M., Tel-Vered, R., Tian, H., and Willner, I. (2011) Optically activated uptake and release of Cu^{2+} or Ag^{+} ions by or from a photoisomerizable monolayer-modified electrode. *Langmuir*, **27** (4), 1380–1386.
25. Zhang, J.J., Riskin, M., Freeman, R., Tel-Vered, R., Balogh, D., Tian, H., and Willner, I. (2011) Electrochemically triggered Au nanoparticles "sponges" for the controlled uptake and release of a photoisomerizable dithienylethene guest substrate. *ACS Nano*, **5** (7), 5936–5944.
26. (a) Wang, Q.C., Qu, D.H., Ren, J., Chen, K.C., and Tian, H. (2004) A lockable light-driven molecular shuttle with a fluorescent signal. *Angew. Chem. Int. Ed.*, **43** (20), 2661–2665; (b) Qu, D.H., Wang, Q.C., and Tian, H. (2005) A half adder based on a photochemically driven [2]rotaxane. *Angew. Chem. Int. Ed.*, **44** (33), 5296–5299; (c) Tian, H. and Wang, Q.C. (2006) Recent progress on switchable rotaxanes. *Chem. Soc. Rev.*, **35** (4), 361–374.
27. Wan, P., Jiang, Y., Wang, Y., Wang, Z., and Zhang, X. (2008) Tuning surface

wettability through photocontrolled reversible molecular shuttle. *Chem. Commun.*, **44**, 5710–5712.

28. Wan, P., Wang, Y., Jiang, Y., Xu, H., and Zhang, X. (2009) Fabrication of reactivated biointerface for dual-controlled reversible immobilization of cytochrome C. *Adv. Mater.*, **21** (43), 4362–4365.
29. Wan, P., Xing, Y., Chen, Y., Chi, L., and Zhang, X. (2011) Host–guest chemistry at interface for photoswitchable bioelectrocatalysis. *Chem. Commun.*, **47** (21), 5994–5996.
30. (a) Zhang, Y., Ren, L., Tu, Q., Wang, X., Liu, R., Li, L., Wang, J.C., Liu, W., Xu, J., and Wang, J. (2011) Fabrication of reversible poly (dimethylsiloxane) surfaces via host–guest chemistry and their repeated utilization in cardiac biomarker analysis. *Anal. Chem.*, **83** (24), 9651–9659; (b) Wang, D., Xie, D., Shi, W., Sun, S., and Zhao, C. (2013) Designing a photoresponsive molecularly imprinted system on a silicon wafer substrate surface. *Langmuir*, **29** (26), 8311–8319.
31. Baron, R., Onopriyenko, A., Katz, E., Lioubashevski, O., Willner, I., Wang, S., and Tian, H. (2006) An electrochemical/photochemical information processing system using a monolayer-functionalized electrode. *Chem. Commun.*, **20**, 2147–2149.
32. Areephong, J., Browne, W.R., Katsonis, N., and Feringa, B.L. (2006) Photo- and electro-chromism of diarylethene modified ITO electrodes—towards molecular based read–write–erase information storage. *Chem. Commun.*, **37**, 3930–3932.
33. Riskin, M., Gutkin, V., Felner, I., and Willner, I. (2008) Photochemical and electrochemical encoding of erasable magnetic patterns. *Angew. Chem. Int. Ed.*, **47** (23), 4416–4420.
34. (a) Namiki, K., Sakamoto, A., Murata, M., Kume, S., and Nishihara, H. (2007) Reversible photochromism of a ferrocenylazobenzene monolayer controllable by a single green light source. *Chem. Commun.*, **44**, 4650–4652; (b) Ivashenko, O., van Herpt, J.T., Feringa, B.L., Rudolf, P., and Browne, W.R. (2013) Electrochemical write and read functionality through oxidative dimerization of spiropyran self-assembled monolayers on gold. *J. Phys. Chem. C*, **117** (36), 18567–18577.
35. Klajn, R., Stoddart, J.F., and Grzybowski, B.A. (2010) Nanoparticles functionalised with reversible molecular and supramolecular switches. *Chem. Soc. Rev.*, **39** (6), 2203–2237.
36. (a) Giljohann, D.A., Seferos, D.S., Daniel, W.L., Massich, M.D., Patel, P.C., and Mirkin, C.A. (2010) In vitro selection of structure-switching signaling aptamers. *Angew. Chem. Int. Ed.*, **49**, 3280–3294; (b) De, M., Ghosh, P.S., and Rotello, V.M. (2008) Applications of nanoparticles in biology. *Adv. Mater.*, **20** (22), 4225–4241; (c) Jung, E., Kim, S., Kim, Y., Seo, S.H., Lee, S.S., Han, M.S., and Lee, S. (2011) A colorimetric high-throughput screening method for palladium-catalyzed coupling reactions of aryl iodides using a gold nanoparticle-based iodide-selective probe. *Angew. Chem. Int. Ed.*, **123** (19), 4478–4481; (d) Luo, W., Zhu, C., Su, S., Li, D., He, Y., Huang, Q., and Fan, C. (2010) Self-catalyzed, self-limiting growth of glucose oxidase-mimicking gold nanoparticles. *ACS Nano*, **4** (12), 7451–7458.
37. (a) Freeman, R., Finder, T., and Willner, I. (2009) Multiplexed analysis of Hg^{2+} and Ag^{+} ions by nucleic acid functionalized CdSe/ZnS quantum dots and their use for logic gate operations. *Angew. Chem. Int. Ed.*, **48** (42), 7818–7821; (b) Yildiz, H.B., Tel-Vered, R., and Willner, I. (2008) Solar cells with enhanced photocurrent efficiencies using oligoaniline-crosslinked Au/CdS nanoparticles arrays on electrodes. *Adv. Mater.*, **18** (21), 3497–3505; (c) Gill, R., Zayats, M., and Willner, I. (2008) Semiconductor quantum dots for bioanalysis. *Angew. Chem. Int. Ed.*, **47** (40), 7602–7625; (d) Ning, Z.J., Voznyy, O., Pan, J., Hoogland, S., Adinolfi, V., Xu, J.X., Li, M., Kirmani, A.R., Sun, J.P., Minor, J., Kemp, K.W., Dong, H.P., Rollny, L., Labelle, A., Carey, G., Sutherland, B., Hill, I.G., Amassian, A., Liu, H., Tang, J., Bakr, O.M., and

Sargent, E.H. (2014) Air-stable n-type colloidal quantum dot solids. *Nat. Mater.*, **13** (8), 822–828.

38. (a) Yang, Y.M., Zhao, Q., Feng, W., and Li, F.Y. (2013) Luminescent chemodosimeters for bioimaging. *Chem. Rev.*, **113** (1), 192–270; (b) Zhou, J., Liu, Z., and Li, F.Y. (2012) Upconversion nanophosphors for small-animal imaging. *Chem. Soc. Rev.*, **41** (3), 1323–1349; (c) Gorris, H.H. and Wolfbeis, O.S. (2013) Photon-upconverting nanoparticles for optical encoding and multiplexing of cells, biomolecules, and microspheres. *Angew. Chem. Int. Ed.*, **52** (13), 3584–3600; (d) Fischer, L.H., Harms, G.S., and Wolfbeis, O.S. (2011) Upconverting nanoparticles for nanoscale thermometry. *Angew. Chem. Int. Ed.*, **50** (20), 4546–4551.

39. (a) Hu, J., Zhang, J., Liu, F., Kittredge, K., Whitesell, J.K., and Fox, M.A. (2001) Competitive photochemical reactivity in a self-assembled monolayer on a colloidal gold cluster. *J. Am. Chem. Soc.*, **123** (7), 1464–1470; (b) Shin, K.H. and Shin, E.J. (2008) Photoresponsive azobenzene-modified gold nanoparticle. *Bull. Korean Chem. Soc.*, **29** (6), 1259.

40. Klajn, R., Bishop, K.J.M., and Grzybowski, B.A. (2007) Light-controlled self-assembly of reversible and irreversible nanoparticle suprastructures. *Proc. Natl. Acad. Sci. U.S.A.*, **104** (25), 10305–10309.

41. (a) Evans, S.D., Johnson, S.R., Ringsdorf, H., Williams, L.M., and Wolf, H. (1998) Photoswitching of azobenzene derivatives formed on planar and colloidal gold surfaces. *Langmuir*, **14** (22), 6436–6440; (b) Sortino, S., Petralia, S., Conoci, S., and Di Bella, S. (2004) Monitoring photoswitching of azobenzene-based self-assembled monolayers on ultrathin platinum films by UV/Vis spectroscopy in the transmission mode. *J. Mater. Chem.*, **14** (5), 811–813; (c) Weidner, T., Brethauer, F., Ballav, N., Motschmann, H., Orendi, H., Bruhn, C., Siemeling, U., and Zharnikov, M. (2008) Correlation between the molecular structure and photoresponse in aliphatic self-assembled monolayers with azobenzene tailgroups. *Langmuir*, **24** (20), 11691–11700.

42. (a) Tamada, K., Akiyama, H., and Wei, T.X. (2002) Photoisomerization reaction of unsymmetrical azobenzene disulfide self-assembled monolayers studied by surface plasmon spectroscopy: influences of side chain length and contacting medium. *Langmuir*, **18** (13), 5239–5246; (b) Akiyama, H., Tamada, K., Nagasawa, J., Abe, K., and Tamaki, T. (2003) Photoreactivity in self-assembled monolayers formed from asymmetric disulfides having para-substituted azobenzenes. *J. Phys. Chem. B*, **107** (1), 130–135.

43. (a) Kalsin, A.M., Kowalczyk, B., Wesson, P., Paszewski, M., and Grzybowski, B.A. (2007) Studying the thermodynamics of surface reactions on nanoparticles by electrostatic titrations. *J. Am. Chem. Soc.*, **129** (21), 6664–6665; (b) Campbell, C.J., Soh, S., and Grzybowski, B.A. (2008) Blocking of disulfide adsorption by coadsorbing omega-functionalized alkane thiols revealed by wet stamping and fluorescence microscopy. *Langmuir*, **24** (20), 11600–11604; (c) Shon, Y.S., Mazzitelli, C., and Murray, R.W. (2001) Unsymmetrical disulfides and thiol mixtures produce different mixed monolayer-protected gold clusters. *Langmuir*, **17** (25), 7735–7741; (d) Choo, H., Cutler, E., and Shon, Y.S. (2003) Synthesis of mixed monolayer-protected gold clusters from thiol mixtures: variation in the tail group, chain length, and solvent. *Langmuir*, **19** (20), 8555–8559.

44. Pollard, M.M., ter Wiel, M.K.J., van Delden, R.A., Vicario, J., Koumura, N., van den Brom, C.R., Meetsma, A., and Feringa, B.L. (2008) Light-driven rotary molecular motors on gold nanoparticles. *Chem. Eur. J.*, **14** (36), 11610–11622.

45. Callari, F., Petralia, S., and Sortino, S. (2006) Highly photoresponsive monolayer-protected gold clusters

by self-assembly of a cyclodextrin-azobenzene-derived supramolecular complex. *Chem. Commun.*, **9**, 1009–1011.
46. Irie, M., Matsuda, K., and Yamaguchi, H. (2007) Excited-state behavior of a fluorescent and photochromic diarylethene on silver nanoparticles. *J. Phys. Chem. C*, **111** (10), 3853–3862.
47. Zhang, J., Whitesell, J.K., and Fox, M.A. (2001) Photoreactivity of self-assembled monolayers of azobenzene or stilbene derivatives capped on colloidal gold clusters. *Chem. Mater.*, **13** (7), 2323–2331.
48. (a) Kudernac, T., van der Molen, S., van Wees, B.J., and Feringa, B.L. (2006) Uni- and bi-directional light-induced switching of diarylethenes on gold nanoparticles. *Chem. Commun.*, **34**, 3597–3599; (b) Dulic, D., van der Molen, S.J., Kudernac, T., Jonkman, H.T., de Jong, J.J.D., Bowden, T.N., van Esch, J., Feringa, B.L., and van Wees, B.J. (2003) One-way optoelectronic switching of photochromic molecules on gold. *Phys. Rev. Lett.*, **91** (20), 207402.
49. (a) Li, J., Speyer, G., and Sankey, O.F. (2004) Conduction switching of photochromic molecules. *Phys. Rev. Lett.*, **93** (24), 248302; (b) Perrier, A., Maurel, F., and Aubard, J. (2007) Theoretical study of the electronic and optical properties of photochromic dithienylethene derivatives connected to small gold clusters. *J. Phys. Chem. A*, **111** (39), 9688–9698.
50. (a) Nishi, H., Asahi, T., and Kobatake, S. (2009) Light-controllable surface plasmon resonance absorption of gold nanoparticles covered with photochromic diarylethene polymers. *J. Phys. Chem. C*, **113** (40), 17359–17366; (b) Nishi, H. and Kobatake, S. (2008) Photochromism and optical property of gold nanoparticles covered with low-polydispersity diarylethene polymers. *Macromolecules*, **41** (11), 3995–4002.
51. (a) Nishi, H., Asahi, T., and Kobatake, S. (2011) Enhanced one-photon cycloreversion reaction of diarylethenes near individual gold nanoparticles. *J. Phys. Chem. C*, **115** (11), 4564–4570; (b) Nishi, H., Asahi, T., and Kobatake, S. (2011) Enhanced photocycloreversion reaction of diarylethene polymers attached to gold nanoparticles in the solid state. *J. Photochem. Photobiol., A*, **221** (2-3), 256–260.
52. Nishi, H., Asahi, T., and Kobatake, S. (2012) Plasmonic enhancement of a photocycloreversion reaction of a diarylethene derivative using individually dispersed silver nanoparticles. *Phys. Chem. Chem. Phys.*, **14** (14), 4898–4905.
53. (a) Raimondo, C., Reinders, F., Soydaner, U., Mayor, M., and Samorì, P. (2010) Light-responsive reversible solvation and precipitation of gold nanoparticles. *Chem. Commun.*, **46** (7), 1147–1149; (b) Sidhaye, D.S., Kashyap, S., Sastry, M., Hotha, S., and Prasad, B.L.V. (2005) Gold nanoparticle networks with photoresponsive interparticle spacings. *Langmuir*, **21** (17), 7979–7984; (c) Housni, A., Zhao, Y., and Zhao, Y. (2010) Using polymers to photoswitch the aggregation state of gold nanoparticles in aqueous solution. *Langmuir*, **26** (14), 12366–12370.
54. Wei, Y., Han, S., Kim, J., Soh, S., and Grzybowski, B.A. (2010) Photoswitchable catalysis mediated by dynamic aggregation of nanoparticles. *J. Am. Chem. Soc.*, **132** (32), 11018–11020.
55. Klajn, R., Wesson, P.J., Bishop, K.J.M., and Grzybowski, B.A. (2009) Writing self-erasing images using metastable nanoparticle "inks". *Angew. Chem. Int. Ed.*, **48** (38), 7035–7039.
56. (a) Asanuma, H., Ito, T., Yoshida, T., Liang, X., and Komiyama, M. (1999) Photoregulation of the formation and dissociation of a DNA duplex by using the cis-trans isomerization of azobenzene. *Angew. Chem. Int. Ed.*, **38** (16), 2393–2395; (b) Kang, H., Liu, H., Phillips, J.A., Cao, Z., Kim, Y., Chen, Y., Yang, Z., Li, J., and Tan, W.H. (2009) Single-DNA molecule nanomotor regulated by photons. *Nano Lett.*, **9** (7), 2690–2696; (c) Zhou, M., Liang, X., Mochizuki, T., and Asanuma, H. (2010) A light-driven DNA nanomachine for the efficient photoswitching of RNA digestion. *Angew. Chem. Int.

Ed., **49** (12), 2167–2170; (d) You, M., Wang, R.W., Zhang, X., Chen, Y., Wang, K., Peng, L., and Tan, W.H. (2010) Photon-regulated DNA-enzymatic nanostructures by molecular assembly. *ACS Nano*, **5** (12), 10090–10095.

57. Yan, Y., Chen, J.I.L., and Ginger, D.S. (2012) Photoswitchable oligonucleotide-modified gold nanoparticles: controlling hybridization stringency with photon dose. *Nano Lett.*, **12** (5), 2530–2536.
58. Peng, T., Dohno, C., and Nakatani, K. (2007) Bidirectional control of gold nanoparticle assembly by turning on and off DNA hybridization with thermally degradable molecular glue. *ChemBioChem*, **8** (5), 483–485.
59. Liu, D., Chen, W., Sun, K., Deng, K., Zhang, W., Wang, Z., and Jiang, X. (2011) Resettable, multi-readout logic gates based on controllably reversible aggregation of gold nanoparticles. *Angew. Chem. Int. Ed.*, **50** (18), 4103–4107.
60. (a) Shao, N., Jin, J., Wang, H., Zheng, J., Yang, R., Chan, W., and Abliz, Z. (2010) Design of bis-spiropyran ligands as dipolar molecule receptors and application to in vivo glutathione fluorescent probes. *J. Am. Chem. Soc.*, **132** (2), 725–736; (b) Shao, N., Wang, H., Gao, X., Yang, R., and Chan, W. (2010) Spiropyran-based fluorescent anion probe and its application for urinary pyrophosphate detection. *Anal. Chem.*, **82** (11), 4628–4636; (c) Sakata, T., Jackson, D.K., Mao, S., and Marriott, G. (2008) Optically switchable chelates: optical control and sensing of metal ions. *J. Org. Chem.*, **73** (1), 227–233.
61. Shiraishi, Y., Shirakawa, E., Tanaka, K., Sakamoto, H., Ichikawa, S., and Hirai, T. (2014) Spiropyran-modified gold nanoparticles: reversible size control of aggregates by UV and visible light irradiations. *ACS Appl. Mater. Interfaces*, **6** (10), 7554–7562.
62. Shiraishi, Y., Tanaka, K., Shirakawa, E., Sugano, Y., Ichikawa, S., Tanaka, S., and Hirai, T. (2013) Light-triggered self-assembly of gold nanoparticles based on photoisomerization of spirothiopyran. *Angew. Chem. Int. Ed.*, **52** (32), 8304–8308.
63. Lu, A.-H., Schmidt, W., Matoussevitch, N., Bonnemann, H., Spliethoff, B., Tesche, B., Bill, E., Kiefer, W., and Schuth, F. (2004) Nanoengineering of a magnetically separable hydrogenation catalyst. *Angew. Chem. Int. Ed.*, **43** (33), 4303–4306.
64. Gupta, M. and Gupta, A.K. (2005) Synthesis and surface engineering of iron oxide nanoparticles for biomedical applications. *Biomaterials*, **26** (18), 3995–4021.
65. (a) Mornet, S., Vasseur, S., Grasset, F., Verveka, P., Goglio, G., Demourgues, A., Portier, J., Pollert, E., and Duguet, E. (2006) Magnetic nanoparticle design for medical applications. *Prog. Solid State Chem.*, **34** (2-4), 237–247; (b) Gleich, B. and Weizenecker, J. (2005) Tomographic imaging using the non-linear response of magnetic particles. *Nature*, **435** (7046), 1214–1217.
66. Hyeon, T. (2003) Chemical synthesis of magnetic nanoparticles. *Chem. Commun.*, **8**, 927–934.
67. (a) Mahendran, V. and Philip, J. (2012) Nanofluid based optical sensor for rapid visual inspection of defects in ferromagnetic materials. *Appl. Phys. Lett.*, **100** (7), 073104; (b) Philip, J., Mahendran, V., and Felicia, L.J. (2013) A simple, in-expensive and ultrasensitive magnetic nanofluid based sensor for detection of cations, ethanol and ammonia. *J. Nanofluids*, **2** (2), 112–119.
68. (a) Mikami, R., Taguchi, M., Yamada, K., Suzuki, K., Sato, O., and Einaga, Y. (2004) Reversible photo-switching of the magnetization of iron oxide nanoparticles at room temperature. *Angew. Chem. Int. Ed.*, **43** (15), 6135–6139; (b) Taguchi, M., Yamada, K., Suzuki, K., Sato, O., and Einaga, Y. (2005) Photoswitchable magnetic nanoparticles of Prussian blue with amphiphilic azobenzene. *Chem. Mater.*, **17** (17), 4554–4559; (c) Suda, M., Nakagawa, M., Iyoda, T., and Einaga, Y. (2007) Reversible photoswitching of ferromagnetic FePt nanoparticles at room temperature. *J. Am. Chem. Soc.*, **129** (17), 5538–5543.
69. Osborne, E.A., Jarrett, B.R., Tu, C., and Louie, A.Y. (2010) Modulation of

T2 relaxation time by light-induced, reversible aggregation of magnetic nanoparticles. *J. Am. Chem. Soc.*, **132** (17), 5934–5935.
70. Yeo, K.M., Gao, C.J., Ahn, K.-H., and Lee, I.S. (2008) Superparamagnetic iron oxide nanoparticles with photoswitchable fluorescence. *Chem. Commun.*, **38**, 4622–4624.
71. Wu, X., Liu, H., Liu, J., Haley, K.N., Treadway, J.A., Larson, J.P., Ge, N., Peale, F., and Bruchez, M.P. (2003) Immunofluorescent labeling of cancer marker Her2 and other cellular targets with semiconductor quantum dots. *Nat. Biotechnol.*, **21** (1), 41–46.
72. (a) Brus, L. (1991) Quantum crystallites and nonlinear optics. *Appl. Phys. A*, **53** (6), 465–474; (b) Alivisatos, A.P. (1996) Semiconductor clusters, nanocrystals, and quantum dots. *Science*, **271** (5251), 933–937; (c) Alivisatos, A.P. (2004) The use of nanocrystals in biological detection. *Nat. Biotechnol.*, **22** (1), 47–52; (d) Nirmal, M. and Brus, L. (1999) Luminescence photophysics in semiconductor nanocrystals. *Acc. Chem. Res.*, **32** (5), 407–414.
73. Xing, Y., Chaudry, Q., Shen, C., Kong, K.Y., Zhau, H.E., Chung, L.W., Petros, J.A., O'Regan, R.M., Yezhelyev, M.V., Simons, J.W., Wang, M.D., and Nie, S. (2007) Bioconjugated quantum dots for multiplexed and quantitative immunohistochemistry. *Nat. Protocols*, **2** (5), 1152–1165.
74. Baron, R., Willner, B., and Willner, I. (2007) Biomolecule-nanoparticle hybrids as functional units for nanobiotechnology. *Chem. Commun.*, **4**, 323–332.
75. Zhu, L., Zhu, M.-Q., Hurst, J.K., and Li, A.D.Q. (2005) Light-controlled molecular switches modulate nanocrystal fluorescence. *J. Am. Chem. Soc.*, **127** (25), 8968–8970.
76. Díaz, S.A., Menéndez, G.O., Etchehon, M.H., Giordano, L., Jovin, T.M., and Jares-Erijman, E.A. (2011) Photoswitchable water-soluble quantum dots: pcFRET based on amphiphilic photochromic polymer coating. *ACS Nano*, **5** (4), 2795–2805.
77. Díaz, S.A., Giordano, L., Jovin, T.M., and Jares-Erijman, E.A. (2012) Modulation of a photoswitchable dual-color quantum dot containing a photochromic FRET acceptor and an internal standard. *Nano Lett.*, **12** (7), 3537–3544.
78. Díaz, S.A., Azcárate, J.C., Giordano, L., Jovin, T.M., and Jares-Erijman, E.A. (2013) Quantum dots as templates for self-assembly of photoswitchable polymers: small, dual-color nanoparticles capable of facile photomodulation. *J. Am. Chem. Soc.*, **135** (8), 3208–3217.
79. Liao, B., Long, P., He, B., Yi, S., Ou, B., Shen, S., and Chen, J. (2013) Reversible fluorescence modulation of spiropyran-functionalized carbon nanoparticles. *J. Mater. Chem. C*, **1** (23), 3716–3721.
80. (a) Haase, M. and Schäfer, H. (2011) Upconverting nanoparticles. *Angew. Chem. Int. Ed.*, **50** (26), 5808–5829; (b) Feng, W., Han, C., and Li, F. (2013) Upconversion-nanophosphor-based functional nanocomposites. *Adv. Mater.*, **25** (37), 5287–5303; (c) Han, S., Deng, R., Xie, X., and Liu, X. (2014) Enhancing luminescence in lanthanide-doped upconversion nanoparticles. *Angew. Chem. Int. Ed.*, **53** (44), 11702–11715.
81. Zhou, Z., Hu, H., Yang, H., Yi, T., Huang, K., Yu, M., Li, F., and Huang, C. (2008) Up-conversion luminescent switch based on photochromic diarylethene and rare-earth nanophosphors. *Chem. Commun.*, **39**, 4786–4788.
82. Boyer, J.-C., Carling, C.-J., Gates, B.D., and Branda, N.R. (2010) Two-way photoswitching using one type of near-infrared light, upconverting nanoparticles, and changing only the light intensity. *J. Am. Chem. Soc.*, **132** (44), 15766–15772.
83. (a) Wu, T., Boyer, J.-C., Barker, M., Wilson, D., and Branda, N.R. (2013) A "plug-and-play" method to prepare water-soluble photoresponsive encapsulated upconverting nanoparticles containing hydrophobic molecular switches. *Chem. Mater.*, **25** (12), 2495–2502; (b) Carling, C.-J., Boyer, J.-C., and Branda, N.R. (2012) Multimodal

fluorescence modulation using molecular photoswitches and upconverting nanoparticles. *Org. Biomol. Chem.*, **10** (30), 6159–6168.

84. Zhang, C., Zhou, H.-P., Liao, L.-Y., Feng, W., Sun, W., Li, Z.-X., Xu, C.-H., Fang, C.-J., Sun, L.-D., Zhang, Y.-W., and Yan, C.-H. (2010) Luminescence modulation of ordered upconversion nanopatterns by a photochromic diarylethene: rewritable optical storage with nondestructive readout. *Adv. Mater.*, **22** (5), 633–637.

85. (a) Chung, T.H., Wu, S.H., Yao, M., Lu, C.W., Lin, Y.S., Hung, Y., Mou, C.Y., Chen, Y.C., and Huang, D.M. (2007) The effect of surface charge on the uptake and biological function of mesoporous silica nanoparticles in 3T3-L1 cells and human mesenchymal stem cells. *Biomaterials*, **28** (19), 2959–2966; (b) Lai, C.Y., Trewyn, B.G., Jeftinija, D.M., Jeftinija, K., Xu, S., Jeftinija, S., and Lin, V.S.Y. (2003) A mesoporous silica nanosphere-based carrier system with chemically removable CdS nanoparticle caps for stimuli-responsive controlled release of neurotransmitters and drug molecules. *J. Am. Chem. Soc.*, **125** (15), 4451–4459; (c) Lu, J., Liong, M., Zink, J.I., and Tamanoi, F. (2007) Mesoporous silica nanoparticles as a delivery system for hydrophobic anticancer drugs. *Small*, **3** (8), 1341–1346; (d) Slowing, I.I., Vivero-Escoto, J.L., Wu, C.W., and Lin, V.S.Y. (2008) Mesoporous silica nanoparticles as controlled release drug delivery and gene transfection carriers. *Adv. Drug Delivery Rev.*, **60** (11), 1278–1288; (e) Torney, F., Trewyn, B.G., Lin, V.S.Y., and Wang, K. (2007) Mesoporous silica nanoparticles deliver DNA and chemicals into plants. *Nat. Nanotechnol.*, **2** (5), 295–300; (f) Coll, C., Bernardos, A., Martínez-Mánez, R., and Sancenón, A.F. (2013) Gated silica mesoporous supports for controlled release and signaling applications. *Acc. Chem. Res.*, **46** (2), 339–349.

86. (a) Liu, R., Zhao, X., Wu, T., and Feng, P. (2008) Tunable redox-responsive hybrid nanogated ensembles. *J. Am. Chem. Soc.*, **130** (44), 14418–14419; (b) Silveira, G.Q., Vargas, M.D., and Ronconi, C.M. (2011) Nanoreservoir operated by ferrocenyl linker oxidation with molecular oxygen. *J. Mater. Chem.*, **21** (16), 6034–6039; (c) Zhang, Z.X., Balogh, D., Wang, F., Tel-Vered, R., Levy, N., Sung, S.Y., Nechushtai, R., and Willner, I. (2013) Light-induced and redox-triggered uptake and release of substrates to and from mesoporous SiO_2 nanoparticles. *J. Mater. Chem. B*, **1** (25), 3159–3166; (d) Nguyen, T.D., Leung, K.C.F., Liong, M., Liu, Y., Stoddart, J.F., and Zink, J.I. (2007) Versatile supramolecular nanovalves reconfigured for light activation. *Adv. Funct. Mater.*, **17** (13), 2101–2110; (e) Khashab, N.M., Trabolsi, A., Lau, Y.A., Ambrogio, M.W., Friedman, D.C., Khatib, H.A., Zink, J.I., and Stoddart, J.F. (2009) Redox- and pH-controlled mechanized nanoparticles. *Eur. J. Org. Chem.*, **11**, 1669–1673; (f) Hernandez, R., Tseng, H.R., Wong, J.W., Stoddart, J.F., and Zink, J.I. (2004) An operational supramolecular nanovalve. *J. Am. Chem. Soc.*, **126** (11), 3370–3371.

87. (a) Liu, J. and Du, X. (2010) pH-and competitor-driven nanovalves of cucurbit[7]uril pseudorotaxanes based on mesoporous silica supports for controlled release. *J. Mater. Chem.*, **20** (18), 3642–3649; (b) Meng, H., Xue, M., Xia, T., Zhao, Y.-L., Tamanoi, F., Stoddart, J.F., Zink, J.I., and Nel, A.E. (2010) Autonomous in vitro anticancer drug release from mesoporous silica nanoparticles by pH-sensitive nanovalves. *J. Am. Chem. Soc.*, **132** (32), 12690–12697; (c) Xue, M., Zhong, X., Shaposhnik, Z., Qu, Y., Tamanoi, F., Duan, X., and Zink, J.I. (2011) pH-operated mechanized porous silicon nanoparticles. *J. Am. Chem. Soc.*, **133** (23), 8798–8801; (d) Park, C., Oh, K., Lee, S.C., and Kim, C. (2007) Controlled release of guest molecules from mesoporous silica particles based on a pH-responsive polypseudorotaxane motif. *Angew. Chem. Int. Ed.*, **46** (9), 1455–1457; (e) Nguyen, T.D., Leung, K.C.F., Liong, M., Pentecost, C.D., Stoddart, J.F., and Zink, J.I. (2006) Construction of a pH-driven supramolecular nanovalve. *Org. Lett.*, **8** (15), 3363–3366.

88. (a) Park, C., Kim, H., Kim, S., and Kim, C. (2009) Enzyme responsive nanocontainers with cyclodextrin gatekeepers and synergistic effects in release of guests. *J. Am. Chem. Soc.*, **131** (46), 16614–16615; (b) Coll, C., Mondragón, L., Martínez-Mánez, R., Sancenón, F., Marcos, D.M., Soto, J., Amorós, P., and Perez-Payé, E. (2011) Enzyme-mediated controlled release systems by anchoring peptide sequences on mesoporous silica supports. *Angew. Chem. Int. Ed.*, **50** (9), 2138–2140; (c) Liu, J., Du, X., and Zhang, X. (2011) Enzyme-inspired controlled release of cucurbit[7]uril nanovalves by using magnetic mesoporous silica. *Chem. Eur. J.*, **17** (3), 810–815.

89. (a) Zhang, Z.X., Balogh, D., Wang, F.A., and Willner, I. (2013) Smart mesoporous SiO_2 nanoparticles for the DNAzyme-induced multiplexed release of substrates. *J. Am. Chem. Soc.*, **135** (5), 1934–1940; (b) Zhang, Z.X., Balogh, D., Wang, F.A., Sung, S.Y., Nechushtai, R., and Willner, I. (2013) Biocatalytic release of an anticancer drug from nucleic-acids-capped mesoporous SiO_2 Using DNA or molecular biomarkers as triggering stimuli. *ACS Nano*, **7** (10), 8455–8468; (c) Zhang, Z.X., Wang, F.A., Balogh, D., and Willner, I. (2014) pH-controlled release of substrates from mesoporous SiO_2 nanoparticles gated by metal ion-dependent DNAzymes. *J. Mater. Chem. B*, **2** (28), 4449–4455; (d) Zhang, Z.X., Wang, F.A., Sohn, Y.S., Nechushtai, R., and Willner, I. (2014) Gated mesoporous SiO_2 nanoparticles using K^+-stabilized G-quadruplexes. *Adv. Funct. Mater.*, **24** (36), 5662–5670.

90. (a) Angelos, S., Choi, E., Vögtle, F., De Cola, L., and Zink, J.I. (2007) Photo-driven expulsion of molecules from mesostructured silica nanoparticles. *J. Phys. Chem. C*, **111** (18), 6589–6592; (b) Lu, J., Choi, E., Tamanoi, F., and Zink, J.I. (2008) Light-activated nanoimpeller-controlled drug release in cancer cells. *Small*, **4** (4), 421–426; (c) Bernardos, A., Mondragón, L., Javakhishvili, I., Mas, N., de la Torre, C., Martínez-Mánez, R., Sancenón, F., Barat, J.M., Hvilsted, S., Orzaez, M., Pérez-Payá, E., and Amorós, P. (2012) Azobenzene polyesters used as gate-like scaffolds in nanoscopic hybrid systems. *Chem. Eur. J.*, **18** (44), 13068–13078; (d) Wang, Y., Li, B., Zhang, L., Song, H., and Zhang, L. (2013) Targeted delivery system based on magnetic mesoporous silica nanocomposites with light-controlled release character. *ACS Appl. Mater. Interfaces*, **5** (1), 11–15; (e) Tarn, D., Ferris, D.P., Barnes, J.C., Ambrogio, M.W., Stoddart, J.F., and Zink, J.I. (2014) A reversible light-operated nanovalve on mesoporous silica nanoparticles. *Nanoscale*, **6** (6), 3335–3343.

91. (a) Croissant, J., Maynadier, M., Gallud, A., N'Dongo, H.P., Nyalosaso, J.L., Derrien, G., Charnay, C., Durand, J.-O., Raehm, L., Serein-Spirau, F., Cheminet, N., Jarosson, T., Mongin, O., Blanchard-Desce, M., Gary-Bobo, M., Garcia, M., Lu, J., Tamanoi, F., Tarn, D., Guardado-Alvarez, T., and Zink, J.I. (2013) Two-photon-triggered drug delivery in cancer cells using nanoimpellers. *Angew. Chem. Int. Ed.*, **52** (51), 13813–14062; (b) Liu, J., Bu, W., Pan, L., and Shi, J. (2013) NIR-triggered anticancer drug delivery by upconverting nanoparticles with integrated azobenzene-modified mesoporous silica. *Angew. Chem. Int. Ed.*, **52** (16), 4375–4379; (c) Mei, X., Yang, S., Chen, D., Li, N., Li, H., Xu, Q., Ge, J., and Lu, J. (2012) Light-triggered reversible assemblies of azobenzene-containing amphiphilic copolymer with β-cyclodextrin-modified hollow mesoporous silica nanoparticles for controlled drug release. *Chem. Commun.*, **48** (80), 10010–10012.

92. Ferris, D.P., Zhao, Y.-L., Khashab, N.M., Khatib, H.A., Stoddart, J.F., and Zink, J.I. (2009) Light-operated mechanized nanoparticles. *J. Am. Chem. Soc.*, **131** (5), 1686–1688.

93. Liu, R., Zhang, Y., and Feng, P. (2009) Multiresponsive supramolecular

nanogated ensembles. *J. Am. Chem. Soc.*, **131** (42), 15128–15129.

94. Wen, Y., Xu, L., Wang, W., Wang, D., Du, H., and Zhang, X. (2012) Highly efficient remote controlled release system based on light-driven DNA nanomachine functionalized mesoporous silica. *Nanoscale*, **4** (15), 4473–4476.

95. Aznar, E., Casasús, R., Garcia-Acosta, B., Marcos, M.D., Martínez-Mánez, R., Sancenón, F., Soto, J., and Amorós, P. (2007) Photochemical and chemical two-channel control of functional nanogated hybrid architectures. *Adv. Mater.*, **19** (17), 2228–2231.

96. Chen, L., Wang, W., Su, B., Wen, Y., Li, C., Zhou, Y., Li, M., Shi, X., Du, H., Song, Y., and Jiang, L. (2014) A light-responsive release platform by controlling the wetting behavior of hydrophobic surface. *ACS Nano*, **8** (1), 744–751.

97. Angelos, S., Yang, Y.-W., Khashab, N.M., Stoddart, J.F., and Zink, J.I. (2009) Dual-controlled nanoparticles exhibiting AND logic. *J. Am. Chem. Soc.*, **131** (32), 11344–11346.

98. Yasuda, S., Nakamura, T., Matsumoto, M., and Shigekawa, H. (2003) Phase switching of a single isomeric molecule and associated characteristic rectification. *J. Am. Chem. Soc.*, **125** (52), 16430–16433.

99. (a) Martin, S., Haiss, W., Higgins, S.J., and Nichols, R.J. (2010) The impact of E-Z photo-isomerization on single molecular conductance. *Nano Lett.*, **10** (6), 2019–2023; (b) Mativetsky, J.M., Pace, G., Elbing, M., Rampi, M.A., Mayor, M., and Samorì, P. (2008) Azobenzenes as light-controlled molecular electronic switches in nanoscale metal–molecule–metal junctions. *J. Am. Chem. Soc.*, **130** (29), 9192–9193; (c) Smaali, K., Lenfant, S., Karpe, S., Oçafrain, M., Blanchard, P., Deresmes, D., Godey, S., Rochefort, A., Roncali, J., and Vuillaume, D. (2010) High on-off conductance switching ratio in optically-driven self-assembled conjugated molecular systems. *ACS Nano*, **4** (4), 2411–2421; (d) Ferri, V., Elbing, M., Pace, G., Dickey, M.D., Zharnikov, M., Samorì, P., Mayor, M., and Rampi, M.A. (2008) Light-powered electrical switch based on cargo-lifting azobenzene monolayers. *Angew. Chem. Int. Ed.*, **47** (18), 3407–3457.

100. Kronemeijer, A.J., Akkerman, H.B., Kudernac, T., van Wees, B.J., Feringa, B.L., Blom, P.W.M., and de Boer, B. (2008) Reversible conductance switching in molecular devices. *Adv. Mater.*, **20** (8), 1467–1473.

101. (a) Matsuda, K., Yamaguchi, H., Sakano, T., Ikeda, M., Tanifuji, N., and Irie, M. (2008) Conductance photoswitching of diarylethene-gold nanoparticle network induced by photochromic reaction. *J. Phys. Chem. C*, **112** (43), 17005–17010; (b) Yamaguchi, H. and Matsuda, K. (2009) Photo-and electrochromic switching of diarylethene-gold nanoparticle network on interdigitated electrodes. *Chem. Lett.*, **38** (10), 946–947; (c) Ikeda, M., Tanifuji, N., Yamaguchi, H., Irie, M., and Matsuda, K. (2007) Photoswitching of conductance of diarylethene-Au nanoparticle network. *Chem. Commun.*, **13**, 1355–1357.

102. van der Molen, S.J., Liao, J., Kudernac, T., Agustsson, J.S., Bernard, L., Calame, M., van Wees, B.J., Feringa, B.L., and Schönenberger, C. (2009) Light-controlled conductance switching of ordered metal-molecule-metal devices. *Nano Lett.*, **9** (1), 76–80.

103. Arramel, P.T.C., Kudernac, T., Katsonis, N., van der Maas, M., Feringa, B.L., and van Wees, B.J. (2013) Reversible light induced conductance switching of asymmetric diarylethenes on gold: surface and electronic studies. *Nanoscale*, **5** (19), 9277–9282.

104. Uchida, K., Yamanoi, Y., Yonezawa, T., and Nishihara, H. (2011) Reversible on/off conductance switching of single diarylethene immobilized on a silicon surface. *J. Am. Chem. Soc.*, **133** (24), 9239–9241.

105. (a) Katsonis, N., Kudernac, T., Walko, M., van der Molen, S.J., van Wees, B.J., and Feringa, B.L. (2006) Reversible conductance switching of single diarylethenes on a gold surface. *Adv. Mater.*, **18** (11), 1397–1400; (b) He,

J., Chen, F., Liddell, P.A., Andréasson, J., Stright, S.D., Gust, D., Moore, T.A., Moore, A.L., Li, J., Sankey, O.F., and Lindsay, S.M. (2005) Switching of a photochromic molecule on gold electrodes: single-molecule measurements. *Nanotechnology*, **16** (6), 695.

106. Kim, Y., Pietsch, T., Erbe, A., Belzig, W., and Scheer, E. (2011) Benzenedithiol: a broad-range single-channel molecular conductor. *Nano Lett.*, **11** (9), 3734–3738.

107. (a) Whalley, A.C., Steigerwald, M.L., Guo, X., and Nuckolls, C. (2007) Reversible switching in molecular electronic devices. *J. Am. Chem. Soc.*, **129** (42), 12590–12591; (b) Cao, Y., Dong, S., Liu, S., Liu, Z., and Guo, X. (2013) Toward functional molecular devices based on graphene–molecule junctions. *Angew. Chem. Int. Ed.*, **52** (14), 3906–4002; (c) Jia, C., Wang, J., Yao, C., Cao, Y., Zhong, Y., Liu, Z., Liu, Z., and Guo, X. (2013) Conductance switching and mechanisms in single-molecule junctions. *Angew. Chem. Int. Ed.*, **52** (33), 8666–8670; (d) Ashraf, M.K., Bruque, N.A., Tan, J.L., Beran, G.J.O., and Lake, R.K. (2011) Conductance switching in diarylethenes bridging carbon nanotubes. *J. Chem. Phys.*, **134** (2), 024524.

108. (a) de Jong, J.J.D., Lucas, L.N., Kellogg, R.M., van Esch, J.H., and Feringa, B.L. (2004) Reversible optical transcription of supramolecular chirality into molecular chirality. *Science*, **304** (5668), 278–281; (b) Irie, M., Kobatake, S., and Horichi, M. (2001) Reversible surface morphology changes of a photochromic diarylethene single crystal by photoirradiation. *Science*, **291** (5509), 1769–1772.

109. Tsuji, Y. and Hoffmann, R. (2014) Frontier orbital control of molecular conductance and its switching. *Angew. Chem. Int. Ed.*, **53** (16), 4093–4181.

110. Tam, E.S., Parks, J.J., Shum, W.W., Zhong, Y., Santiago-Berríos, M.B., Zheng, X., Yang, W., Chan, G.K.-L., Abruña, H.D., and Ralph, D.C. (2011) Single-molecule conductance of pyridine-terminated dithienylethene switch molecules. *ACS Nano*, **5** (6), 5115–5123.

111. (a) Kim, Y., Hellmuth, T.J., Sysoiev, D., Pauly, F., Pietsch, T., Wolf, J., Erbe, A., Huhn, T., Groth, U., Steiner, U.E., and Scheer, E. (2012) Charge transport characteristics of diarylethene photoswitching single-molecule junctions. *Nano Lett.*, **12** (7), 3736–3742; (b) Briechle, B.M., Kim, Y., Ehrenreich, P., Erbe, A., Sysoiev, D., Huhn, T., Groth, U., and Scheer, E. (2012) Current-voltage characteristics of single-molecule diarylethene junctions measured with adjustable gold electrodes in solution. *Beilstein J. Nanotechnol.*, **3** (1), 798–808.

112. Irie, M. (2000) Diarylethenes for memories and switches. *Chem. Rev.*, **100** (5), 1685–1716.

113. Meng, F., Hervault, Y.-M., Norel, L., Costuas, K., Van Dyck, C., Geskin, V., Cornil, J., Hng, H.H., Rigaut, S., and Chen, X. (2012) Photo-modulable molecular transport junctions based on organometallic molecular wires. *Chem. Sci.*, **3** (10), 3113–3138.

114. Taniguchi, M., Nojima, Y., Yokota, K., Terao, J., Sato, K., Kambe, N., and Kawai, T. (2006) Self-organized interconnect method for molecular devices. *J. Am. Chem. Soc.*, **128** (47), 15062–15063.

115. Roldan, D., Kaliginedi, V., Cobo, S., Kolivoska, V., Bucher, C., Hong, W., Royal, G., and Wandlowski, T. (2013) Charge transport in photoswitchable dimethyldihydropyrene-type single-molecule junctions. *J. Am. Chem. Soc.*, **135** (16), 5974–5977.

7
Hybrid Organic/Photochromic Approaches to Generate Multifunctional Materials, Interfaces, and Devices

Emanuele Orgiu and Paolo Samorì

7.1
Introduction

Silicon industry's greatest strength has always relied on its ability to downscale device dimensions and generate low-cost functions regardless of the relatively high cost per area. Conversely, "van der Waals solids" relying on weak interactions such as organic semiconductors (OSs) have always been thought of complementing this requirement for low-cost production as well as large-area applications by virtue of their characteristic chemical versatility and easy processability. Hence, OSs hold a great potential for the integration in the everyday flexible electronics [1–22] as they feature a number of fundamental properties such as light emission [23–45], charge transport [45–74], and photovoltaic response [75–96].

So far, synthetic pathways and processing of organic semiconductors have greatly improved and generated a variety of possible candidate molecules/processes that are mature enough to meet the initially envisaged applications [97–112].

Nevertheless, the initial desire for reducing the device size is ultimately getting to an end for the Si-based electronics because of the limitations coming from the photolithographic steps involved in the production process of the devices. In this regard, organic and polymer electronics, that is, the electronics based on organic and polymeric semiconductors, cannot complement its inorganic counterpart and the relentless need of miniaturization unless new challenging avenues are pursued.

A new strategy can be envisaged in which molecular functions are integrated into organic semiconductors via a controlled combination with an additional molecular component featuring supplementary functions. This is the case of photochromic systems [113–118] that are small organic molecules capable of undergoing reversible isomerization upon light irradiation at definite wavelengths between (at least) two (meta)stable states exhibiting markedly different properties at the molecule level. In these molecular switches, the specific isomeric state is *selected* upon exposure to a light stimulus with a proper wavelength. The most common families of photochromic molecules, depicted in Figure 7.1,

Photochromic Materials: Preparation, Properties and Applications, First Edition.
Edited by He Tian and Junji Zhang.

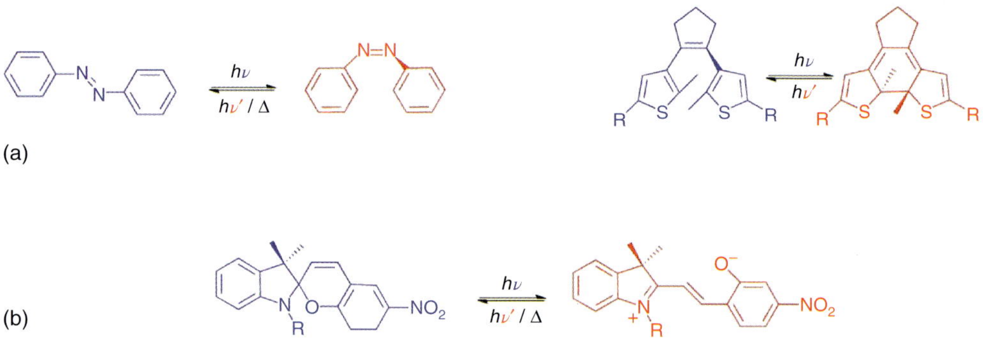

Figure 7.1 Photochromic molecules encompassed in this review, featuring two independently addressable states that lead to reversible structural changes such as geometry/steric hindrance and electronic changes such as dipole moments, π-conjugation, HOMO–LUMO gap, and redox potential. All these changes occurring on the molecular level affect macroscopic properties such as shape, aggregation behavior, conductance of the resulting materials, or interface containing these photoswitches. In particular, the sketch depicts (a) azobenzenes, (b) diarylethenes, and (c) spiropyrans.

are azobenzenes (AZOs) [119–128], diarylethenes (DAEs) [129–140], and spiropyrans(SPs)/-oxazines [141, 142]. In addition, the photochromic family also encompasses fulgides [143], imidazole dimers [144], and stilbenes that are not considered in this chapter because they have not yet been exploited in combination with organic semiconductors.

In the last two decades, photochromic molecules have been shown to represent key building blocks for both optical data storage [132, 145–149] and optical switching [150, 151]. The optical readout relies on the opportunity of sensing the differences in fluorescence [152–154] and absorption, even in the infrared (IR) region [155], of the two isomeric states. Nevertheless, optical readout is a typically destructive process because the molecule's state is varied upon light exposure.

An alternative method relies on the measurement of the electrical characteristics in molecular junctions where the isomerization of the photochromic molecule gives rise to a binary conductance [156–161]. However, the fabrication of molecular junctions requires high precision and resolution necessitating time-consuming processes, bulky equipment, highly controlled processing environments as well as advanced and costly deposition/fabrication techniques. Moreover, neat films of photochromic molecules could be certainly exploited for large-area electronics via cheap production solutions if they did not exhibit moderate switching ratios [162] or low charge carrier mobilities [163]. These evidences seem to suggest that the huge potential held by photochromic molecules can be better harnessed when they are combined with organic or polymeric semiconductors, which bring in (i) the low-cost manufacturing, (ii) the ease of processability, (iii) a variety of different device geometries and interfaces that can be used all at the same time in order to generate multifunctional devices, and (iv) a proper semiconductor behavior of the materials, which is required in three-terminal device configurations.

This chapter reviews recent strategies that have been pursued by marrying the exceptional (opto)electronic characteristics of organic and polymeric semiconductors with the photoresponsive nature of photochromic molecules, in an effort to construct bifunctional or multifunctional devices.

7.1.1 Tuning the Charge Injection in Organic-Based Devices by Means of Photochromic Molecules

Controlling and optimizing charge injection is one of the major problems in organic-based devices [164–166]. Two possible approaches could be undertaken to solve this problem: (i) the chemisorption of self-assembled monolayers (SAMs) on the injecting electrodes or (ii) the introduction of a physisorbed interlayer between injecting electrode and charge transporting material. The former method is a universal strategy to modify various physicochemical properties of surfaces [167] including their surface wettability and metal work function. This can make it possible to improve charge injection at the metal–semiconductor interface by either reducing the interfacial energy mismatch between these materials or enhancing the molecular order and degree of crystallinity within the semiconductor layer assembling on top of the SAMs, the latter being the key to optimize the transfer integral and to promote the charge transport within the active semiconducting material.

The introduction of a photochromic interlayer between the electrode and the semiconductor is a viable method to confer a bifunctional nature to such interfaces, that is, by changing surface properties such as wettability or work function by irradiation at given wavelengths, as previously reported [156, 160, 168–170]. In this framework, it has been recently shown how a photomodulable bistability of the charge injection in an AZO-based SAM chemisorbed onto Au source and drain electrodes can be attained [171]. Organic Field-Effect Transistor (OFETs) in bottom-gate bottom-contact configuration using an air-stable perylene diimide (PDIF-CN2) as semiconducting layer were tested in order to provide a proof of concept of the suggested working principle. The thiol-functionalized biphenyl AZO derivatives utilized within this work were previously shown to form highly ordered and tightly packed SAMs on Au(111), with isomerization yields as high as 96% [172]. Such a large yield of isomerization could be obtained owing to the strong intermolecular interactions between adjacent molecules within the 2D crystalline domains [173]. In Crivillers *et al.*'s work, this azobenzene-based self-assembled monolayer (AZO-SAM) is chemisorbed on the gold source and drain electrodes of an OFET to optically switch the injection/extraction of charges at the electrode/semiconductor interface (Figure 7.2a). In such configuration, the device irradiation needed to induce the photoisomerization of the AZO-SAM has to be performed from the top of the device. Thus, the semiconductor could act a masking light-absorbing layer. As a consequence, the electroactive material was selected in view of two features: (i) its absorption spectra not overlapping the characteristic absorption bands of the chosen AZO and (ii) its electrical

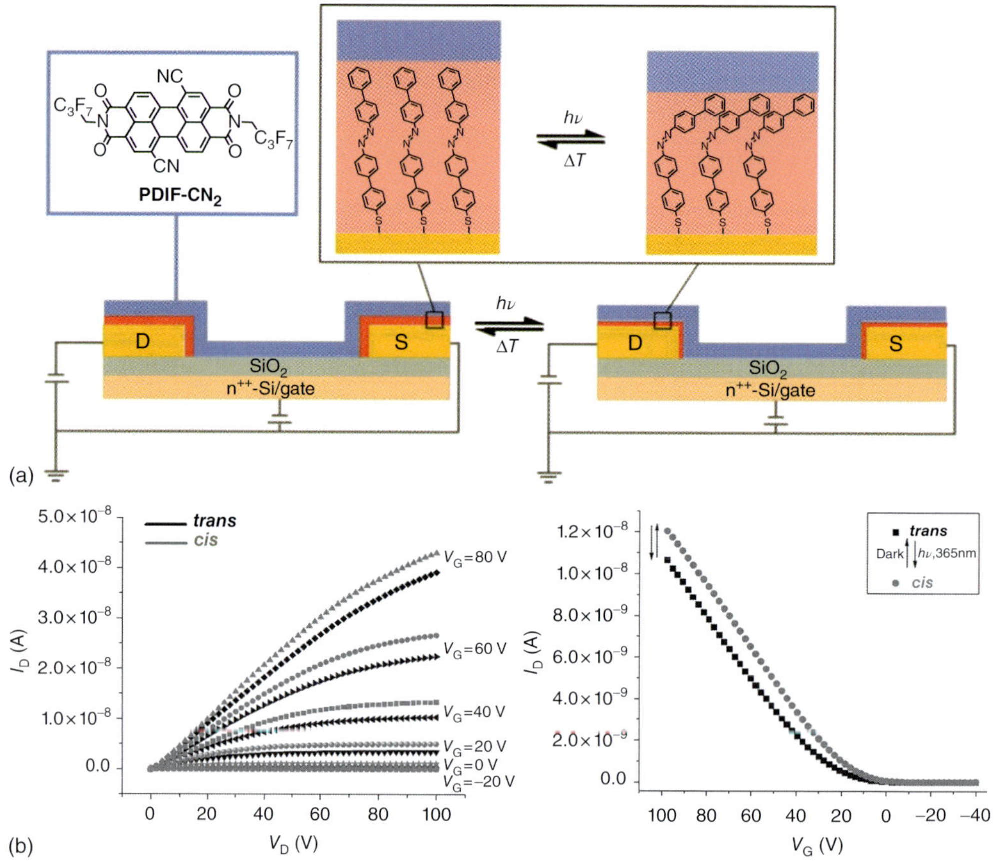

Figure 7.2 (a) Representation of the working principle of the AZO-SAM Field-Effect Transistor (FET) device. The semiconductor is a perylene diimide derivative, namely PDIF-CN2. The AZO-SAM chemisorbed on gold source and drain injecting electrodes undergoes reversible isomerization upon exposure to UV light (*trans*-to-*cis*) and switches back (*cis*-to-*trans*) after thermal recovery. As a result, the charge injection and ultimately the drain current are photomodulated. ($W = 10$ mm, $L = 10$ μm, film thickness 8–10 nm). (b) Transfer (right) and output (left) curves acquired after an *in situ* switching from of the AZO-SAM from *trans* to *cis*. The semiconductor is a spin-coated PDIF-CN2 film. (Reproduced with permission [171]. 2011, Wiley.)

properties being very little sensitive to exposure to ultraviolet (UV) and visible light. The *trans*-AZO isomer spectra show an intense $\pi-\pi^*$ transition band at ~365 nm and a weaker n$-\pi^*$ band around 450 nm while the *cis* isomer is characterized by a more pronounced n$-\pi^*$ transition band at ~450 nm. Upon illumination of a *trans*-SAM with UV light, the isomerization from *trans* to *cis* occurs. The *cis*-to-*trans* isomerization can be triggered by irradiation into the n$-\pi^*$ band of the *cis* form upon white light exposure. Nevertheless, PDIF-CN2 thin films show reduced absorption features in the 350–410 nm range and an intense absorption band that peaks at ~550 nm. Considering its optical features,

PDIF-CN2 was chosen as an ideal candidate material for the AZO-SAM FET experiment.

Initially, the AZO-SAMs were chemisorbed in *trans* form and successively exposed to UV light (365 nm) during 90 min. A thermal back-reaction was achieved by just storing the sample in the dark for 24 h. The photoinduced *cis*-to-*trans* isomerization that could be attained upon exposure to white light was deliberately not carried out because of the photogeneration of charge carriers, which may render it complicated to separate the different contributions to the current coming from both the back-isomerization reaction and the illumination with the white light.

The output and transfer characteristics in *cis*-SAM-based FETs displayed a 20% enhancement in the maximum source–drain current ($I_{D,max}$) as portrayed in Figure 7.2b. This change was accompanied by a decrease in threshold voltage and a ΔV_{TH} (*trans* to *cis*) of 9.2 and 7.1 V in the linear and the saturation regime. Such a variation confirmed the suggested working principle: The lower thickness of the SAM in the *cis* form leads to a reduced tunneling barrier/thickness, which results in higher drain current. The optical modulation process was found to be reversible up to four reversible cycles with negligible switching fatigue. In order to nail down the switching mechanisms, two blank experiments were carried out: a PDIF-CN2 solution was spin cast on both bare gold and undecanethiol-SAM-treated gold electrodes and irradiated with UV light for 90 min. In both cases, no changes were detected by means of electrical characterization supporting the hypothesis that the AZO-SAM is the photoresponsive part of the device. Furthermore, PDIF-CN2 displays a weak absorption in the UV, thus ruling out a possible role played by the photoexcitation of the semiconductor.

In addition to the difference in tunneling resistance of the *cis*-AZO and *trans*-AZO, the change of the major electrical FET parameters could relate to (i) a difference in film morphology at the semiconductor/AZO-SAM interface and (ii) a change in the work function (WF) of the AZO-SAM owing to the photoisomerization. The *trans*- and *cis*-AZO-SAM show different hydrophobic properties, with static water contact angle values of $87.2° \pm 1.3°$ and $70.6° \pm 1.9°$, respectively. The different wettability may play a major role on the interfacial film microstructure, thus making charge injection more or less favorable in the two isomeric states of the SAM. In order to sort out this issue, two additional and separate *trans* and *cis* devices were fabricated. PDIF-CN2 solutions were spin-coated onto two different samples having *trans*- and *cis*-AZO-SAM-modified electrodes, respectively.

Electrical characterization (output and transfer curves) exhibited markedly higher drain current for the *cis*-AZO transistor, thus supporting the findings of the *in situ* experiment. The I_D (*cis*)/I_D (*trans*) increase amounted to 240%, which the authors attributed to the potentially improved interfacial semiconductor/AZO-SAM morphology of PDIF-CN2 films separately spin-coated onto the two isomeric forms of the AZO-SAM. Atomic Force Microscopy (AFM) imaging performed on regions of PDIF-CN2 films spin-coated over AZO-SAM-functionalized Au electrodes did not revealed significant change except for an almost negligible roughness enhancement. Molecular reorganization and the reduced injecting area of the source electrode [58], where a low number of AZOs interact with the organic layer, in a bottom-contact architecture are likely to increase the device fatigue reported in their work. In addition to the reduced cyclability issue that may affect photochromic molecular systems, in this specific case a molecular rearrangement of the semiconductor at the interface occurs, leading to a more static interface, which was accompanied by a lower current modulation efficiency after repeated cycles. In order to discard a possible work function difference between *trans*- and *cis*-AZO-SAM, Kelvin probe measurements showed only a subtle difference of about 70 meV between the two isomeric states of the films. Thus, the observed differences in output current could not be attributed to a significant work function shift, leading to a more favorable energy-level alignment. In spite of the low values recorded in FET devices, the difference in mobility ($\mu_{trans} - \mu_{cis}$) further reflects an improved charge injection in transistor with *cis*-AZO-functionalized electrodes.

Another remarkable approach that uses photochromic molecules interlayer in order to finely modulate the charge injection into an organic semiconductor is reported by Zacharias *et al.* [174]. In their work, they describe a solution-processed multilayered organic light-emitting diode (OLED) device where a photochromic layer of a dithienylenethene can be photoswitched reversibly. Interestingly, ON-to-OFF ratio as large as 10^3 was achieved for both electroluminescence and current density. Zacharias *et al.* tested three different configurations showed in Figure 7.3a: (i) incorporation of what the authors call compound 1 (oxetane-functionalized dithienylethene, XDTE, Figure 7.3a) into electroluminescent polymer, which we call device of type “blend”; (ii) a thin layer of compound 1 between the anode and the emissive layer, here called device of type “bilayered”; and (iii) a further hole-transporting layer sandwiched between PEDOT:PSS and 1, “trilayered” device.

A marked decrease in electroluminescence efficiency was measured in device of “blend” type even upon addition of small quantities (0.5 wt%) of XDTE. Moreover, no significant isomerization could be observed under UV light irradiation. Conversely, “bilayered” devices would display electroluminescence and ON-to-OFF ratio of ~5. In “trilayered” devices, an extra hole-transporting layer such as a triphenylamine dimer (XTPD) derivative was added to the structure. The authors reported on seven different XTPD derivatives (2–8) with HOMO level ranging from −5.17 down to −5.56 eV.

The highest ON-to-OFF on single XTPD layers reached about 200 and is obtained when the HOMO level of the XTPD was close in energy to the work function of PEDOT:PSS. Conversely, the switching ratio was found to decrease with lowering the HOMO level of the XTPD, with ON-to-OFF ~1 in the worst case (lowest XTPD HOMO level).

The switching ratio could be further improved by modifying the thickness of the XDTE layer. In particular, OLED devices with a double layer formed by XTPD derivative 2 and 3 (16-nm thick) and a 40-nm thick XDTE layer displayed ON-to-OFF ratio of 3000. The devices' emission onset was about 6 V in the OFF state but was reduced to 2.7 V by UV irradiation (ON state). Exposure to a monochromatic

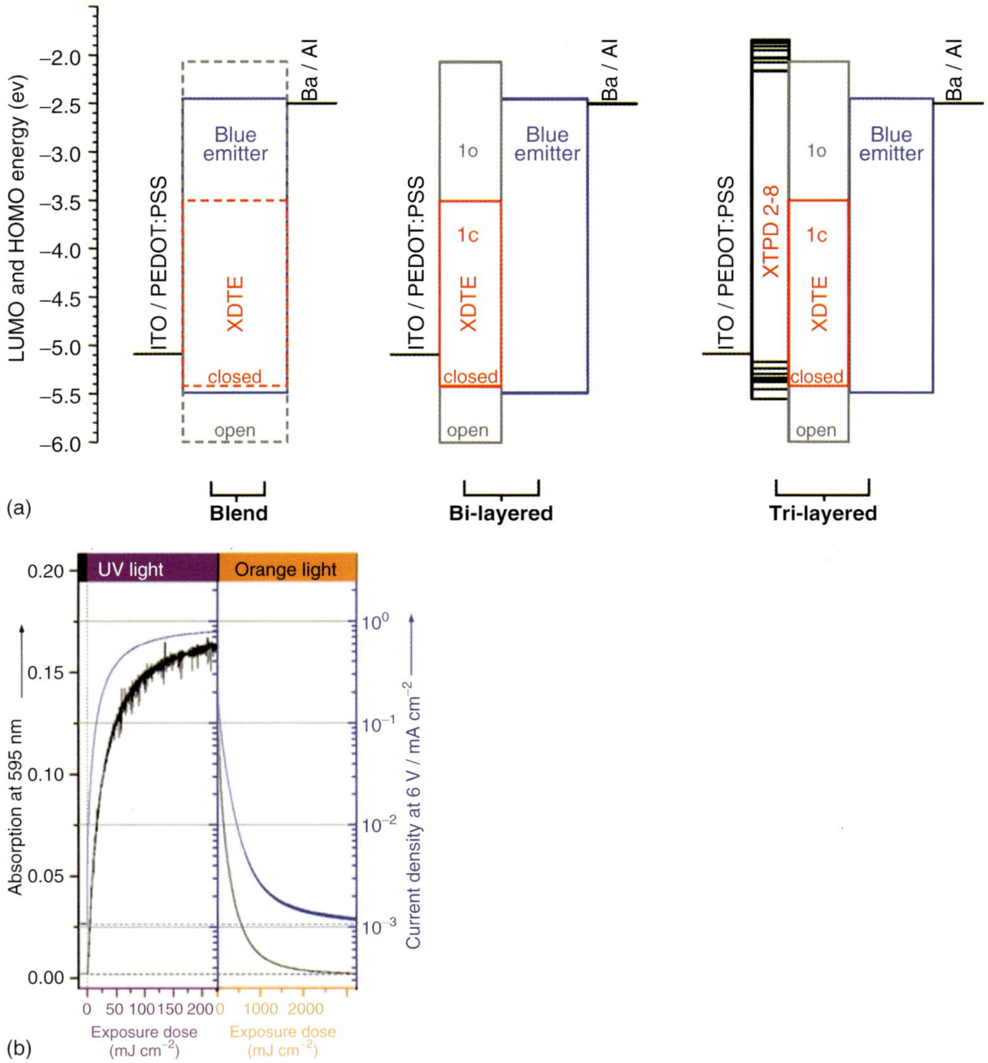

Figure 7.3 (a) Different device structures tested in Zacharias *et al*. The *y*-axis indicates the energy of the HOMO and LUMO levels of each single component of the multilayered structures encompassed in this study. In particular, (left) XDTE was blended with the emitting material, device of type "blend." (center) XDTE was employed as an interlayer, device of type "bilayered." (right) Various hole-injecting materials 2–8 and their combinations were used as an interlayer between XDTE and the ITO/PEDOT electrode, device of type "trilayered." (b) Comparative plot of an optimized device comprising the following layers: PEDOT (35 nm)/2 (8 nm)/3 (8 nm)/1 (40 nm)/blue emitting polyspirofluorene (70 nm)/Ba(4 nm)/Al (150 nm), in which absorption and current density are simultaneously recorded under irradiation with UV (312 nm, left) and orange (590 nm, right) light sources. (Panel (a) is readapted from Ref. [175]. Reproduced with permission [174]. 2009, Wiley.)

590 nm light would make the devices switch back to the OFF state. OLED devices featuring electroluminescence up to 0.8 cd A^{-1} ($\lambda_{max} = 428$ nm) and constant with the bias were measured.

In order to understand the working principle of their OLED devices, the authors have employed a reference device where the photochromic layer is replaced by the XTPD 6 that exhibits a HOMO energy of ~5.4 eV, therefore energetically close to that of XDTE. By utilizing this configuration, the electroluminescence obtained was of 2.1 cd A^{-1} ($\lambda_{max} = 456$ nm). Regardless of the reduced efficiency of the electron-blocking layer at the interface between the blue-emitting polyspirofluorene and 1, the lower efficiency of the switchable device is owing to the blue-shifted emission that stems from a different location of the emission zone. Finally, the electron-blocking material and the position of the emission area were improved by using a further XTPD layer (7) between layer 1 and the blue-emitting layer.

The switching dynamics were monitored by recording current density and the photochromic layer absorption in a functioning device using a reflection set-up. Following Lambert–Beer law, at a wavelength of 595 nm, the absorption should be directly proportional to the concentration of compound 1c. The absorption increase featuring an exponential trend (with a rate constant or the ring closure amounting to $k_{closed} = 20\ J^{-1}\ cm^2$) was observed upon irradiation with a 312-nm light source as displayed in the left part of Figure 7.3b. The back-reaction, from ON to OFF state, was obtained by irradiation with an orange light source centered at $\lambda = 590$ nm and featured a rate constant $k_{open} = 3.4\ J^{-1}\ cm^2$ Interestingly, the OLEDs incorporating XDTE tested in Zacharias *et al.* showed a remarkably high resistance to fatigue, indicating that the switching properties of the pristine photochromic layer are nearly unchanged once incorporated in the device.

By and large, the use of photochromic molecules at the electrode/active layer interface allows one to optically tailor the work function of the injecting electrode while changing surface properties such as the wettability. Furthermore, the fraction of photochromic molecules involved in the SAM or in the interlayer can be modulated by selecting the proper irradiation power and duration at a given wavelength, paving the way to multilevel-injection interfaces where ideally a number of intermediate levels could be separately addressed by optical stimuli.

Conversely, a number of open questions are yet to be addressed: (i) how the mechanical constraint exerted on both chemisorbed SAMs and films of photochromic molecular systems by the presence of an upper layer affects the isomerization yield of single molecules and, ultimately, the ensemble behavior, (ii) the switching fatigue experienced during device operation, (iii) the need for additional chemistry steps in order to vary the structure of the photochromic molecule when it is employed as an interlayer since this function requires solubility in solvents orthogonal to those of the underlying layer or the possibility to cross-link the whole photochromic layer for the deposition of an upper layer on it, and (iv) how the ensemble effect (cooperative or collective effect) could be favorable to the isomerization yield of photochromic SAMs.

7.2
Tuning the Polaronic Transport in Organic Semiconductors by Means of Photochromic Molecules

In view of their increasingly higher field-effect mobilities [11, 20], organic semiconductors hold great promise for the generation of a new electronics era benefitting from an ideally infinite number of possibilities to tune their optical, mechanical, and transport properties [20]. This number is as large as the solutions that synthetic chemistry can offer to the scientific and industrial community.

The functionalization of organic semiconductors with different side groups allows the tuning of numerous characteristics including the solubility, assembly capacity, and optical properties. Likewise one or more photoresponsive bistable photochromic groups have been covalently grafted to organic semiconductors in order to confer to them a bifunctional nature, by adding a selective response to distinct wavelengths. This concept resulted in a number of examples of dyads, multiads, or polymers incorporating photochromic units, as presented in Section 7.2.1.

Nonetheless, the above-mentioned approach brings in all the complexity of the synthesis of macromolecules that adds many and sometimes complex synthetic steps, thus it is time consuming and consequently does not hold potential for upscaling. Moreover, the semiconducting properties of the organic semiconductor as much as the isomerization of the photochromic moiety included in the macromolecule could be hindered by the effect that other component can exert. In addition, driving the self-assembly of these sophisticated (macro)molecules toward highly ordered supramolecular architectures, which is crucial for efficient charge transport in organic active layers, may not be trivial owing to their multiple functionalities. Most likely because of these reasons, hitherto no examples of photochromic molecule/polymeric semiconductor exhibiting both functions at the same time were reported so far.

Conversely, the solution processability of most of the organic semiconductors can be exploited for the realization of molecular devices where both functionalities are embedded by either fabricating multilayered devices, where an interlayer of photochromic molecules is sandwiched, or blending photochromic molecules and organic semiconductors. Examples of both cases together with advantages and disadvantages of these two approaches are discussed in Sections 7.2.2 and 7.2.3, respectively.

7.2.1
Photochromic Molecules and Organic Semiconductors Incorporated in Dyads, Multiads, and Polymers

In the attempt to bring organic (semi)conducting molecules and photochromic molecules together, several important studies were reported demonstrating that these two varieties of molecules could be mutually advantageous: dyads, multiads, and polymers decorated with photochromic units were synthesized and tested leading to major changes in conductivity and optical properties upon light irradiation [150, 157, 159, 176–182].

Among the various excellent works, recently Pärs *et al.* reported on the possibility to covalently tether an organic semiconducting molecule such a perylene bisimide (PBI) to a dithinylcyclopentene (DCP) [183]. The molecular system would act as an optical transistor operating under ambient conditions (Figure 7.4). In this work, the DCP unit photoisomerizes from the open form to the closed form upon exposure to UV light (280–310 nm) while the ring opening occurs under visible light (500–650 nm) exposure. In order to exploit the PBI–DCP–PBI triad as an optically gated molecular system in which the fluorescence is finely tuned, the authors carried out a controlled series of irradiating cycles by alternating light pulses at 300 nm (conversion of the DCP unit to the closed form) and 514 nm (probing of the fluorescence in the PBI units). Thanks to this irradiation scheme, the fluorescence intensity of the PBI units could be tuned by energy transfer from the PBI to the DCP unit in the closed form. In particular, the change in fluorescent intensity (I) can be either ON (I_{ON}) or OFF (I_{OFF}) depending on whether DCP is in its open or closed isomeric state, respectively. Clearly, the OFF level of intensity corresponds to a nonzero intensity. The switching ratio can be defined as

$$\frac{I_{final} - I_{initial}}{I_{ON}} = \frac{\Delta I}{I_{ON}} \tag{7.1}$$

where I_{final} and $I_{initial}$ are related to I_{ON}/I_{OFF} depending on the specific sequence performed. Furthermore, the fatigue resistance of the triads was probed by illuminating the sample at 635 nm while a wavelength of 514 nm would probe the fluorescence. This sequence was repeated five times per switching cycle, while a "300–514 nm" sequence was then used to induce the back isomerization of DCP to the closed form. The outcome of a 3000 cycles-long experiment is displayed in Figure 7.4 revealing a fluorescence contrast drop from 0.83 down to 0.38. In

Figure 7.4 (a) Schematic sketch of the molecular triad consisting of two perylene bisimide (PBI) units that are covalently linked to a dithienylcyclopentene (DCP). Top: closed form and bottom: open form. (b) Absorption spectra of the PBI–DCP–PBI triad dissolved in toluene at a concentration of 1.5×10^{-6} mol l^{-1} for the open (red) and the closed (blue) form of the DCP unit. For comparison, the dashed line shows the absorption spectrum of pure PBI in the same solvent at a concentration of 1.5×10^{-6} mol l^{-1}. (c) Contrast of the modulation of the fluorescence intensity as a function of the number of switching cycles that each consists of 2×5 conversion/probe sequences. Each data point corresponds to the average over 50 switching cycles. The illumination intensities (exposure times) were 96 mW cm^{-2} (250 ms) at 514 nm, 130 mW cm^{-2} (250 ms) at 300 nm, and 43 W cm^{-2} (250 ms) at 635 nm. The kinks after about 1000 and 2500 switching cycles reflect slight readjustments of the fourth-harmonic generation due to drifts during the long-term experiment. The line serves as a guide for the eye. Insets bottom: modulation of the fluorescence intensity at the beginning and the end of the experiment. Inset top: transistor analogy identifying the S_1 and S_0 state of PBI as source and drain, the conversion beams as gate voltage of different polarity and the optical pumping as external circuit, respectively. (Reproduced with permission [183]. 2011, Wiley.)

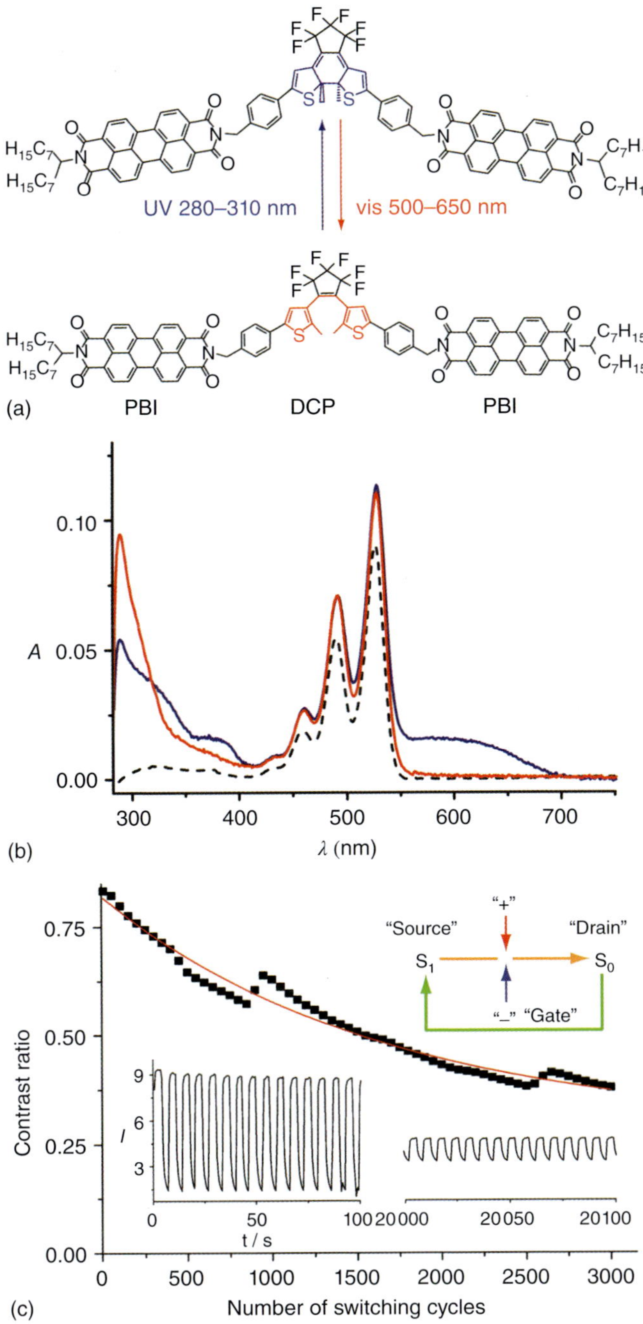
UV 280–310 nm
vis 500–650 nm
(a)
PBI
DCP
PBI
A
0.10
0.05
0.00
300
400
500
600
700
λ (nm)
(b)
Contrast ratio
0.75
0.50
0.25
0.00
0
500
1000
1500
2000
2500
3000
Number of switching cycles
"Source"
"Drain"
"+"
"–" "Gate"
S1
S0
I
9
6
3
0
50
100
20000
20050
20100
t / s
(c)

addition, the two insets show the variation of the fluorescence intensity in the first and last 100 s of the experiment where, respectively, a high and then low intensities are measured. In this work, the authors propose an analogy of their optical molecular model to a real three-terminal devices where the gate function is exerted by the conversion beam while the source and drain would be the S_1 and S_0 states of the PBI units, as they ensure the flow of fluorescence photons. Following their analogy, the sign of the gate "optical electrode" would swing between 635 and 300 nm in such a way that, like in a transistor, the amount of electrons between "source" (S_1) and "drain" (S_0) could be controlled, as shown in the top inset of Figure 7.4. Another example of modulation of the fluorescence operated via a DAE dyad based on perylene diimide is reported by Tan *et al.* [179].

Alongside to the undoubtedly disruptive examples of integration of photochromic molecules into dyads, a remarkable effort has also been devoted to the synthesis of photochromic conjugated conductive polymers as, in principle, any photochromic switching unit conjugated within a π-electron polymer would lead to a modulable conductivity [175, 177, 178, 180, 181, 184, 185]. However, despite the notable advances, these materials suffer from a slow or sometimes even irreversible switching mechanism along with low electrical conductivities.

It is fair to point out that the above-mentioned dyad/multiad strategy often requires complicated and multistep synthetic paths, hampering their easy production and upscalability and therefore their actual technological exploitation. By and large, the ability to photochemically tune materials properties such as the conductivity or the fluorescence represents an overall advantage of photochromic molecules, but their integration in real devices such as transistors requires their combination with suitable semiconductors.

7.2.2 The Multilayer Approach

The use of photochromic molecules in electronic devices dates back to the late 1980s when Tachibana *et al.* the integration of AZO derivatives in multilayered Langmuir–Blodgett structures could offer tunable bistable conductivity under light irradiation [186, 187]. Unfortunately, this approach revealed a noticeable thermal instability of the light-responsive compounds, which would lead to undesired changes in the device output. It is only in 2001 that a number of experiments carried out by Tsujioka *et al.* brought to the attention of the scientific community the possibility of using DAE derivative compounds in multilayered photoswitchable bipolar memory devices [188–192]. Nevertheless, these devices could initially reach ON-to-OFF current ratio below 100 owing to a photostationary state with <15% of molecules in the closed form. A possible reason for this not outstanding result can be put forward: upon thermal evaporation, the molecules in the open form are "frozen" in that specific isomeric state, which makes it difficult for them to photoisomerize when embedded in a densely packed supramolecular assembly. This limited solid-state conversion could somehow be overcome in a later study where an ON-to-OFF current ratio of about 100 in electrical switching (photostationary state featuring >90% of closed isomers) was measured for a different DAE derivative. Although the authors did not nail down the reasons for such a remarkably high conversion yield, the superior switching capacity of the new derivative, where the side triphenylamine are replaced by benzothiazole units, could be the reason for the improved performance [193].

An insightful analysis of the transport in a multilayer diode incorporating a photochromic layer [194] has been recently published by Meerholz and coworkers [195]. In this work, a cross-linkable dithienylethene (XDTE) serves as a photo-modulable hole-injection barrier layer in an OLED leading to remarkable changes in both luminance and current density. By means of *in situ* reflectance absorption spectroscopy, all the molecules in the closed form are monitored so that all the output current variations can be associated with fractions of closed isomers, X, ranging from 0 (all of the molecules are in the open form) to 1 (all of the molecules are in the closed form). Interestingly, the switching can be attained using either optical or electrical stimuli, leading to two different morphologies in the film incorporating the molecules in the closed state. Optical irradiation generates a quasi-isotropic distribution of the molecules in the XDTE film while filament-like morphology dominates the assembly when electrical signals are applied. The authors introduce the electrical closing of the XDTE in view of the observed electrochemical ring closure observed in solution, which likely creates an alternative charge transport pathway across the film. Figure 7.5 provides evidence for the two different types of output device current modulation, further evidencing the fine and achievable control. A subtle control over the output current is also attained when a careful combination of electrical and optical stimuli is employed. These findings demonstrate that the photochromic diode can attain a variety of continuous current levels, rendering a good prototype of a multilevel memory device.

Nevertheless, a multilayer approach is demanding in terms of device processing, as it must ensure minimal interlayer penetration. Apart from using orthogonal solvents severely limiting the choice of the molecular components, a viable method for preparing multilayered structures is via cross-linking of successive layers [174, 195] or by depositing the photoswitchable layer by means of a PDMS stamp at the top of an already prepared device [196]. Unfortunately, the former way requires additional synthetic effort to impart the capacity to undergo cross-link to the chosen compounds. Furthermore, since the switching efficiency can be reduced by the presence of overlayers, a different geometry is desirable where the interaction between light and photochromic molecules is maximized and not by any absorptive top molecule, eventually reducing the number of photons reaching the component that should undergo photoisomerization.

7.2.3 The Blending Approach

A practical way of combining the photoswitching capability of photochromic molecules and the charge transport characteristics of organic semiconductors

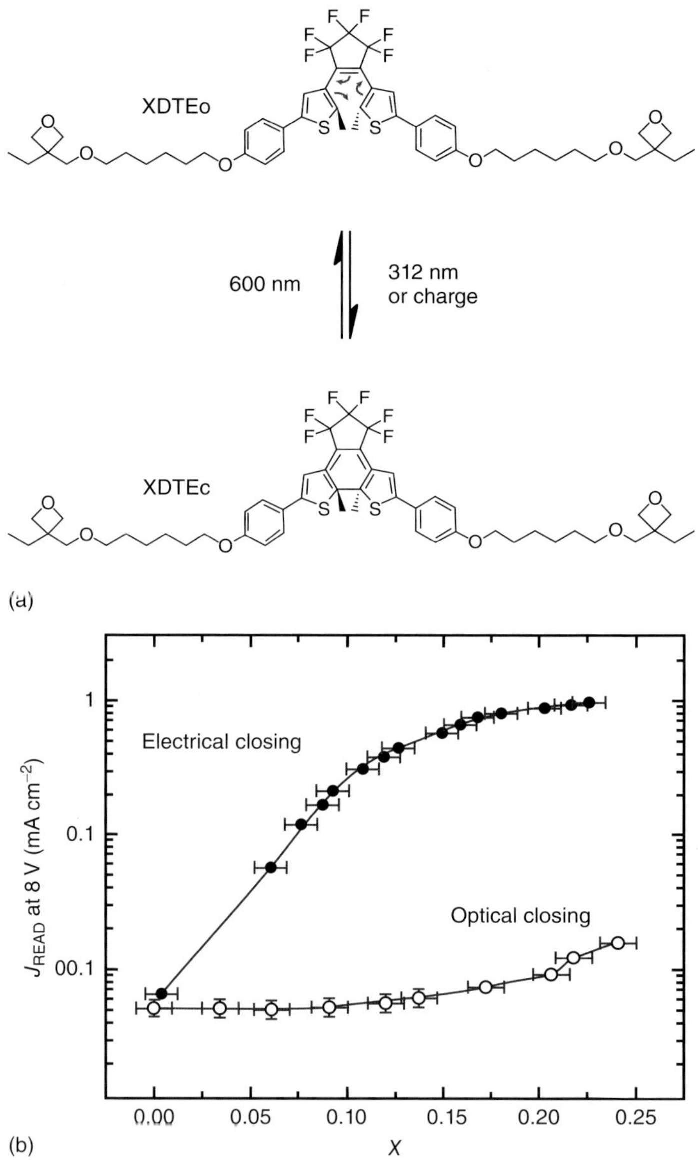
XDTEo
600 nm
312 nm
or charge
XDTEc
(a)
Electrical closing
Optical closing
J_READ at 8 V (mA cm^-2)
1
0.1
00.1
0.00
0.05
0.10
0.15
0.20
0.25
X
(b)

Figure 7.5 (a) The cross-linkable dithienylethene (XDTE) used in Ref. [56]. (b) Comparison of the current density (at 8 V) measured as a function of closed isomer fraction in the XDTE interlayer with electrical or optical stimuli. The "electrical closing" transport relies on the formation of electrically induced conductive filaments within the film, while in the "optical closing" case the closed isomers are quasi-isotropically distributed in the XDTE layer. (Reproduced with permission [195]. 2013, Wiley.)

is indeed by creating a binary mixture. In 2009, Crispin *et al.* [197] proposed a model supported by theoretical calculations suggesting the use of a variety of DAEs blended with organic semiconductors in order to realize optically switchable two-terminal devices. A few years earlier, experiments were already carried out by different research groups [198–203], showing the possibility to introduce light-controlled traps in a polymer film in order to modulate the output current in a diode. The bicomponent film relied on the use of SP (6-nitro-1′,3′,3′,-trimethylspiro[2*H*-1-benzopyran-2,2′-indoline]) as photochromic molecule and either poly[2-methoxy-5(2′-ethylhexyloxy)-1,4-phenylenevinylene] (MEH-PPV) [199, 200, 202] or poly[methyl(phenyl)silylene] (PMPSi) [198] as polymer semiconductor. The analysis of the generated trapping levels due to the presence of dipolar species in the blend was tested in the framework of the Space-Charge-Limited Current (SCLC) [198–200] and by simulations and modeling [202]. In the former case, the comparison of the $J-V$ curves before and after irradiation provided different curve shapes that could be ascribed to SP being in the high-energy-gap state or in the low-energy-gap state, thus acting or not as trapping centers for charges. The physics behind this experiment is dramatically different from what was presented in the previous section where the photochromic molecules would form a separate layer while interacting with the semiconductor mostly changing their resistivity across the device.

Optically triggered conformational changes of SPs in a semiconductor polymer matrix were employed more recently by Li *et al.* [204] in order to tune the channel conductance in an OFET. In this case, SPs, which exhibit a very huge variation in dipole moment in the two distinct isomeric forms, are expected to cause a significant change in the energy landscape of charges transported within the poly(3-hexylthiophene) (P3HT) layer. The device scheme is displayed in Figure 7.6a where also a possible, yet unproven, assembly of the two interacting molecules, that is, P3HT and SP, is depicted. Upon UV irradiation for about 3 min, the low-conductance state of the FET channel was changed gradually into a higher conductance state when an increase in output current is recorded (Figure 7.6b). The back-reaction was performed under white light and it required about 8 min for the devices to restore the initial drain current value. Significantly, the devices showed long-term stability and they could be tested in air up to 3 h without giving evidence of degradation. As an alternative to the proposed mechanism, the authors suggest that charge transfer between the P3HT backbone and the phenoxide unit of the SP can occur. The photogenerated phenoxide

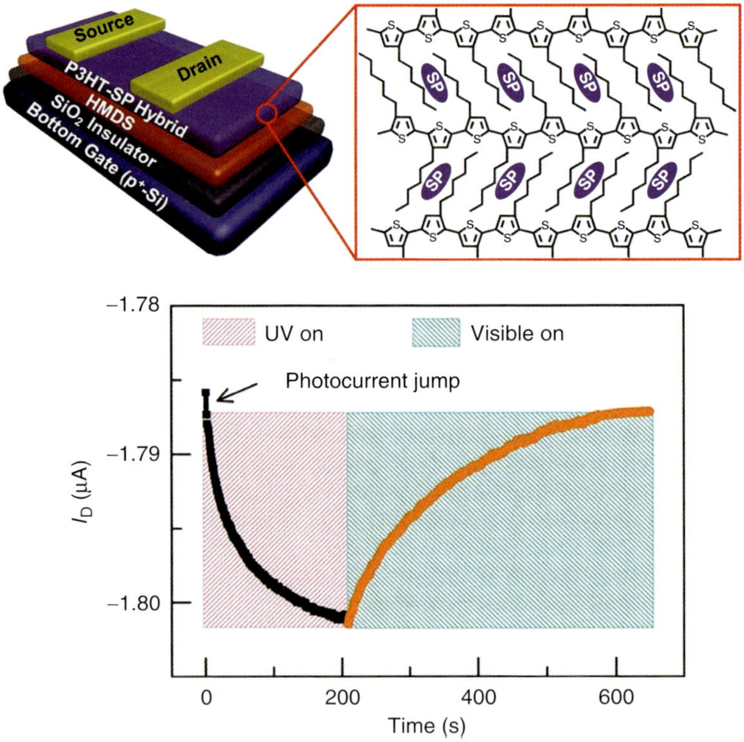

Figure 7.6 (a) Device layout employed by Li *et al.*: a top-contact bottom-gate transistor with HMDS-functionalized silicon oxide as the dielectric layer and gold source and drain electrodes as the injecting electrodes. The bicomponent film is a blend of P3HT and SP. The SP molecules are thought of being randomly dispersed within the interdigitated P3HT side chains and close to the polythiophene backbone. (b) An example of a switching cycle: the drain current versus time plot highlights how the device is able to respond with a different output upon irradiation with a different wavelength. (Reproduced with permission [204]. 2012, RCS.)

group in the SP-open form can act as a charge trap that can generate a decrease in mobility. Ishiguro *et al.* recently showed that the approach based on the combination of SPs with a polymer semiconductor as a mixed active layer in an FET device can be expanded also to other polymers such as poly(triarylamine) (PTAA) [205].

We have recently reported a different strategy that relies on the insertion of phototunable energetic levels in the bandgap of P3HT by using DAEs derivatives [206]. In this work, DAEs were selected as the photochromic molecules because they offer the significant advantages of high fatigue resistance, thermal stability of both switching states, and feature radically different electronic properties, including HOMO and LUMO energy levels and redox characteristics. Bifunctional organic thin-film transistors (OTFTs) were fabricated by using blends of DAEs with a polymeric semiconductor acting as the electroactive material (see Figure 7.7a).

The dual functionality in these OTFTs was achieved by engineering the energy levels of the blend through the incorporation of the DAE phototunable energy levels in the polymer matrix in order to tune the charge transport. Hence, blending the semiconducting material with a suitable DAE should make it possible to selectively control and modulate the charge carriers within the film as a result of light irradiation at different wavelengths. A schematic of the energy-level engineering in the above-mentioned bicomponent films is depicted in Figure 7.7b, where two DAE derivatives in the open (DAE_1o) or in the closed (DAE_1c) form feature different energy levels that are supposed to act differently on the hole transport by generating deeper (DAE_1c) or shallower (DAE_2c) energy-trapping levels in the band gap of P3HT. Dynamic characterization of DAE_1o/P3HT-based FET devices was performed upon irradiation of the transistor over second-scale cycles (Figure 7.7c) while switching the light source between UV (at a fixed wavelength) and visible light, respectively. This experiment clearly showed that a dynamic control of the output current of the device could be performed optically in a reversible manner on a timescale of seconds without exhibiting fatigue for several irradiation cycles. Significantly, the response time of the device, which represents the current variation of the output current in response to a single 3-ns-long laser pulse centered at 310 nm, was found to fall within the timescale of a few microseconds, a truly relevant result paving the way toward exploitation in real electronics.

Inorganic–organic hybrid structures can also be exploited to assemble multicomponent materials incorporating photochromic molecules. Among various systems, metallic nanoparticles represent ordered nanostructured scaffolds that can be easily synthesized with a high precision as monodisperse objects. In this framework, Raimondo *et al.* decorated gold nanoparticles (AuNPs) with a chemisorbed SAM of AZO and studied the interparticle propensity to undergo aggregation as a function of the isomeric state of the AZO changed by light irradiation at specific wavelength [207]. In a later set of experiments, these photoresponsive nanoparticles were incorporated in the polymer semiconductor via blending in the attempt to provide optical response to the FET [208]. In particular, P3HT was blended with AuNPs coated with a chemisorbed AZO-based SAM, the latter acting as trap charges in the device channel (Figure 7.8, top panel). Through a light-induced isomerization between the *trans* and *cis* forms of the AZO molecules chemically tethered to the AuNPs, a variation in the tunneling barrier is expected that sets the efficiency of the trapping process in the semiconductor film. The electrical characterization of the FETs upon irradiation with UV light was investigated on devices having different AuNPs/P3HT ratios; it revealed significant changes in field-effect mobility and threshold voltage between devices with azobenzobenzene-coated AuNPs with respect to AuNPs with a non-photoresponsive coating or pristine P3HT both used as a reference (Figure 7.8, bottom panel). An increase in drain current was observed that reached saturation within a few minutes although bias stress signature started to appear after 10 min inducing a decrease in the I_D. By and large, this approach makes it possible to impart a dual functionality to an organic/nanoparticles

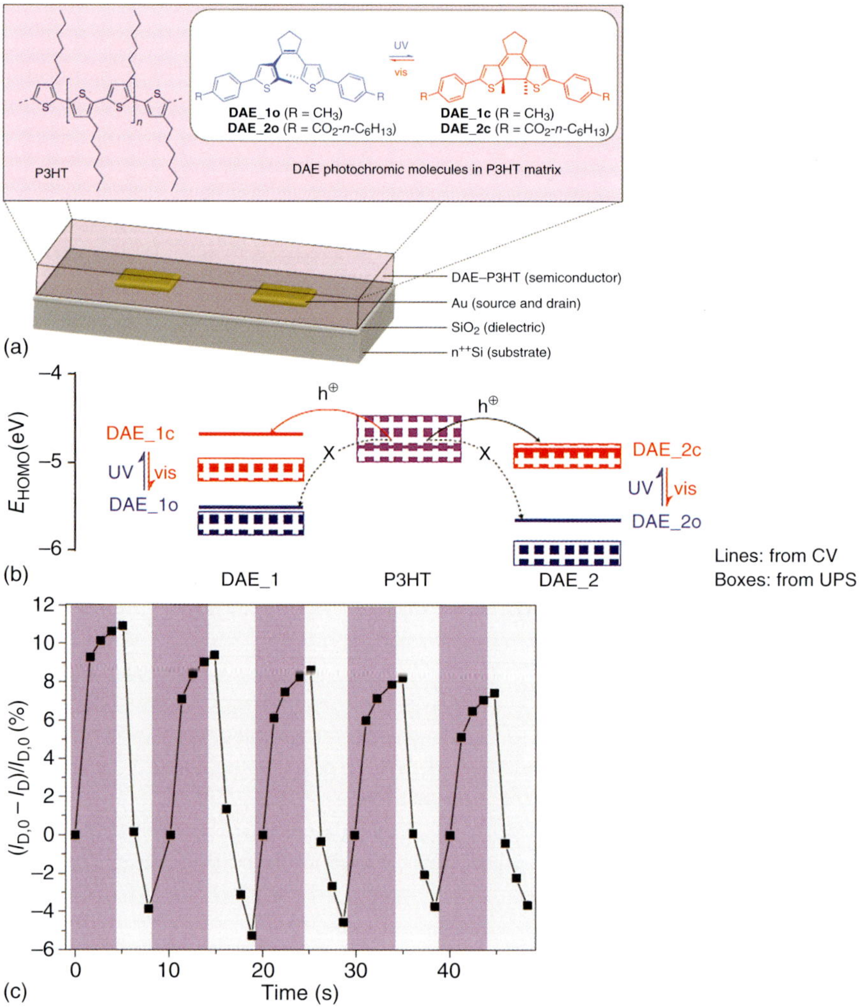

Figure 7.7 (a) General scheme of the tested organic thin-film transistors including chemical formulae of the molecules employed in the bicomponent film, namely the photochromic molecules DAE_1 and DAE_2 in their two isomeric states and the P3HT semiconducting polymer. (b) Energy diagram shows the HOMO levels of the employed components and depicts the photomodulation mechanism occurring in the device (UV, ultraviolet; Vis, visible). (c) Dynamic switching of an OTFT made from DAE_1o (20 wt% in P3HT) under several irradiation cycles with ultraviolet ($\lambda_{irr} = 365$ nm, $P_i = 62$ mW cm^2) and white light ($\lambda_{irr} > 400$ nm, $P_i = 5.06$ mW cm^2). (Reproduced with permission [206]. 2012, Nature Publishing Group.)

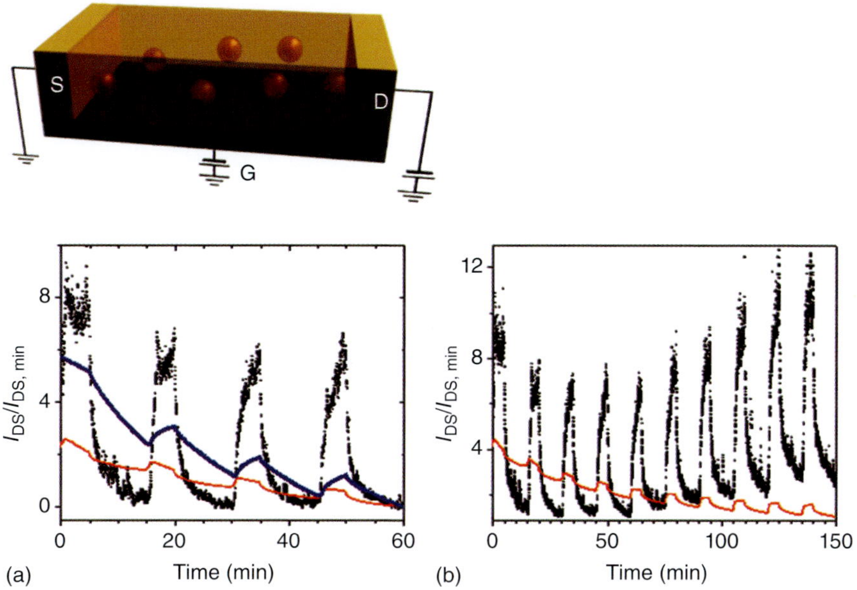

Figure 7.8 (Top panel) Device scheme employed in the study: gold nanoparticles with various coatings are blended with P3HT, which acts as the semiconductor layer for the bottom-contact bottom-gate transistor. (Bottom panel) Photoresponse cycles versus time of s-AZONPs/P3HT (in black), OPE-NPs/P3HT (in red), and P3HT (in blue). For the latter, the $I_D/I_{D,min}$ was multiplied by a factor of 100 for a better comparison. (a) 4 Cycles and (b) 10 cycles. ($V_G - V_{TH} = -4$ V, $V_D = -10$ V, $L = 5$ μm). (Reproduced with permission [208]. 2012, US National Academy of Science.)

(NPs)-based FET through the gating of the drain current both electrically (by means of the gate electrode), as in a conventional transistor, and optically (via optical stimuli).

In summary, multifunctional devices such as photoswitchable transistors and memories were realized via blending of the organic semiconductor with different types of photochromic molecules, which make this approach more appealing for device mass production than a multilayered approach. Yet, numerous challenges must be faced in the future including (i) control over phase segregation in bicomponent films, (ii) how the crystallization of the organic polymer is affected by the presence of a photochromic molecule in the mixture, (iii) dewetting and miscibility in a common solvent, (iv) extending the approach to small semiconducting molecules interacting with photochromic molecules given that only polymers were considered up to now, and (v) how the presence of injected charges within the film may affect the isomerization mechanisms and its yield given that this may later depend on the specific photochromic molecule/organic semiconductor pair considered.

7.3
Photoresponsive Dielectric Interfaces and Bulk

It is widely recognized that the physics of organic-based transistors is governed by interfacial properties of different nature occurring at both the electrode/semiconductor and the dielectric/semiconductor interface [58, 209, 210]. In these devices, polaron transport occurs in the active layer within the very first few nanometers in the vicinity of the dielectric layer. This implies that even minimal chemical changes at the interface could induce major variations in the overall device performances. Thus, a viable approach to attain a bistable nature in an OFET is through the modification of the semiconductor/dielectric interface with photochromic molecules whose different isomeric state would lead to changes in the energetic landscape offered to the charge carriers by introducing phototunable variations in the interfacial dipole.

In the work by Zhang *et al.* [211, 212], the interfacial properties are controlled by functional SAMs of SPs that have been often employed to modify the morphology of organic semiconductors [213, 214]. In this work, SAMs of SPs confer a multifunctional nature to the OFET given that a significant change in dipole moment occurs when a UV light induces the switching between their neutral form, closed-SP (6.4 D), and the zwitterionic colored form, open-SP (13.9 D) as determined by density functional theory calculations [196]. Pentacene-based OFETs were prepared in bottom-gate top-contact geometry on a heavily doped p-type silicon wafer with a 300-nm-thick layer of thermally grown SiO_x. SAMs of SP were prepared on silicon wafer substrates by silanizing first the SiO_x with APTMS and successively tethering an SP carboxylic acid to the surface through covalent amide bond formation with the aid of DCC (carbo diimide dehydrating-activating agent). XPS and FTIR measurements proved the effective immobilization of the SPs on the surface.

Evidence of the SAM switching was provided through (i) ultraviolet–visible (UV–Vis) absorption spectra that revealed a large characteristic difference in the absorption band at wavelength $\lambda = 563$ nm upon UV ($\lambda = 365$ nm) and visible light ($\lambda > 520$ nm) irradiation, proving the reversible switching between closed SP and open SP; (ii) the calculated percent conversion (indicated with x_e) of SP molecules from closed SP to open SP at the photostationary state was of about 84%; and (iii) light-induced variation of the water contact angle of 12–15° suggesting that the surface is more polar after UV irradiation, as one would expect when SP-SAM is in the open form.

However, about 10-fold lower mobility measured on devices with SP-SAM-functionalized oxide with respect to pristine SiO_2 suggested that some interfacial trapping related to both the presence of SP molecules and unreacted amine groups on the surface has to be invoked to explain such a decrease. In addition, a large (negative) shift in threshold voltage was measured but could not be attributed to any variation in capacitance of the SAM/SiO_2 structure upon irradiation. Overall, large and reversible variations in drain current in response to UV and white light irradiation were measured over ~100 SP-functionalized devices.

Aiming at providing evidence for the switching reversibility, the authors performed a number of short irradiation time measurements. The devices consistently exhibited long term and reversible operational stability. Figure 7.9a clearly shows several switching cycles of the drain current recorded as a function of time. The measurements were performed over a period of ~3 h and for ~200 cycles; the devices showed a good switching behavior without appreciable degradation when operated in ambient atmosphere. Moreover, two important merit figures were used to quantify the response to irradiation: the current change ratios (P) and the responsivity (R) expressed in A/W:

$$P = \frac{I_{light}}{I_{dark}} = \frac{|I_l - I_{dark}|}{I_{dark}} \tag{7.2}$$

$$R = \frac{I_{light}}{P_{ill}} = \frac{|I_l - I_{dark}|}{LWI_{ill}} \tag{7.3}$$

where I_{dark} and I_{light} are, respectively, the drain current measured in dark and under light irradiation, I_l is the drain current under illumination, P_{ill} is the illumination power incident on the device channel, I_{ill} is the light power intensity, L and W are, respectively, the channel length and width of the device. Noteworthy, the device photoresponsivity was found being bias-dependent. Figure 7.9b shows the responsivity data of the device at several gate voltages for a constant drain bias. In the best cases, R as high as ~400 A W^{-1} and P up to ~450 were extracted at very low effective light power density of 7.4 μW cm^{-2}. The reported values were found to exceed those previously measured in most of the organic-based phototransistors and are comparable with those of amorphous silicon (R = 300 A W^{-1} and P = 1000), which makes this finding truly relevant for technological exploitation. Regarding the switching mechanisms, the authors considered several possibilities. First, the effect of the molecular dipole change (13.9 D for open SP and 6.4 D for closed SP) should be taken into account. If one considers the relationship between electric field across a SAM and its molecular dipole:

$$E_{SAM} = \frac{N\mu_{mol}}{\varepsilon d_{mol}} \tag{7.4}$$

where ε is the dielectric constant, and N and d_{mol} are the area density and height of the SP-SAM, respectively, it appears that the voltage variation across the monolayer in response to a change in molecular dipole would amount to 1.1 – 1.7 V. An additional change in dielectric constant upon isomerization could be taken into account that would further explain the authors' findings.

A charge-transfer mechanism between pentacene and photogenerated phenoxide ion groups (open-SP form) could be invoked as a possible mechanism. These latter groups could act as trap centers, thus leading to a decrease in charge mobility and a threshold voltage shift toward more negative values, which did not occur in the experiments. To exclude the possibility of interfacial charge transfer, a poly(methyl methacrylate) (PMMA) interlayer was used to cover the SP layer, but this attempt resulted in device performances comparable to those without interlayer.

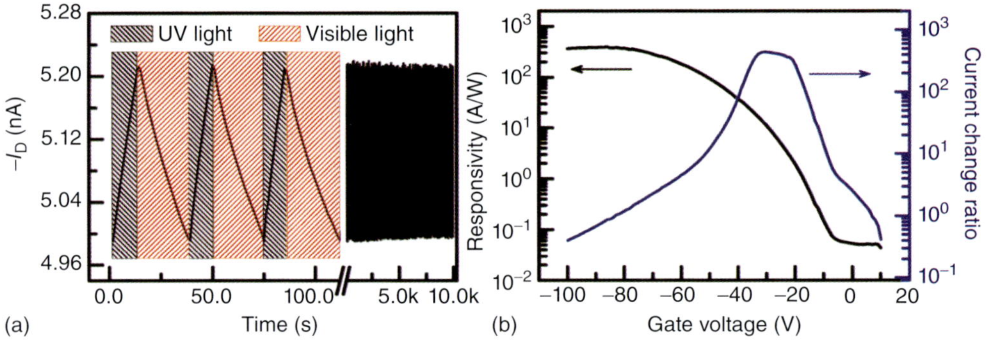

Figure 7.9 (a) Plot of I_D versus time for a pentacene-based FET with SP-SAM-functionalized silicon oxide. The switching experiment lasted for about 3 h (first three cycles expanded for clarity) revealing a reversible photoswitching upon irradiation with UV and visible light. $V_D = -30$ V; $V_G = -15$ V. (b) Responsivity (*R*) and photosensitivity (*P*) as a function of the gate voltage (effective irradiance power = 7.4 μW cm^{-2}) for the previously mentioned device (drain voltage was kept at −100 V). (Reproduced with permission [211]. 2011, ACS.)

Another intriguing approach to confer a bistable nature in OFET was reported by Tseng *et al.* [215] who integrated both aggregated clusters and SAM of AZO derivatives at the dielectric/pentacene interface. In their work, SAMs of several substituted AZO derivatives indicated with AZO-Sil-R (with R = CF_3, H, $C_{12}H_{25}$, CH_3) were chemisorbed onto SiO_x (Figure 7.10a). Each type of SAM features a different dipole moment depending on the head group attached. In addition, discrete clusters of multiple layers of AZO-functionalized acid molecules indicated as compound AZO-acid-R (with R = CF_3, CH_3) could also lead to bistability in the pentacene film surrounding them (Figure 7.10b). Before evaporation of the organic semiconductor, the SAMs formation was characterized by ellipsometry and attenuated total reflectance infrared spectroscopy (ATR-IR). Furthermore, pentacene films deposited on the SAMs were investigated by both atomic force microscopy and X-ray diffraction. These experiments revealed a polycrystalline nature in all the films grown on the AZO monolayers, and the diffraction peak intensities were found to slightly differ depending on the specific terminal functional group that could be ascribed to the difference in surface energy. OFETs in staggered geometry were prepared on the pentacene-AZO composite films by thermal evaporation of gold source and drain electrodes through a shadow mask. All the devices exhibited p-type behavior, which is expected when pentacene acts as the active layer. Field-effect mobility values would fall within the 0.27–0.35 cm^2 V^{-1} s^{-1} range for devices prepared on different SAM-modified substrates, with the exception of the film deposited on the mixed monolayer of AZO-Sil-CF_3/C_{10}, which exhibited lower mobilities when compared to those recorded on a single-component SAM. The mobility was higher in OFET where the pentacene film was deposited onto the $C_{12}H_{25}$-terminated AZO- SAM likely owing to the different crystallinity of the active layer, as confirmed by X-ray and AFM data. The threshold voltage showed remarkable shifts to more positive

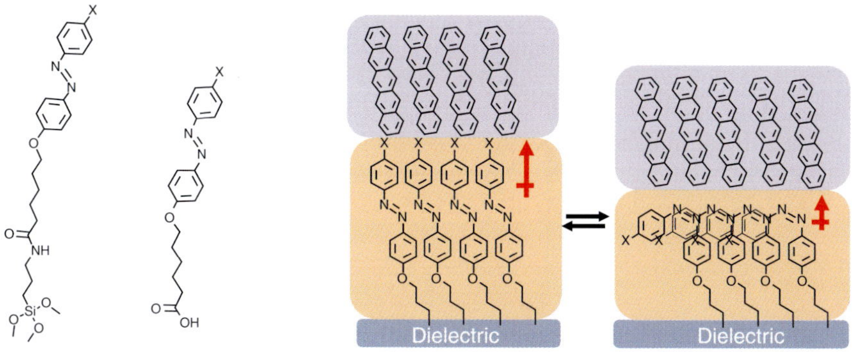

Figure 7.10 (a) (left) Chemical structures of several substituted azobenzene derivatives indicated with AZO-Sil-R (with R = CF_3, H, $C_{12}H_{25}$, CH_3) chemisorbed in SAM form onto silicon oxide layers. Each type of SAM features a different dipole moment depending on the head group attached. In addition, (right) discrete clusters of multiple layers of azobenzene-carrying acid molecules, indicated as compound AZO-acid-R (with R = CF_3, CH_3), could also lead to bistability when acting within the bulk of the semiconductor (pentacene). (b) Schematics of the switching principle presented in the work by Tseng *et al.* When the azobenzene SAM undergoes isomerization, the dipole of the molecular backbone is changed leading to a different dipolar interaction with the pentacene upper layer. (Panel (a) is readapted from Ref. [214]. Reproduced with permission [215]. 2012, ACS.)

voltages in devices with both AZO-Sil-CF_3 monolayer and mixed layer-assembled SiO_2 surface while shifts to higher negative threshold voltage values were measured in devices bearing AZO-Sil-H and AZO-Sil-CH_3 monolayer-assembled surfaces when compared to the OTS-modified surfaces that were used as a reference. In fact, a fluorinated surface bears an intrinsically electronegative nature leading to more positive field-effect threshold voltages, whereas electron-rich surfaces need a more negative bias to compensate the higher amount of positive charge accumulated.

Upon UV-induced *trans*-to-*cis* isomerization of AZO derivatives, the system undergoes a variation in both direction and magnitude of the overall molecular dipole moment of the SAM chemisorbed on SiO_x (Figure 7.10b). The effect of such photoisomerization on the $I_D - V_G$ curves was evaluated. Importantly, the authors performed control experiments in FETs with no AZO-SAM, which confirmed the photochemical stability of the pentacene films, evidenced through the negligible variations in current upon UV irradiation.

Discrete-cluster interfaces led to better memory window and faster responses than those recorded in SAM-functionalized devices, a result that the authors attributed to a better contact and more efficient charge transfer between pentacene and the trapping centers. In addition, the supposedly better contact would lead to a faster discharging of the trapping sites, which results therefore in shorter retention times. A similar approach to introduce controlled deep trapping levels by using a layer of thermally evaporated DAEs (instead of SPs) at the pentacene/PMMA interface was employed by Yoshida *et al.* [216]. In their

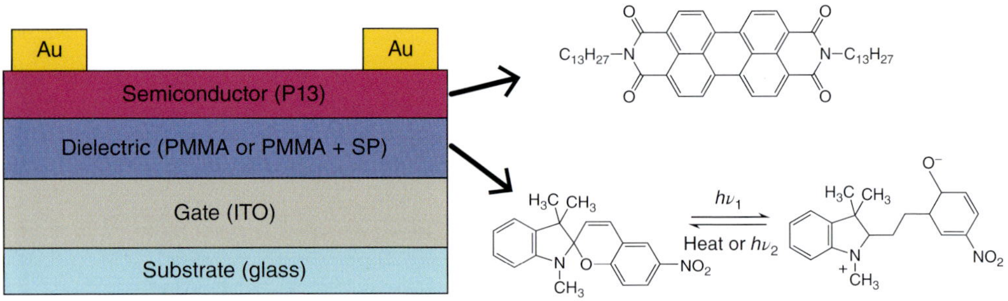

Figure 7.11 Sketch of the bottom-gate top-contact FET structure employed in Lutzyk *et al.* [217]. The active layer is an n-type semiconductor (PTCDI-$C_{13}H_{27}$). The gate is either PMMA or PMMA blended with spiropyrans. (Reproduced with permission [217]. 2011, ACS.)

work, a rewritable memory is realized by photoswitching the energy levels of a DAE derivative sandwiched between the polymer dielectric and the active layer.

Aiming at phototuning the dielectric properties in order to modulate the drain current, Lutsyk *et al.* [217] integrated SP derivatives in a PMMA matrix, which serves as the gate insulator in an FET device (Figure 7.11). Their findings rely on the reversible modification of the field-effect mobility owing to phototunable formation of trap states in the organic semiconductor in the vicinity of highly polar molecules such as SP or MC. According to the literature, deep trap states can be formed when the dipole moment is as high as 10D [197, 218]. An ITO layer acts as the bottom gate for the whole structure. Successively, a 1 : 0.1 wt% solution of PMMA and SP (solvent:ethyl acetate) was spin-coated to form the insulating gate layer. Both the organic semiconductor and the electrodes were thermally evaporated.

Upon UV irradiation, an increase in drain current is observed already after 1-min exposure of the device to the light source. This effect was attributed to the isomerization of SP to MC. The MC isomer switches back to the SP form either by white light irradiation or slow thermal recovery. The former photoreaction is undoubtedly faster than the latter, but the authors chose the back-reaction in dark in order to rule out the contribution to the drain current coming from the generation of photocarriers in the organic layer upon irradiation with visible light. This would be instrumental to make a proof of principle of their theory. The relaxation time at ambient temperature lasted 16 h during which the drain current value recovers its initial value. Despite the fact that kinetics is not considered within the manuscript, the authors found that the isomerization process was reversible. The threshold voltage shift that stems from the SP isomerization into MC in the PMMA matrix was of about 5 V and differed of 7 and 12 V, respectively, from devices where only PMMA is used as the dielectric. Furthermore, addition of SP to PMMA led to changes in the device transconductance by ~8%, whereas a further reversible ~2% increase was measured upon irradiation with UV light.

These results apparently contradict previous theoretical prediction [197, 203, 219] and experimental evidence [198, 199, 220] where current was found to increase upon SP to MC isomerization in a semiconductor matrix. However, this trend is consistent with a previous experimental observation [221], which used pentacene as the active layer and SPs/PMMA as the dielectric. Nevertheless, the switching mechanism could be explained in two possible ways: (i) formation of charged states originated by dipoles in the vicinity of the dielectric/semiconductor interface or (ii) changes of the dielectric bulk properties of the insulator.

The studies presented earlier represent an exciting improvement for the normal functioning of OFETs because of the added functionalities brought into play by the photochromic molecules integrated in the dielectric bulk and interface. Nevertheless, some issues are yet to be further understood. Though several hypotheses such as charge transfer were put forward, a clear understanding of the physics of the photochromic SAM/organic semiconductor interface still needs to be unraveled. Furthermore, the relationship between the change in steric hindrance of the photochromic molecule upon isomerization and the mechanical constraint exerted by the dielectric polymer is still an open question.

7.4
Conclusions and Future Outlooks

This chapter describes a number of attempts combining the world of photochromic molecules and organic electronics in order to generate multifunctional devices. Thanks to the features added by the photoresponsive nature of these bistable molecular switches new routes can be mapped out that are likely to drive the present and future organic electronics community toward devices exhibiting novel and multiple functionalities.

In particular, optical switching of the output current has been achieved and proven in combination with organic semiconductors by introducing controlled photomodulable trap levels either in the semiconductor bulk or at the interface between dielectric and active layer, leading to a novel generation of fast optical response FETs and multilevel memories with ON-to-OFF ratios approaching 10^4.

Furthermore, by tuning the process of charge injection with light, a different amount of charge can be transferred from an injecting electrode into the active layer of an organic-based transistor or an OLED either by changing the tunneling barrier of a photoresponsive SAM or using an interlayer featuring phototunable energy levels.

By and large, organic-based devices' performances are strongly interface dependent; therefore, exploiting photochromic molecules to tune not only the bulk transport properties of the active layer but also those of the metal/semiconductor and dielectric/semiconductor interfaces will open up a numerous opportunities for the generation of bifunctional up to multifunctional devices.

However, this work also attempts to draw a roadmap while aspiring to put in evidence the most significant issues that are yet to be further understood. In spite of the suggested hypothesis of charge transfer, a more nuanced study on the physics of charge transport is needed to disentangle the mutual effect that the semiconductor and photochromic component exert in a binary blend or when they share a large area interface. Furthermore, how a charge injected into the semiconductor can affect the isomerization process of the photochromic molecules is still an open question that will require careful spectroscopic investigation as well as modeling/simulation activity. The role of the light power density on the isomerization of the photochromic molecule in the solid state also needs to be properly tackled.

To date, the role of the very many intermolecular interactions on the switching efficiency in each specific semiconductor/photochromic molecule pair needs to be unraveled. This could be achieved by an in-depth photochemical investigation of the isomerization yield in solution at different concentrations compared to that recorded in films. In case of ordered assemblies, the influence of intermolecular interactions can be reproducible by designing systems and assemblies thereof that can undergo isomerization in processes that are either cooperative or collective in nature. By exploiting these characteristics, a high photochemical yield and improved faster time response to the optical stimuli could be certainly achieved.

To gain deeper insight into the working principles of molecular systems incorporating both organic semiconductors and photochromic molecules, further fundamental research is required. Optimized materials, a deep and all-embracing understanding of their photophysical and photochemical processes, and more nuanced design rules for the devices will pave the way toward a novel generation of multifunctional, low-cost, large-area, and flexible electronics.

Acknowledgments

We are grateful to Dr Nuria Crivillers, Dr Corinna Raimondo, Dr Andrea Liscio, Dr Vincenzo Palermo as well as Prof. Stefan Hecht, Prof. Marcel Mayor, Prof. Franco Cacialli, Dr David Beljonne, and Dr Jérôme Cornil and their research teams for the joint research activity on combining photochromic molecules and organic semiconductors that was an essential source of inspiration for this chapter. This work was supported by the EC through the ERC project SUPRA-FUNCTION (GA-257305), as well as the Agence Nationale de la Recherche through the LabEx project CSC, and the International Center for Frontier Research in Chemistry (icFRC).

References

1. Katz, H.E., Lovinger, A.J., Johnson, J., Kloc, C., Siegrist, T., Li, W., Lin, Y.Y., and Dodabalapur, A. (2000) A soluble and air-stable organic semiconductor with high electron mobility. *Nature*, **404** (6777), 478–481.
2. Crone, B., Dodabalapur, A., Lin, Y.Y., Filas, R.W., Bao, Z., LaDuca,

A., Sarpeshkar, R., Katz, H.E., and Li, W. (2000) Large-scale complementary integrated circuits based on organic transistors. *Nature*, **403** (6769), 521–523.
3. Rogers, J.A., Bao, Z., Baldwin, K., Dodabalapur, A., Crone, B., Raju, V.R., Kuck, V., Katz, H., Amundson, K., Ewing, J., and Drzaic, P. (2001) Paper-like electronic displays: large-area rubber-stamped plastic sheets of electronics and microencapsulated electrophoretic inks. *Proc. Natl. Acad. Sci. U.S.A.*, **98** (9), 4835–4840.
4. Meijer, E.J., De Leeuw, D.M., Setayesh, S., Van Veenendaal, E., Huisman, B.H., Blom, P.W.M., Hummelen, J.C., Scherf, U., and Klapwijk, T.M. (2003) Solution-processed ambipolar organic field-effect transistors and inverters. *Nat. Mater.*, **2** (10), 678–682.
5. Zschieschang, U., Klauk, H., Halik, M., Schmid, G., and Dehm, C. (2003) Flexible organic circuits with printed gate electrodes. *Adv. Mater.*, **15** (14), 1147–1151.
6. Arias, A.C., Ready, S.E., Lujan, R., Wong, W.S., Paul, K.E., Salleo, A., Chabinyc, M.L., Apte, R., Street, R.A., Wu, Y., Liu, P., and Ong, B. (2004) All jet-printed polymer thin-film transistor active-matrix backplanes. *Appl. Phys. Lett.*, **85** (15), 3304–3306.
7. Ling, M.M. and Bao, Z.N. (2004) Thin film deposition, patterning, and printing in organic thin film transistors. *Chem. Mater.*, **16** (23), 4824–4840.
8. McCulloch, I. (2005) Thin films – rolling out organic electronics. *Nat. Mater.*, **4** (8), 583–584.
9. Street, R.A., Wong, W.S., Ready, S.E., Chabinyc, I.L., Arias, A.C., Limb, S., Salleo, A., and Lujan, R. (2006) Jet printing flexible displays. *Mater. Today*, **9** (4), 32–37.
10. Slinker, J.D., DeFranco, J.A., Jaquith, M.J., Silveira, W.R., Zhong, Y.W., Moran-Mirabal, J.M., Craighead, H.G., Abruna, H.D., Marohn, J.A., and Malliaras, G.G. (2007) Direct measurement of the electric-field distribution in a light-emitting electrochemical cell. *Nat. Mater.*, **6** (11), 894–899.
11. Facchetti, A. (2007) Semiconductors for organic transistors. *Mater. Today*, **10** (3), 28–37.
12. Klauk, H., Zschieschang, U., Pflaum, J., and Halik, M. (2007) Ultralow-power organic complementary circuits. *Nature*, **445** (7129), 745–748.
13. Berggren, M., Nilsson, D., and Robinson, N.D. (2007) Organic materials for printed electronics. *Nat. Mater.*, **6** (1), 3–5.
14. Cho, J.H., Lee, J., Xia, Y., Kim, B., He, Y., Renn, M.J., Lodge, T.P., and Frisbie, C.D. (2008) Printable ion-gel gate dielectrics for low-voltage polymer thin-film transistors on plastic. *Nat. Mater.*, **7** (11), 900–906.
15. Yan, H., Chen, Z.H., Zheng, Y., Newman, C., Quinn, J.R., Dötz, F., Kastler, M., and Facchetti, A. (2009) A high-mobility electron-transporting polymer for printed transistors. *Nature*, **457** (7230), 679–686.
16. Rauch, T., Boberl, M., Tedde, S.F., Furst, J., Kovalenko, M.V., Hesser, G.N., Lemmer, U., Heiss, W., and Hayden, O. (2009) Near-infrared imaging with quantum-dot-sensitized organic photodiodes. *Nat. Photonics*, **3** (6), 332–336.
17. Simon, D.T., Kurup, S., Larsson, K.C., Hori, R., Tybrandt, K., Goiny, M., Jager, E.H., Berggren, M., Canlon, B., and Richter-Dahlfors, A. (2009) Organic electronics for precise delivery of neurotransmitters to modulate mammalian sensory function. *Nat. Mater.*, **8** (9), 742–746.
18. Rogers, J.A., Someya, T., and Huang, Y.G. (2010) Materials and mechanics for stretchable electronics. *Science*, **327** (5973), 1603–1607.
19. Mannsfeld, S.C.B., Tee, B.C.K., Stoltenberg, R.M., Chen, C.V.H.H., Barman, S., Muir, B.V.O., Sokolov, A.N., Reese, C., and Bao, Z.N. (2010) Highly sensitive flexible pressure sensors with microstructured rubber dielectric layers. *Nat. Mater.*, **9** (10), 859–864.
20. Arias, A.C., MacKenzie, J.D., McCulloch, I., Rivnay, J., and Salleo, A. (2010) Materials and applications for large area electronics: solution-based approaches. *Chem. Rev.*, **110** (1), 3–24.

21. Pfattner, R., Mas-Torrent, M., Bilotti, I., Brillante, A., Milita, S., Liscio, F., Biscarini, F., Marszalek, T., Ulanski, J., Nosal, A., Gazicki-Lipman, M., Leufgen, M., Schmidt, G., Molenkamp, L.W., Laukhin, V., Veciana, J., and Rovira, C. (2010) High-performance single crystal organic field-effect transistors based on two dithiophene-tetrathiafulvalene (DT-TTF) Polymorphs. *Adv. Mater.*, **22** (37), 4198–4203.
22. Cavallini, M., D'Angelo, P., Criado, V.V., Gentili, D., Shehu, A., Leonardi, F., Milita, S., Liscio, F., and Biscarini, F. (2011) Ambipolar multi-stripe organic field-effect transistors. *Adv. Mater.*, **23** (43), 5091–5097.
23. Burroughes, J.H., Bradley, D.D.C., Brown, A.R., Marks, R.N., Mackay, K., Friend, R.H., Burns, P.L., and Holmes, A.B. (1990) Light-emitting-diodes based on conjugated polymers. *Nature*, **347** (6293), 539–541.
24. Greenham, N.C., Moratti, S.C., Bradley, D.D.C., Friend, R.H., and Holmes, A.B. (1993) Efficient light-emitting-diodes based on polymers with high electron-affinities. *Nature*, **365** (6447), 628–630.
25. Berggren, M., Inganas, O., Gustafsson, G., Rasmusson, J., Andersson, M.R., Hjertberg, T., and Wennerstrom, O. (1994) Light-emitting-diodes with variable colors from polymer blends. *Nature*, **372** (6505), 444–446.
26. Burrows, P.E., Shen, Z., Bulovic, V., McCarty, D.M., Forrest, S.R., Cronin, J.A., and Thompson, M.E. (1996) Relationship between electroluminescence and current transport in organic heterojunction light-emitting devices. *J. Appl. Phys.*, **79** (10), 7991–8006.
27. Baldo, M.A., O'Brien, D.F., You, Y., Shoustikov, A., Sibley, S., Thompson, M.E., and Forrest, S.R. (1998) Highly efficient phosphorescent emission from organic electroluminescent devices. *Nature*, **395** (6698), 151–154.
28. Grice, A.W., Bradley, D.D.C., Bernius, M.T., Inbasekaran, M., Wu, W.W., and Woo, E.P. (1998) High brightness and efficiency blue light-emitting polymer diodes. *Appl. Phys. Lett.*, **73** (5), 629–631.
29. Kim, J.S., Granstrom, M., Friend, R.H., Johansson, N., Salaneck, W.R., Daik, R., Feast, W.J., and Cacialli, F. (1998) Indium-tin oxide treatments for single- and double-layer polymeric light-emitting diodes: the relation between the anode physical, chemical, and morphological properties and the device performance. *J. Appl. Phys.*, **84** (12), 6859–6870.
30. Friend, R.H., Gymer, R.W., Holmes, A.B., Burroughes, J.H., Marks, R.N., Taliani, C., Bradley, D.D.C., Dos Santos, D.A., Bredas, J.L., Logdlund, M., and Salaneck, W.R. (1999) Electroluminescence in conjugated polymers. *Nature*, **397** (6715), 121–128.
31. Aziz, H., Popovic, Z.D., Hu, N.X., Hor, A.M., and Xu, G. (1999) Degradation mechanism of small molecule-based organic light-emitting devices. *Science*, **283** (5409), 1900–1902.
32. Mitschke, U. and Bauerle, P. (2000) The electroluminescence of organic materials. *J. Mater. Chem.*, **10** (7), 1471–1507.
33. Ho, P.K.H., Kim, J.S., Burroughes, J.H., Becker, H., Li, S.F.Y., Brown, T.M., Cacialli, F., and Friend, R.H. (2000) Molecular-scale interface engineering for polymer light-emitting diodes. *Nature*, **404** (6777), 481–484.
34. Scherf, U. and List, E.J.W. (2002) Semiconducting polyfluorenes – towards reliable structure-property relationships. *Adv. Mater.*, **14** (7), 477–487.
35. List, E.J.W., Guentner, R., de Freitas, P.S., and Scherf, U. (2002) The effect of keto defect sites on the emission properties of polyfluorene-type materials. *Adv. Mater.*, **14** (5), 374–378.
36. Muller, C.D., Falcou, A., Reckefuss, N., Rojahn, M., Wiederhirn, V., Rudati, P., Frohne, H., Nuyken, O., Becker, H., and Meerholz, K. (2003) Multi-colour organic light-emitting displays by solution processing. *Nature*, **421** (6925), 829–833.
37. Chen, C.T. (2004) Evolution of red organic light-emitting diodes: materials

and devices. *Chem. Mater.*, **16** (23), 4389–4400.

38. Forrest, S.R. (2004) The path to ubiquitous and low-cost organic electronic appliances on plastic. *Nature*, **428** (6986), 911–918.
39. Anthopoulos, T.D., Frampton, M.J., Namdas, E.B., Burn, P.L., and Samuel, I.D.W. (2004) Solution-processable red phosphorescent dendrimers for light-emitting device applications. *Adv. Mater.*, **16** (6), 557–560.
40. Zaumseil, J., Friend, R.H., and Sirringhaus, H. (2006) Spatial control of the recombination zone in an ambipolar light-emitting organic transistor. *Nat. Mater.*, **5** (1), 69–74.
41. Muccini, M. (2006) A bright future for organic field-effect transistors. *Nat. Mater.*, **5** (8), 605–613.
42. Cicoira, F. and Santato, C. (2007) Organic light emitting field effect transistors: advances and perspectives. *Adv. Funct. Mater.*, **17** (17), 3421–3434.
43. Walzer, K., Maennig, B., Pfeiffer, M., and Leo, K. (2007) Highly efficient organic devices based on electrically doped transport layers. *Chem. Rev.*, **107** (4), 1233–1271.
44. Reineke, S., Lindner, F., Schwartz, G., Seidler, N., Walzer, K., Luessem, B., and Leo, K. (2009) White organic light-emitting diodes with fluorescent tube efficiency. *Nature*, **459** (7244), 234–238.
45. Capelli, R., Toffanin, S., Generali, G., Usta, H., Facchetti, A., and Muccini, M. (2010) Organic light-emitting transistors with an efficiency that outperforms the equivalent light-emitting diodes. *Nat. Mater.*, **9** (6), 496–503.
46. Assadi, A., Svensson, C., Willander, M., and Inganas, O. (1988) Field-effect mobility of poly(3-hexylthiophene). *Appl. Phys. Lett.*, **53** (3), 195–197.
47. Dodabalapur, A., Torsi, L., and Katz, H.E. (1995) Organic transistors: 2-dimensional transport and improved electrical characteristics. *Science*, **268** (5208), 270–271.
48. Torsi, L., Dodabalapur, A., Rothberg, L.J., Fung, A.W.P., and Katz, H.E. (1996) Intrinsic transport properties and performance limits of organic field-effect transistors. *Science*, **272** (5267), 1462–1464.
49. Bao, Z.N., Feng, Y., Dodabalapur, A., Raju, V.R., and Lovinger, A.J. (1997) High-performance plastic transistors fabricated by printing techniques. *Chem. Mater.*, **9** (6), 1299–1301.
50. Sirringhaus, H., Brown, P.J., Friend, R.H., Nielsen, M.M., Bechgaard, K., Langeveld-Voss, B.M.W., Spiering, A.J.H., Janssen, R.A.J., Meijer, E.W., Herwig, P., and de Leeuw, D.M. (1999) Two-dimensional charge transport in self-organized, high-mobility conjugated polymers. *Nature*, **401** (6754), 685–688.
51. Katz, H.E., Bao, Z.N., and Gilat, S.L. (2001) Synthetic chemistry for ultrapure, processable, and high-mobility organic transistor semiconductors. *Acc. Chem. Res.*, **34** (5), 359–369.
52. Pappenfus, T.M., Chesterfield, R.J., Frisbie, C.D., Mann, K.R., Casado, J., Raff, J.D., and Miller, L.L. (2002) A pi-stacking terthiophene-based quinodimethane is an n-channel conductor in a thin film transistor. *J. Am. Chem. Soc.*, **124** (16), 4184–4185.
53. Stutzmann, N., Friend, R.H., and Sirringhaus, H. (2003) Self-aligned, vertical-channel, polymer field-effect transistors. *Science*, **299** (5614), 1881–1884.
54. Halik, M., Klauk, H., Zschieschang, U., Schmid, G., Ponomarenko, S., Kirchmeyer, S., and Weber, W. (2003) Relationship between molecular structure and electrical performance of oligothiophene organic thin film transistors. *Adv. Mater.*, **15** (11), 917–922.
55. Halik, M., Klauk, H., Zschieschang, U., Schmid, G., Dehm, C., Schutz, M., Maisch, S., Effenberger, F., Brunnbauer, M., and Stellacci, F. (2004) Low-voltage organic transistors with an amorphous molecular gate dielectric. *Nature*, **431** (7011), 963–966.
56. Brédas, J.L., Beljonne, D., Coropceanu, V., and Cornil, J. (2004) Charge-transfer and energy-transfer processes in pi-conjugated oligomers and polymers: a molecular picture. *Chem. Rev.*, **104** (11), 4971–5003.

57. Newman, C.R., Frisbie, C.D., da Silva, D.A., Bredas, J.L., Ewbank, P.C., and Mann, K.R. (2004) Introduction to organic thin film transistors and design of n-channel organic semiconductors. *Chem. Mater.*, **16** (23), 4436–4451.
58. Dinelli, F., Murgia, M., Levy, P., Cavallini, M., Biscarini, F., and de Leeuw, D.M. (2004) Spatially correlated charge transport in organic thin film transistors. *Phys. Rev. Lett.*, **92** (11), 116802.
59. Chesterfield, R.J., McKeen, J.C., Newman, C.R., Ewbank, P.C., da Silva, D.A., Bredas, J.L., Miller, L.L., Mann, K.R., and Frisbie, C.D. (2004) Organic thin film transistors based on N-alkyl perylene diimides: charge transport kinetics as a function of gate voltage and temperature. *J. Phys. Chem. B*, **108** (50), 19281–19292.
60. Chua, L.L., Zaumseil, J., Chang, J.F., Ou, E.C.W., Ho, P.K.H., Sirringhaus, H., and Friend, R.H. (2005) General observation of n-type field-effect behaviour in organic semiconductors. *Nature*, **434** (7030), 194–199.
61. Sirringhaus, H. (2005) Device physics of solution-processed organic field-effect transistors. *Adv. Mater.*, **17** (20), 2411–2425.
62. Anthopoulos, T.D., Setayesh, S., Smits, E., Colle, M., Cantatore, E., de Boer, B., Blom, P.W.M., and de Leeuw, D.M. (2006) Air-stable complementary-like circuits based on organic ambipolar transistors. *Adv. Mater.*, **18** (14), 1900–1904.
63. Briseno, A.L., Mannsfeld, S.C.B., Ling, M.M., Liu, S., Tseng, R.J., Reese, C., Roberts, M.E., Yang, Y., Wudl, F., and Bao, Z. (2006) Patterning organic single-crystal transistor arrays. *Nature*, **444** (7121), 913–917.
64. Coropceanu, V., Cornil, J., da Silva, D.A., Olivier, Y., Silbey, R., and Brédas, J.L. (2007) Charge transport in organic semiconductors. *Chem. Rev.*, **107** (4), 926–952.
65. Cornil, J., Brédas, J.L., Zaumseil, J., and Sirringhaus, H. (2007) Ambipolar transport in organic conjugated material. *Adv. Mater.*, **19** (14), 1791–1799.
66. Zaumseil, J. and Sirringhaus, H. (2007) Electron and ambipolar transport in organic field-effect transistors. *Chem. Rev.*, **107** (4), 1296–1323.
67. Finlayson, C.E., Friend, R.H., Otten, M.B.J., Schwartz, E., Cornelissen, J.J.L.M., Nolte, R.L.M., Rowan, A.E., Samorì, P., Palermo, V., Liscio, A., Peneva, K., Müllen, K., Trapani, S., and Beljonne, D. (2008) Electronic transport properties or ensembles of perylene-substituted poly-isocyanopeptide arrays. *Adv. Funct. Mater.*, **18** (24), 3947–3955.
68. Rivnay, J., Jimison, L.H., Northrup, J.E., Toney, M.F., Noriega, R., Lu, S.F., Marks, T.J., Facchetti, A., and Salleo, A. (2009) Large modulation of carrier transport by grain-boundary molecular packing and microstructure in organic thin films. *Nat. Mater.*, **8** (12), 952–958.
69. Dabirian, R., Palermo, V., Liscio, A., Schwartz, E., Otten, M.B.J., Finlayson, C.E., Treossi, E., Friend, R.H., Calestani, G., Müllen, K., Nolte, R.J.M., Rowan, A.E., and Samorì, P. (2009) The relationship between nanoscale architecture and charge transport in conjugated nanocrystals bridged by multichromophoric polymers. *J. Am. Chem. Soc.*, **131** (20), 7055–7063.
70. Shehu, A., Quiroga, S.D., D'Angelo, P., Albonetti, C., Borgatti, F., Murgia, M., Scorzoni, A., Stoliar, P., and Biscarini, F. (2010) Layered distribution of charge carriers in organic thin film transistors. *Phys. Rev. Lett.*, **104** (24), 246602.
71. Bronstein, H., Chen, Z., Ashraf, R.S., Zhang, W., Du, J., Durrant, J.R., Tuladhar, P.S., Song, K., Watkins, S.E., Geerts, Y., Wienk, M.M., Janssen, R.A.J., Anthopoulos, T., Sirringhaus, H., Heeney, M., and McCulloch, I. (2011) Thieno 3,2-b thiophene-diketopyrrolopyrrole-containing polymers for high-performance organic field-effect transistors and organic photovoltaic devices. *J. Am. Chem. Soc.*, **133** (10), 3272–3275.
72. Lezama, I.G., Nakano, M., Minder, N.A., Chen, Z.H., Di Girolamo, F.V., Facchetti, A., and Morpurgo, A.F. (2012) Single-crystal organic charge-transfer interfaces probed using

Schottky-gated heterostructures. *Nat. Mater.*, **11** (9), 788–794.
73. Brondijk, J.J., Roelofs, W.S.C., Mathijssen, S.G.J., Shehu, A., Cramer, T., Biscarini, F., Blom, P.W.M., and de Leeuw, D.M. (2012) Two-dimensional charge transport in disordered organic semiconductors. *Phys. Rev. Lett.*, **109** (5), 056601.
74. Cornil, J., Verlaak, S., Martinelli, N., Mityashin, A., Olivier, Y., Van Regemorter, T., D'Avino, G., Muccioli, L., Zannoni, C., Castet, F., Beljonne, D., and Heremans, P. (2013) Exploring the energy landscape of the charge transport levels in organic semiconductors at the molecular scale. *Acc. Chem. Res.*, **46** (2), 434–443.
75. Yu, G., Gao, J., Hummelen, J.C., Wudl, F., and Heeger, A.J. (1995) Polymer photovoltaic cells: enhanced efficiencies via a network of internal donor-acceptor heterojunctions. *Science*, **270** (5243), 1789–1791.
76. Halls, J.J.M., Walsh, C.A., Greenham, N.C., Marseglia, E.A., Friend, R.H., Moratti, S.C., and Holmes, A.B. (1995) Efficient photodiodes from interpenetrating polymer networks. *Nature*, **376** (6540), 498–500.
77. Schmidt-Mende, L., Fechtenkötter, A., Müllen, K., Moons, E., Friend, R.H., and MacKenzie, J.D. (2001) Self-organized discotic liquid crystals for high-efficiency organic photovoltaics. *Science*, **293** (5532), 1119–1122.
78. Svensson, M., Zhang, F.L., Veenstra, S.C., Verhees, W.J.H., Hummelen, J.C., Kroon, J.M., Inganas, O., and Andersson, M.R. (2003) High-performance polymer solar cells of an alternating polyfluorene copolymer and a fullerene derivative. *Adv. Mater.*, **15** (12), 988–991.
79. Peumans, P., Uchida, S., and Forrest, S.R. (2003) Efficient bulk heterojunction photovoltaic cells using small-molecular-weight organic thin films. *Nature*, **425** (6954), 158–162.
80. Hoppe, H. and Sariciftci, N.S. (2004) Organic solar cells: an overview. *J. Mater. Res.*, **19** (7), 1924–1945.
81. Yang, F., Shtein, M., and Forrest, S.R. (2005) Controlled growth of a molecular bulk heterojunction photovoltaic cell. *Nat. Mater.*, **4** (1), 37–41.
82. Coffey, D.C. and Ginger, D.S. (2006) Time-resolved electrostatic force microscopy of polymer solar cells. *Nat. Mater.*, **5** (9), 735–740.
83. Peet, J., Kim, J.Y., Coates, N.E., Ma, W.L., Moses, D., Heeger, A.J., and Bazan, G.C. (2007) Efficiency enhancement in low-bandgap polymer solar cells by processing with alkane dithiols. *Nat. Mater.*, **6** (7), 497–500.
84. Kim, J.Y., Lee, K., Coates, N.E., Moses, D., Nguyen, T.Q., Dante, M., and Heeger, A.J. (2007) Efficient tandem polymer solar cells fabricated by all-solution processing. *Science*, **317** (5835), 222–225.
85. Gunes, S., Neugebauer, H., and Sariciftci, N.S. (2007) Conjugated polymer-based organic solar cells. *Chem. Rev.*, **107** (4), 1324–1338.
86. Blom, P.W.M., Mihailetchi, V.D., Koster, L.J.A., and Markov, D.E. (2007) Device physics of polymer: fullerene bulk heterojunction solar cells. *Adv. Mater.*, **19** (12), 1551–1566.
87. Palermo, V., Otten, M.B.J., Liscio, A., Schwartz, E., de Witte, P.A.J., Castriciano, M.A., Wienk, M.M., Nolde, F., De Luca, G., Cornelissen, J.J.L.M., Janssen, R.A.J., Müllen, K., Rowan, A.E., Nolte, R.J.M., and Samorì, P. (2008) The relationship between nanoscale architecture and function in photovoltaic multichromophoric arrays as visualized by Kelvin probe force microscopy. *J. Am. Chem. Soc.*, **130** (44), 14605–14614.
88. Thompson, B.C. and Frechet, J.M. (2008) Polymer-fullerene composite solar cells. *Angew. Chem.*, **47** (1), 58–77.
89. Campoy-Quiles, M., Ferenczi, T., Agostinelli, T., Etchegoin, P.G., Kim, Y., Anthopoulos, T.D., Stavrinou, P.N., Bradley, D.D., and Nelson, J. (2008) Morphology evolution via self-organization and lateral and vertical diffusion in polymer:fullerene solar cell blends. *Nat. Mater.*, **7** (2), 158–164.

90. Moulé, A.J. and Meerholz, K. (2008) Controlling morphology in polymer-fullerene mixtures. *Adv. Mater.*, **20** (2), 240–245.

91. Brédas, J.L., Norton, J.E., Cornil, J., and Coropceanu, V. (2009) Molecular understanding of organic solar cells: the challenges. *Acc. Chem. Res.*, **42** (11), 1691–1699.

92. Roncali, J. (2009) Molecular bulk heterojunctions: an emerging approach to organic solar cells. *Acc. Chem. Res.*, **42** (11), 1719–1730.

93. Lee, M.R., Eckert, R.D., Forberich, K., Dennler, G., Brabec, C.J., and Gaudiana, R.A. (2009) Solar power wires based on organic photovoltaic materials. *Science*, **324** (5924), 232–235.

94. Brabec, C.J., Gowrisanker, S., Halls, J.J., Laird, D., Jia, S., and Williams, S.P. (2010) Polymer-fullerene bulk-heterojunction solar cells. *Adv. Mater.*, **22** (34), 3839–3856.

95. Helgesen, M., Sondergaard, R., and Krebs, F.C. (2010) Advanced materials and processes for polymer solar cell devices. *J. Mater. Chem.*, **20** (1), 36–60.

96. Facchetti, A. (2013) Polymer donor-polymer acceptor (all-polymer) solar cells. *Mater. Today*, **16** (4), 123–132.

97. Miao, Q., Lefenfeld, M., Nguyen, T.Q., Siegrist, T., Kloc, C., and Nuckolls, C. (2005) Self-assembly and electronics of dipolar linear acenes. *Adv. Mater.*, **17** (4), 407–412.

98. Anthony, J.E. (2006) Functionalized acenes and heteroacenes for organic electronics. *Chem. Rev.*, **106** (12), 5028–5048.

99. Schmidt, R., Ling, M.M., Oh, J.H., Winkler, M., Konemann, M., Bao, Z.N., and Wurthner, F. (2007) Core-fluorinated rerylene bisimide dyes: air stable n-channel organic semiconductors for thin film transistors with exceptionally high on-to-off current ratios. *Adv. Mater.*, **19** (21), 3692–3695.

100. Palermo, V. and Samorì, P. (2007) Molecular self-assembly across multiple length scales. *Angew. Chem. Int. Ed.*, **46** (24), 4428–4432.

101. Sergeyev, S., Pisula, W., and Geerts, Y.H. (2007) Discotic liquid crystals: a new generation of organic semiconductors. *Chem. Soc. Rev.*, **36** (12), 1902–1929.

102. McCulloch, I., Heeney, M., Chabinyc, M.L., DeLongchamp, D., Kline, R.J., Coelle, M., Duffy, W., Fischer, D., Gundlach, D., Hamadani, B., Hamilton, R., Richter, L., Salleo, A., Shkunov, M., Sporrowe, D., Tierney, S., and Zhong, W. (2009) Semiconducting thienothiophene copolymers: design, synthesis, morphology, and performance in thin-film organic transistors. *Adv. Mater.*, **21** (10-11), 1091–1109.

103. Li, C., Liu, M.Y., Pschirer, N.G., Baumgarten, M., and Müllen, K. (2010) Polyphenylene-based materials for organic photovoltaics. *Chem. Rev.*, **110** (11), 6817–6855.

104. Gsanger, M., Oh, J.H., Konemann, M., Hoffken, H.W., Krause, A.M., Bao, Z.N., and Wurthner, F. (2010) A crystal-engineered hydrogen-bonded octachloroperylene diimide with a twisted core: an n-channel organic semiconductor. *Angew. Chem. Int. Ed.*, **49** (4), 740–743.

105. Lee, W.Y., Oh, J.H., Suraru, S.L., Chen, W.C., Wurthner, F., and Bao, Z.N. (2011) High-mobility air-stable solution-shear-processed n-channel organic transistors based on core-chlorinated naphthalene diimides. *Adv. Funct. Mater.*, **21** (21), 4173–4181.

106. De Luca, G., Pisula, W., Credgington, D., Treossi, E., Fenwick, O., Lazzerini, G.M., Dabirian, R., Orgiu, E., Liscio, A., Palermo, V., Müllen, K., Cacialli, F., and Samorì, P. (2011) Non-conventional processing and post-processing methods for the nanostructuring of conjugated materials for organic electronics. *Adv. Funct. Mater.*, **21** (7), 1279–1295.

107. Figueira-Duarte, T.M. and Müllen, K. (2011) Pyrene-based materials for organic electronics. *Chem. Rev.*, **111** (11), 7260–7314.

108. Diao, Y., Tee, B.C.-K., Giri, G., Xu, J., Kim, D.H., Becerril, H.A., Stoltenberg, R.M., Lee, T.H., Xue, G., Mannsfeld, S.C.B., and Bao, Z. (2013) Solution coating of large-area organic semiconductor thin films with aligned single-crystalline domains. *Nat. Mater.*, **12**, 665–671.

109. Facchetti, A. (2013) Organic semiconductors: made to order. *Nat. Mater.*, **12**, 598–600.

110. Treat, N.D., Nekuda Malik, J.A., Reid, O., Yu, L., Shuttle, C.G., Rumbles, G., Hawker, C.J., Chabinyc, M.L., Smith, P., and Stingelin, N. (2013) Microstructure formation in molecular and polymer semiconductors assisted by nucleation agents. *Nat. Mater.*, **12**, 628–633.

111. Nielsen, C.B., Turbiez, M., and McCulloch, I. (2013) Recent advances in the development of semiconducting DPP-containing polymers for transistor applications. *Adv. Mater.*, **25** (13), 1859–1880.

112. McCulloch, I. (2013) Organic electronics. *Adv. Mater.*, **25** (13), 1811–1812.

113. Feringa, B.L., Jager, W.F., and Delange, B. (1993) Organic materials for reversible optical-data storage. *Tetrahedron*, **49** (37), 8267–8310.

114. Bouas-Laurent, H. and Durr, H. (2001) Organic photochromism. *Pure Appl. Chem.*, **73** (4), 639–665.

115. Russew, M.-M. and Hecht, S. (2010) Photoswitches: from molecules to materials. *Adv. Mater.*, **22**, 3348–3360.

116. Irie, M. (2000) Photochromism: memories and switches – introduction. *Chem. Rev.*, **100** (5), 1683–1684.

117. Peters, M.V., Stoll, R.S., Kuehn, A., and Hecht, S. (2008) Photoswitching of basicity. *Angew. Chem. Int. Ed.*, **47** (32), 5968–5972.

118. Minkin, V.I. (2004) Photo-, thermo-, solvato-, and electrochromic spiroheterocyclic compounds. *Chem. Rev.*, **104** (5), 2751–2776.

119. Shinkai, S. (1987) Switch-functionalized systems in biomimetic chemistry. *Pure Appl. Chem.*, **59** (3), 425–430.

120. Liu, Z.F., Hashimoto, K., and Fujishima, A. (1990) Photoelectrochemical information-storage using an azobenzene derivative. *Nature*, **347** (6294), 658–660.

121. Ikeda, T. and Tsutsumi, O. (1995) Optical switching and image storage by means of azobenzene liquid-crystal films. *Science*, **268** (5219), 1873–1875.

122. Nagele, T., Hoche, R., Zinth, W., and Wachtveitl, J. (1997) Femtosecond photoisomerization of cis-azobenzene. *Chem. Phys. Lett.*, **272** (5-6), 489–495.

123. Archut, A., Vogtle, F., De Cola, L., Azzellini, G.C., Balzani, V., Ramanujam, P.S., and Berg, R.H. (1998) Azobenzene-functionalized cascade molecules: photoswitchable supramolecular systems. *Chem. Eur. J.*, **4** (4), 699–706.

124. Tamai, N. and Miyasaka, H. (2000) Ultrafast dynamics of photochromic systems. *Chem. Rev.*, **100** (5), 1875–1890.

125. Hugel, T., Holland, N.B., Cattani, A., Moroder, L., Seitz, M., and Gaub, H.E. (2002) Single-molecule optomechanical cycle. *Science*, **296** (5570), 1103–1106.

126. Alemani, M., Peters, M.V., Hecht, S., Rieder, K.-H., Moresco, F., and Grill, L. (2006) Electric field-induced isomerization of azobenzene by STM. *J. Am. Chem. Soc.*, **128** (45), 14446–14447.

127. Zeitouny, J., Aurisicchio, C., Bonifazi, D., De Zorzi, R., Geremia, S., Bonini, M., Palma, C.-A., Samorì, P., Listorti, A., Belbakra, A., and Armaroli, N. (2009) Photoinduced structural modifications in multicomponent architectures containing azobenzene moieties as photoswitchable cores. *J. Mater. Chem.*, **19** (27), 4715–4724.

128. Smaali, K., Lenfant, S., Karpe, S., Ocafrain, M., Blanchard, P., Deresmes, D., Godey, S., Rochefort, A., Roncali, J., and Vuillaume, D. (2010) High on-off conductance switching ratio in optically-driven self-assembled conjugated molecular systems. *ACS Nano*, **4** (4), 2411–2421.

129. Gilat, S.L., Kawai, S.H., and Lehn, J.M. (1993) Light-triggered electrical and optical switching devices. *J. Chem. Soc., Chem. Commun.*, (18), 1439–1442.

130. Tsivgoulis, G.M. and Lehn, J.M. (1995) Photonic molecular devices – reversibly photoswitchable fluorophores for nondestructive readout for optical memory. *Angew. Chem., Int. Ed. Engl.*, **34** (10), 1119–1122.

131. Gilat, S.L., Kawai, S.H., and Lehn, J.M. (1995) Light-triggered molecular devices: photochemical switching of optical and electrochemical properties

in molecular wire type diarylethene species. *Chem. Eur. J.*, **1** (5), 275–284.

132. Stellacci, F., Bertarelli, C., Toscano, F., Gallazzi, M.C., Zotti, G., and Zerbi, G. (1999) A high quantum yield diarylethene-backbone photochromic polymer. *Adv. Mater.*, **11** (4), 292–295.
133. Irie, M. (2000) Diarylethenes for memories and switches. *Chem. Rev.*, **100** (5), 1685–1716.
134. Irie, M., Kobatake, S., and Horichi, M. (2001) Reversible surface morphology changes of a photochromic diarylethene single crystal by photoirradiation. *Science*, **291** (5509), 1769–1772.
135. Murguly, E., Norsten, T.B., and Branda, N.R. (2001) Nondestructive data processing based on chiroptical 1,2-dithienylethene photochromes. *Angew. Chem. Int. Ed.*, **40** (9), 1752–1755.
136. Tian, H. and Yang, S. (2004) Recent progresses on diarylethene based photochromic switches. *Chem. Soc. Rev.*, **33** (2), 85–97.
137. Kobatake, S., Takami, S., Muto, H., Ishikawa, T., and Irie, M. (2007) Rapid and reversible shape changes of molecular crystals on photoirradiation. *Nature*, **446** (7137), 778–781.
138. Tian, H. and Wang, S. (2007) Photochromic bisthienylethene as multifunction switches. *Chem. Commun.*, (8), 781–792.
139. Arai, R., Uemura, S., Irie, M., and Matsuda, K. (2008) Reversible photoinduced change in molecular ordering of diarylethene derivatives at a solution–HOPG interface. *J. Am. Chem. Soc.*, **130** (29), 9371–9379.
140. Yagai, S., Ishiwatari, K., Lin, X., Karatsu, T., Kitamura, A., and Uemura, S. (2013) Rational design of photoresponsive supramolecular assemblies based on diarylethene. *Chem. Eur. J.*, **19** (22), 6971–6975.
141. Raymo, F.M. and Tomasulo, M. (2005) Electron and energy transfer modulation with photochromic switches. *Chem. Soc. Rev.*, **34** (4), 327–336.
142. Yildiz, I., Deniz, E., and Raymo, F.M. (2009) Fluorescence modulation with photochromic switches in nanostructured constructs. *Chem. Soc. Rev.*, **38** (7), 1859–1867.
143. Yokoyama, Y. (2000) Fulgides for memories and switches. *Chem. Rev.*, **100** (5), 1717–1739.
144. Kishimoto, Y. and Abe, J. (2009) A fast photochromic molecule that colors only under UV light. *J. Am. Chem. Soc.*, **131** (12), 4227–4229.
145. Parthenopoulos, D.A. and Rentzepis, P.M. (1989) 3-Dimensional optical storage memory. *Science*, **245** (4920), 843–845.
146. Kawata, S. and Kawata, Y. (2000) Three-dimensional optical data storage using photochromic materials. *Chem. Rev.*, **100** (5), 1777–1788.
147. Irie, M. (1989) Advances in photochromic materials for optical-data storage media. *Jpn. J. Appl. Phys., Part 1*, **28**, 215–219.
148. Myles, A.J. and Branda, N.R. (2002) 1,2-Dithienylethene photochromes and non-destructive erasable memory. *Adv. Funct. Mater.*, **12** (3), 167–173.
149. Yao, B.L., Wang, Y.L., Menke, N., Lei, M., Zheng, Y., Ren, L.Y., Chen, G.F., Chen, Y., and Fan, M.G. (2005) Optical properties and applications of photochromic fulgides. *Mol. Cryst. Liq. Cryst.*, **430**, 211–219.
150. Huang, Y.Y., Liang, W., Poon, J.K.S., Xu, Y., Lee, R.K., and Yariv, A. (2006) Spiro-oxazine photochromic fiber optical switch. *Appl. Phys. Lett.*, **88** (18), 181102.
151. Raymo, F.M. (2002) Digital processing and communication with molecular switches. *Adv. Mater.*, **14** (6), 401–414.
152. Tsujioka, T. and Irie, M. (1999) Theoretical study of signal-to-noise ratio on near-field photochromic memory with fluorescence readout. *Appl. Opt.*, **38** (23), 5066–5072.
153. Tsujioka, T. and Irie, M. (1998) Fluorescence readout of near-field photochromic memory. *Appl. Opt.*, **37** (20), 4419–4424.
154. Lim, S.J., Seo, J., and Park, S.Y. (2006) Photochromic switching of excited-state intramolecular proton-transfer (ESIPT) fluorescence: a unique route to high-contrast memory switching and nondestructive readout. *J. Am. Chem. Soc.*, **128** (45), 14542–14547.

155. Uchida, K., Saito, M., Murakami, A., Nakamura, S., and Irie, M. (2003) Non-destructive readout of the photochromic reactions of diarylethene derivatives using infrared light. *Adv. Mater.*, **15** (2), 121–125.
156. Dulic, D., van der Molen, S.J., Kudernac, T., Jonkman, H.T., de Jong, J.J.D., Bowden, T.N., van Esch, J., Feringa, B.L., and van Wees, B.J. (2003) One-way optoelectronic switching of photochromic molecules on gold. *Phys. Rev. Lett.*, **91** (20), 207402.
157. van der Molen, S.J., Liao, J.H., Kudernac, T., Agustsson, J.S., Bernard, L., Calame, M., van Wees, B.J., Feringa, B.L., and Schonenberger, C. (2009) Light-controlled conductance switching of ordered metal-molecule-metal devices. *Nano Lett.*, **9** (1), 76–80.
158. Li, J., Speyer, G., and Sankey, O.F. (2004) Conduction switching of photochromic molecules. *Phys. Rev. Lett.*, **93** (24), 248302.
159. Speyer, G., Li, J., and Sankey, O.F. (2004) Investigations of conductance of photochromic switching molecules. *Phys. Status Solidi B*, **241** (10), 2326–2330.
160. Kronemeijer, A.J., Akkerman, H.B., Kudernac, T., van Wees, B.J., Feringa, B.L., Blom, P.W.M., and de Boer, B. (2008) Reversible conductance switching in molecular devices. *Adv. Mater.*, **20** (8), 1467–1473.
161. Whalley, A.C., Steigerwald, M.L., Guo, X., and Nuckolls, C. (2007) Reversible switching in molecular electronic devices. *J. Am. Chem. Soc.*, **129** (42), 12590–12591.
162. Yassar, A., Garnier, F., Jaafari, H., Rebiere-Galy, N., Frigoli, M., Moustrou, C., Samat, A., and Guglielmetti, R. (2002) Light-triggered molecular devices based on photochromic oligothiophene substituted chromenes. *Appl. Phys. Lett.*, **80** (23), 4297–4299.
163. Hayakawa, R., Higashiguchi, K., Matsuda, K., Chikyow, T., and Wakayama, Y. (2013) Optically and electrically driven organic thin film transistors with diarylethene photochromic channel layers. *ACS Appl. Mater. Interfaces*, **5** (9), 3625–3630.
164. Kahn, A., Koch, N., and Gao, W.Y. (2003) Electronic structure and electrical properties of interfaces between metals and pi-conjugated molecular films. *J. Polym. Sci., Part B: Polym. Phys.*, **41** (21), 2529–2548.
165. Koch, N., Kahn, A., Ghijsen, J., Pireaux, J.J., Schwartz, J., Johnson, R.L., and Elschner, A. (2003) Conjugated organic molecules on metal versus polymer electrodes: demonstration of a key energy level alignment mechanism. *Appl. Phys. Lett.*, **82** (1), 70–72.
166. Duhm, S., Heimel, G., Salzmann, I., Glowatzki, H., Johnson, R.L., Vollmer, A., Rabe, J.P., and Koch, N. (2008) Orientation-dependent ionization energies and interface dipoles in ordered molecular assemblies. *Nat. Mater.*, **7** (4), 326–332.
167. Love, J.C., Estroff, L.A., Kriebel, J.K., Nuzzo, R.G., and Whitesides, G.M. (2005) Self-assembled monolayers of thiolates on metals as a form of nanotechnology. *Chem. Rev.*, **105** (4), 1103–1169.
168. Crivillers, N., Osella, S., Van Dyck, C., Lazzerini, G.M., Cornil, D., Liscio, A., Di Stasio, F., Mian, S., Fenwick, O., Reinders, F., Neuburger, M., Treossi, E., Mayor, M., Palermo, V., Cacialli, F., Cornil, J., and Samorì, P. (2013) Large work function shift of gold induced by a novel perfluorinated azobenzene-based self-assembled monolayer. *Adv. Mater.*, **25** (3), 432–436.
169. Crivillers, N., Liscio, A., Di Stasio, F., Van Dyck, C., Osella, S., Cornil, D., Mian, S., Lazzerini, G.M., Fenwick, O., Orgiu, E., Reinders, F., Braun, S., Fahlman, M., Mayor, M., Cornil, J., Palermo, V., Cacialli, F., and Samorì, P. (2011) Photoinduced work function changes by isomerization of a densely packed azobenzene-based SAM on Au: a joint experimental and theoretical study. *Phys. Chem. Chem. Phys.*, **13** (32), 14302–14310.
170. Klajn, R. (2010) Immobilized azobenzenes for the construction of photoresponsive materials. *Pure Appl. Chem.*, **82** (12), 2247–2279.

171. Crivillers, N., Orgiu, E., Reinders, F., Mayor, M., and Samorì, P. (2011) Optical modulation of the charge injection in an organic field-effect transistor based on photochromic self-assembled-monolayer-functionalized electrodes. *Adv. Mater.*, **23** (12), 1447–1452.
172. Elbing, M., Blaszczyk, A., von Haenisch, C., Mayor, M., Ferri, V., Grave, C., Rampi, M.A., Pace, G., Samorì, P., Shaporenko, A., and Zharnikov, M. (2008) Single component self-assembled monolayers of aromatic azo-biphenyl: influence of the packing tightness on the SAM structure and light-induced molecular movements. *Adv. Funct. Mater.*, **18** (19), 2972–2983.
173. Pace, G., Ferri, V., Grave, C., Elbing, M., von Hanisch, C., Zharnikov, M., Mayor, M., Rampi, M.A., and Samorì, P. (2007) Cooperative light-induced molecular movements of highly ordered azobenzene self-assembled monolayers. *Proc. Natl. Acad. Sci. U.S.A.*, **104** (24), 9937–9942.
174. Zacharias, P., Gather, M.C., Koehnen, A., Rehmann, N., and Meerholz, K. (2009) Photoprogrammable organic light-emitting diodes. *Angew. Chem. Int. Ed.*, **48** (22), 4038–4041.
175. Yang, J.H. and Ng, M.K. (2006) Synthesis of a photochromic conjugated polymer incorporating spirobenzopyran in the backbone. *Synthesis (Stuttgart)*, **18**, 3075–3079.
176. Guo, X.F., Zhang, D., Gui, Y., Wax, M.X., Li, J.C., Liu, Y.Q., and Zhu, D.B. (2004) Reversible photoregulation of the electrical conductivity of spiropyran-doped polyaniline for information recording and nondestructive processing. *Adv. Mater.*, **16** (7), 636–640.
177. Harvey, C.P. and Tovar, J.D. (2011) Main-chain photochromic conducting polymers. *Polym. Chem. (UK)*, **2** (12), 2699–2706.
178. Marsella, M.J., Wang, Z.Q., and Mitchell, R.H. (2000) Backbone photochromic polymers containing the dimethyldihydropyrene moiety: toward optoelectronic switches. *Org. Lett.*, **2** (19), 2979–2982.
179. Tan, W.J., Li, X., Zhang, J.J., and Tian, H. (2011) A photochromic diarylethene dyad based on perylene diimide. *Dyes Pigm.*, **89** (3), 260–265.
180. Wagner, K., Byrne, R., Zanoni, M., Gambhir, S., Dennany, L., Breukers, R., Higgins, M., Wagner, P., Diamond, D., Wallace, G.G., and Officer, D.L. (2011) A multiswitchable poly(terthiophene) bearing a spiropyran functionality: understanding photo- and electrochemical control. *J. Am. Chem. Soc.*, **133** (14), 5453–5462.
181. Logtenberg, H., van der Velde, J.H.M., de Mendoza, P., Areephong, J., Hjelm, J., Feringa, B.L., and Browne, W.R. (2012) Electrochemical switching of conductance with diarylethene-based redox-active polymers. *J. Phys. Chem. C*, **116** (45), 24136–24142.
182. Klajn, R. (2014) Spiropyran-based dynamic materials. *Chem. Soc. Rev.*, **43** (1), 148–184.
183. Pars, M., Hofmann, C.C., Willinger, K., Bauer, P., Thelakkat, M., and Kohler, J. (2011) An organic optical transistor operated under ambient conditions. *Angew. Chem. Int. Ed.*, **50** (48), 11405–11408.
184. Choi, H., Lee, H., Kang, Y.J., Kim, E., Kang, S.O., and Ko, J. (2005) Photochromism and electrical transport characteristics of a dyad and a polymer with diarylethene and quinoline units. *J. Org. Chem.*, **70** (21), 8291–8297.
185. Kawai, T., Nakashima, Y., and Irie, M. (2005) A novel photoresponsive pi-conjugated polymer based on diarylethene and its photoswitching effect in electrical conductivity. *Adv. Mater.*, **17** (3), 309–314.
186. Tachibana, H., Komizu, H., Nakamura, T., Matsumoto, M., Tanaka, M., Manda, E., Kawabata, Y., and Kato, T. (1989) Conductive langmuir-blodgett film containing mesogenic unit – phase-transition accompanied with the change in conductivity. *Chem. Lett.*, (5), 841–844.
187. Tachibana, H., Nakamura, T., Matsumoto, M., Komizu, H., Manda, E., Niino, H., Yabe, A., and Kawabata, Y. (1989) Photochemical switching in

conductive langmuir-blodgett films. *J. Am. Chem. Soc.*, **111** (8), 3080–3081.

188. Tsujioka, T., Hamada, Y., Shibata, K., Taniguchi, A., and Fuyuki, T. (2001) Nondestructive readout of photochromic optical memory using photocurrent detection. *Appl. Phys. Lett.*, **78** (16), 2282–2284.
189. Tsujioka, T. and Irie, M. (2010) Electrical functions of photochromic molecules. *J. Photochem. Photobiol., C*, **11** (1), 1–14.
190. Tsujioka, T. and Kondo, H. (2003) Organic bistable molecular memory using photochromic diarylethene. *Appl. Phys. Lett.*, **83** (5), 937–939.
191. Tsujioka, T. and Masuda, K. (2003) Electrical carrier-injection and transport characteristics of photochromic diarylethene films. *Appl. Phys. Lett.*, **83** (24), 4978–4980.
192. Tsujioka, T., Shimizu, M., and Ishihara, E. (2005) Organic bistable memory characteristics with a photochromic diarylethene layer. *Appl. Phys. Lett.*, **87** (21), 213506.
193. Tsujioka, T., Sasa, T., and Kakihara, Y. (2012) Nonvolatile organic memory based on isomerization of diarylethene molecules by electrical carrier injection. *Org. Electron.*, **13** (4), 681–686.
194. Shallcross, R.C., Zacharias, P., Koehnen, A., Koerner, P.O., Maibach, E., and Meerholz, K. (2013) Photochromic transduction layers in organic memory elements. *Adv. Mater.*, **25** (3), 469–476.
195. Shallcross, R.C., Körner, P.O., Maibach, E., Köhnen, A., and Meerholz, K. (2013) A photochromic diode with a continuum of intermediate states: towards high density multilevel storage. *Adv. Mater.*, **25** (34), 4807–4813.
196. Shen, Q., Cao, Y., Liu, S., Steigerwald, M.L., and Guo, X. (2009) Conformation-induced electrostatic gating of the conduction of spiropyran-coated organic thin-film transistors. *J. Phys. Chem. C*, **113** (24), 10807–10812.
197. Jakobsson, F.L.E., Marsal, P., Braun, S., Fahlman, M., Berggren, M., Cornil, J., and Crispin, X. (2009) Tuning the energy levels of photochromic diarylethene compounds for opto-electronic switch devices. *J. Phys. Chem. C*, **113** (42), 18396–18405.
198. Nespurek, S., Sworakowski, J., Combellas, C., Wang, G., and Weiter, M. (2004) A molecular device based on light controlled charge carrier mobility. *Appl. Surf. Sci.*, **234** (1-4), 395–402.
199. Andersson, P., Robinson, N.D., and Berggren, M. (2005) Switchable charge traps in polymer diodes. *Adv. Mater.*, **17** (14), 1798–1803.
200. Andersson, P., Robinson, N.D., and Berggren, M. (2005) Diodes based on blends of molecular switches and conjugated polymers. *Synth. Met.*, **150** (3), 217–221.
201. Weiter, M., Navratil, J., Vala, M., and Toman, P. (2009) Photoinduced reversible switching of charge carrier mobility in conjugated polymers. *Eur. Phys. J. Appl. Phys.*, **48** (1), 10401.
202. Weiter, M., Vala, M., Zmeskal, O., Nespurek, S., and Toman, P. (2007) A molecular photosensor based on photoswitching of charge carrier mobility. *Macromol. Symp.*, **247**, 318–325.
203. Toman, P. and Nespurek, S. (2008) Modeling of hole transport in poly[2-methoxy-5-(2′-Ethylhexyloxy)-p-phenylene vinylene] doped with polar additives. *Mol. Cryst. Liq. Cryst.*, **496**, 25–38.
204. Li, Y.R., Zhang, H.T., Qi, C.M., and Guo, X.F. (2012) Light-driven photochromism-induced reversible switching in P3HT-spiropyran hybrid transistors. *J. Mater. Chem.*, **22** (10), 4261–4265.
205. Ishiguro, Y., Hayakawa, R., Chikyow, T., and Wakayama, Y. (2013) Optical switching of carrier transport in polymeric transistors with photochromic spiropyran molecules. *J. Mater. Chem. C*, **1** (17), 3012–3016.
206. Orgiu, E., Crivillers, N., Herder, M., Grubert, L., Patzel, M., Frisch, J., Pavlica, E., Duong, D.T., Bratina, G., Salleo, A., Koch, N., Hecht, S., and Samorì, P. (2012) Optically switchable transistor via energy-level phototuning in a bicomponent organic semiconductor. *Nat. Chem.*, **4** (8), 675–679.

207. Raimondo, C., Reinders, F., Soydaner, U., Mayor, M., and Samorì, P. (2010) Light-responsive reversible solvation and precipitation of gold nanoparticles. *Chem. Commun.*, **46** (7), 1147–1149.

208. Raimondo, C., Crivillers, N., Reinders, F., Sander, F., Mayor, M., and Samorì, P. (2012) Optically switchable organic field-effect transistors based on photoresponsive gold nanoparticles blended with poly(3-hexylthiophene). *Proc. Natl. Acad. Sci. U.S.A.*, **109** (31), 12375–12380.

209. Kobayashi, S., Nishikawa, T., Takenobu, T., Mori, S., Shimoda, T., Mitani, T., Shimotani, H., Yoshimoto, N., Ogawa, S., and Iwasa, Y. (2004) Control of carrier density by self-assembled monolayers in organic field-effect transistors. *Nat. Mater.*, **3** (5), 317–322.

210. Kagan, C.R., Afzali, A., Martel, R., Gignac, L.M., Solomon, P.M., Schrott, A.G., and Ek, B. (2003) Evaluations and considerations for self-assembled monolayer field-effect transistors. *Nano Lett.*, **3** (2), 119–124.

211. Zhang, H., Guo, X., Hui, J., Hu, S., Xu, W., and Zhu, D. (2011) Interface engineering of semiconductor/dielectric heterojunctions toward functional organic thin-film transistors. *Nano Lett.*, **11** (11), 4939–4946.

212. Zhang, J.J., Zou, Q., and Tian, H. (2013) Photochromic materials: more than meets the eye. *Adv. Mater.*, **25** (3), 378–399.

213. Virkar, A., Mannsfeld, S., Oh, J.H., Toney, M.F., Tan, Y.H., Liu, G.Y., Scott, J.C., Miller, R., and Bao, Z. (2009) The role of OTS density on pentacene and C-60 nucleation, thin film growth, and transistor performance. *Adv. Funct. Mater.*, **19** (12), 1962–1970.

214. Lee, H.S., Kim, D.H., Cho, J.H., Hwang, M., Jang, Y., and Cho, K. (2008) Effect of the phase states of self-assembled monolayers on pentacene growth and thin-film transistor characteristics. *J. Am. Chem. Soc.*, **130** (32), 10556–10564.

215. Tseng, C.-W., Huang, D.-C., and Tao, Y.-T. (2012) Electric bistability induced by incorporating self-assembled monolayers/aggregated clusters of azobenzene derivatives in pentacene-based thin-film transistors. *ACS Appl. Mater. Interfaces*, **4** (10), 5483–5491.

216. Yoshida, M., Suemori, K., Uemura, S., Hoshino, S., Takada, N., Kodzasa, T., and Kamata, T. (2010) Development of field-effect transistor-type photorewritable memory using photochromic interface layer. *Jpn. J. Appl. Phys.*, (4), 49.

217. Lutsyk, P., Janus, K., Sworakowski, J., Generali, G., Capelli, R., and Muccini, M. (2011) Photoswitching of an n-type organic field effect transistor by a reversible photochromic reaction in the dielectric film. *J. Phys. Chem. C*, **115** (7), 3106–3114.

218. Nespurek, S. and Sworakowski, J. (2001) Molecular current modulator consisting of conjugated polymer chain with chemically attached photoactive side groups. *Thin Solid Films*, **393** (1-2), 168–176.

219. Toman, P., Bartkowiak, W., Nespurek, S., Sworakowski, J., and Zalesny, R. (2005) Quantum-chemical insight into the design of molecular optoelectrical switch. *Chem. Phys.*, **316** (1-3), 267–278.

220. Vala, M., Weiter, M., Zmeskal, O., Nespurek, S., and Toman, P. (2008) Light-induced change of charge carrier mobility in semiconducting polymers. *Macromol. Symp.*, **268**, 125–128.

221. Shen, Q.A., Wang, L.J., Liu, S., Cao, Y., Gan, L., Guo, X.F., Steigerwald, M.L., Shuai, Z.G., Liu, Z.F., and Nuckolls, C. (2010) Photoactive gate dielectrics. *Adv. Mater.*, **22** (30), 3282–3287.

8
Photochromic Bulk Materials

Masakazu Morimoto, Seiya Kobatake, Masahiro Irie, Hari Krishna Bisoyi, Quan Li, Sheng Wang, and He Tian

8.1
Photochromic Polymers

When photochromic chromophores are incorporated into polymer backbones or side groups, photoirradiation brings about changes in various properties of polymer solutions and solids. These polymers having the chromophores are named photochromic polymers. The photochromic polymers are useful for various types of applications, such as photochromic glasses, ultraviolet (UV) sensors, optical waveguides, optical memories, holographic recording media, nonlinear optics, and so on [1, 2]. Upon photoirradiation, the polymers reversibly change their physical and chemical properties, such as polymer chain conformation, shape of polymer gels, surface wettability, membrane potential, membrane permeability, pH, solubility, sol–gel transition temperature, and phase separation temperature of polymer blends. Researches on these photochromic polymers in 1967–2000 were reviewed by Irie [3–6]. In this section, recent development carried out after 2000 is described.

8.1.1
Glass Transition Temperature

In most cases, the reactivity of the photochromic chromophores in polymer matrices differs from that in solutions. The movement of the molecules depends on the glass transition temperature (T_g) and the free volume in the polymer matrices. Therefore, the photochromic reactivities and thermal decoloration rates are influenced by T_g of the polymer matrices [7]. A matrix effect for spirooxazine, spiropyran, and azobenzene was summarized in the literature [8]. The fluorescence photoswitching of a photochromic diarylethene derivative in polymer matrices has been studied at a single-molecule level to reveal the matrix effect [9, 10]. When the molecules are embedded in soft polymer matrices having a low T_g near room temperature, such as poly(*n*-butyl methacrylate), the cyclization and

Photochromic Materials: Preparation, Properties and Applications, First Edition.
Edited by He Tian and Junji Zhang.

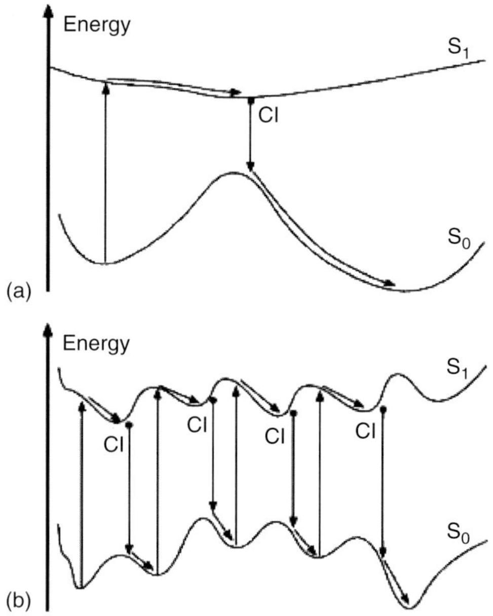

Figure 8.1 A schematic diagram of the potential energy surface of diarylethene (a) in the gas phase and (b) in the polymer matrix. (Reprinted with permission from [11]. Copyright 2007, American Chemical Society.)

cycloreversion quantum yields of each molecule are constant. On the other hand, in high T_g (~100 °C) rigid polymer matrices, such as Zeonex or poly(methyl methacrylate), the quantum yields are not constant at the single-molecule level. The quantum yields of the molecule increase with an increasing number of absorbed photons. In other words, the molecule remembers the number of absorbed photons and undergoes the reaction after absorbing a certain number of photons. The environmental effect is explained by a multilocal minima model, as shown in Figure 8.1 [11]. The polymer chains surrounding the molecule work as steric hindrance. The steric hindrance modifies the potential energy surfaces of both ground and excited states. In the absence of such steric hindrance as in low T_g matrices, the molecule undergoes one-step photoreaction, which leads to a constant quantum yield. In high T_g matrices, the multilocal potential energy surfaces prevent the one-step photoreaction and require multistep photoexcitations to reach the final reaction process. It requires many steps for the molecule to undergo the reaction from open- (or closed-) to closed- (or open-) isomers. Therefore, the quantum yields depend on the number of excitation. The result provides a new insight into the photoreaction of molecules in polymer matrices.

Photochromic reactions change T_g of solid materials composed of photochromic chromophores. As a result, physical and chemical properties can be changed. Tsujioka *et al.* [12] found that metal atoms are selectively deposited on a diarylethene amorphous film depending on the photoisomerization state.

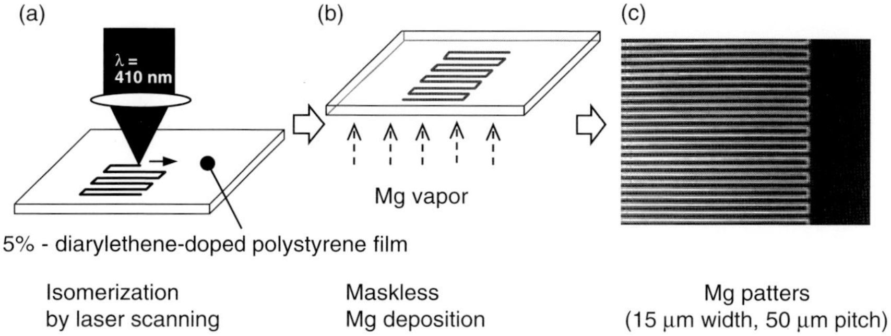

Figure 8.2 Fine metal patterning process on 5% diarylethene-doped polystyrene film. (a) Fine colored patterns on the surface was prepared by violet laser spot scanning. (b) Mg vapor was evaporated to the surface without a shadow mask. (c) The fine metal patterns were obtained with 15 μm widths and 50 μm pitches corresponding to the photoirradiated colored pattern. (Adapted with permission from [15]. Copyright 2010, The Royal Society of Chemistry.)

Metal vapor atoms are deposited selectively on a UV-irradiated colored surface but not on an uncolored surface. Upon UV irradiation, the diarylethenes undergo photocyclization reactions and change the color from colorless to blue. At the same time, T_g of the film increases from 32 to 95 °C. The colored surface has a higher T_g, and metal can deposit on the colored hard surface. The mobility of the polymer chain can be controlled by doping diarylethene molecules into polymer matrices [13, 14]. The uncolored state on 5% diarylethene-doped polystyrene film did not form Mg-deposited film, while the photogenerated colored state formed Mg metal film [15]. Fine Mg patterning process by scanning a laser spot followed by evaporating Mg has been demonstrated as shown in Figure 8.2. An uncolored diarylethene-doped polystyrene film was prepared on a glass substrate. The laser spot with a wavelength of 410 nm was scanned on the film, and colored parallel lines with a pitch of 50 μm were obtained. Mg was evaporated to the surface at a deposition rate of 1 nm s^{-1} without a shadow mask. Parallel fine Mg patterns with a width of 15 μm were successfully obtained.

8.1.2 Fluorescence

Reversible photoisomerization between two isomers can modulate the fluorescence of a fluorescent dye attached to a photochromic chromophore. In most cases, the colorless form is fluorescent, while the photogenerated colored isomer quenches the fluorescence of the dye via a FRET mechanism. The fluorophores are embedded in the polymer matrices or are incorporated into polymer backbones or side groups. Scheme 8.1 shows typical synthetic routes of photochromic polymers in which photochromic chromophores and fluorescent dyes are incorporated into polymer backbones or side groups [16–19]. Park *et al.* [17] prepared aggregation-induced enhanced emission (AIEE)-based

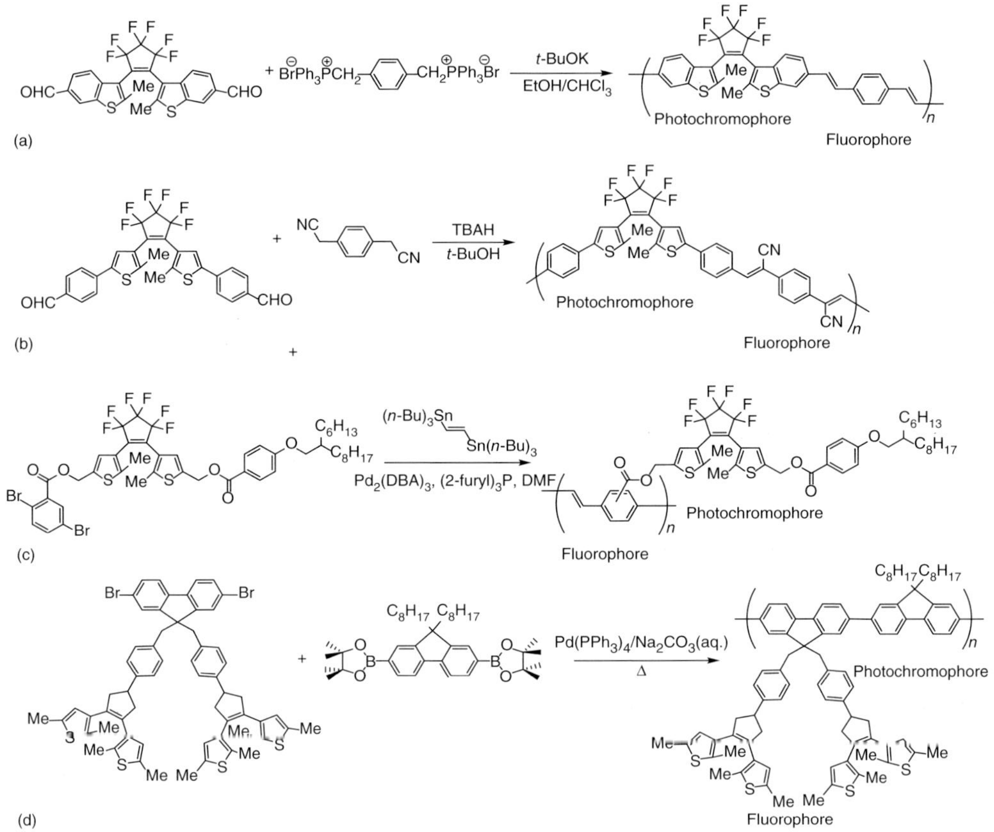

Scheme 8.1 Typical synthetic methods of fluorescent photochromic polymers: (a) the Wittig polycondensation reaction [16], (b) the Knoevenagel polycondensation reaction [17], (c) the Still polycondensation reaction [18], and (d) Pd-catalyzed Suzuki coupling reaction [19].

fluorescent photochromic polymer (Scheme 8.1b) showing strong fluorescence in the neat polymer film. Fluorescence switching ratio of more than 10 times was observed by alternating irradiation with UV and visible light. Harbron *et al.* [20, 21] prepared photochromic chromophore-doped conjugated polymer nanoparticles. Scheme 8.2 shows the structure of photochromic chromophores and conjugated polymers. When the diarylethene was used as the photochromic chromophores, the nanoparticles exhibited efficient thermally irreversible fluorescence photomodulation in aqueous suspension [21]. In the open-ring form of the diarylethene, the nanoparticles exhibit a bright fluorescence. Upon UV irradiation, the open-ring form converts to the visible-absorbing closed-ring form, which causes an efficient fluorescence quenching via the FRET mechanism. Li *et al.* [22] prepared photoswitchable polymer nanoparticles by emulsion polymerization through embedding an iridium(III) complex as a fluorophore and

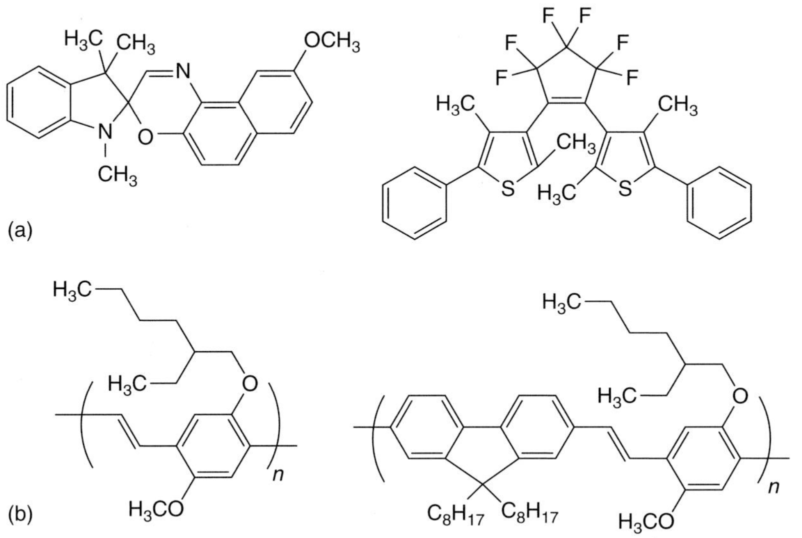

Scheme 8.2 The structure of (a) photochromic chromophores and (b) conjugated polymers.

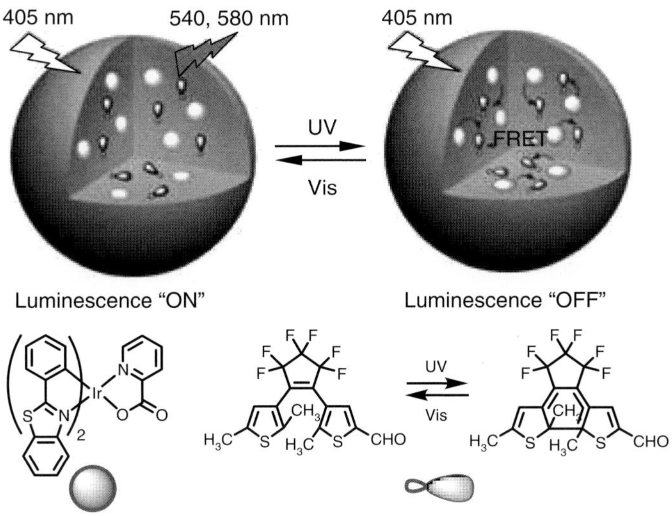

Figure 8.3 Schematic illustration of photoswitchable polymer nanoparticles embedded with an Ir(III) complex and a diarylethene derivative. (Adapted with permission from [22]. Copyright 2012, Springer.)

a diarylethene derivative as a photochrome, as shown in Figure 8.3. The polymer nanoparticles show a reversible luminescence modulation with diarylethene switching using UV and visible light in aqueous solution and even in living cells. The polymer nanoparticles displayed fast intracellular luminescence modulation by photoswitching, indicating that the nanoparticles can be used as photoswitchable luminescence probes for a diversity of bioimaging applications. Raymo *et al.* [23] prepared a cyclodextrin-based polymer in which two entrapped species can be operated independently upon photoirradiation. The fluorescence of one component enables the visualization of the labeled cells, and the release of nitric oxide from the other induces their mortality. Hurst *et al.* [24] have synthesized polymer nanoparticles showing reversible fluorescence photoswitching and demonstrated that on-and-off switching can be elicited with alternating UV (<400 nm) and visible (>450 nm) light. Using liposomes, these photoswitchable nanoparticles can be delivered into living cells and their presence is confirmed by fluorescence imaging. These high-contrast, photoswitchable nanoparticles have potential utility as smart inks, time-sensitive displays, and biological markers. Biesalski *et al.* [25] prepared light-switchable polymer networks attached to solid substrates. Benzophenone monomer was introduced to the spiropyran-functional polymer as photo-cross-linkable groups by free radical copolymerization. The benzophenone photo-cross-linkable groups were immobilized on a glass surface. Surface-attached polymer networks bearing the photofunctional spiropyran were prepared as shown in Figure 8.4.

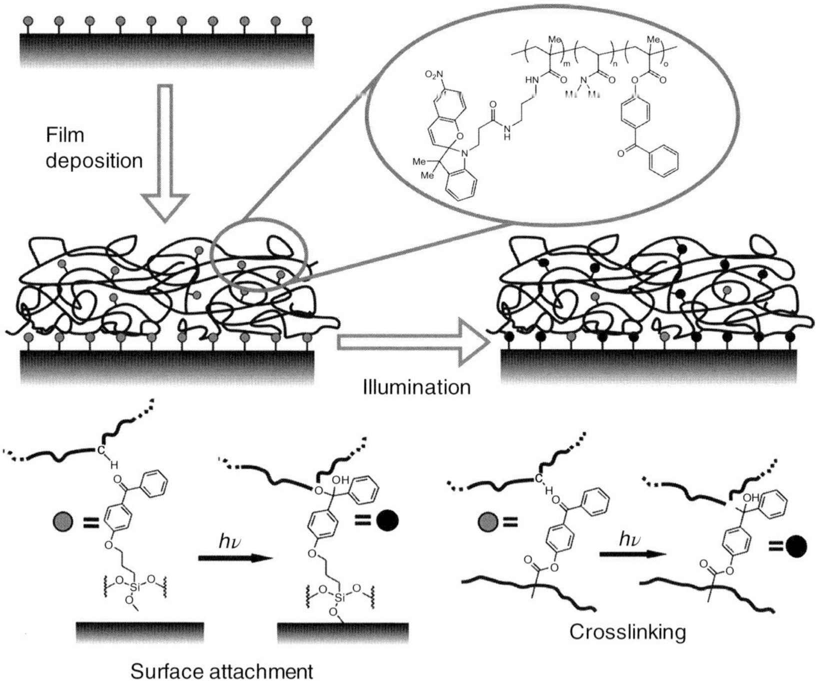

Figure 8.4 Schematic illustration of the preparation of the surface-attached, photoswitchable polymer networks. (Adapted with permission from [25]. Copyright 2013, American Chemical Society.)

8.1.3
Conductivity

π-Conjugated photochromic conducting polymers can modulate the electrical conductivity by alternating irradiation with UV and visible light. A photochromic conducting polymer composed of diarylethene and fluorene (Scheme 8.3a) was prepared by Pd-catalyzed Suzuki coupling and showed optical memory

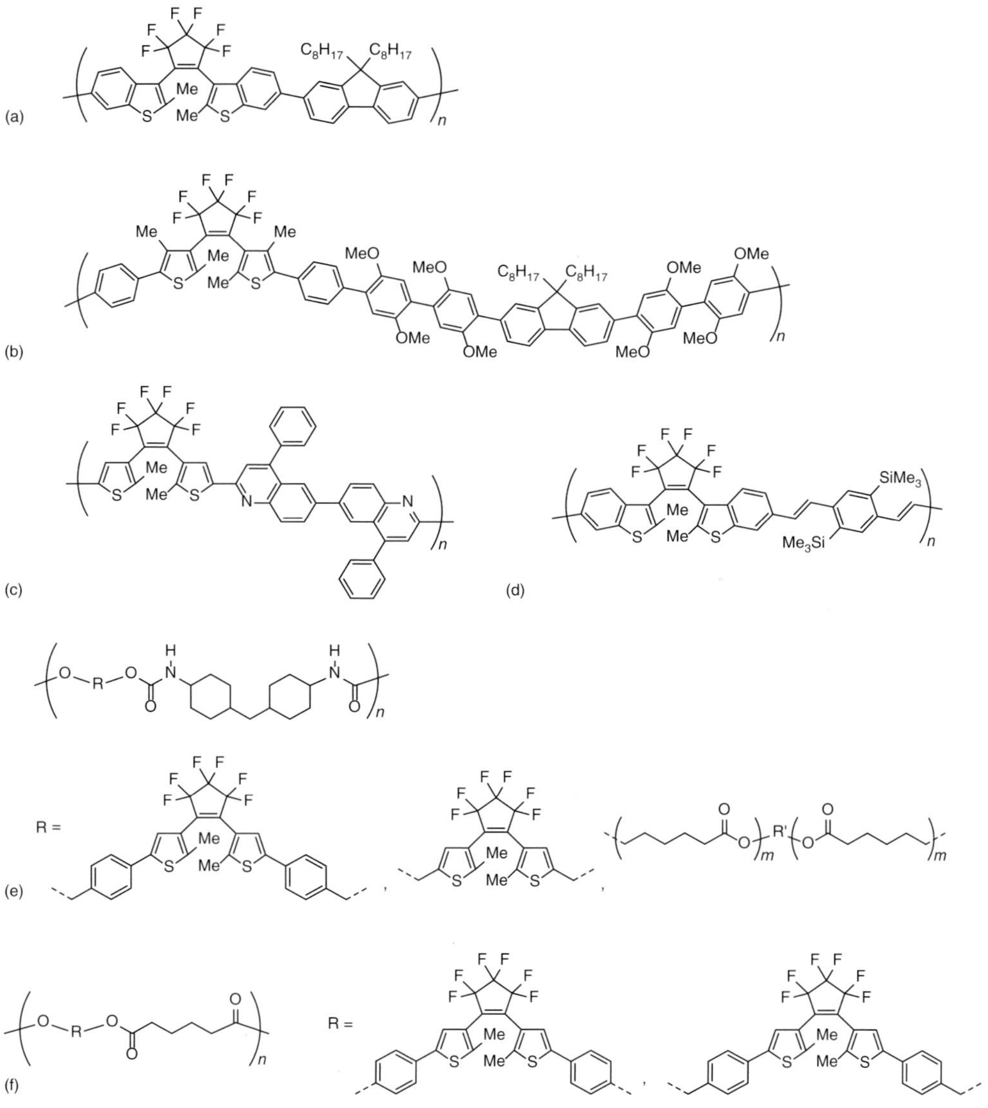

Scheme 8.3 The structure of diarylethene polymers.

effect in current–voltage characteristics as well as in the fluorescence emission [26, 27]. Another type of photochromic conducting polymer composed of diarylethene and fluorene (Scheme 8.3b) was also prepared and showed high degree of conductivity modulation upon alternating irradiation with UV and visible light [28]. Kim *et al.* [29] have synthesized a π-conjugated photochromic polymer incorporating bis(phenylquinoline) unit (Scheme 8.3c). Good transport characteristics were observed in the colored form of the diarylethenes with phenylquinoline group. When the photochromic diarylethene unit is the open-ring form, it showed an irreversible oxidation peak at 1.1 V. Upon irradiation with UV light, the peak was shifted to ~1.2 V. The current for a photocell containing the pure polymer film upon irradiation with 366 nm light increases twice larger than that of the open form at 2 V. Kim and Lee [30] have achieved photoelectrical switching of a diarylethene polymer connected to trimethylsilyl-modified *p*-phenylenevinylene units (Scheme 8.3d). The electrical properties of a photocell containing the polymer can be modulated with UV and visible light. The electrical change was persistent and reversible, thereby allowing the photopatterning of an electrode that can modulate electrical properties by photoirradiation.

A nitrospiropyran-containing methacrylate polymer was prepared for controlling contact electrification with light [31]. The polymer was demonstrated to switch the sign and magnitude of contact electrification with light, such as electrostatic self-assembly or actuation.

Polyurethanes having photochromic diarylethenes in the main chain (Scheme 8.3e) was used for multistate optical memories [32] and optical metrology [33]. Polyesters having diarylethenes (Scheme 8.3f) were also used for volume phase holographic grating [34], light-controlled resistance modulation [35], and photoresponsive actuators [36].

8.1.4
Living Radical Polymerization

Vinyl monomers having photochromic chromophores can be used for not only conventional radical polymerization but also living radical polymerization. Reversible addition–fragmentation chain transfer (RAFT) polymerization is one of the living radical polymerization that can control the molecular weight and the end groups of the polymer. Pan *et al.* [37] prepared photochromic polymeric vesicles by RAFT polymerization-induced self-assembly and reorganization. The resulting vesicles show photochromic behavior different from that of their free polymer chains in DMF, and the vesicles exhibited stronger fluorescence and excellent photostability due to confinement of conformational flexibility of the polymer chains in the aggregates (Figure 8.5).

Kobatake *et al.* [38–41] fabricated gold nanoparticles covered with a diarylethene polymer that was prepared by RAFT polymerization of a

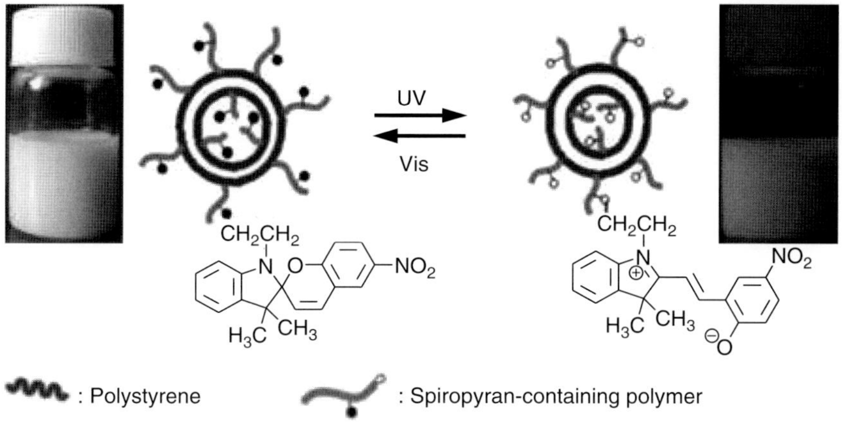

Figure 8.5 Schematic illustration of photoresponsive polymeric vesicles. (Reprinted with permission from [37]. Copyright 2011, John Wiley and Sons.)

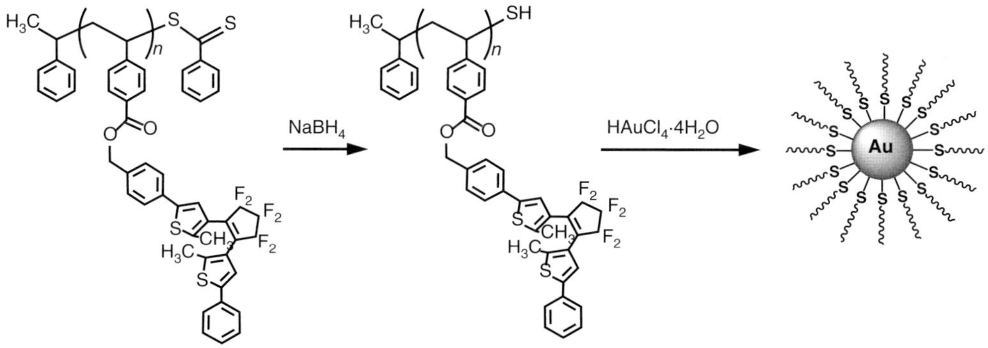

Scheme 8.4 Synthetic route of gold nanoparticle covered with a photochromic diarylethene polymer.

diarylethene-pendant styrene monomer, followed by reduction of dithiobenzoate group to SH, as shown in Scheme 8.4. The diarylethene chromophores in the vicinity of the gold nanoparticles undergo faster photocycloreversion reaction than those of the outside area in the polymer shell corresponding to a nonenhanced area. Thermoresponsive poly(*N*-isopropylacrylamide) with a diarylethene chromophore at the end group was also synthesized by RAFT polymerization (Scheme 8.5) [42]. The resulting polymer has a lower critical solution temperature at 33 °C and exhibits photochromism in water upon photoirradiation. The localized surface plasmon resonance band of gold nanoparticles covered with the polymer exhibited a bathochromic shift with heating and photoirradiation.

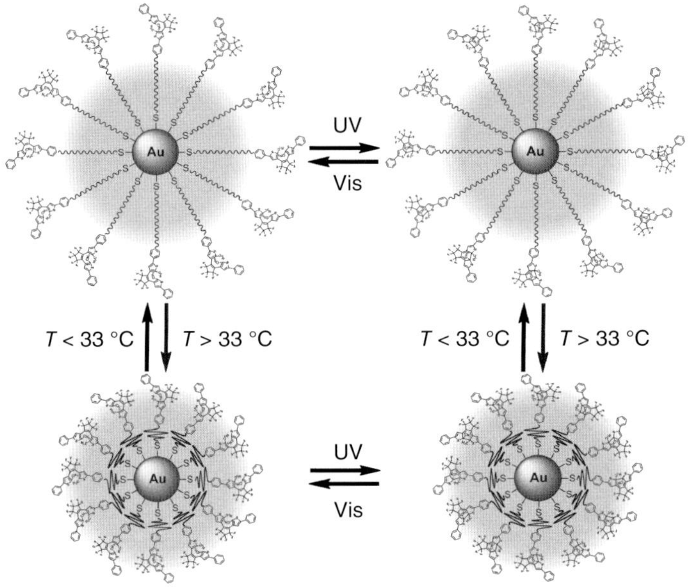

Scheme 8.5 Schematic illustration for photoswitching and thermoswitching of gold nanoparticle covered with poly(*N*-isopropylacrylamide) attached with a diarylethene chromophore at the end group. (Reprinted with permission from [42]. Copyright 2013, Elsevier.)

8.1.5 Surface Relief Grating

In 1995, massive motion of materials on the surface at room temperature (well below T_g) under irradiation with interfering polarized light was reported [43, 44]. Azobenzene undergoes isomerization from *trans* to *cis* form upon irradiation with UV light and reverses upon heating or irradiation with visible light. During the isomerization, the molecule undergoes a large structure change and thus the configurational change of azobenzene molecule brings about photoinduced orientation. Irradiation of polymer films having pendant azobenzene chromophores for a period longer than required for photoinduced orientation produces a relief grating on the film surface. The grating with depths of up to 1 μm can be obtained. A variety of single beams create different surface deformations. The surface relief gratings can find various applications such as one-step holographic image storage, the orientation layers in a liquid crystal cell, and so on [45].

8.1.6 Photomechanical Effect

Photoresponsive polymers having photochromic chromophores give rise to molecular motions by the photoisomerization and thereby can be deformed by photoirradiation. This photomechanical effect was first demonstrated in solution by using polymers having azobenzene chromophores in the main chain in the beginning of the 1980s [46, 47]. It is well known that the distance between 4 and 4′ carbons of azobenzene decreases from 0.90 to 0.55 nm upon photoisomerization from the *trans* to *cis* form. The length change was used as a driving force to control the conformation of polymer chains by incorporating the residues into polymer backbone, as shown in Scheme 8.6. Intrinsic viscosity, $[\eta]$, of polymer (**1**) in *N*,*N*-dimethylacetamide was found to decrease from 1.22 to 0.50 dl g^{-1} by irradiation with UV light and to return to the initial value in the dark. Upon alternate irradiation with UV and visible light, the viscosity reversible changed as much as 60%. Before UV irradiation the polymer has a rodlike extended conformation, while the *trans*-to-*cis* isomerization kinks the polymer chains, resulting in a compact conformation and a decrease in the viscosity. The amount of viscosity change was found to depend on the length of the flexible methylene chains in the polymer backbone. The viscosity changes of polymers (**2**), (**3**), (**4**), and (**5**) were 63, 41, 20, and 4%, respectively. The methylene chains act as strain absorbers and suppress the conformational changes. A detailed study of photoinduced conformational changes using a time-resolved light scattering method revealed that the conformational changes take place in milliseconds time range in solution [48].

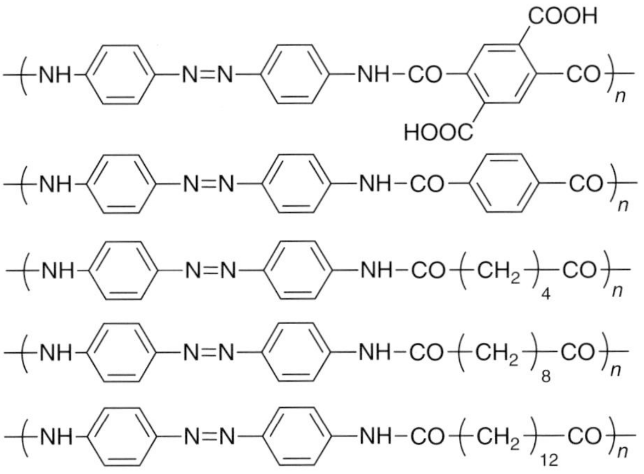

(1)

(2)

(3)

(4)

(5)

Scheme 8.6 Photoresponsive polymers having azobenzene residues in the main chain.

Although it is, in principle, possible to amplify the photostimulated conformational changes of polymer chains in solution to a macroscopic change in the size of polymer gels or films, it is hard to link the molecular-scale events to macroscopic shape changes. Self-standing films composed of polymers having azobenzene chromophores in the main chain were prepared, and their photoresponsive behavior was examined in dry conditions as well as in the solvent-swollen state. In contrast to expectation, any appreciable change of their shape was not observed upon photoirradiation. The reason is that the strain energy produced in polymer chains by the geometrical structure changes of azobenzene units is readily released by the free volume of the polymer matrices. The direct conversion of the geometrical structure changes of azobenzene chromophores to a macroscale deformation was failed.

Although many polymer systems showing photostimulated deformation have been reported, the deformation is limited to a few percent and most of them are ascribed to photoheating effect [4]. If the deformation is larger than 10%, it is safe to say that it is due to photochemical effect. One photodeformable material that shows a large deformation is a polyacrylamide gel containing a small amount of triphenylmethane leucohydroxide or leucocyanide groups [4].

In 2001, Finkelmann *et al.* [49] reported the first real photochemical contraction of liquid crystalline elastomers in the dry system. The structure of elastomer having pendant mesogens is shown in Scheme 8.7. The azobenzene pendant groups isomerize from *trans* to *cis* form upon irradiation with 365 nm light and revert to the initial *trans* form with a relaxation time of some 100 s in the dark. Upon irradiation with 365 nm light, the elastomer contracts as much as 22% at 313 K, while it reverts to the initial length in 250 min after switching off the light. The shape change of the elastomer is ascribed to order–disorder phase transition from the nematic to the isotropic phase. The nematic elastomer is known to contract upon heating above the phase transition temperature from the nematic to the isotropic phase. When azobenzene chromophores are incorporated into the elastomer, the phase transition temperature shifts depending on the ratio of *trans* and *cis* forms. The phase transition temperature change induced by photoirradiation causes the contraction isothermally. This is the mechanism of the photoinduced shape change of the elastomer. Based on this mechanism, various photoresponsive liquid crystalline elastomers having azobenzene in the main chain or side group have also been prepared and their shape changes upon irradiation with polarized light were studied [50–58].

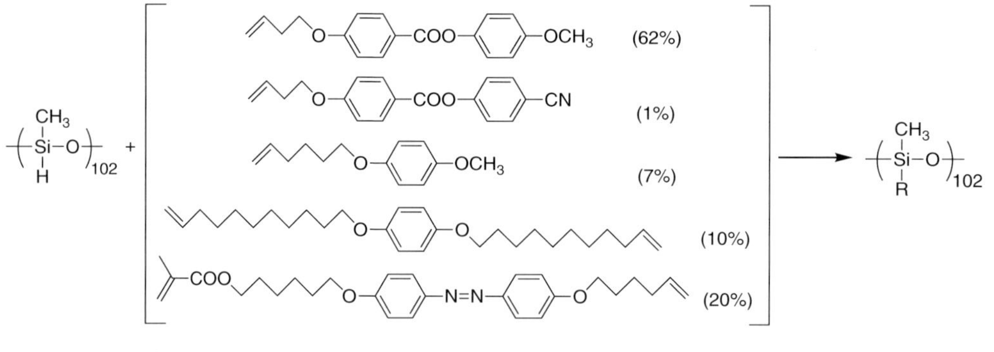

Scheme 8.7 Chemical structure of elastomer having azobenzene chromophore synthesized from poly(oxy(methylsilylene)).

8.2 Single-Crystalline Photoswitches

8.2.1 Crystalline-State Photochromic Materials

Photochromic molecules, which undergo photochromic reactions in the crystalline state, are rare. For example, most of azobenzene and spirobenzopyran derivatives cannot undergo photochromic reactions in crystals. The reason is that they require large geometrical structure changes during the photochromic reactions, and the large geometrical structure change is prohibited in crystals. Examples of crystalline-state photochromic molecules are paracyclophanes [59], triarylimidazole dimer [60, 61], diphenylmaleronitrile [62], aziridine [63], 2-(2,4-dinitrobenzyl)pyridine [64, 65], *N*-salicylideneaniline [66, 67], and triazene [68]. However, in the case of these molecules, photogenerated colored states are thermally unstable at room temperature. In contrast, diarylethene derivatives bearing heterocyclic aryl groups undergo thermally irreversible and fatigue-resistant photochromic reactions in the single-crystalline phase [2, 6, 69–73]. This section focuses on the single-crystalline photochromism of diarylethene derivatives.

8.2.2 Photochromic Diarylethene Single Crystals

Photochromic reactions of the diarylethene derivatives are based on photoinduced electrocyclic reactions of the central hexatriene moiety. The molecules reversibly isomerize between open- and closed-ring isomers by cyclization and cycloreversion reactions upon irradiation with UV and visible light. Generally, the open-ring isomers are colorless and have no absorption bands in the visible region, while the closed-ring isomers are colored due to their extended π-conjugation. Diarylethene derivatives exhibit outstanding photochromic performance with thermal irreversibility, fatigue resistance, and rapid response. Both of the isomers are thermally stable in the dark at room temperature [74, 75]. The coloration/decoloration cycles can be repeated more than 30 000 times [76–79]. Time-resolved spectroscopy revealed that the cyclization and cycloreversion reactions complete in less than several picoseconds after photoexcitation [78–83].

Figure 8.6 shows photographs of diarylethene single crystals. Upon irradiation with UV light, the colorless single crystals turn to yellow, red, blue, or green, depending on chemical structures of the component molecules. The coloration is due to the formation of the corresponding closed-ring isomers in the single crystals. The colors remain stable in the dark at room temperature. Upon irradiation with visible light, the colored crystals revert to the colorless crystals. The reversible color changes can be repeated many times.

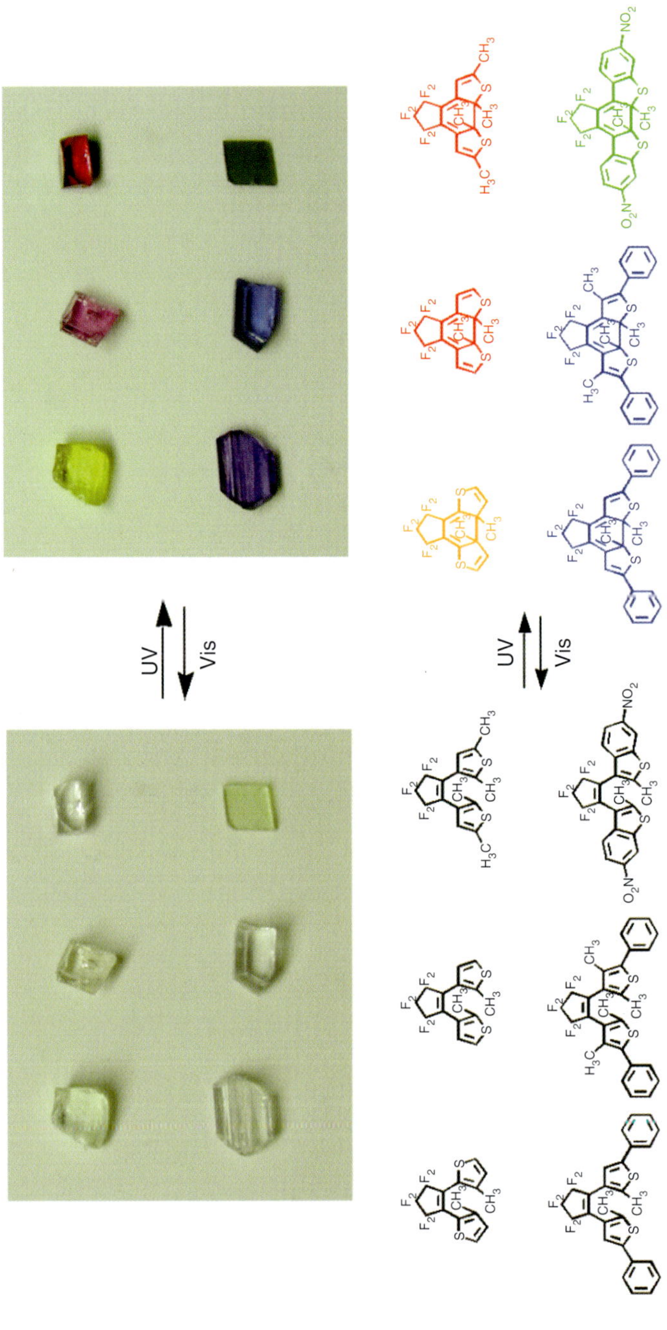

Figure 8.6 Photographs of photochromic diarylethene single crystals.

8.2.3 *In situ* X-ray Crystallographic Analysis of Photoisomerization Reaction

Photochromic reactions of the diarylethene molecules in the single-crystalline phase were followed by *in situ* X-ray crystallographic analysis [84, 85]. X-ray analysis was carried out for a UV-irradiated single crystal of diarylethene **6**. Photogenerated closed-ring isomer in the UV-irradiated single crystal was observed as a disordered structure. Figure 8.7a illustrates the molecular structures of the

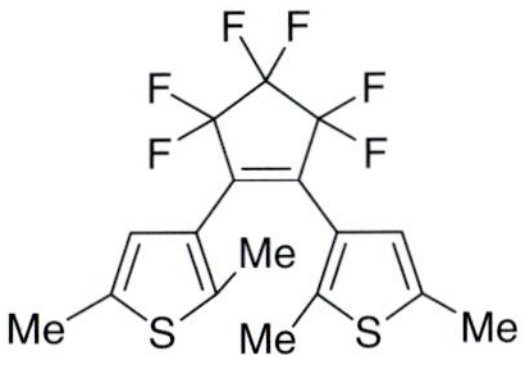

6

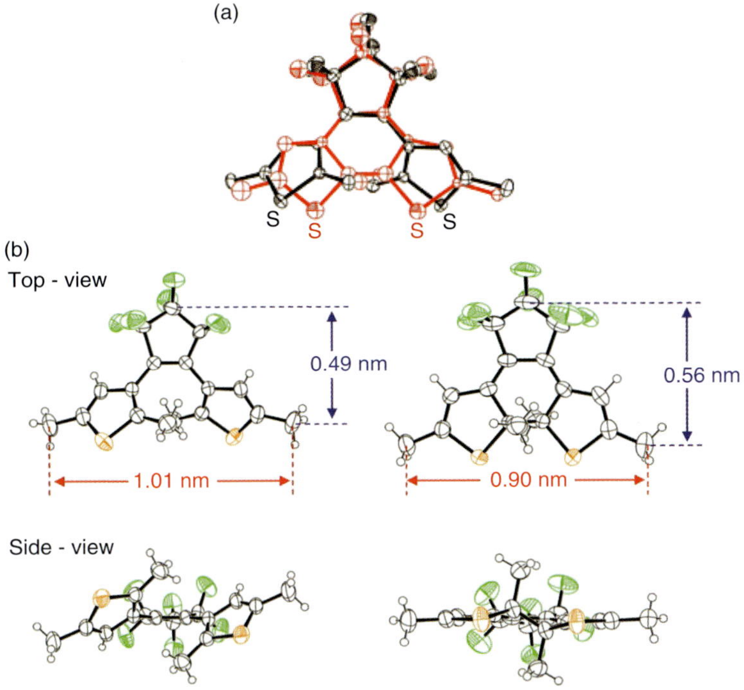

Figure 8.7 (a) Geometrical change of **6** upon photocyclizaiton revealed by *in situ* X-ray crystallographic analysis. Black and red molecules indicate open-ring and photogenerated closed-ring isomers, respectively. Hydrogen atoms are omitted for clarity. (b) Top and side views of geometrical structures of the open- and closed-ring isomers. (Adapted with permission from [70]. Copyright 2005, The Royal Society of Chemistry.)

open-ring isomer and photogenerated closed-ring isomer. As can be seen, the sulfur and the reacting carbon atoms significantly move to undergo the cyclization reaction, while other atoms remain at the original positions even after the reaction. This small structural change allows the molecule to undergo the photoisomerization reaction in the crystal lattice. Figure 8.7b shows the structural difference between the open- and closed-ring isomers in more detail. The top view indicates that the base width of the triangle decreases from 1.01 to 0.90 nm, and the height increases from 0.49 to 0.56 nm upon photocyclization. The side view shows that the thickness of the molecule becomes thinner when the molecule undergoes the photocyclization reaction. Such a small but distinct change of the molecular geometry upon photoisomerization induces shape changes and photomechanical performance of bulk single crystals as described later.

8.2.4 Photoisomerization Quantum Yields

The diarylethene molecules can adopt two types of conformations in solution, so-called parallel and antiparallel conformations, as shown in Scheme 8.8. The two conformers interconvert each other, and in most cases they exist in the ratio of 50 : 50 in the equilibrium state. Photoinduced conrotatory electrocyclic reaction of the diarylethene takes place only from the antiparallel conformation. In general, therefore, photocyclization quantum yield of the diarylethene derivatives in solution is less than 0.5.

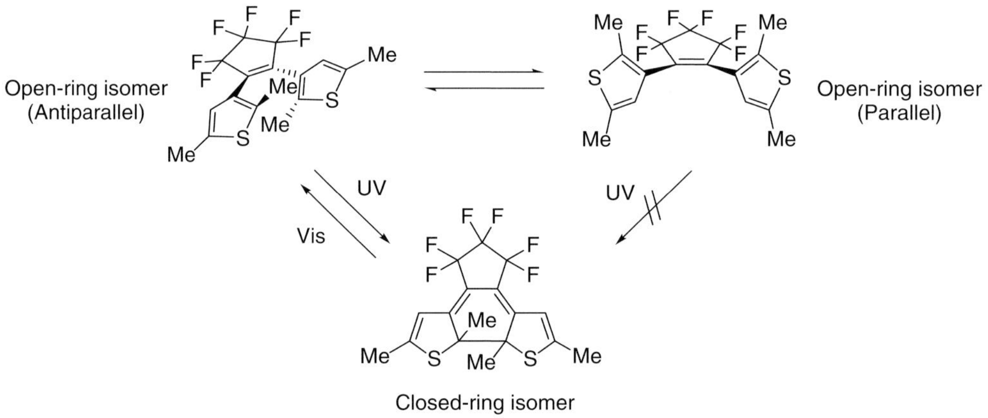

Scheme 8.8 Conformational change and photoisomerization reaction of diarylethene.

In crystals, on the other hand, interconversion between parallel and antiparallel conformations cannot take place. In most cases, diarylethene molecules are packed in the photoreactive antiparallel conformation. Diarylethene crystals can undergo photocyclization with a very high cyclization quantum yield. Figure 8.8 shows a correlation between cyclization quantum yields in diarylethene crystals

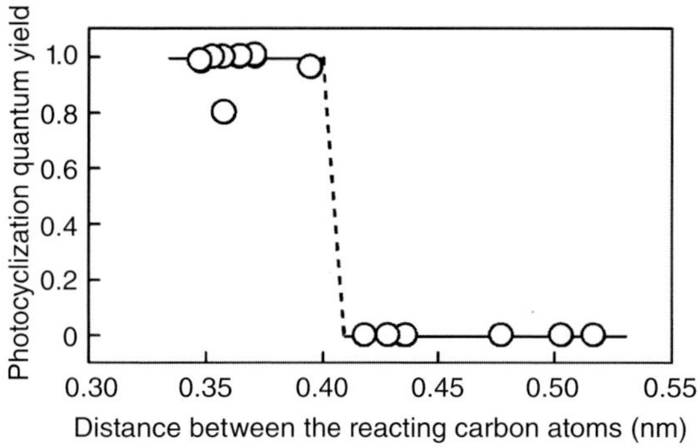

Figure 8.8 Relationship between photocyclization quantum yield and distance between reacting carbon atoms for 14 diarylethene single crystals. (Adapted with permission from [86]. Copyright 2002, The Royal Society of Chemistry.)

and the distance between reacting carbon atoms of the diarylethene molecules [86, 87]. When the molecules are packed in the antiparallel conformation and the distance between the reacting carbon atoms is less than 0.40 nm, the cyclization quantum yield becomes close to unity (100%). This means that photon energy absorbed by the single crystal is almost quantitatively used for the cyclization reaction. However, when the distance is larger than 0.42 nm, the photocyclization reaction in crystals is suppressed.

8.2.5 Multicolor Photochromism of Multicomponent Crystals

Single-component photochromic systems reversibly interconvert between two states, "colorless" and "colored," originating from two corresponding isomers. In contrast, multicomponent systems composed of different kinds of photochromic molecules can undergo reversible multistate switching between more than two states by the combination of two states of each component. For example, four states ($=2^2$) can be produced in a two-component photochromic system, and eight states ($=2^3$) in a three-component system. If a multicomponent photochromic crystal composed of three kinds of photochromic molecules that exhibit three primary colors, such as yellow, red, and blue, can be prepared, the crystal is expected to show a full range of colors upon photoirradiation.

A full-color photochromic crystal has been prepared by incorporating three kinds of diarylethene molecules, **6**–**8**, exhibiting red, yellow, and blue, respectively, into a single crystal [88, 89]. Recrystallization of a mixture of **6**–**8** (molar ratio **6** : **7** : **8** = 1 : 0.4 : 0.5) from an acetonitrile solution gave a single crystal composed of **6**–**8** in the molar ratio of 97.4 : 2.4 : 0.2. Figure 8.9 shows a photograph of the three-component crystal. Upon irradiation with appropriate wavelength of light, the colorless crystal turned to yellow, red, and blue. These colors are ascribed to the formation of closed-ring isomers of **6**–**8** in the crystal. The colors remained

Figure 8.9 Photograph of partially colored three-component crystal of **6·7·8**. (Reprinted with permission from [89]. Copyright 2003, American Chemical Society.)

stable in the dark at room temperature. Upon irradiation with visible light, all of the colors were completely bleached.

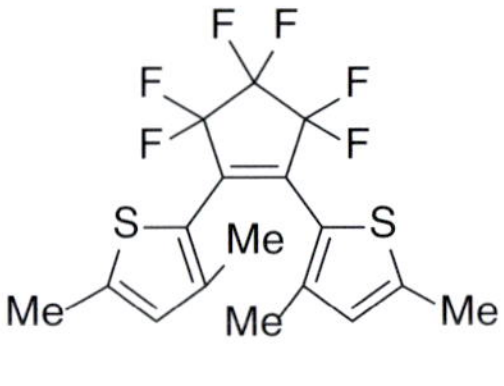

7

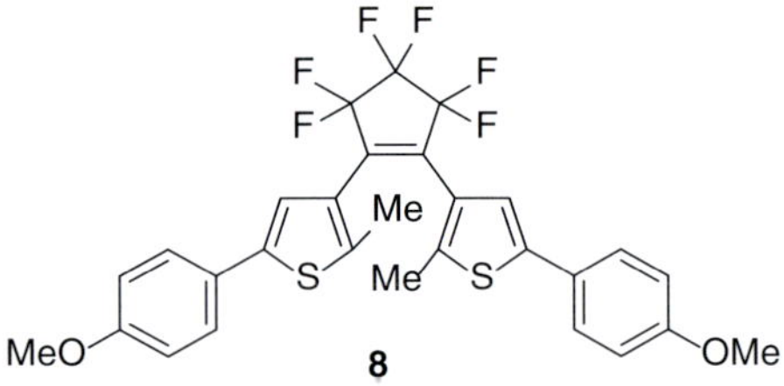

8

Another full-color photochromic crystal has also been prepared by using diarylethenes **9–11** bearing oxazole, thiazole, and thiophene rings as aryl groups and exhibiting yellow, red, and blue, respectively [90, 91]. The sizes and geometries of the three molecules are quite similar to each other. Therefore, the crystal

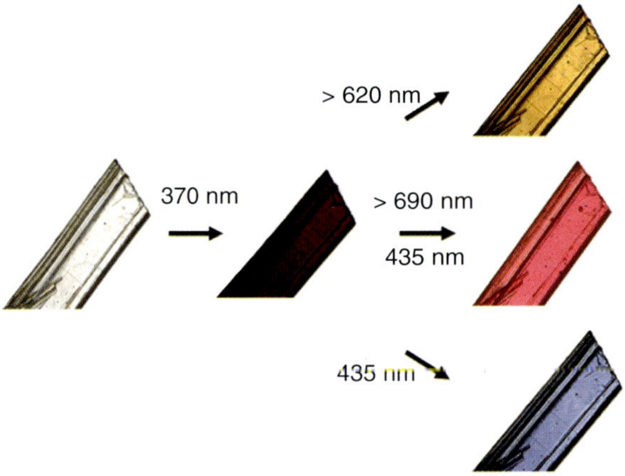

Figure 8.10 Color changes of three-component crystal of **9·10·11**. (Adapted with permission from [91]. Copyright 2007, American Chemical Society.)

can contain three-component molecules in almost equal amounts. As shown in Figure 8.10, the three-component crystal exhibited yellow, red, and blue upon irradiation with light of appropriate wavelengths.

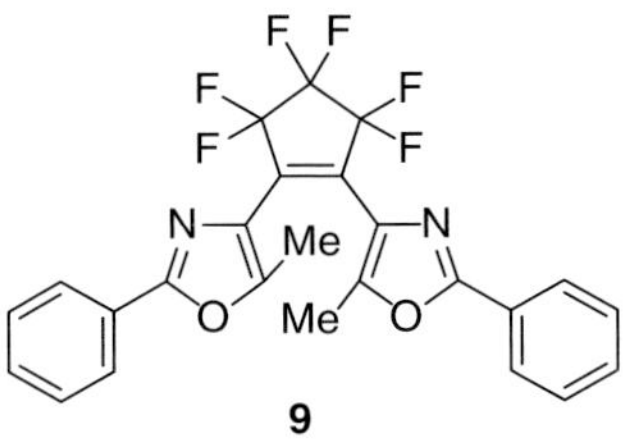

9

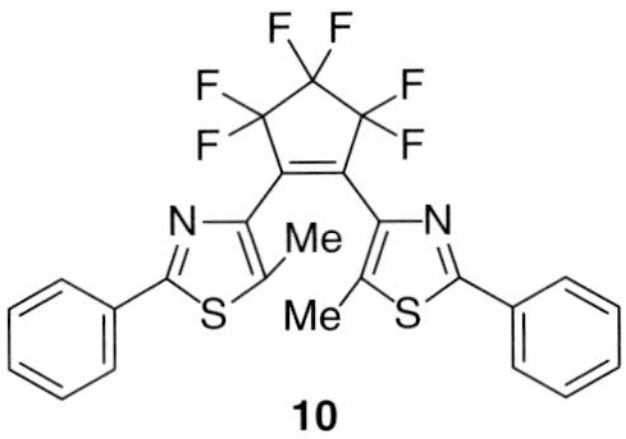

10

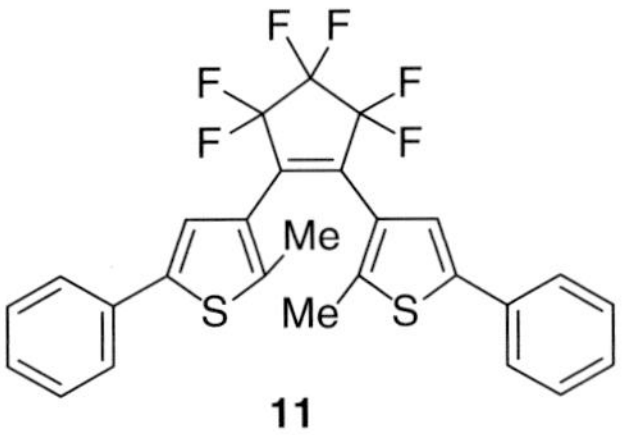

11

8.2.6
Nanoperiodic Structures Fabricated by Photochromic Reactions

In the multicolor photochromic crystals composed of three kinds of the diarylethenes described above the dopant molecules were substitutionally incorporated into the lattice of the host crystal. When molecules bear substituents that can interact intermolecularly, the stoichiometry of the component molecules can be definitely controlled. Here, crystal structures and photochromic performance of stoichiometric 1 : 1 co-crystals composed of **12** bearing pentafluorophenyl groups and **13** or **14** bearing naphthyl or phenyl groups, respectively, are described [92, 93].

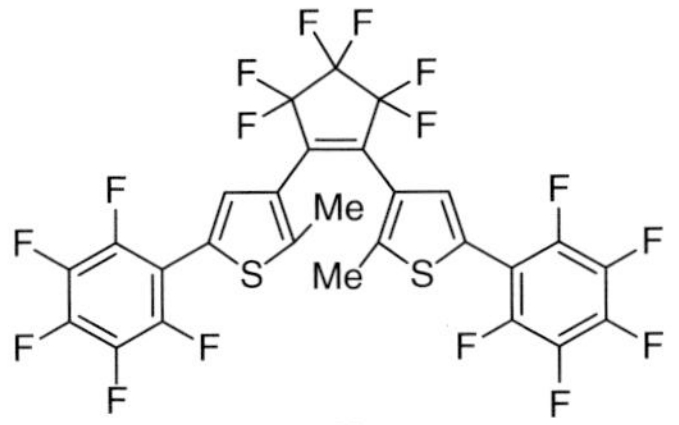

12

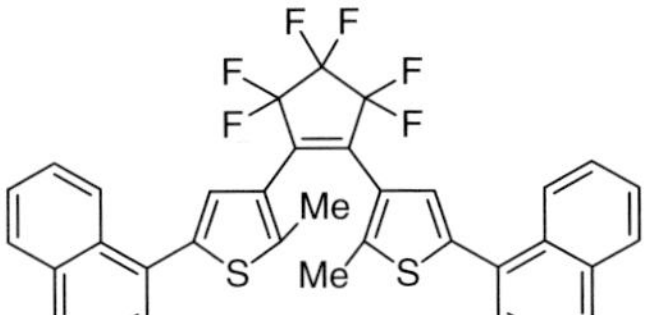

13

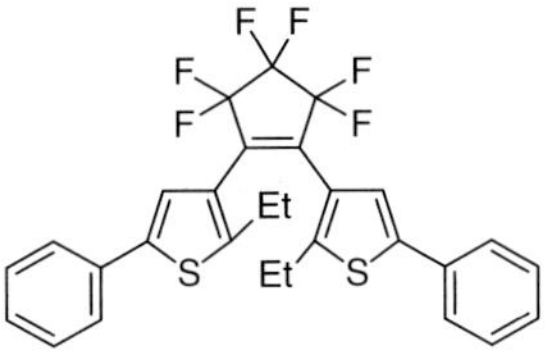

14

The co-crystal of **12**·**13** was prepared by recrystallization of a 1 : 1 (molar ratio) mixture of **12** and **13** from hexane. Figure 8.11a shows molecular packing diagrams in the co-crystal. The crystal is composed of **12** and **13** in the molar ratio of 1 : 1. The pentafluorophenyl groups of **12** and the naphthyl groups of **13** are stacked by intermolecular aryl–perfluoroaryl interactions. The diagrams viewed from the *a*-, *b*-, and *c*-axes indicate that **12** and **13** molecules are packed in a 3D alternating arrangement to form a mosaiclike structure. The co-crystal of **12** and **14** was also prepared by recrystallization of a 1 : 1 mixture of the two components. Figure 8.11b shows molecular packing diagrams in the co-crystal. The composition ratio in the crystal was **12** : **14** = 1 : 1. The crystal has a layered structure in which molecular layers of **12** and **14** are alternately stacked. The thickness of each layer is about 0.65 nm.

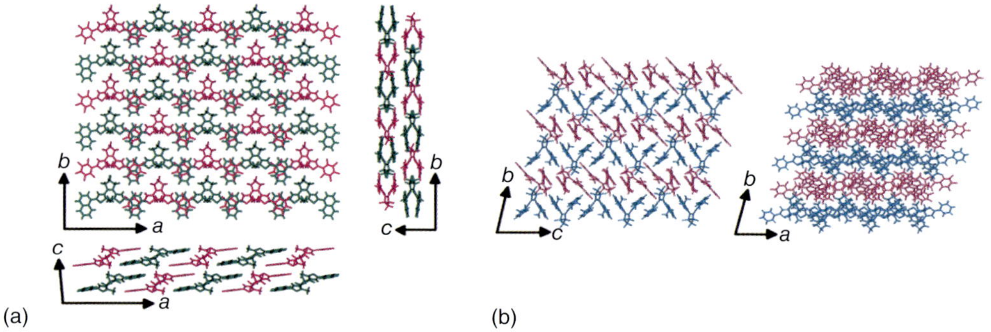

Figure 8.11 Molecular packing diagrams of co-crystals of **12·13** (a) and **12·14** (b). Red, green, and blue molecules indicate **12–14**, respectively. ((a) Adapted with permission from. [93]. Copyright 2004, The Japan Chemical Journal Forum and Wiley Periodicals, Inc. (b) Adapted from [92] by permission of the Royal Society of Chemistry (RSC) on behalf of the European Society for Photobiology, the European Photochemistry Association, and RSC. Copyright 2003.)

The two-component co-crystals underwent photochromism. Upon irradiation with UV light, the colorless crystals of **12·13** and **12·14** turned green and blue, respectively. The colors were bleached upon irradiation with visible light. The conversion ratio from the open- to closed-ring isomers of the two components was monitored by high performance liquid chromatography (HPLC). In the co-crystals of **12·13** and **12·14**, **13** and **14** selectively underwent photocyclization to yield the corresponding closed-ring isomers, while cyclization of **12** was significantly suppressed. The highly selective photocyclization of **13** and **14** was also confirmed by *in situ* X-ray crystallographic analysis.

Figure 8.12 shows schematic illustrations of photochromic reactions in the co-crystals of **12·13** and **13·14** with well-controlled nanostructures. As a result of the selective photocyclizations, the open- and closed-ring isomers, which have different refractive indices, are arranged periodically at the molecular level in the UV-irradiated crystals. A similar nanostructure can be fabricated by highly selective photocyclization reactions in a single-component diarylethene crystal, which contains photoreactive and photoinactive conformers in the ratio of 1 : 1 [94]. Such photoreversible periodic changes in refractive indices in the crystalline nanostructures have potential applications as a new type of photonic device.

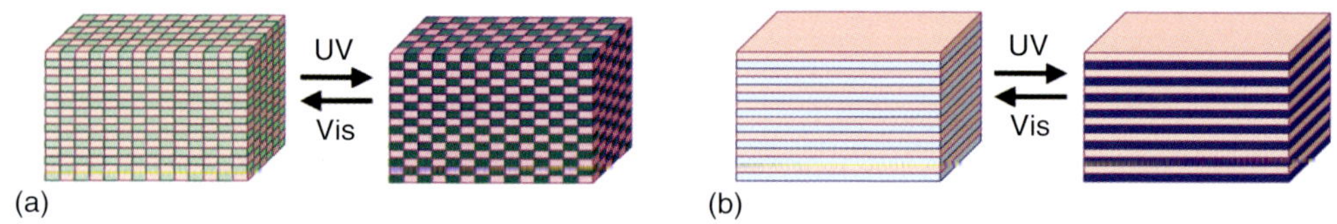

Figure 8.12 Schematic illustrations of photochromic reactions in co-crystals **12·13** (a) and **12·14** (b). Red, green, and blue areas indicate **12–14**, respectively. (Adapted with permission from [93]. Copyright 2004, The Japan Chemical Journal Forum and Wiley Periodicals, Inc.)

8.2.7 Photoinduced Shape Changes and Mechanical Performance

In a densely packed molecular crystal, strain energy is considered to directly affect the shape of the bulk crystal. Figure 8.13a shows the color and shape changes of a single crystal of diarylethene **14** with the size of about 10 μm [95, 96]. Upon irradiation with UV light, the crystal transformed the shape from square to lozenge accompanying the color change from colorless to blue. Upon irradiation with visible light, the blue color was completely bleached and the shape of the crystal returned to the initial state. Figure 8.14a shows the molecular-packing diagrams of **14**. The shape deformation from square to lozenge indicates that the crystal contracts along the *c*-axis and expands along the *b*-axis. The crystal lattice deformation was also observed by *in situ* X-ray crystallographic analysis. Figure 8.14b shows the geometrical structure change of the diarylethene molecule upon photoisomerization. Upon photocyclization, the height of the molecule increases from 0.61 to 0.73 nm, which is the direction of the *b*-axis. On the other hand, the thickness of the molecules is reduced. The cofacial packing of the diarylethene molecules along the *c*-axis, as shown in Figure 8.14a, indicates that the thin layers of closed-ring isomers allow the molecules to be stacked one by one, resulting in the contraction along the *c*-axis. The molecular structure change of the diarylethene molecules directly affects the crystal shape.

Figure 8.13b shows another type of crystal shape deformation observed for diarylethene **10** [95]. Upon irradiation with UV light, the rectangular crystal contracts along the long axis of the crystal and reverts to the initial shape upon irradiation with visible light. When the crystal has a rodlike shape, it bends

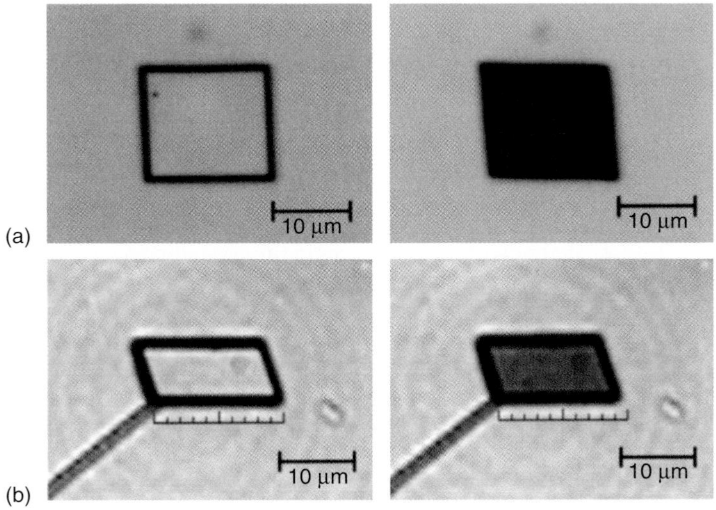

Figure 8.13 Deformation of diarylethene crystals of (a) **14** and (b) **10** upon UV (365 nm) and visible ($\lambda > 500$ nm) light irradiation. (Adapted with permission from [95]. Copyright 2007, Macmillan Publishers Ltd.)

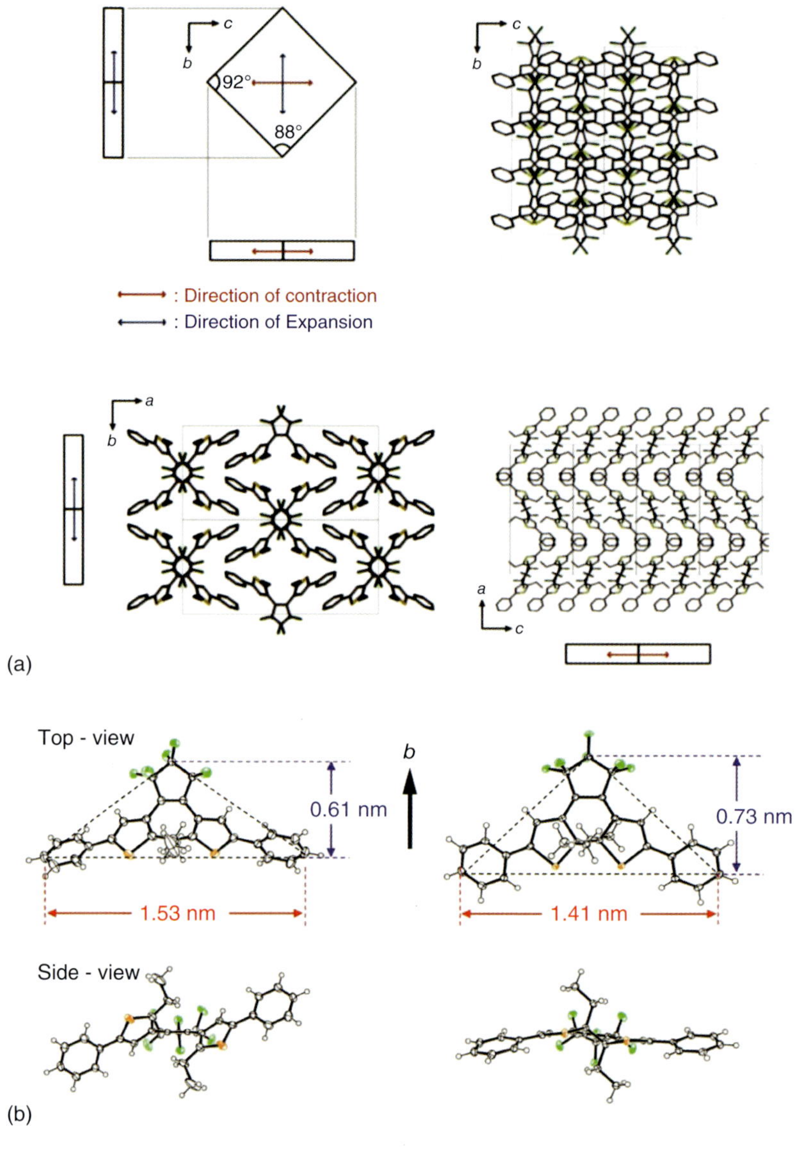

Figure 8.14 (a) Molecular packing of single crystal of **14**. The red arrows indicate the direction of contraction and blue arrows indicate the direction of expansion of the crystal upon UV irradiation. (b) Geometrical structures of the open- and closed-ring isomers of **14** in crystals. (Adapted from [73] by permission of the Royal Society of Chemistry (RSC) on behalf of the European Society for Photobiology, the European Photochemistry Association, and RSC. Copyright 2010.)

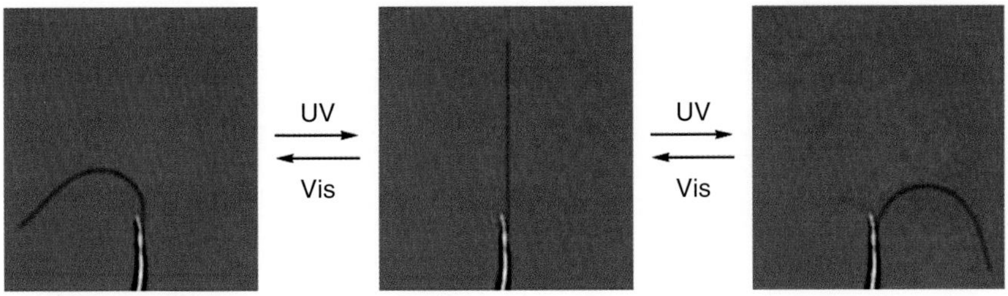

Figure 8.15 Photoreversible crystal shape change of a mixed crystal of **15** and **16**. (Reprinted with permission from [97]. Copyright 2012, WILEY-VCH Verlag GmbH & Co. KGaA, Weinheim.)

toward the direction of incident UV light. The bending motion is due to the gradient in the extent of the photoisomerization from the surface caused by the high absorbance of the crystal. The shrinkage of the irradiated part of the crystal induces the bending as observed in bimetals. Upon irradiation with visible light, the bent rodlike crystal returned to the initial straight one.

Two-component mixed crystals composed of diarylethenes **15** and **16** were found to exhibit superior fatigue resistant property [97]. Recrystallization of a mixture of equimolar amounts of **15** and **16** gave rodlike crystals, which contain the two components in the ratio of **15** : **16** = 63 : 37. The crystal showed reversible bending more than 1000 times upon irradiation with UV and visible light. A long rodlike mixed crystal with the length of 3 mm showed curling into a hairpin shape upon irradiation with UV light, as shown in Figure 8.15. The crystal remained its crystallinity even after the curling and reverted to the original straight shape upon irradiation with visible light. A noteworthy feature of the molecular crystal actuators is that they can work even at extremely low temperature. The rodlike crystal exhibited photoinduced reversible bending at 4.6 K, and the bending was found to take place in less than 5 μs after the excitation with a UV laser pulse.

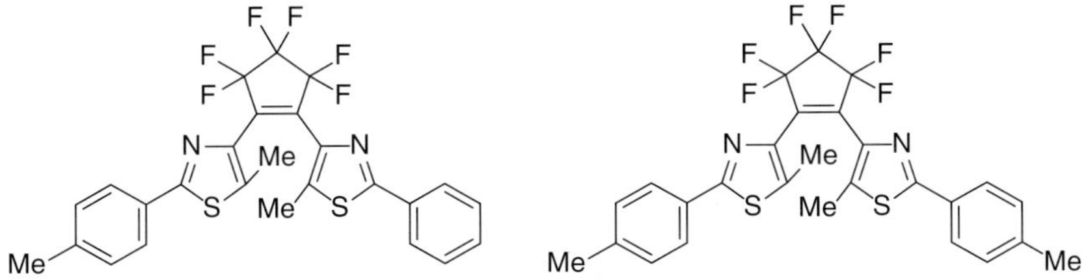

15 **16**

A two-component co-crystal of **13·Np**^{F} also exhibited photoinduced reversible bending [98]. Upon irradiation with UV light, the rectangular platelike crystal turned its color from colorless to blue and bent moving away from the UV light source. Upon irradiation with visible light, the blue color disappeared and the crystal recovered the original straight shape. The rectangular platelike co-crystal

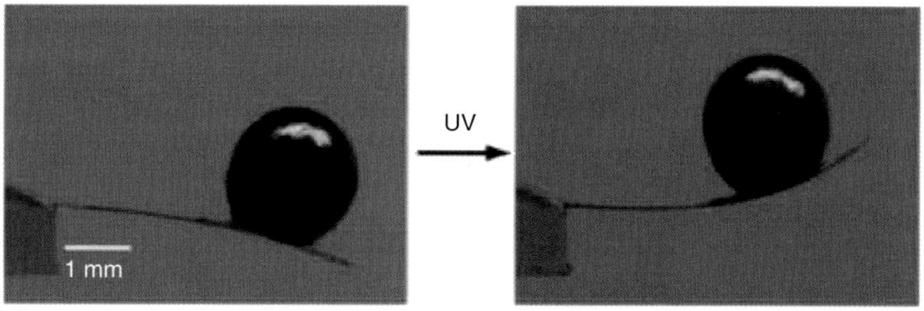

Figure 8.16 Photomechanical work of a molecular crystal cantilever made of a co-crystal of **13** and **Np^F**. (Reprinted with permission from [98]. Copyright 2010, American Chemical Society.)

was fixed at the edge of a glass plate to prepare a cantilever arm and a lead ball was loaded onto the crystal, as shown in Figure 8.16. Upon irradiation with UV light, the cantilever arm (0.17 mg) lifted the metal ball (46.77 mg) as high as 0.95 mm. The amount of the work performed by the cantilever arm is as large as 0.43 μJ. The photogenerated maximum stress was estimated to be 44 MPa by the analysis of the deformation based on structural mechanics. *In situ* X-ray crystallographic analysis revealed that the macroscopic deformation of the crystal is due to the anisotropic deformation of the crystal lattice induced by the geometrical structure changes of the diarylethene molecules upon photoisomerization. The molecular crystals link the geometrical structure changes of molecules in the nanoscopic molecular world to the mechanical movement of the bulk materials and perform the mechanical work in the macroscopic real world.

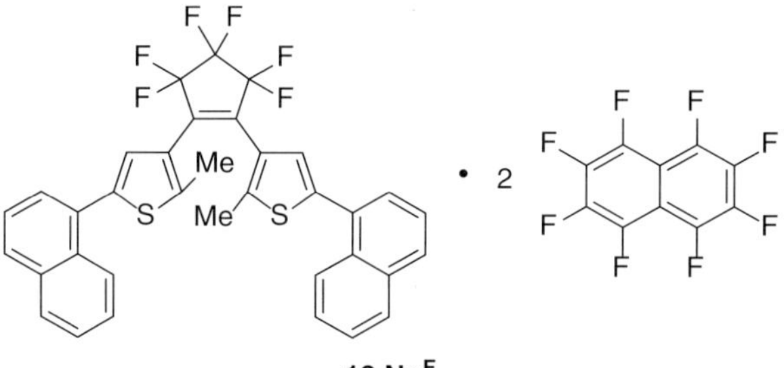

13·Np^F

A thin single crystal of diarylethene **17** was found to twist upon irradiation with UV light, as shown in Figure 8.17 [99]. Upon irradiation with visible light, the twisted crystal recovered the original shape. The reversible twisting could be repeated more than 30 cycles upon alternate irradiation with UV and visible light.

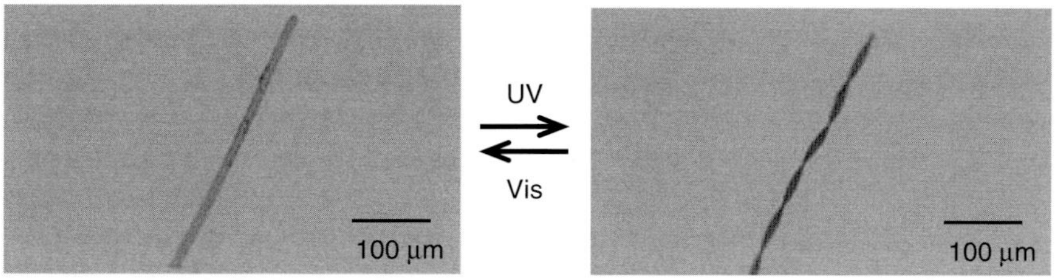

Figure 8.17 Photoreversible twisting of a single crystal of **17**. (Reprinted with permission from [99]. Copyright 2013, WILEY-VCH Verlag GmbH & Co. KGaA, Weinheim.)

The twisting movement of the crystal is attributed to the anisotropic deformation of the unit cell upon photoisomerization of the diarylethene molecule.

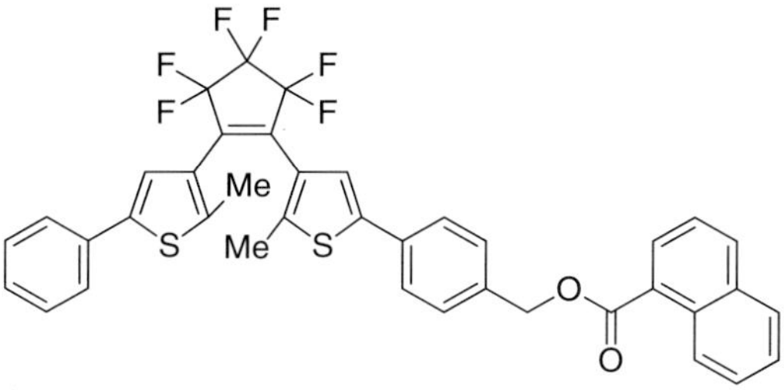

17

Photoinduced shape deformation of bulk molecular crystals was also observed for other diarylethenes [100, 101], azobenzene [102], salicylideneaniline [103], furylfulgide [104], anthracene derivatives [105–111], and [2+2] photocycloaddition systems [112, 113].

8.3 Photochromic Liquid Crystals

8.3.1 Introduction

Liquid crystals (LCs) are fluid phases with direction dependent (i.e., anisotropic) physical properties. They usually flow like ordinary liquids such as water, alcohol, and toluene. However, the molecules constituting the LC phases possess some degree of orientational and positional order, which results in different values of physical properties when measured from different directions. In other words, LCs may simultaneously share the flow properties of isotropic liquids and the anisotropic properties of crystalline solids. Thermodynamically, LCs are situated between the crystalline solid state and the liquid state (Figure 8.18). For this reason, LCs are considered as an intermediate state or mesophase of matter. The molecules in the LC phase exhibit more ordering than in liquids but less

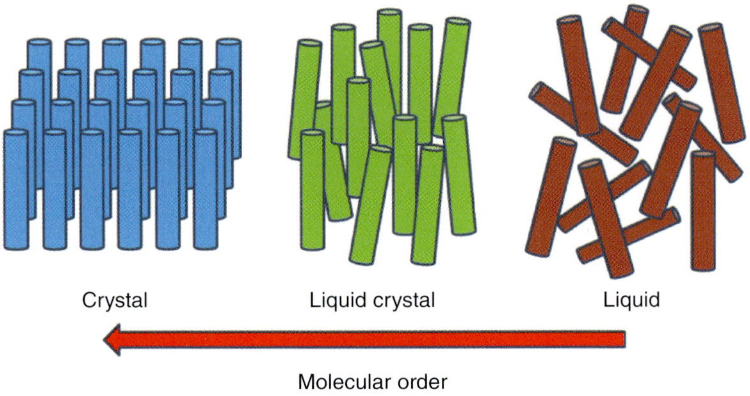

Figure 8.18 Molecular organization in crystal, LC, and liquid states of matter. In the solid crystal phase, the molecules possess both long-range positional order and orientational order; in the LC phase, the molecules possess long-range orientational order but short-range positional order; and in the liquid phase, there is no orientational or positional order of the molecules. In the figure, the molecular organization in the nematic phase of an LC is depicted.

ordering than crystalline solids. This mesophase is referred to as the *fourth state of matter* after solid, liquid, and gaseous states of matter. The constituents of a mesophase are called *mesogens* , and the nature of these mesogens can be organic, inorganic, and organometallic. Owing to their dynamic nature, LCs can be switched by applying different external stimuli. The unique combination of order and dynamics renders LCs as appealing soft materials from both scientific and technological point of view [114–130]. LCs self-assemble into various structures with the help of many different types of molecular interactions such as van der Waals, dipolar and quadrupolar interaction, charge transfer, $\pi-\pi$ interaction, metal coordination, and hydrogen bonding. LCs are known to play a critical role in living systems and biology. For instance, life would not be possible to exist without the ordered and dynamic self-assembly of lipids into bilayers within the cell membrane, an order and self-structuration typical of LCs. Several biological molecules such as lipids, carbohydrates, proteins, and nucleic acids have been found to exist in various liquid crystalline phases under certain conditions.

LCs have become quintessential materials in our daily life in the form of flat panel liquid crystal display (LCD) devices. LCD devices using LCs as the active switching components such as mobile phones, laptop and television monitors, and projectors have dramatically changed the way we present information in the twenty-first century. However, the beyond display applications of LCs are numerous, and new applications are steadily emerging. LCs can potentially be used as functional materials for electron and ion transporting, sensory, and optical materials. LCs are used as templates for the fabrication of mesoporous and nanoscale materials. Recently, their biomedical and diagnostic applications such as in controlled drug delivery, protein binding, phospholipid labeling, and microbe detection have been demonstrated. Moreover, photochemically, thermally, or mechanically induced structure changes of LCs have been used for the fabrication of multifunctional stimuli-responsive materials and devices.

LCs can be classified into different categories in many distinct ways. However, the most commonly used classification of LCs is under two categories: thermotropic LCs and lyotropic LCs. Thermotropic LCs can be obtained by the variations of temperature or pressure, whereas lyotropic LCs are fabricated by dissolving amphiphilic materials in suitable solvents where self-assembled structures behave as the building blocks of LC phases. Lyotropic LCs are stabilized by both concentration of the solute and temperature. If some compounds are able to display both thermotropic and lyotropic LC phases, they are referred to as *amphotropic* . Depending on the shape of the molecules forming thermotropic LC phases, they can be subdivided into calamitic (rodlike), discotic (disklike), and bent-core (bananalike) LCs (Figure 8.19).

Depending on the supramolecular organization of the molecules in the LC phases, they can be classified into nematic, smectic, columnar, B phase, and so on. The most commonly observed LC phase structures exhibited by rodlike, disklike, and bent-core molecules are shown in Figure 8.20.

LC phase behaviors of materials can be characterized by polarizing optical microscopy (POM), differential scanning calorimetry (DSC), and X-ray diffraction (XRD) studies. These three techniques are complementary to each other in identifying the LC phase; hence, they are often used in combination to establish the identities of different mesophases and the corresponding supramolecular organizations.

POM is generally employed to mark phase transitions in an LC material by heating on a hot stage. The presence of fluidity, birefringence, and sometimes

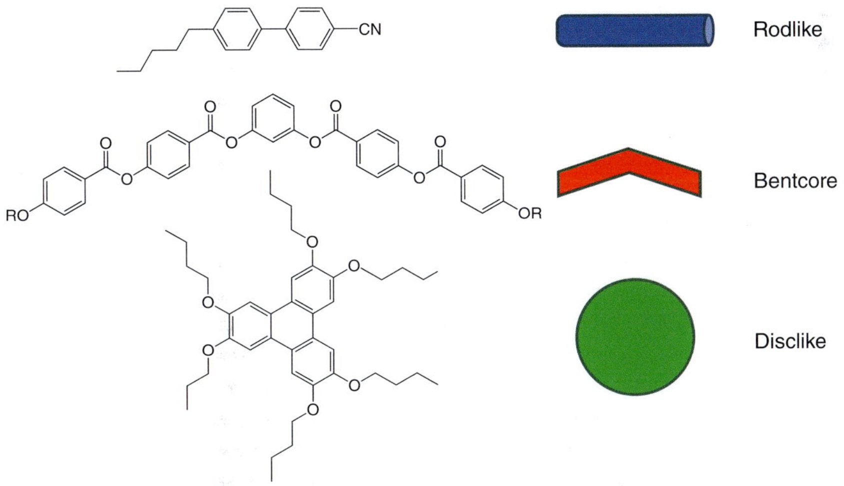

Figure 8.19 Typical molecular shapes of thermotropic LCs of rodlike, bent-core, and discotics.

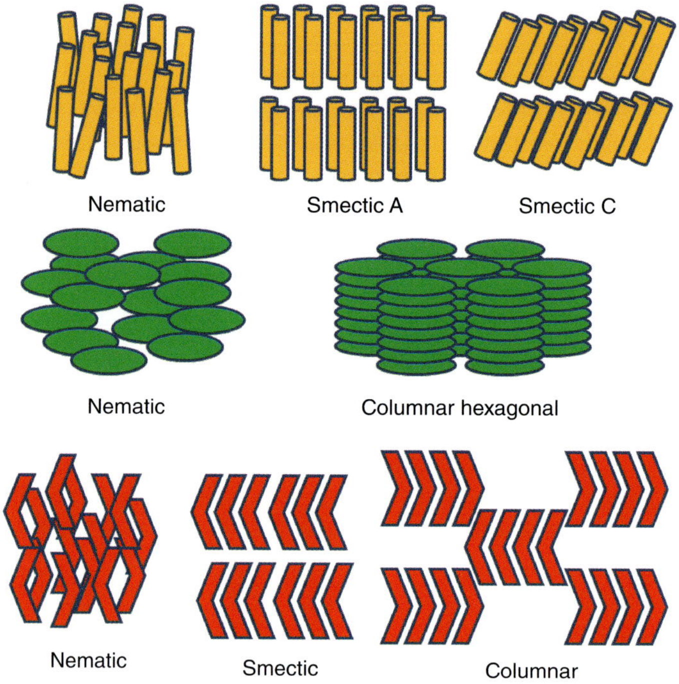

Figure 8.20 Commonly observed LC phase structures of rodlike, disklike, and bent-core molecules.

characteristic textures gives strong indication of the LC phases. DSC is used to determine the temperatures of phase transitions and enthalpy changes related to each transition. With increasing temperature, transitions are typically observed for transformations between the crystalline, LC, and isotropic liquid states. The temperature at which the crystalline solid phase transforms into an LC phase is referred to as the *melting temperature* (T_m), whereas the phase transition from mesophase to isotropic liquid is termed as the *clearing transition* and the temperature at which it occurs is referred to as *clearing temperature* (T_c). Though this technique cannot identify the type of LC phase, the magnitude of enthalpy change gives some information about the degree of molecular ordering within a mesophase. The supramolecular organization and the corresponding packing parameters in each LC phase can be probed in detail by XRD studies. This technique allows one to obtain a detailed insight into the various microstructures adopted in the self-assembly of the mesogens in the mesophase.

Photochromic materials undergo reversible photoisomerization between at least two stable forms having distinct light absorption profiles [6, 131–139]. The most commonly studied photochromic systems and their photochemical interconversions are shown in Figure 8.21. During the reversible photoisomerization, some physical properties of photochromic compounds such as absorption spectra, fluorescence emission, conjugation, electron conductivity, dipole interaction,

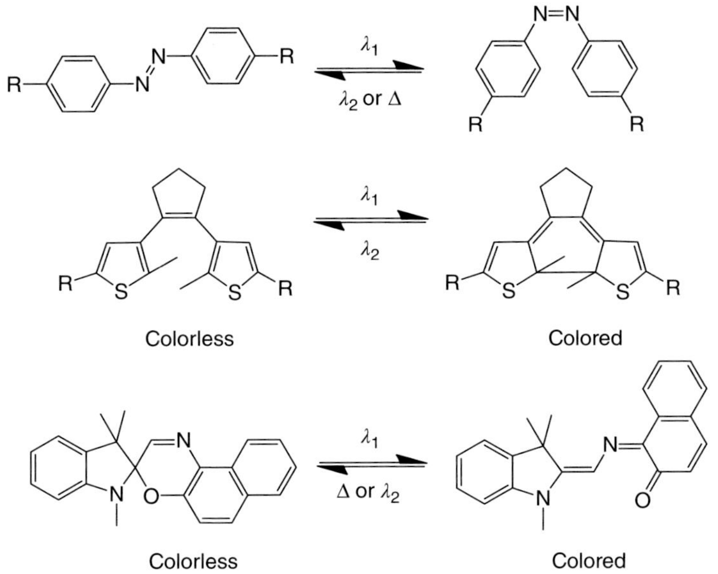

Figure 8.21 Molecular structures and photoisomerization of commonly studied photochromic molecular switches. In the figure, λ_1 is the wavelength responsible for driving the molecule from state 1 (initial) to state 2 (photoisomer) while λ_2 drives the system from state 2 to state 1. Δ represents thermal relaxation.

and geometrical shape may be tuned by light irradiation. Photochromic materials are very promising and possess great potential in several scientific research fields, ranging from chemistry, physics, and materials science to nanotechnology.

Combining the properties of photochromic molecules and LCs might yield versatile functional materials with novel and/or enhanced properties for practical applications. In order to pursue the above-mentioned hypothesis, different types of photochromic LC materials have been designed, synthesized, and studied. Another goal in preparing photochromic LCs is to obtain a material responding reversibly to both an electric field and light irradiation. The most straightforward design of such a material would be a hybrid molecule composed of mesogenic and photochromic units covalently linked to each other. However, photochromic LCs have been realized either by chemical linking of photochromic groups with mesogenic moieties or by physical mixing (doping) of photochromic molecules with LC hosts. Promising results have been obtained by both the approaches. In the following, we present the different classes of photochromic LCs.

8.3.2
Spiropyran- and Spirooxazine-Based Photochromic Liquid Crystals

Spiropyrans and spirooxazines are an interesting family of photochromic materials due to their unique properties such as excellent photofatigue resistance, strong photocoloration, and fast thermal relaxation. The colorless ring-closed spiroforms of spiropyran and spirooxazine can be transformed into the colored ring-opened merocyanine form upon irradiation with UV light, whereas its reverse process occurs thermally in the dark or photochemically by irradiation with visible light. A unique feature of spiropyran is its significantly increased dipole moment after photoisomerization from ring-closed spiropyran form to ring-open charge-separated zwitterionic merocyanine form. The physical and chemical properties of the two forms of spiropyran and spirooxazine are distinctly different; therefore, the thermally reversible photochromic switching has been the basis for the intelligent materials with applications in smart devices. Liquid crystalline spiropyran and spirooxazine derivatives have been designed and synthesized as both molecular and polymeric materials.

Cabrera *et al.* synthesized photochromic LC polysiloxanes **18** containing spiropyran groups (Scheme 8.9) [140, 141]. The phenylbenzoate copolymers with the spiropyran side chains give a mesophase in which the clearing temperature decreases with increasing spiropyran content in the macromolecules. These copolymers were found to exhibit interesting photochromic behavior. Red-, blue-, and yellow-colored polymer films could be obtained by photoirradiation with suitable wavelength at appropriate temperatures. The thermochromic spiropyran-to-merocyanine dye conversion occurs on heating the copolymers. The copolymer films cast from solution acquire a pink color at room temperature and exhibit strong birefringence. Irradiation with visible light (>500 nm) brings about a pale yellow color, while irradiation of the yellow film with UV light (365 nm) results in a deep red color. A yellow film irradiated with UV light at −20 °C turns blue and can be converted back to yellow by irradiation with visible light. The possibility of controlling the formation of the primary colors with light and temperature may make it possible to tailor these LC polymers for new applications in imaging technology. Prior to this, they synthesized hybrid molecules containing spiropyran and mesogenic groups; however, these materials show some peculiar phase behavior that was named as quasi-LC. Subsequently, Yitzchaik *et al.* reported acrylic copolymers **19** with mesogenic and spiropyran side chains, which showed photochromic mesophases [142]. Upon irradiation with UV light, these copolymers were found to strongly absorb visible light. The color disappears on irradiation with visible light or thermally. Interestingly in these copolymers, two distinct sites were observed: mesogenic domains and amorphous sites. Main chains and photochromic side chains are presumably located in the amorphous site that expands and causes the mesophase to disappear as the spiropyran content increases in the polymer. Shragina *et al.* have reported photochromic LCs **20** and **21** containing spironaphthoxazine substituted with mesogenic group. The compound spiroindoline-naphthoxazine **21** with mesogenic substituent showed LC properties. Natarajan *et al.* synthesized first example of photochromic cholesteric LC siloxane compound **22** [143] containing spiropyran group. This LC compound was observed to exhibit selective reflection property of cholesteric LCs as well as photochromism in thin films, in fibers, and in solution.

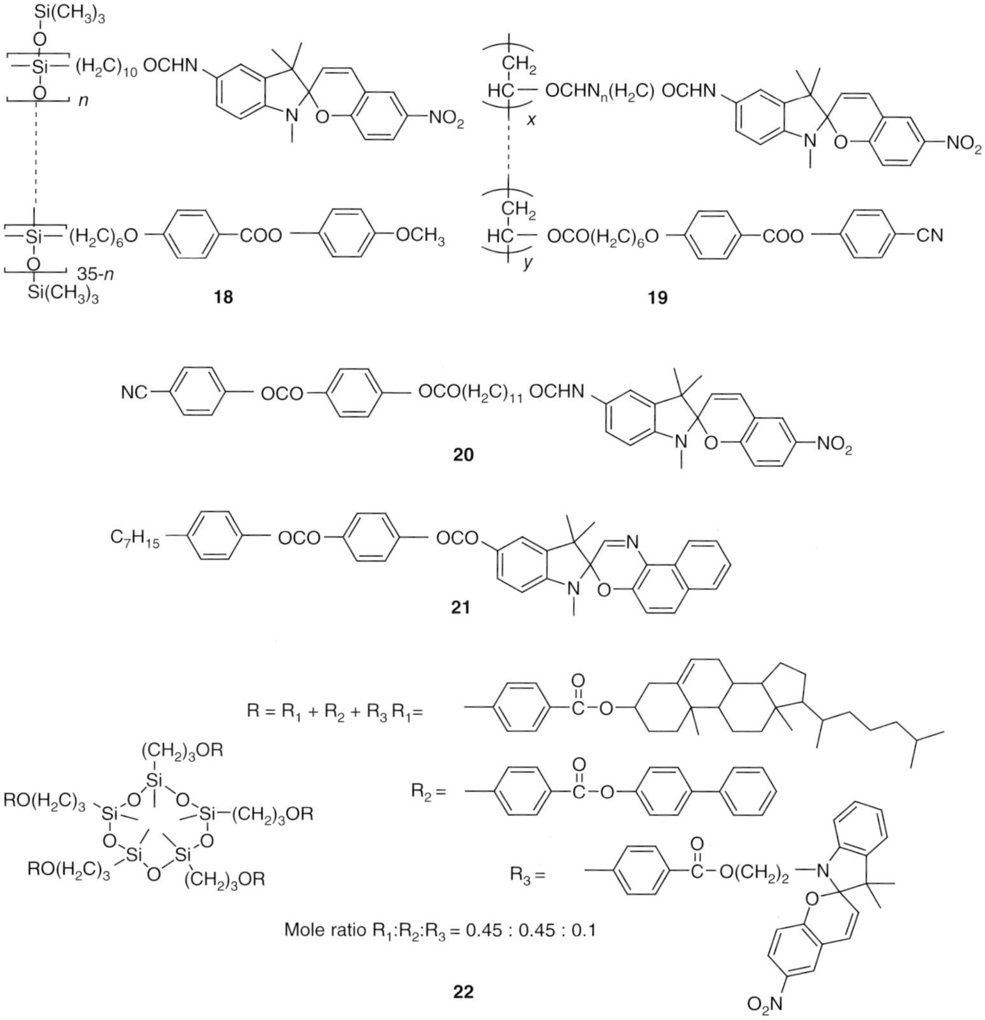

Scheme 8.9 Spiropyran- and spirooxazine-based photochromic LC polymers and oligomers.

Hattori and Uryu designed and synthesized polymerizable photochromic LC materials **23–25** containing biphenylene as the mesogenic moiety (Scheme 8.10) [144]. These hybrid photochromic compounds displayed metastable mesophases as revealed by POM, DSC, and XRD studies. It has been presumed that the appearance of mesophase on cooling was caused by reorientation of the spirooxazine moieties in the molten state. Photochromic polymerizable acrylates containing spirooxazine moieties with a chiral substituent were also prepared, which yielded photochromic chiral LC systems **26** and **27** [145]. The photochromic acrylates containing both an undecamethylene group and a (2*S*,3*S*)-2-chloro-3-methylpentanoyloxy group or a (−)-menthoxyacetoxy group

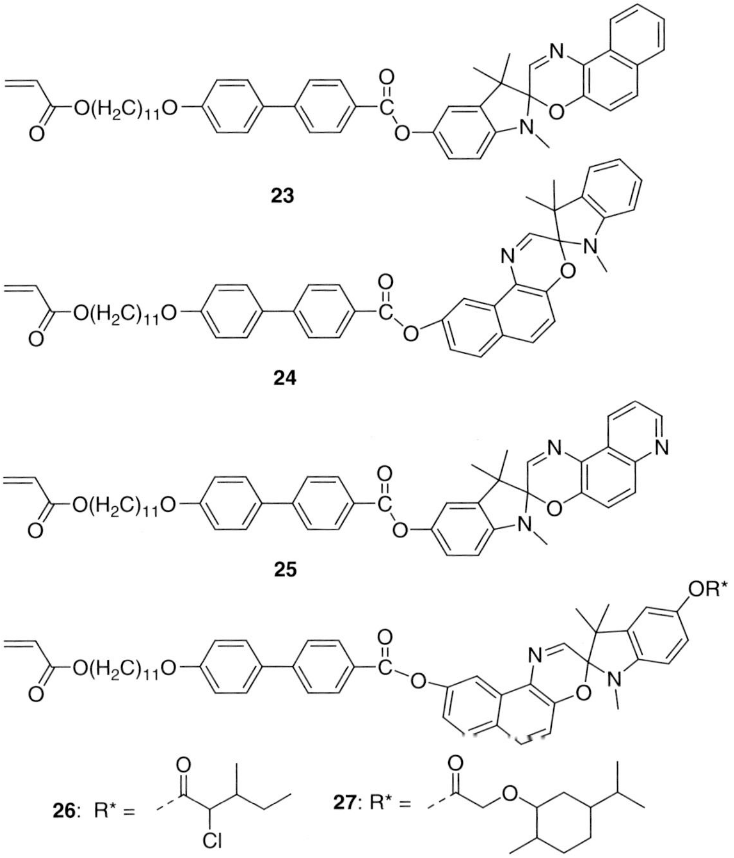

Scheme 8.10 Polymerizable spirooxazine-based LCs.

show a supercooled mesophase. Interestingly, the latter compound was found to reflect right-handed blue light at room temperature.

Keum *et al.* have synthesized and studied the LC phase behavior of various spiropyran-based materials (Scheme 8.11) [146, 147]. These compounds **28–30** were found to exhibit metastable monotropic nematic LC phase as studied by DSC and POM.

Although spirooxazines have a spirocarbon atom as a chiral center, they usually exist as racemates. Even if enantiomers are separated, each enantiomer is racemized by thermal and optical interconversions. Thus, spirooxazines are not usually able to act as chiral dopants for LC hosts. To use spirooxazines as chiral dopants, their synthesis with an additional chiral auxiliary is required. Accordingly, Li *et al.* synthesized axially chiral spirooxazines (Scheme 8.12), which exhibit high helical twisting power (HTP) [148]. Photoresponsive cholesteric LCs were fabricated by doping these molecular switches **31–34** into commercially available nematic LC hosts. Among these chiral dopants, the compound **31** with bridged binaphthalene

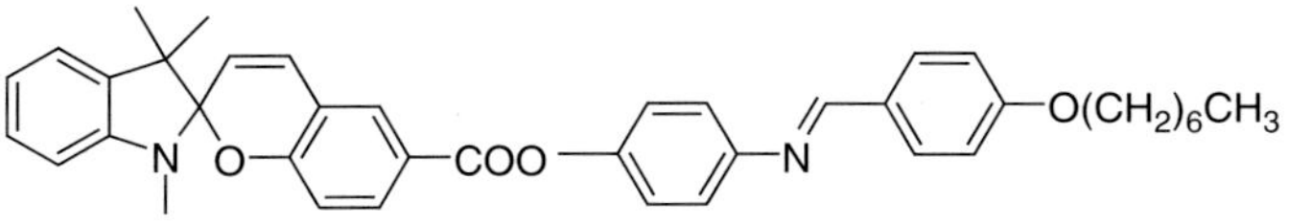

28

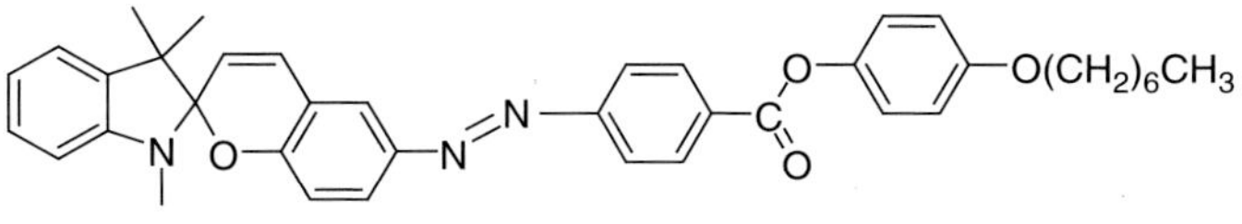

29

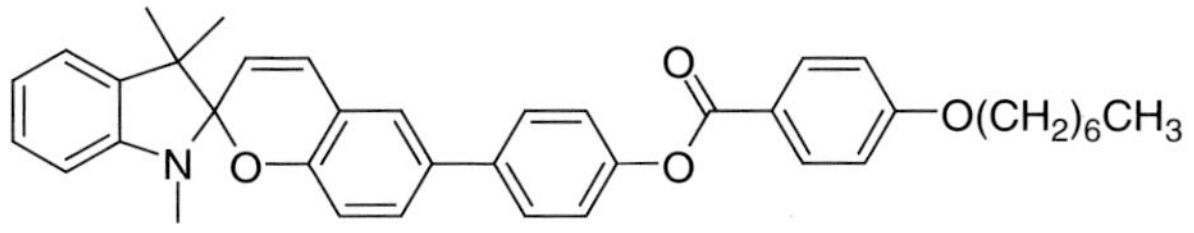

30

Scheme 8.11 Photochromic spiropyran-based LCs.

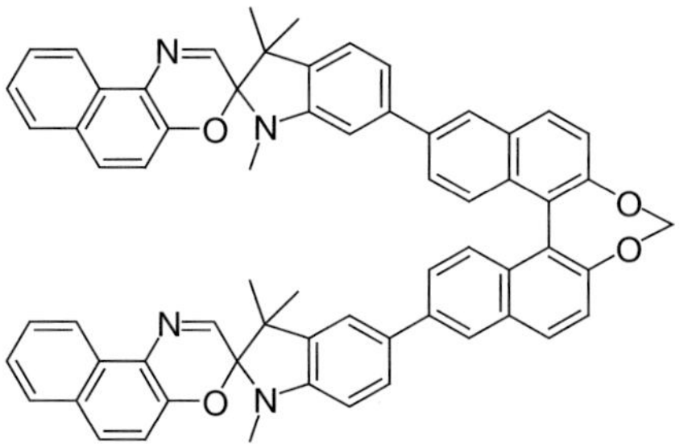

31

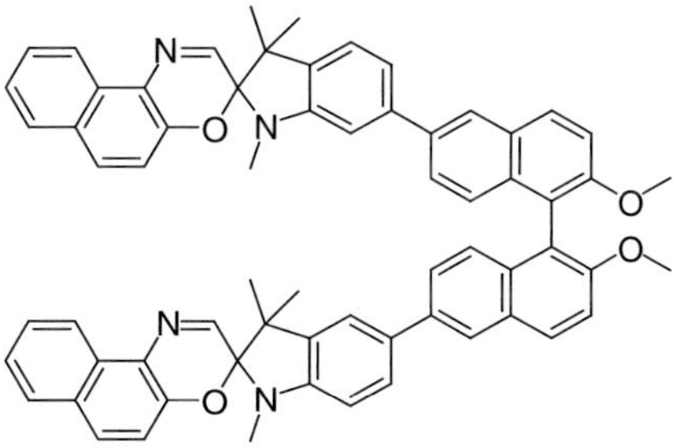

32

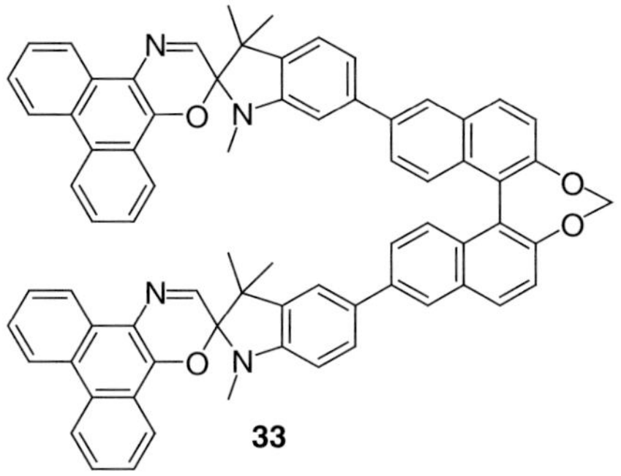

33

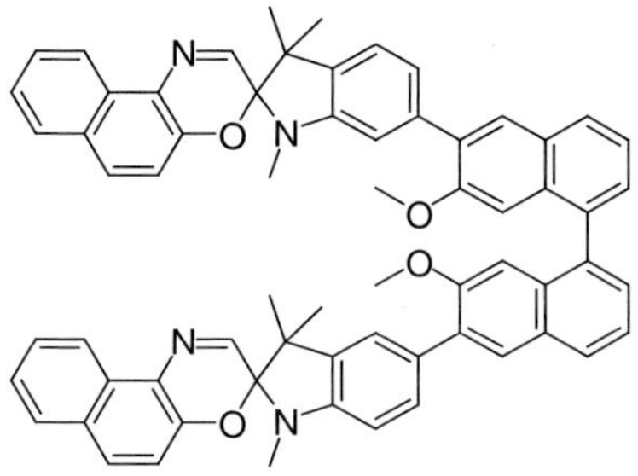

34

Scheme 8.12 Spirooxazine-based molecular switches for induction of photoresponsive cholesteric LCs.

group showed the largest HTP value in E7. Interestingly, the HTP for compounds **31** and **33** increases upon irradiation with UV light, while the HTP of other compounds decreases under the same condition. This observation has been attributed to more compatibility of the rodlike merocyanine form for compounds **31** and **33** compared to the other compounds. Photochromic reverse-mode polymer dispersed LCs have also been fabricated by doping spirooxazine-based photochromic compounds [149, 150].

8.3.3
Diarylethene-Based Photochromic Liquid Crystals

Dithienylcyclopentenes are considered to be the most promising photochromic materials for optical memory and photoswitching applications because of their excellent fatigue resistance and superior thermal stability of both photoisomers, fast photocyclization, and electrical conductivity [151]. They undergo a reversible photocyclization reaction between colorless ring-opened and colored ring-closed forms. Upon irradiation with UV light, they can transform from the colorless open-ring form to the colored closed-ring form. The reverse process is thermally stable and occurs only by visible light irradiation. Since the physical and chemical properties of the two isomeric forms are different, optically reversible switching has been the basis for generating new functional materials from these photochromic moieties. Various liquid crystalline materials based on dithienylcyclopentene scaffold with and without mesogen substitution has been reported.

Mehl *et al.* made a modular approach toward photochromic LCs by using 1,2-bis(2-methylbenzo[*b*]thiophen-3-yl)hexafluorocyclopentene system connected to two cyanobiphenyl groups via flexible spacers with 10 methylene units (Scheme 8.13) [152–156]. The system is shown in Scheme 8.13. This is very versatile approach, which can lead to the systematic investigations of the influence

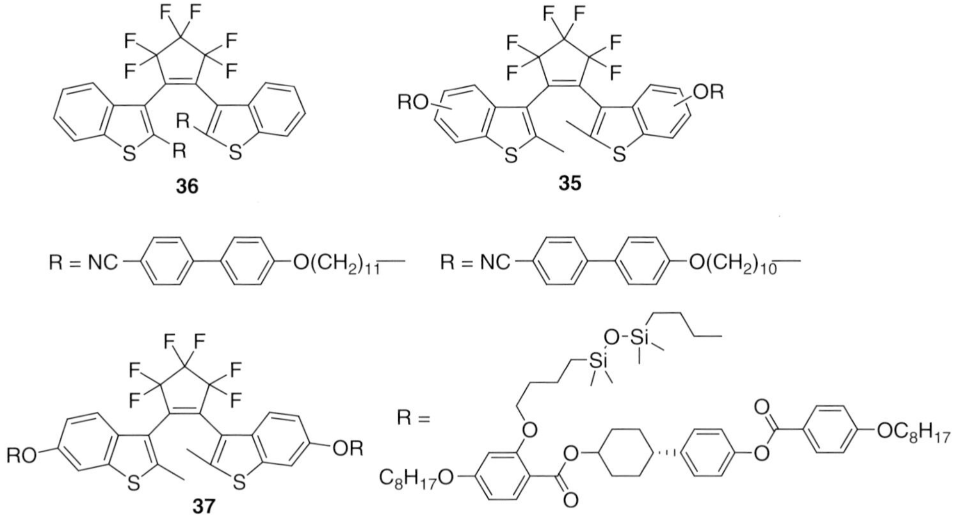

Scheme 8.13 Mesogen-functionalized LC diarylethenes.

of the photochromic group, the spacer lengths, and the position of mesogens on the properties of such systems. POM revealed broken focal conic defect in conjunction with schlieren textures, indicating the formation of a smectic C phase in compound **35** [152]. The presence of nematic phase before the material transforms into isotropic state was confirmed by the observation of a schlieren texture, with two and four brush defects. The sample in the photostationary state (which contains both open-ring and closed-ring isomers) exhibited a different phase behavior. It displays only the nematic phase between the crystalline and isotropic state. Repeated heating and cooling cycles of the samples did not have any influence on the transition temperatures and their associated enthalpies, suggesting the thermal stability of the system. Different structural isomers of compound **35** have also been realized by changing the position of the mesogens [153]. The LC phase behavior for all of these systems is broadly similar in terms of the type of phase structures. However, upon photoirradiation, the phase behaviors of the compounds in photostationary states are significantly different. This system shows that combining individual functionalities of a photochromic core and mesogens linked by flexible spacers allows the design of materials where mesomorphic phase structures and stability can be modulated.

The LC phase behavior of the open-ring isomers **36** and its photostationary state was investigated using POM and DSC [154]. Compound **36** melts at 140.1 °C and on cooling from the isotropic liquid exhibits a nematic phase at 96.1 °C. Irradiation with UV light alters these properties significantly. In the photostationary state, the reduction in intramolecular flexibility due to ring closure of the system reduces the stability of the nematic phase while the melting point is less altered. In this compound, it was shown that the introduction of cyanobiphenyl groups in 2,2′-positions via alkyl spacers in a diarylethene derivative enhances the rate of photoswitching without altering the other photochromic properties. This also affords a new alternative in the design of photochromic LCs based on diarylethene derivatives.

Frigoli and Mehl have designed and synthesized photoswitchable fluids exhibiting room temperature nematic phases as shown in Scheme 8.13 [155]. The compound **37** has a nematic range over 50 °C including room temperature and undergoes a glass transition at about 3 °C. A new series of photochromic unconventional LC has been reported by Mehl *et al.* as shown in Scheme 8.14 [156]. The thermotropic phase behavior of the open-ring isomers were investigated by DSC and POM. When cooled from isotropic state, the compounds **38** and **39** exhibit monotropic nematic phases identified from their typical schlieren textures. The stability of the nematic phases are drastically reduced in the photostationary states, which has been attributed to the reduction of flexibility of the central cores. Surprisingly, compound **40** does not exhibit any mesomorphism at all.

Chen *et al.* reported photochromic glassy LCs as a new class of functional optical materials (Scheme 8.15) [157]. These materials were synthesized by functionalizing diarylethene core with nematogens. Depending on the type and number of nematogens, three distinct types of thermotropic LC phase behavior have been observed. Compound **41** on cooling from isotropic state exhibited

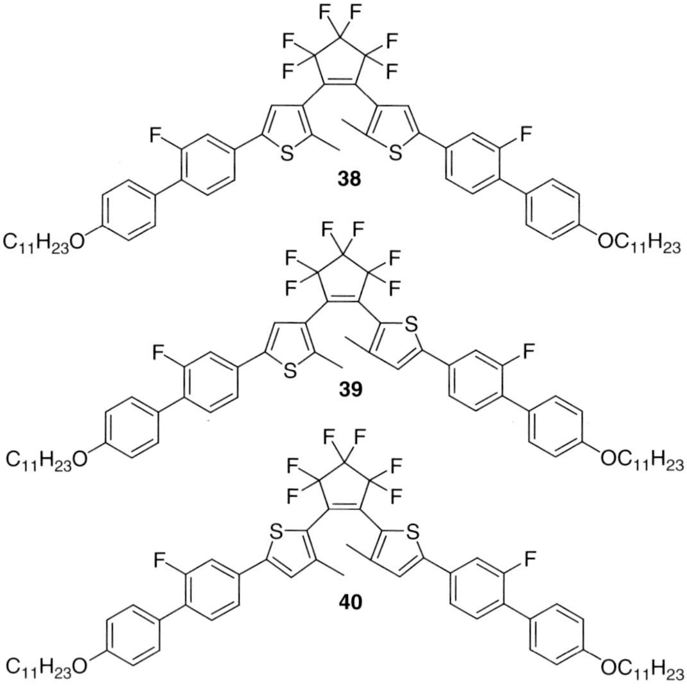

Scheme 8.14 Liquid crystalline photochromic diarylethene derivatives.

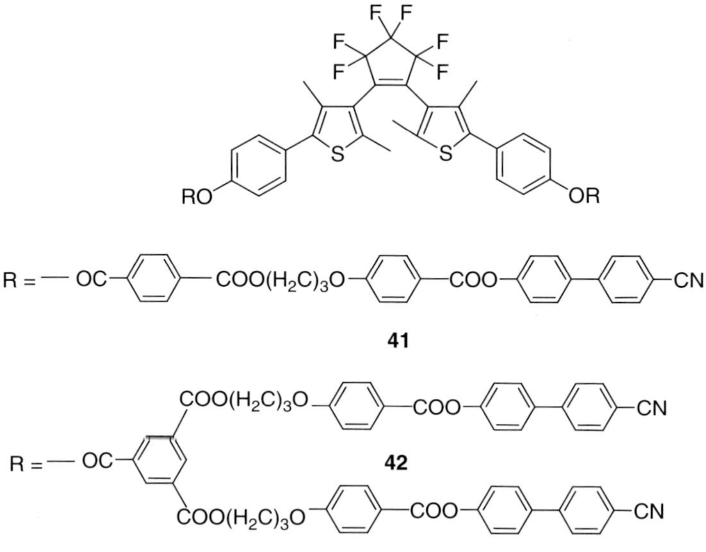

Scheme 8.15 Diarylethene-based photochromic glassy LCs.

nematic phase that transformed into glassy LC film with smectic (lamellar) ordering. However, compound **42** containing more number of mesogens resulted in a nematic glass. Therefore, morphologically stable glassy nematic LCs comprising a dithienylethene core has been successfully designed and synthesized.

Li *et al.* reported mesogenic diarylethene compounds **43–45**, which show chiral nematic phases over broad temperature range as evidenced by typical oily streak textures (Scheme 8.16) [158]. These compounds do not crystallize but transform into photochromic glassy state. These compounds can be used as chiral dopants to induce photoresponsive cholesteric LCs by adding them into commercially available nematic LC hosts.

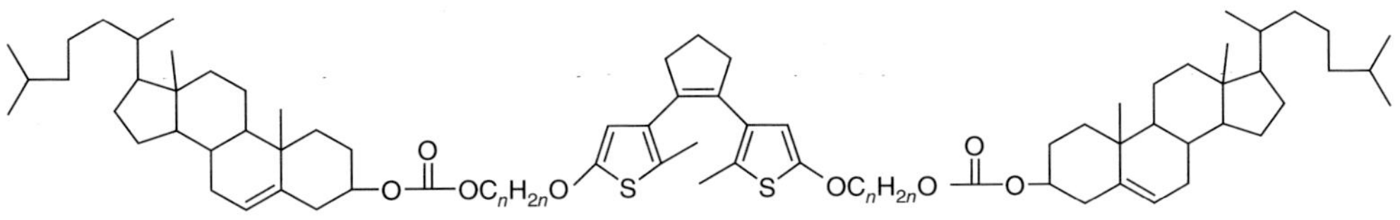

43: $n = 6$; **44**: $n = 9$; **45**: $n = 11$

Scheme 8.16 Cholesterol-substituted LC diarylethene derivatives.

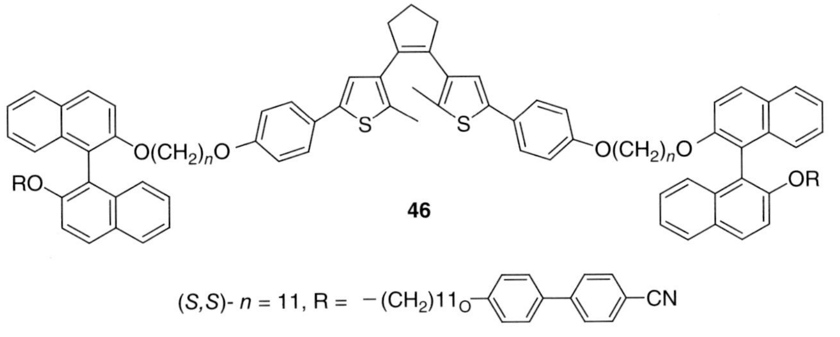

Scheme 8.17 Axially chiral photochromic diarylethene derivatives.

Li *et al.* synthesized mesogenic axially chiral diarylethenes **46**, which possess very high HTP (Scheme 8.17) [159]. These were employed as chiral dopants for the fabrication of cholesteric LCs whose pitch could be modulated over wide range. This enables reflection color tuning over a wide wavelength range. Subsequently compounds **47–50** were designed and synthesized that furnished unprecedented results (Scheme 8.18) [160, 161]. Interestingly, each member of the chiral dopants with bridged binaphthyl groups containing different alkylenedioxy bridges displays unique characteristics with respect to their helicity induction capabilities in nematic LCs. Compound **47** with a methylenedioxy bridge exhibits dramatic increase in HTP upon UV irradiation, thus enabling a blue shift of the reflection

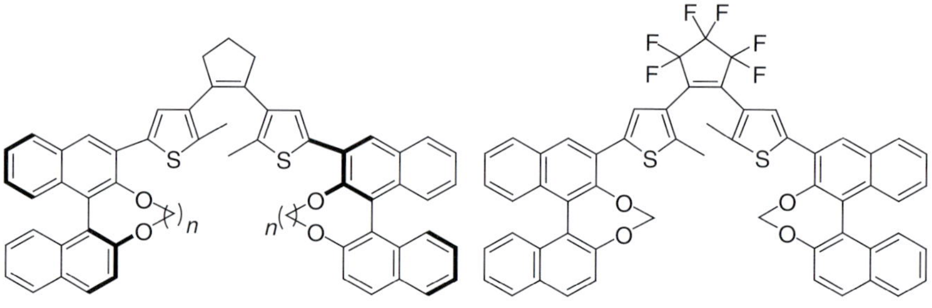

(*S,S*)-**47,48,49,50** : *n* = 1,2,3,4 (*S,S*) or (*R,R*)-**51**

Scheme 8.18 Chemical structures of axially chiral photochromic diarylethene dopants.

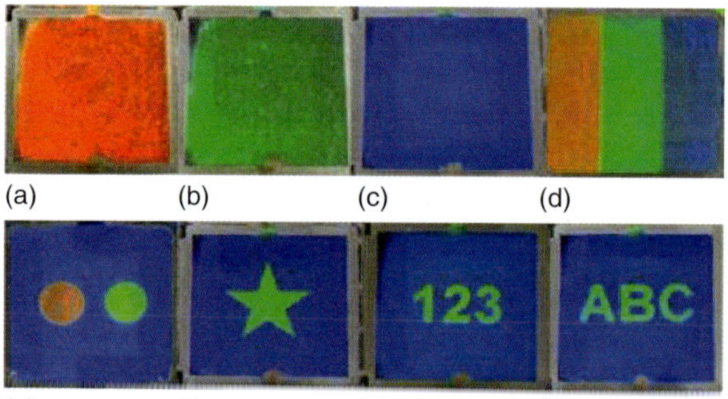

(a) (b) (c) (d)
(e) (f) (g) (h)

Figure 8.22 Thermally stable red, green, and blue reflection colors in cholesteric films containing **47**. (Reproduced with permission from [160]. Copyright 2012, American Chemical Society.)

color. The three primary colors red, green, and blue could be attained in a single film by varying the UV irradiation time while masking different areas of a cholesteric film (Figure 8.22). Compounds **48** and **49** were found to possess much higher HTPs than **47** but experience a dramatic decrease in HTP during photoisomerization that leads to a red shift of the selective reflection upon UV irradiation. Surprisingly compound **50** with butylenedioxy bridge causes light-driven reversible handedness inversion of the cholesteric LCs fabricated from three different nematic LC hosts. Right-handed helices were induced in all the hosts before UV irradiation, but upon UV irradiation the right-handed helices unwound, yielding nematic phases that upon further irradiation afforded left-handed helices as revealed by POM studies (Figure 8.23) [161]. The change of dihedral angle of the binaphthalene groups leading to conformational alterations has been attributed to these interesting observations. Very recently, the compound **51**with axial chirality has been reported, which possesses superior thermal stability and high HTP in different nematic hosts [162].

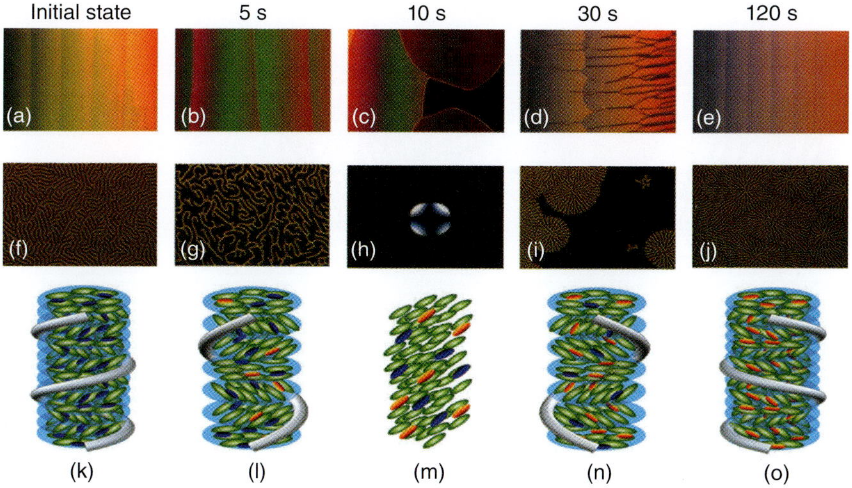

Figure 8.23 Polarizing optical microscopic demonstration of handedness inversion in cholesteric films containing **50**. (Reproduced with permission from [161]. Copyright 2013, Wiley-VCH GmbH & Co.)

To investigate the effect of photoisomerization of dithienylethene dopants on the spontaneous polarization of a ferroelectric liquid crystal (FLC) host, compounds **52** and **53** have been synthesized keeping in mind their compatibility with the lamellar structure of the chiral SmC phase (Scheme 8.19) [163–165]. Reversible modulation of the spontaneous polarization could be achieved by photocyclization of the compound by irradiating with UV and visible light sequentially (Figure 8.24). The spontaneous polarization photomodulation cycle could be repeated a number of times without any sign of photochemical degradation.

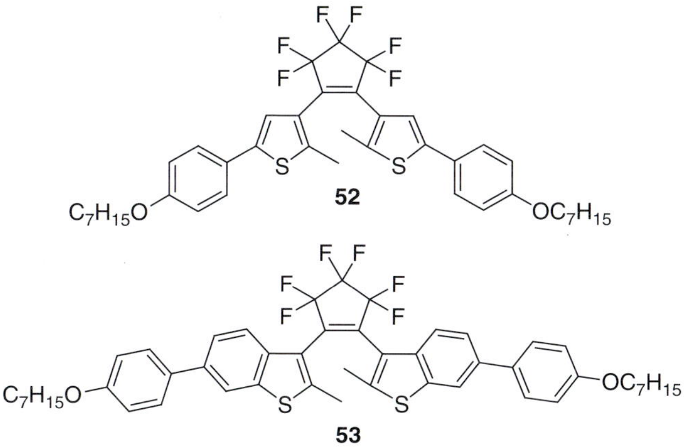

Scheme 8.19 Photoresponsive diarylethenes used in ferroelectric LCs.

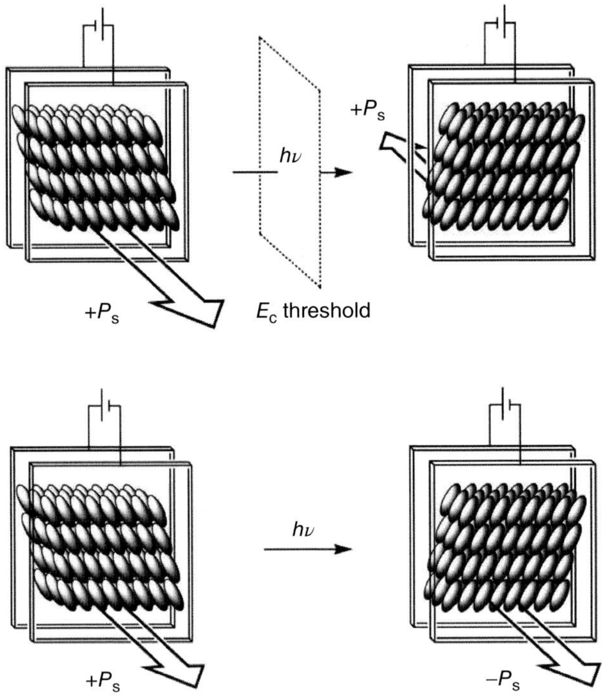

Figure 8.21 Modulation of spontaneous polarization in ferroelectric LCs enabled by photochromic materials. (Reproduced with permission from [165]. Copyright 2004, The Royal Society of Chemistry.)

The photomodulation of spontaneous polarization has been attributed to the loss of conformational flexibility of the dopant upon photocyclization.

8.3.4 Azobenzene-Based Photochromic Liquid Crystals

Azobenzenes are widely studied and a well-known family of photochromic compounds that undergoes *trans*–*cis* isomerization upon UV irradiation. The *cis* isomer can be driven back to *trans* form either by visible light irradiation or by thermal relaxation. The dramatic shape change from the rodlike structure of *trans* form to bent-shaped *cis* form has been the basis of numerous studies including optical imaging, optical memory storage and retrieval, and so on. Different varieties of LC polymeric and low molar mass materials based on azobenzene photoswitch entity have been synthesized and studied. In mesomorphic photochromic azobenzene derivatives, the properties such as absorption wavelength, mesophase morphologies, and transition temperatures can be controlled through their molecular structure engineering.

Functional cross-linked photomobile LC polymers containing achiral and chiral rodlike azobenzene derivatives have been systematically explored by

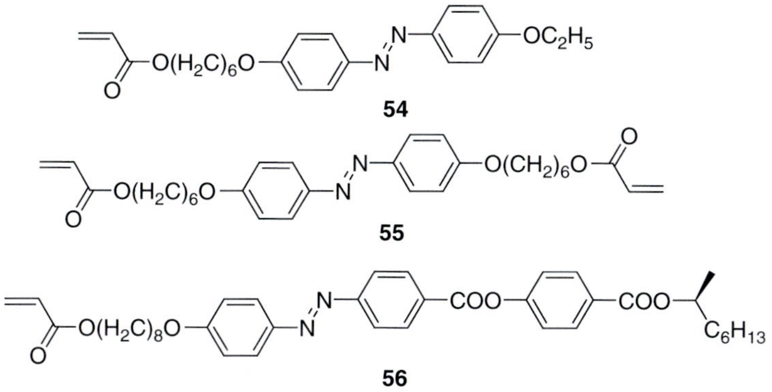

Scheme 8.20 Azobenzene-based polymerizable photochromic LC monomers.

Ikeda *et al.* (Scheme 8.20) [50, 166]. Reversible phase transitions in these LC polymers induced by photochemical isomerization of side-chain and main-chain azobenzene chromophores have been observed. They have shown that a large variety of cross-linked LC polymers can be prepared by changing the structures of the main chains and mesogens (core, spacer, and tail) and the position of the mesogens. It has been demonstrated that LC polymers based on azobenzene derivative **54** and the cross-linker **55** exhibit remarkable directional bending on irradiation with polarized light (Figure 8.25). LC polymers composed of chiral azobenzene derivative **56** showed smectic phases with ferroelectric properties

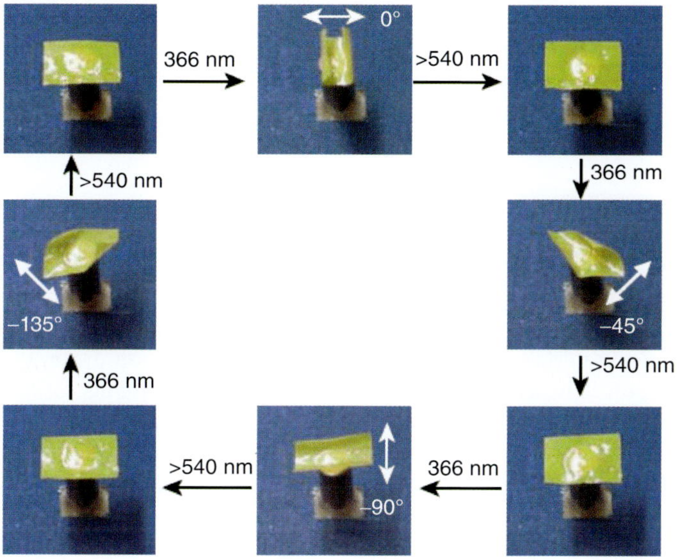

Figure 8.25 Precise control of the bending direction of a polydomain cross-linked LC polymer film by linearly polarized light. White arrows indicate directions of linearly polarized light. (Reproduced with permission from [50]. Copyright 2003, Nature Publishing Group.)

[166]. FLC polymer thin films with a high LC order could be prepared by photopolymerization under an applied electric field. It has been observed that the concentration and location of photochromic moieties play an important role during the process.

Side-chain LC polymers **57** and **58** with photoresponsive azobenzene groups (Scheme 8.21) have been targeted for the development of optical and holographic data storage due to their ease of photoreversion between *trans* and *cis* states [167, 168]. Even though some of the final polymers are not LC, light irradiation with suitable wavelength is found to induce ordered domains in the polymer matrices. The incorporated side-chain mesogens act in a cooperative manner, thereby enhancing refractive indices and birefringence of the materials. LC polymers that form glasses on cooling produce optically transparent films with high anisotropy. Cholesteric polymers containing azobenzene moieties have also been used to control the pitch lengths of the helical superstructures.

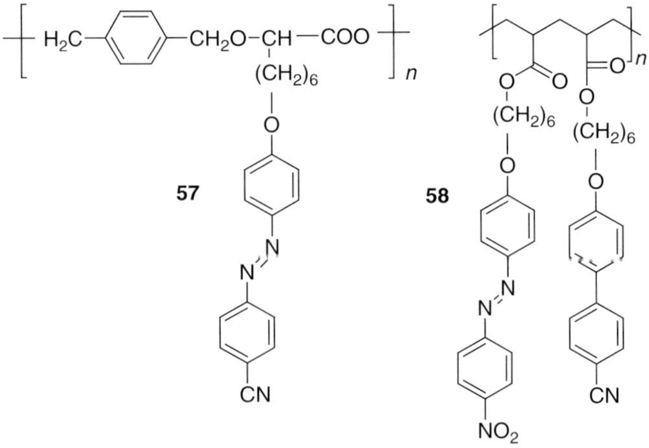

Scheme 8.21 Azobenzene-based photochromic LC polymers.

Norikane *et al.* designed and synthesized LC macrocyclic compound **61** tethered by multiple azobenzene units bearing alkoxy chains (Scheme 8.22) [169]. The molecular structures promote the formation of columnar and lamellar phases. These photochromic materials exhibit isothermal phase transition from LC to isotropic state upon light irradiation due to significant shape change of the molecules. Triphenylene-based discotic LCs bearing six rodlike azobenzene peripheral moieties linked through alkyl chains have been synthesized as shown in Scheme 8.22 [170, 171]. These disk–rod hybrid compounds exhibit discotic nematic phase behavior though discotic compounds are more favorable for columnar phase formation. However, compound **60** shows smectic A and rectangular columnar phases. Upon light irradiation, this compound was found

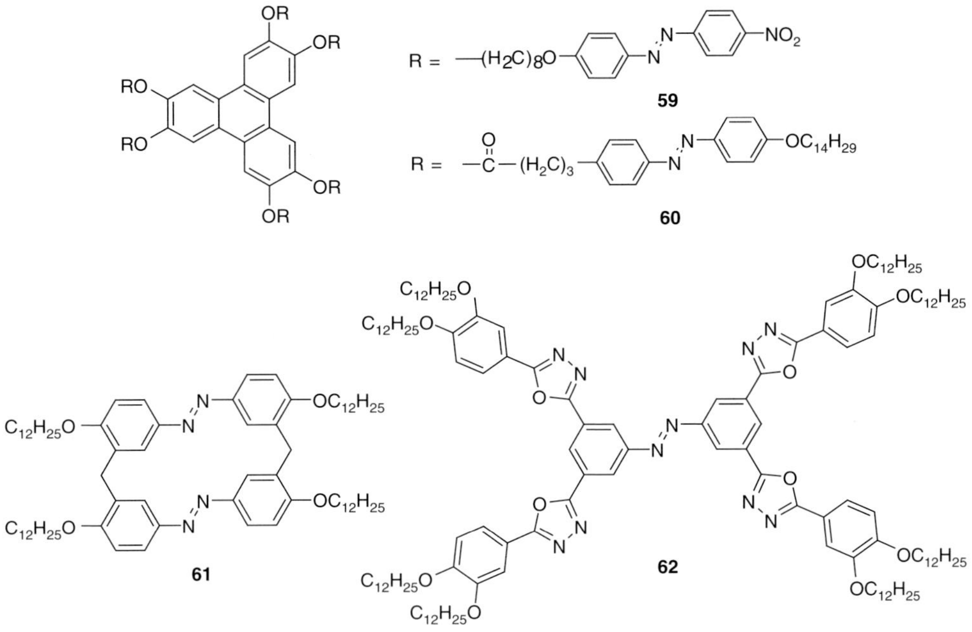

Scheme 8.22 Azobenzene-based discotic photochromic LCs.

to exhibit interesting reversible shape alteration between disk and rod in the LC phase and undergo isothermal phase transition into the isotropic state. Discotic azo compound **62** was prepared by Westphal *et al.* [172]. The formation of hexagonal columnar mesophase in the compound has been established by POM, DSC, and XRD studies. Upon photoisomerization, the compound transforms to isotropic phase. Such photoisomerizable materials are promising candidates for controlling the conductivity in electro-optic devices.

Several bent-core LCs containing azobenzene side wings have been synthesized and studied by several groups (Scheme 8.23) [173–182]. Prasad *et al.* reported that the compound **63** exhibits uniaxial and biaxial nematic phases in addition to higher order smectic phases. Subsequently, Li *et al.* synthesized compounds **64** and **65** and found that they exhibit exclusively nematic and chiral nematic phases. Interestingly, compound **65** exhibited unprecedented thermally, electrically, and photochemically controlled pitch modulation in the cholesteric phase. Azobenzene-containing polymerizable bent-core LCs **66, 67** have been designed and synthesized. Chiral LC phases with interesting properties have been observed in achiral bent-core compounds **68, 69** containing azobenzene side wings and additional substituents at the central core. Bent-shaped compound **70** has been found to exhibit enantiomeric excess in the LC phase when irradiated with circularly polarized light.

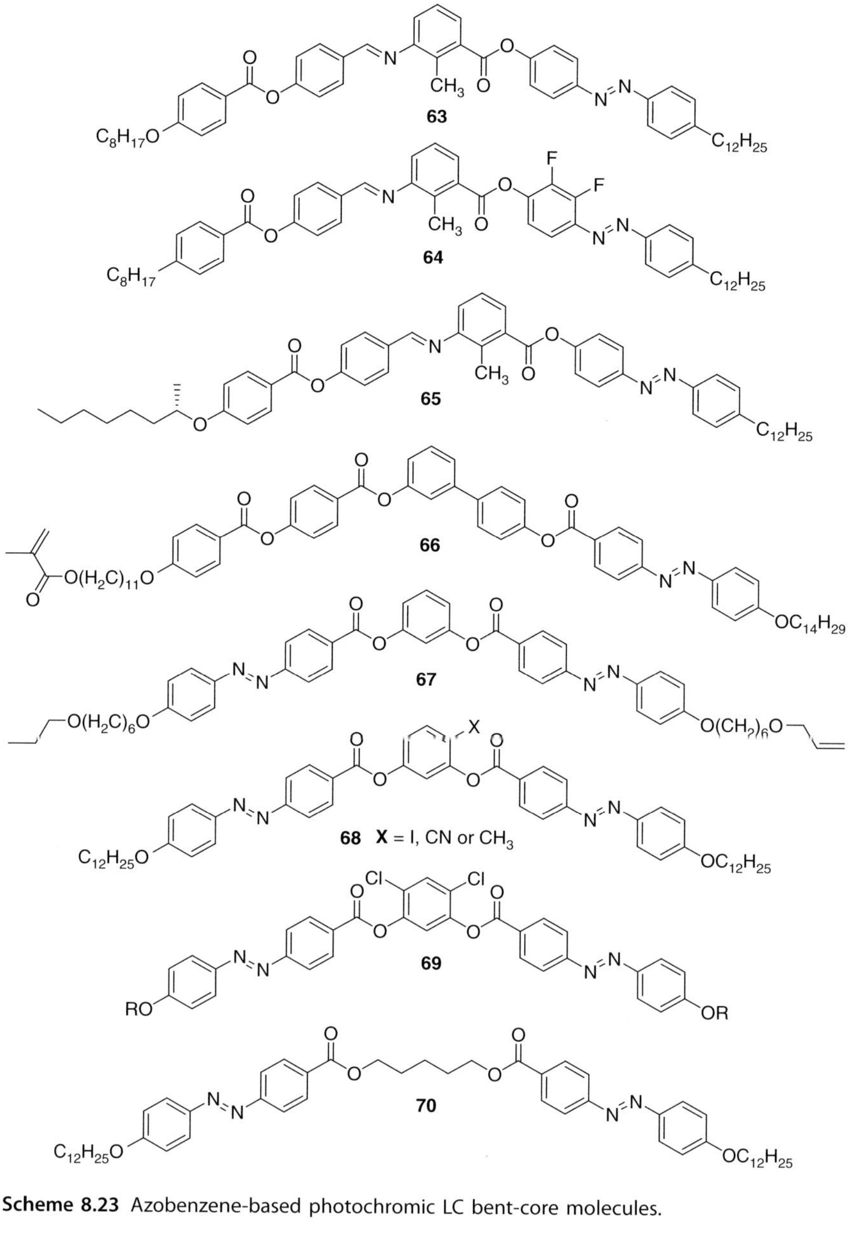

Scheme 8.23 Azobenzene-based photochromic LC bent-core molecules.

Li *et al.* designed and synthesized the mesogenic azoarene **71**, which undergoes *trans–cis* reversible photoisomerization by visible light (Scheme 8.24) [183]. As a chiral dopant, it efficiently induces photoresponsive cholesteric LCs because of its high HTP. Since the HTP of this compound shows a large change in its value between the initial and photostationary states, the pitch of the cholesteric LC

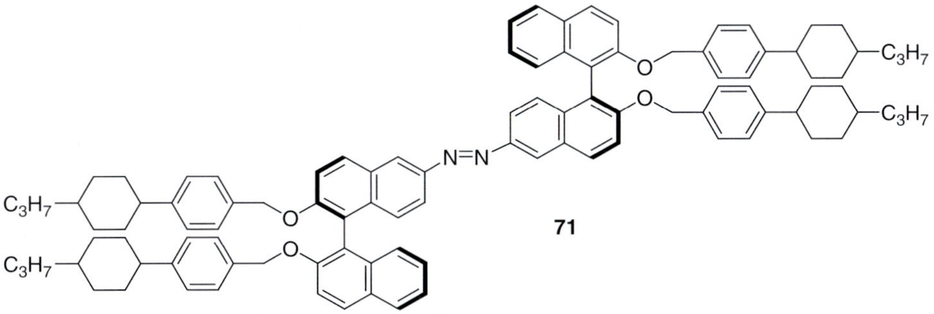

71

Scheme 8.24 Visible light-driven photochromic azobenzene-based chiral switch.

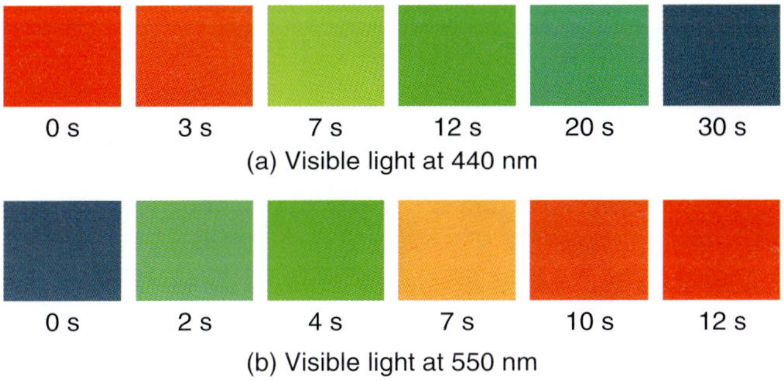

Figure 8.26 Full-range reflection color obtained by doping the mesogenic photochromic material **71** in a nematic LC host. (Reproduced with permission from [183]. Copyright 2012, American Chemical Society.)

could be tuned over a wide range that results reversible reflection colors across the visible spectrum (Figure 8.26).

New chiral LCs **72** containing cholesteryl group and azobenzene moieties have been synthesized and studied (Scheme 8.25) [184]. A sequence of interesting mesophases was observed in these materials, and reversible photoresponsive properties were demonstrated by UV irradiation. When doped into nematic LC hosts, they were able to induce cholesteric phases.

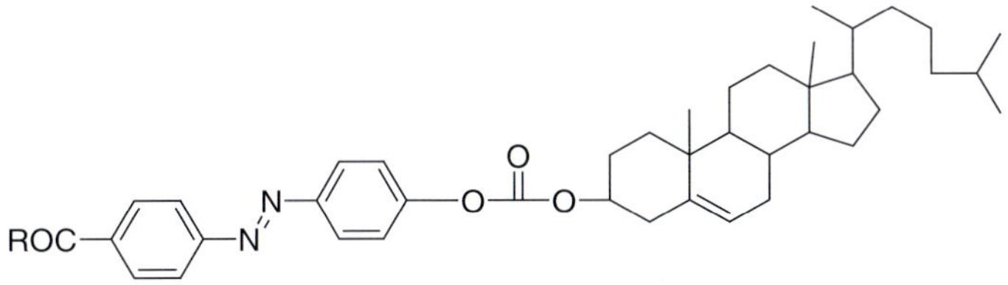

72 R = CH_3 or CH_3CH_2

Scheme 8.25 Cholesterol containing photochromic azobenzene LCs.

Axially chiral mesogenic azoarene **73–75** have been designed and synthesized by Li *et al.* (Scheme 8.26) [185, 186]. These molecular switches do not exhibit

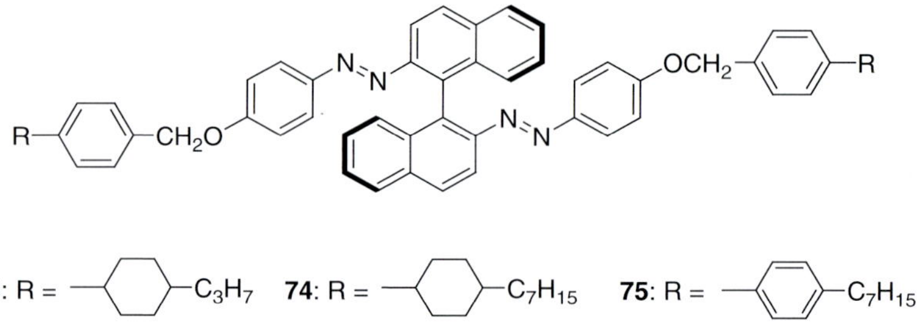

Scheme 8.26 Chemical structures of axially chiral photochromic azoarenes.

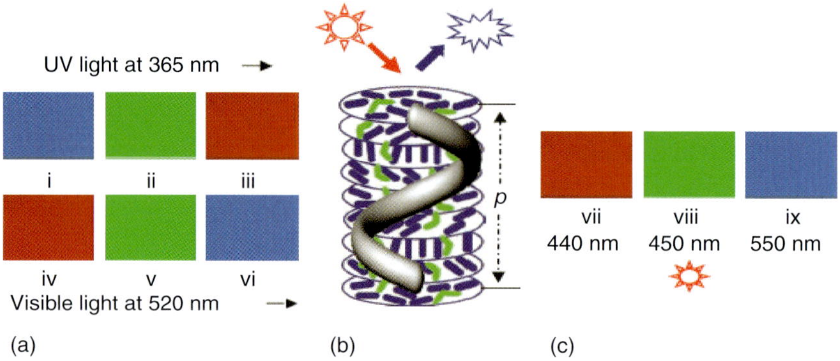

Figure 8.27 Dynamic and photostationary red, green, and blue reflection colors achieved in cholesteric LCs using compound **74**. (Reproduced with permission from [186]. Copyright 2011, Wiley-VCH Verlag GmbH & Co.)

any mesomorphism themselves but act as very efficient chiral dopants to induce photoresponsive cholesteric LCs. Due to their very high HTPs, a small quantity of these materials are needed to fabricate cholesteric LCs. Moreover, since the HTP values of these dopants have very large difference between the initial and final states, they enable wide tunability of the reflection color of the cholesteric phases. Notably, the compound **73** furnishes cholesteric LC with fast and reversible phototuning of reflection color across the entire visible region. Dynamic control of red, green, and blue reflection colors have been leveraged in photoresponsive cholesteric LCs by using compound **74** as a light-driven chiral molecular switch (Figure 8.27). Moreover, photostationary red, green, and blue colors have been achieved by irradiation with different wavelength of visible light. Multistimuli-responsive photodisplays have been demonstrated by using these materials in nematic LCs. Similarly, mesogenic azoarene derivatives **76** and **77** having both tetrahedral and axial chirality have been synthesized (Scheme 8.27) [187]. Due to better compatibility of these compounds in nematic hosts, photoresponsive cholesteric LCs have been fabricated. The reflection colors of these cholesteric LCs could be reversible tuned by light irradiation.

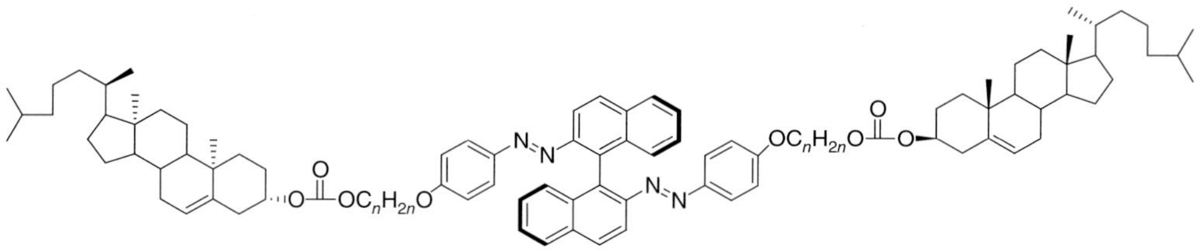

76 $n = 9$; **77** $n = 11$

Scheme 8.27 Azoarenes with tetrahedral and axial chirality.

8.3.5 Other Photochromic Liquid Crystals

New biindenylidenedione derivatives **78** containing two biphenyl substituents have been synthesized and characterized (Scheme 8.28) [188]. All of them exhibit photochromic properties, and the photochromic state may be returned to the original form thermally. Compound **78** displays monotropic smectic A mesophase, which has been confirmed by DSC and POM investigations. These are the first examples of liquid crystalline materials exhibiting light-induced radical behavior. Reversible thermal and photochemical switching of LC phases and luminescence in diphenylbutadiene-based mesogenic dimers consisting of a cholesterol moiety linked to a diphenylbutadiene chromophore through flexible alkyl chains have been studied by Abraham *et al.* Photoinduced *cis*–*trans* isomerization of the butadiene chromophore in these materials causes an isothermal phase transition from the smectic to the cholesteric phase [189]. Moreover, by photochemically controlling the *cis*:*trans* isomer ratio, the color of the cholesteric film of **79** can be tuned over the entire visible region.

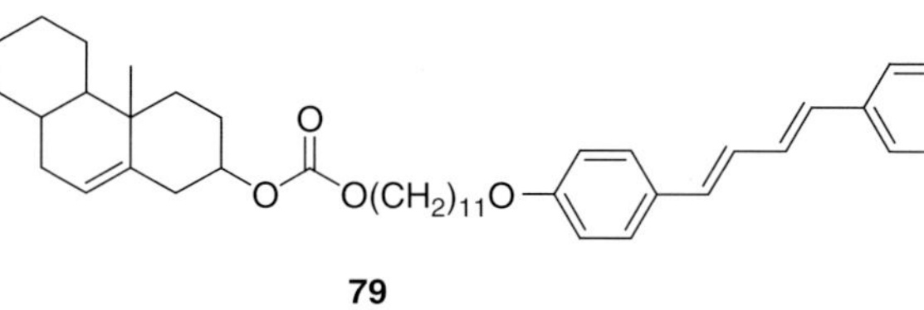

78

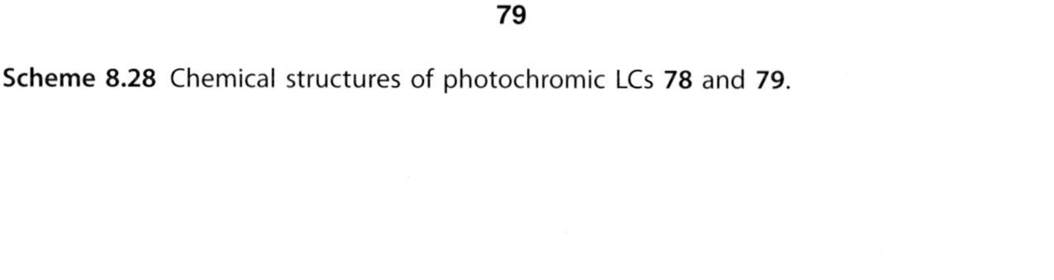

79

Scheme 8.28 Chemical structures of photochromic LCs **78** and **79**.

Over the recent years, in addition to photochromic LCs, different kinds of photoluminescent LCs have been synthesized and studied (Scheme 8.29) [190–200]. Such LC molecules deliberately incorporate photoluminescent moieties such as triazine, oxadiazole, thiadiazole, anthracene, and pyrene moieties into their chemical structure. Therefore, these materials exhibit photoluminescence in their solid and LC states as well as in solutions.

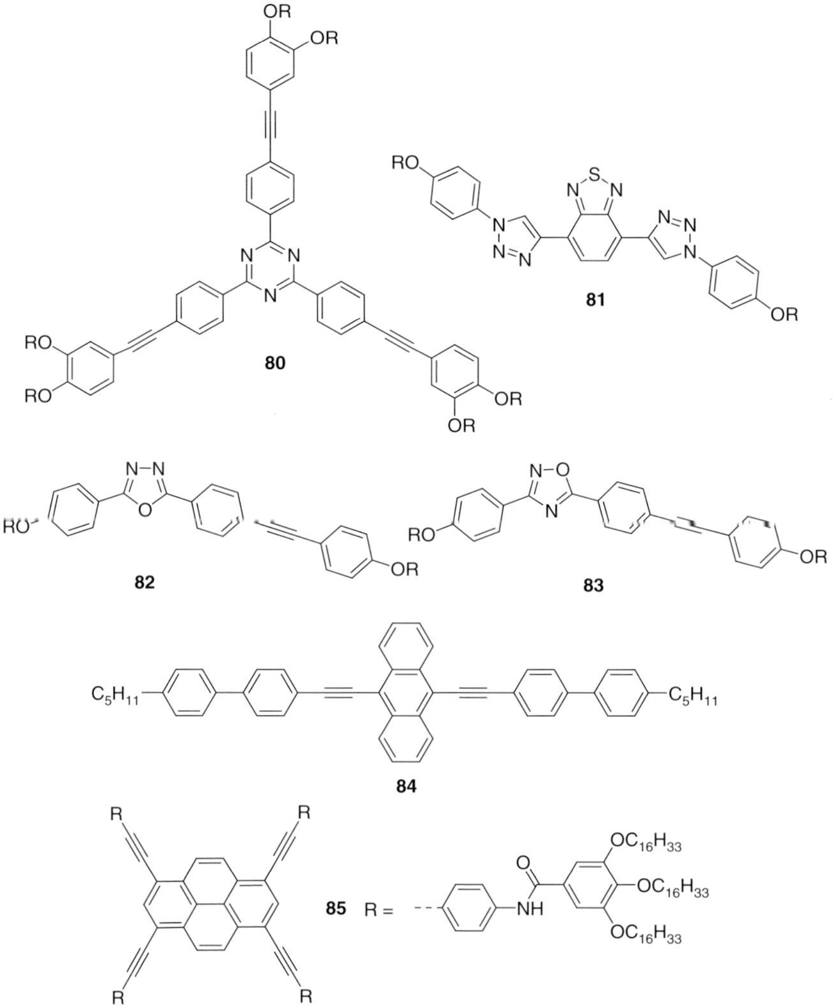

Scheme 8.29 Molecular structures of different types of photoluminescent LCs.

8.3.6
Conclusions and Outlook

Photochromic LCs based on a variety of photochromic scaffolds such as spiropyrans, spirooxazines, diarylethenes, and azobenzene derivatives have been designed and synthesized to study the synergetic effect of two classes of functional materials, that is, photochromic materials and LCs. Copolymers containing different quantities of spirooxazines display mesophase behavior and interesting photochromic properties. Chiral and achiral spirooxazine derivatives carrying mesogens have been synthesized. Photochromic spirooxazine-based chiral molecular switches have been realized that efficiently induce cholesteric phases in commercially available nematic LCs. Diarylethene-based photochromic LCs displaying distinct phase behaviors have been reported. However, appealing results have been achieved by designing photochromic diarylethene-based chiral molecular switches and doping them in nematic LC hosts. Photoresponsive cholesteric LCs exhibiting red, green, and blue reflection colors as well as light-driven helix inversion have been demonstrated. By doping diarylethene-based molecular switches, the spontaneous polarization of FLCs has been modulated. Azobenzene-based cross-linked LC polymers and copolymers have been demonstrated to show remarkable photomechanical properties and optical memory effects. In addition, discotic and bent-core LCs containing azobenzene moieties produce liquid crystalline materials with fascinating properties. Similar to the diarylethene photoswitches, azobenzene-based chiral molecular switches have enabled the fabrication of photoresponsive cholesteric LCs with promising properties. Recently, cholesteric LCs containing photochromic chiral molecular switches have been found to exhibit appealing applicable properties such as omnidirectional lasing and circularly polarized reflections when investigated in the form of spherical microshells and microdroplets [201, 202]. Combining nanomaterials such as upconversion nanoparticles with photochromic cholesteric LCs has also been shown to widen their scope in the context of pitch modulation and handedness inversion by using near-infrared (NIR) light irradiation [203–206]. Synthesis of materials exhibiting both photochromic and LC properties is challenging since their combination usually results in the loss of one of these properties. Overall, the photochromic LCs are interesting materials with promising properties [207, 208], but this area needs more systematic attention to explore these intriguing materials. Taking into account their present achievements and problems yet to be addressed, photochromic LCs still show a great potential in the area of advanced functional materials.

8.4 Photochromic Gels

8.4.1 Introduction

Organogels have attracted much attention in supramolecular chemistry and materials science in recent years [209–212]. They consist of 3D networks that are formed by self-assembly through noncovalent interactions, such as hydrogen bonding, hydrophobic–hydrophobic interactions, π–π stacking, and metal–ligand coordination. In the process of gelation, the gelator molecules self-assemble through noncovalent interactions to form fibrous architectures on the nanometer scale, which, in turn, builds up a micrometer-scale entangled 3D network entrapping or immobilizing organic solvent molecules, preventing the organic solvent molecules from flowing [213, 214]. Of particular focus on gelators, materials science are "smart gels," which show reversible changes in morphology or physical properties in response to various external stimuli, such as temperature, light, force, and chemical or sound [215]. Such responsive systems are highly desirable in thermo- and mechano-responsive sensor materials or applications like drug delivery or catalysis, or nano- and mesoscopic assemblies with interesting optical and electronic properties, and so on. So far, many functional gels have specific applications in light harvesting; energy and charge transfer; photo, enzyme, and switches; and chiral or molecular shape selection [216].

Among the stimuli, light is particularly attractive because it is noninvasive and can be delivered instantaneously to a precise location. The photoresponsive gels, which can change their properties upon light irradiation, is therefore currently a focus of attention. Several photoresponsive gels have been proposed so far with photoswitching units such as azobenzene, spiropyran, and spirooxazine, diarylethene, salicylideneaniline, imidazole, and stilbene [217, 218]. The organogels created with these photoresponsive properties such as sol–gel phase transition, gel–precipitation transition, color or fluorescence change, and morphological change by photochromic reaction of these moieties upon light irradiation. They are photoresponsive gels, namely, photochromic gels Those gels as photofunctional soft materials have generated enormous interest for basic scientific and technological in recent years. They have demonstrated that these gels may have potential applications in a number of areas including nanomaterials and delivery or modification agents for paints, inks, cleaning agents, cosmetics, polymers, drugs by photon controlled [219, 220]. The most commonly used photoresponsive chromophores for the design of photochromic gels are azobenzene, diarylethene, and spiropyran, spirooxazine, 2*H*-chromenes. salicylideneaniline, and so on.

8.4.2
Azobenzene Gels

Azobenzene is famous photochrome that undergoes *trans–cis* isomerization when irradiated with light tuned to an appropriate wavelength. The reverse *cis–trans* isomerization can be driven by light or occurs thermally in the dark. Its photochromic properties make it an ideal component of numerous molecular devices and functional materials. Azobenzene as photochromic building block in molecular design has been utilized as a light-triggered switch in a variety of polymers, surface-modified materials, molecular probes and sensor, molecular machines, optical storage devices, and metal-ion chelators. The change in geometry upon isomerization orients the molecules to perform a task, modulates interactions that change the structure of the bulk material, changes the spectroscopic properties, or moves a substituent that blocks or unblocks activity [221]. In photochromic gels fields, azobenzene derivatives were first designed to develop photochromic gels. The reversible photoinduced *trans–cis* isomerization of azobenzene unit significantly influences the gelation ability of the associated gelators. However, the large volume and polarity changes associated with *trans*-to-*cis* isomerization can induce considerable variation in the properties even with partial isomerization yield. Due to the large steric repulsion of the *cis* isomers, they have relatively weak association constants resulting in the destruction of the self-assembly. In generally, azobenzenes combined with steroid group, sugar, bisurea, semicarbazide groups, and melamine are developed to design the photochromic gels.

Shinkai and coworkers have made significant contributions to azobenzene-derived gelators. They designed and synthesized a lot of azobenzene with steroid skeleton gelators. They investigated the sol–gel phase transition as well as the chirality of supramolecular stacks using CD spectroscopy and have revealed the formation of supramolecular architectures of different shapes and sizes by SEM and TEM [222].

Nobuyuki Tamaoki groups reported azobenzenes **86a–c** (Figure 8.28) functionalized with two urethane moieties linked to two cholesteryl ester units form

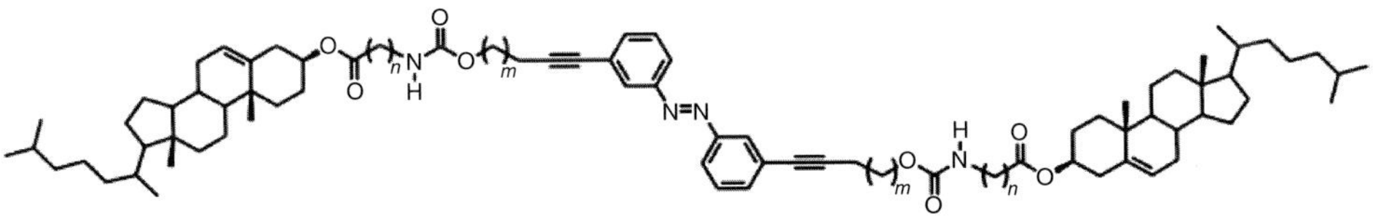

86a: $m = 3, n = 1$; **86b:** $m = 2, n = 2$; **86c:** $m = 3, n = 2$;

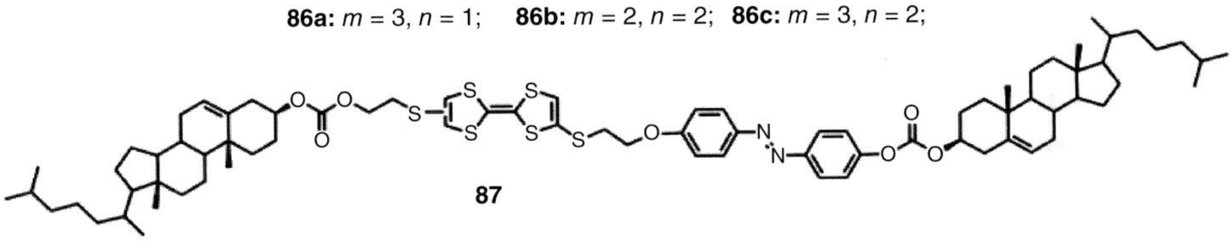

87

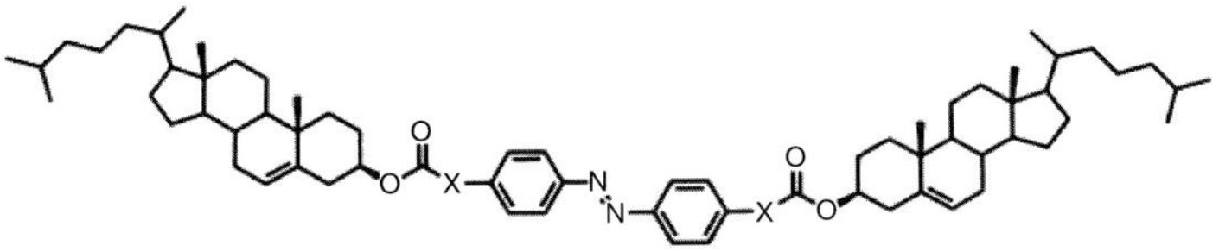

88a: X = 0; **88b:** X = $-O(CH_2)_2O-$; **88c:** $-O(CH_2)_6O-$

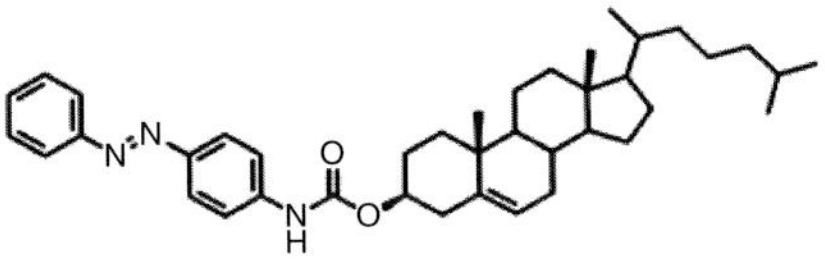

89

Figure 8.28 Chemical structures of photochromic gels based on azobenzene with teroid skeletons.

gels that exhibit sol–gel phase transitions upon photoirradiation as a result of *trans–cis* isomerization of the azobenzene units. During the sol–gel phase transitions, hydrogen bonds that are partly responsible for stabilizing the gels are broken or reformed [223]. In a bischolesterol-functionalized gelator **87**, with azobenzene moiety and a redox activity TTF moiety, exhibits both photo and redox control of the electronic properties [224]. Recently, a series of photoresponsive dicholesterol-linked azobenzene gelators **88a–c** with different spacer lengths have been reported [225]. Their gelation properties were affected by the spacer-link groups and cosolvent dramatically. Accordingly, azobenzene gelator **88b** with a spacer of two methylene units is found to be the best gelator. The addition of methanol is found to modulate the speed of gelation, photoresponsive property, and the morphology of the gel nanostructures [226]. A cholesterol imide based on azobenzene gelator **89** exhibited reversible sol–gel transition via photoisomerization of the azobenzene unit upon irradiation with UV and visible lights [227].

In addition, Shinkai *et al.* have reported the photochromic hydrogelators based on azobenzene with sugar bolaamphiphiles [228]. These bolaamphiphiles consist of two solvophilic aminophenyl sugar skeletons as the chiral aggregate forming site and a solvophobic azobenzene moiety as the π–π stacking site. The sugar-coated nanofibers formed by a combination of azobenzene and disaccharide lactones **90a–c** in Figure 8.29 provided a bioactive interface for cell attachment and

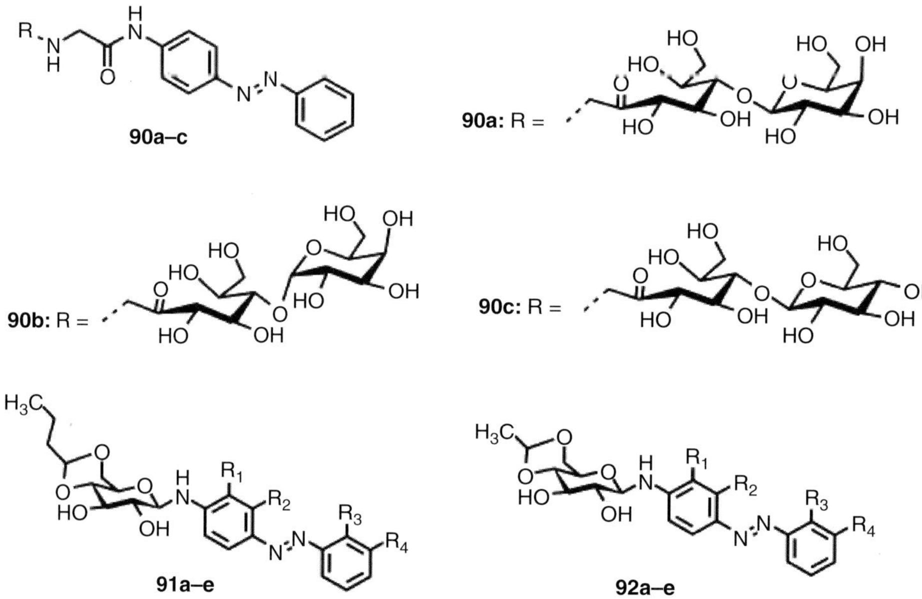

a: $R_1 = R_2 = R_3 = R_4 = H$
b: $R_1 = F, R_2 = H, R_3 = F, R_4 = H$
c: $R_1 = Cl, R_2 = H, R_3 = Cl, R_4 = H$
d: $R_1 = H, R_2 = Cl, R_3 = H, R_4 = Cl$
e: $R_1 = Br, R_2 = H, R_3 = Br, R_4 = H$

Figure 8.29 Chemical structures of photochromic gels based on azobenzene with sugar skeletons.

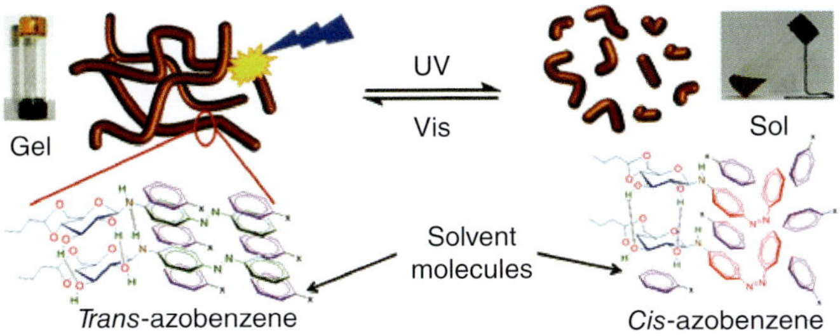

Figure 8.30 Illustration of a phase-selective gelation for the removal of small amounts of toxic solvents from water under light irradiation. (Adapted with permission from [230]. Copyright 2011, American Chemical Society.)

exhibited lectin binding [229]. In addition, the hydrogels exhibited a reversible sol–gel transition in response to temperature and UV irradiation.

The latest report on a photochromic gelator pertains to a sugar-based amphiphilic system containing an azobenzene moiety **91** and **92** [230]. The partial *trans–cis* isomerization of the azobenzene moiety allows photoinduced chopping of the entangled long fibers to short fibers, resulting in controlled fiber length and gel–sol transition. These gelators facilitate phase-selective gelation of aromatic solvents from aqueous emulsion. Such a phase-selective gelation is useful for the removal of small amounts of toxic solvents from water as shown in Figure 8.30.

Ikeda's groups have reported a chiral azobenzene **93** containing a cyclic syn-carbonate moiety that self-assembles to form a photoswitchable organogel [231]. They have also demonstrated photoresponsive properties of anisotropic gels composed of **94** and a room temperature nematic LC 4-cyano-4′-pentyl-biphenyl [232]. (Figure 8.31) The photoinduced gel–sol transition of the physical gel upon UV irradiation at room temperature leads to the *trans–cis* photoisomerization, which induces the transition from nematic LC gel (Figure 8.31b) to a cholesteric sol phase (Figure 8.31c). Subsequent visible light irradiation or keeping the sol at room temperature causes *cis–trans* photoisomerization of the azobenzene moiety leading to regelation. During this process, the cholesteric molecular alignment in the sol phase (Figure 8.31c) behaves as a template for the aggregation of gelators into cholesteric LC gel (Figure 8.31d). The polarity change of the azobenzenes in **94** should be a key factor to the induction of photostimulated on–off switching of the hydrogen bonding, which eventually led to the gel–sol transition [233, 234].

The azobenzene-appended melamine **95** and the complementary H-bonding barbiturate (**A**) or cyanurate (**B**) derivatives (Figure 8.32) have been exploited for the preparation of photoresponsive rosette assemblies [235]. In aliphatic solvents, a rosette possessing the sterically bulky tridodecyloxyphenyl substituent in the barbiturate component **A** does not hierarchically organize into higher order columnar aggregates. However, the sterically nondemanding *N*-dodecylcyanurate **B** results hierarchically organized elongated columnar fibrous aggregates in

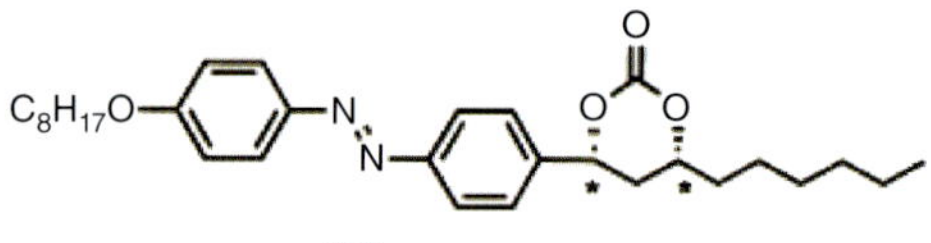

93

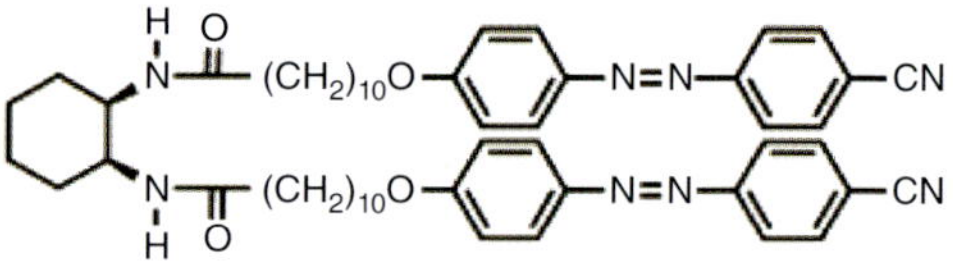

94

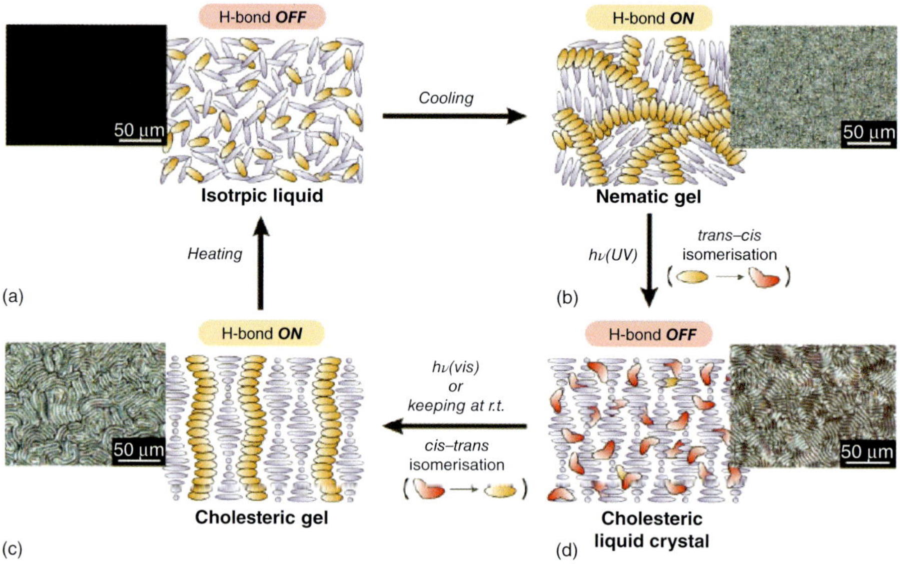

Figure 8.31 POM pictures and the respective schematic illustration of photoinduced structural changes in LC physical gels consisting of 4-cyano-4′-pentyl-biphenyl containing 3 wt% of **94**: (a) isotropic liquid state at 120 °C; (b) nematic gel state at room temperature; (c) cholesteric LC phase (LC sol state) at room temperature after UV irradiation of the nematic gel for 15 min; and (d) cholesteric gel state at room temperature after keeping the cholesteric LC phase. (Adapted with permission from [233]. Copyright 2003, Wiley-VCH.)

cyclohexane, which eventually leads to the formation of an organogel. Dynamic light scattering and UV–vis studies revealed that the dissociation and the reformation of columnar aggregates could be controlled by the *trans–cis* isomerization of the azobenzene moiety.

The gelator **96** showed a positive CD spectrum in polar solvents while it is inverted to a negative spectrum in nonpolar solvents (Figure 8.33). The gelation exhibits interesting responsive to temperature, photoirradiation, and polarity of the solvents. The chiroptical switching properties of a gel with nanotubular morphology formed by the coassembly of a lipid gelator and an azobenzene derivative are also reported [236].

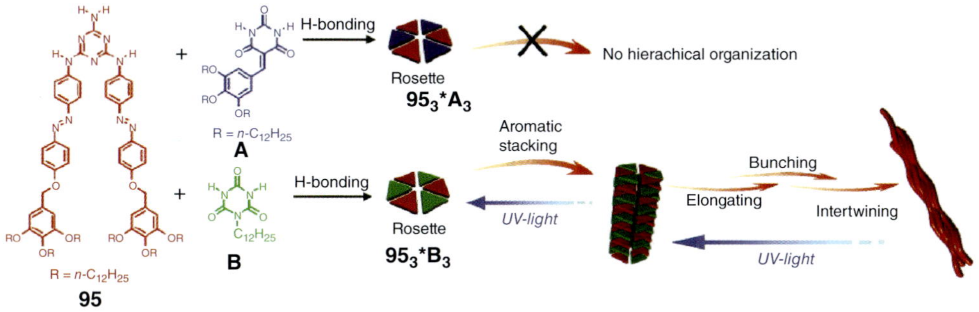

Figure 8.32 Schematic representation showing aggregation of the melamine–azobenzene conjugate (**95**) with barbiturate (**A**) and cyanurate (**B**) derivatives. Rosette $\mathbf{95_3{\cdot}B_3}$ can hierarchically organize into intertwined fibers (red arrows). The fiber formation can be regulated by light (blue arrows). (Adapted with permission from [235]. Copyright 2005, American Chemical Society.)

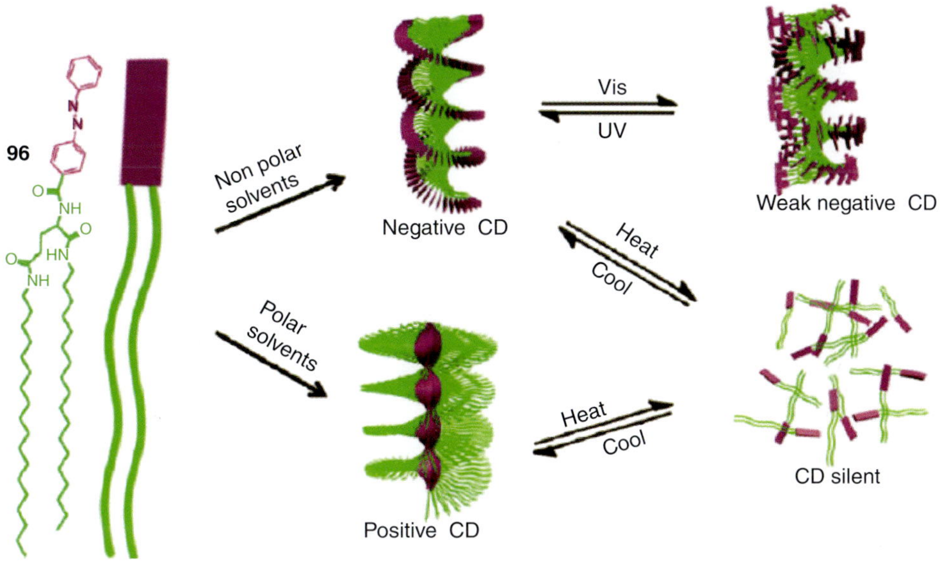

Figure 8.33 Illustration of the multichannel supramolecular chiroptical switches composed by **96**. (Adapted with permission from [236]. Copyright 2012, American Chemical Society.)

8.4.3 Spiropyran and Spirooxazine Gels

Spiropyrans and spirooxazines are two well-known photochromic compounds with the spiroring structure. The photochromic reaction of spiropyrans and closely related spirooxazines involves reversible photochemical cleavage of the C–O bond of the spirounit. Because of their interesting photochromic properties, they have been investigated for the design of optical memories, switches, and displays [134, 237]. Despite numerous studies on spiropyrans, spirooxazines, and their transition metal complex systems known in solutions, corresponding studies on their photochromic organogels as well as those of their related chromene systems are rare. The first report on the spiropyran gel is reported by Hachisako groups [238]. They have determined the critical aggregation concentrations of organogelators by following the kinetics of the thermal merocyanine–spiropyran isomerization of L-glutamic acid-derived lipids with spiropyran head groups **97a–c** (Figure 8.34).

A photochromic spiropyran with a cholesterol moiety gel **98** was reported, which dissolved in carbon disulfide (CS2) and benzyl alcohol (BA) and formed a homogeneous gel at room temperature [239]. The gels exhibit purple color due to the strong H-bonding interaction between the merocyanine form of **98** and BA molecules. UV–vis light irradiation successfully promote rapid and reversible color change between purple and yellow by photoisomerization of the spiropyran moiety.

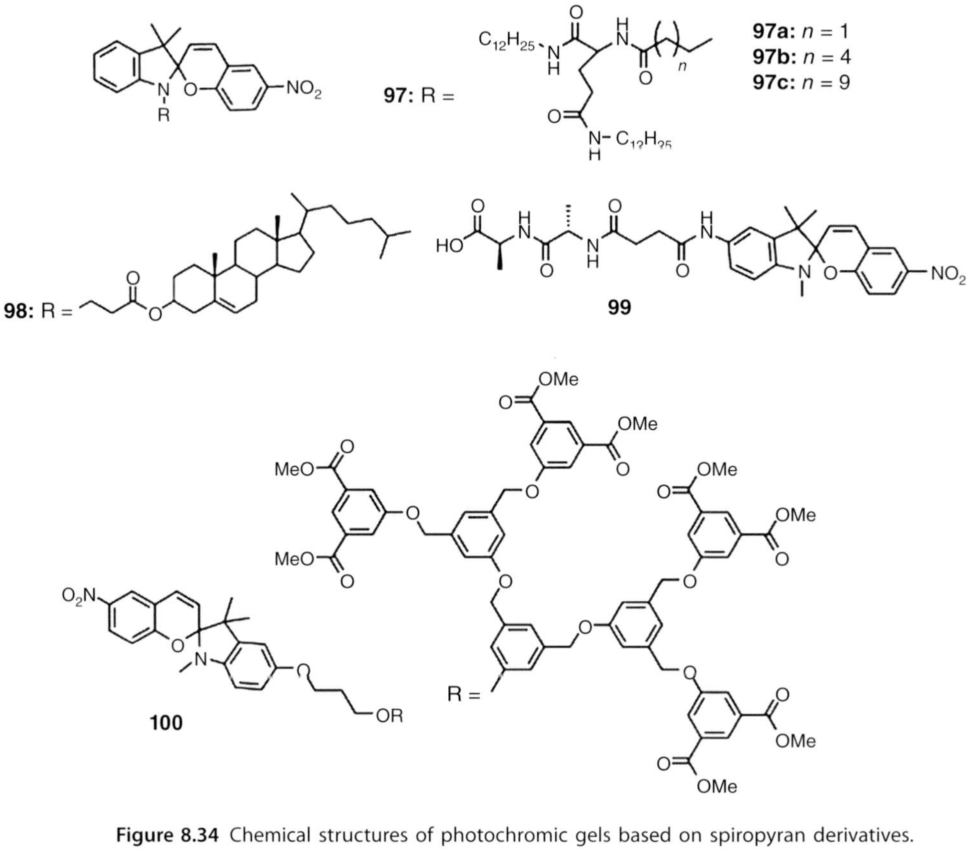

Figure 8.34 Chemical structures of photochromic gels based on spiropyran derivatives.

The spiropyran-linked dipeptide molecule **99** is reported to form a hydrogel upon photoisomerization into a merocyanine form [240]. In addition, the hydrogel is found to undergo gel-to-sol transition upon interaction with vancomycin due to the strong interaction of the latter with peptide unit of the gelator, thereby weakening the supramolecular interactions responsible for the gelation. The spiropyran-derived dendron molecule **100** shows different self-assembling properties depending on the processing conditions [241]. For instance, cooling of a hot **100** solution of 80–30 °C followed by UV irradiation to the open merocyanine leads to the formation of nano-/microsized particles, while cooling into 0 °C results in fibrous morphology and gel. Visible light irradiation of the gel caused merocyanine–spiropyran isomerization and gel-to-sol transition.

Doping of spiropyran into an organogel system based on 4-*tert*-butyl-1-phenylcyclohexanol was found to increase the lifetime of the photomerocyanine and also stabilize the organic photochromic material [242]. Recently, Raghavan's groups have reported photorheological properties of a gel prepared from a lecithin/sodium deoxycholate reverse micelles doped with spiropyran [243]. The gel consisting of long wormlike reverse micelles changed its dimension upon UV-induced isomerization of spiropyran to merocyanine and caused about 10-fold decrease in the viscosity. When the UV irradiation was switched off, merocyanine was reverted back to spiropyran form and the viscosity recovered its initial value. This cycle can be repeated several times without loss of response.

Spirooxazines belong to another interesting class of materials that is known to show photochromic and acidichromic properties. Spirooxazines have been shown to possess high fatigue resistance and excellent photostability and pH stability [244]. A series of photochromic spirooxazines of gallic acid **101a,b** and cholesterol derivatives **102** form photoresponsive gels [245] (Figure 8.35). In the presence of *p*-toluenesulfonic acid, the gelation ability of these molecules is further enhanced owing to an acid-mediated ring opening of the photochromic moiety. Kinetic studies revealed that the rate of bleaching of the open to closed form is much slower in the gel state when compared to that in the solution.

8.4.4 Diarylethenes Gels

Diarylethenes derivatives are distinct class of photochromic system that can undergo a reversible ring-closure reaction upon irradiation with UV and visible light, which are the most promising compounds because of their excellent fatigue resistance and thermal stability. The open-ring and closed-ring isomers of the diarylethenes differ from each other not only in their absorption but also in various physical and chemical properties, such as luminescence, refractive index, oxidation/reduction potentials, chiral properties, magnetic interactions, and so on [246]. Feringa's groups have conducted extensive studies on photoresponsive as well as chiroptical properties of organogels derived from dithienylethene-based molecules **103** and **104** (Figure 8.35) [247–251]. One of these photochromic organogelator **103a** showed the reversible optical

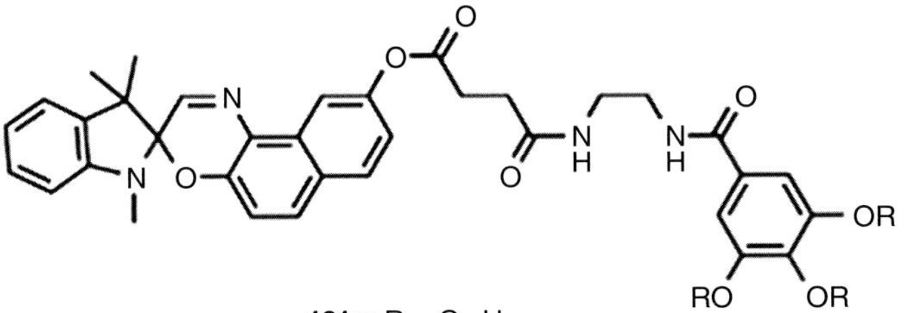

101a: R = $C_{12}H_{25}$
101b: R = $C_{16}H_{33}$

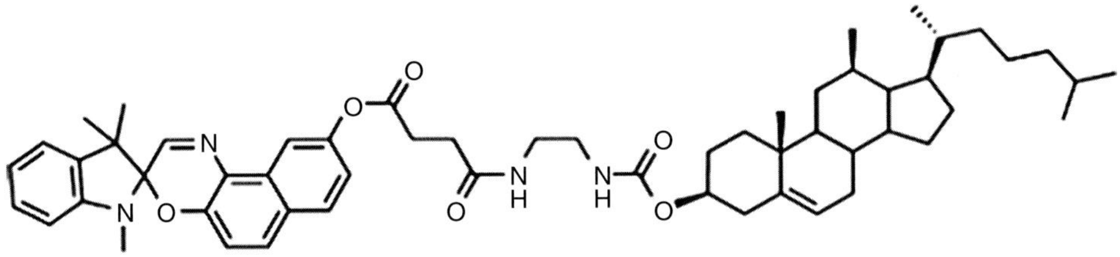

102

Figure 8.35 Chemical structures of photochromic gels based on spirooxazine derivatives.

transcription of supramolecular chirality into molecular chirality. The optical switching between different supramolecular chiral aggregates and the interplay of molecular and supramolecular chirality in these systems is attractive for designing molecular memory systems and smart functional materials (Figure 8.36).

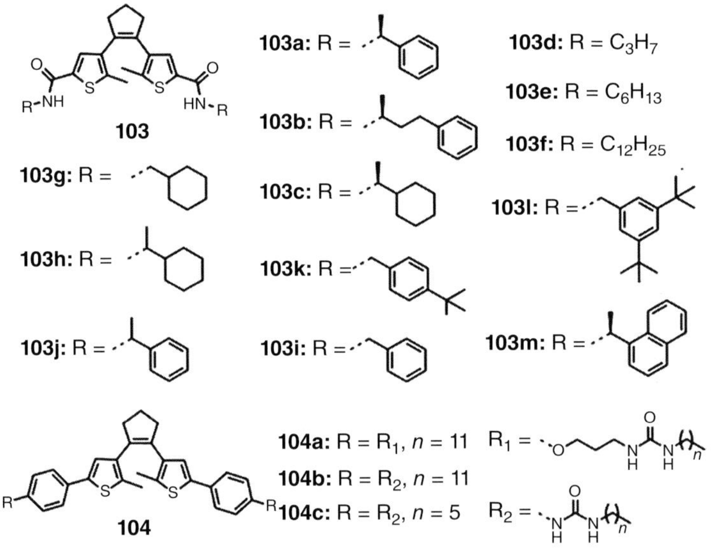

Figure 8.36 Chemical structures of photochromic diarylethene gels with photoresponsive and chiroptical properties.

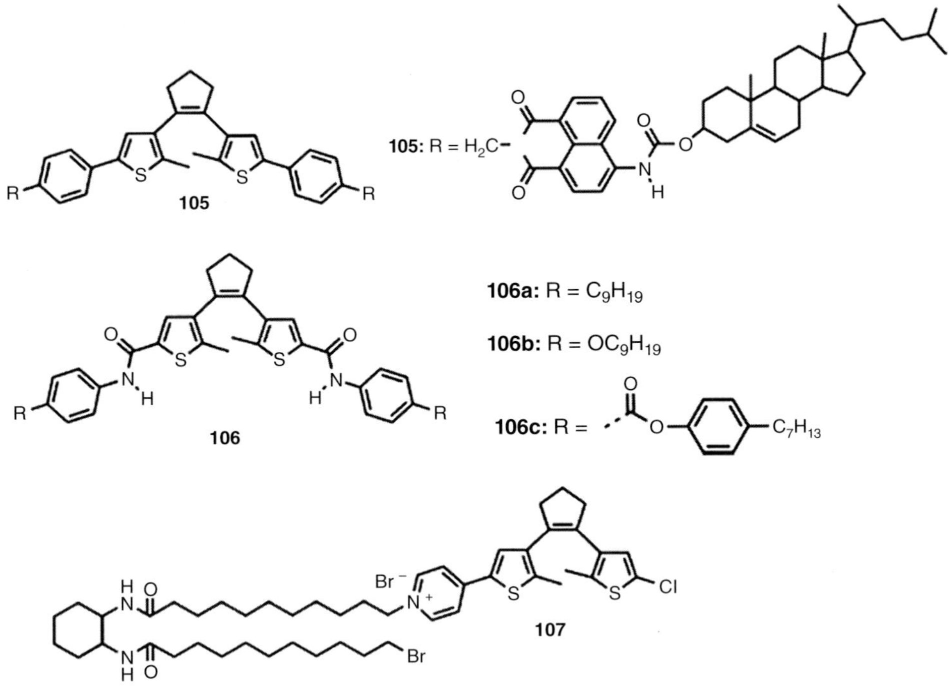

Figure 8.37 Chemical structures of photochromic fluorescent dithienylethene gels.

Diarylethenes are also found in a variety of π-gelators 20–22 (Figure 8.37), which show reversible photoinduced changes in the gelation and electronic properties with multiswitch function. For example, Tian groups have reported a multiple switching fluorescent photochromic organogel based on bisthienylethene-bridged naphthalimides **105** by light, thermal, fluoride anions, and proton stimuli (Figure 8.38) [252]. It exhibits excellent photochromic properties and defined thermo-reversible properties in organogel system, in the same time, by taking advantage of the F^-, and proton can induce obviously different absorption and fluorescence spectra of **105** under sequential alternating UV–vis light irradiation, a complicated multiple switch is realized, which makes it promising application in the fields of opto- and electronic smart materials, logic gate, nanomachines, fluorescence sensors, and other molecular photonic devices.

Li's groups synthesized a series of photochromic gels **106** based on dithienylcyclopentene amides with a phenylene unit as a bridge between the amide and long alkyl chain [253]. These gelator molecules were found to be able to induce gelation in apolar solvents to form entangled networks driven by intermolecular hydrogen bonding along with π–π interactions. Their excellent reversible photochromism and the thermally reversible property in sol–gel transition were exhibited. A fluorescent organogel based on photochromic dithienylethene **107** was also reported, whose optimal excitation wavelength results in little structural change in both open and closed isomers of diarylethene, thus presenting the first

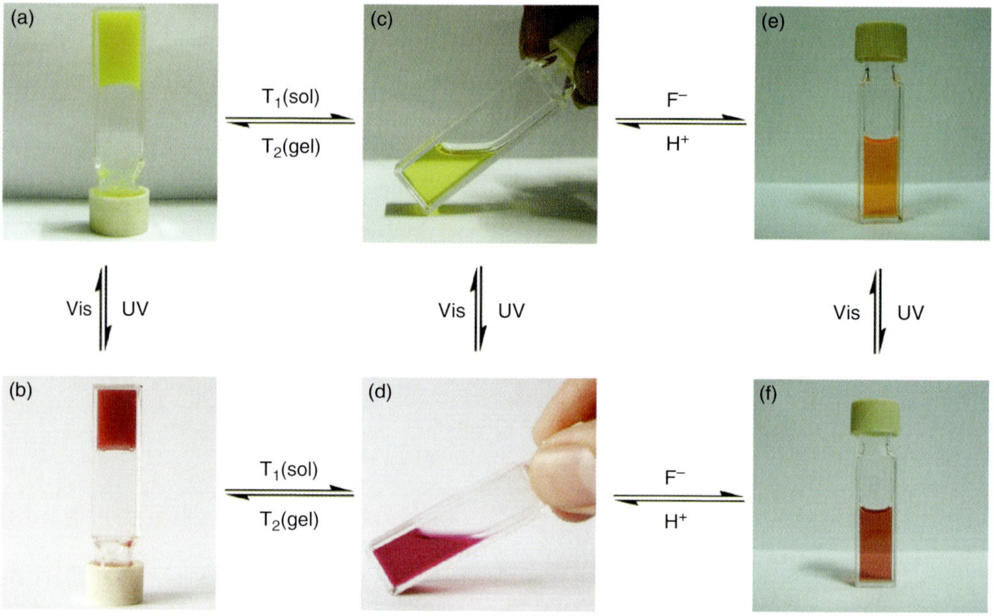

Figure 8.38 Multiple switching images of compound **105** in the cooperative effect of light, thermal, fluoride anions, proton. (a) Gel(open); (b) gel(closed); (c) sol(open); (d) sol(closed); (e) sol(open) + F^-; and (f) sol(closed) + F^-. (Adapted with permission from [252]. Copyright 2006, The Royal Society of Chemistry.)

example of a fluorescent switch with nondestructive readout ability in the gel state [254].

Recently, some diarylethene gels **108–111** are reported as showed in Figure 8.39. A new photochromic dithienylethene functionalized with a hydrophobic cholesterol unit and a hydrophilic poly(ethylene glycol)-modified pyridinium group **108** was demonstrated by light-controlled formation of vesicles and supramolecular organogels [255]. The incorporation of dithienylethene allows for photochemical control over vesicle formation in water and for light-controlled self-assembly of organogel fibers in apolar aromatic solvents. These features demonstrate that light control of a supramolecular structure can be achieved in aqueous as well as organic media, and that this ability can be present in a single molecule. This opens the way toward the effective development of new strategies in soft nanotechnology for applications in controlled chemical release systems.

A novel multiresponsive gelator based on a tetrapeptide–dithienylcyclopentene conjugate **109** was designed and synthesized [256]. This organogel exhibits a smart multistimuli-responsive behavior upon exposure to external stimuli such as temperature, light, chemicals, and mechanical force. Moreover, in the presence of catechol, this gelator forms a more robust organogel, with a dramatic change of the assembly behavior and rheological properties. Feringa's groups reported

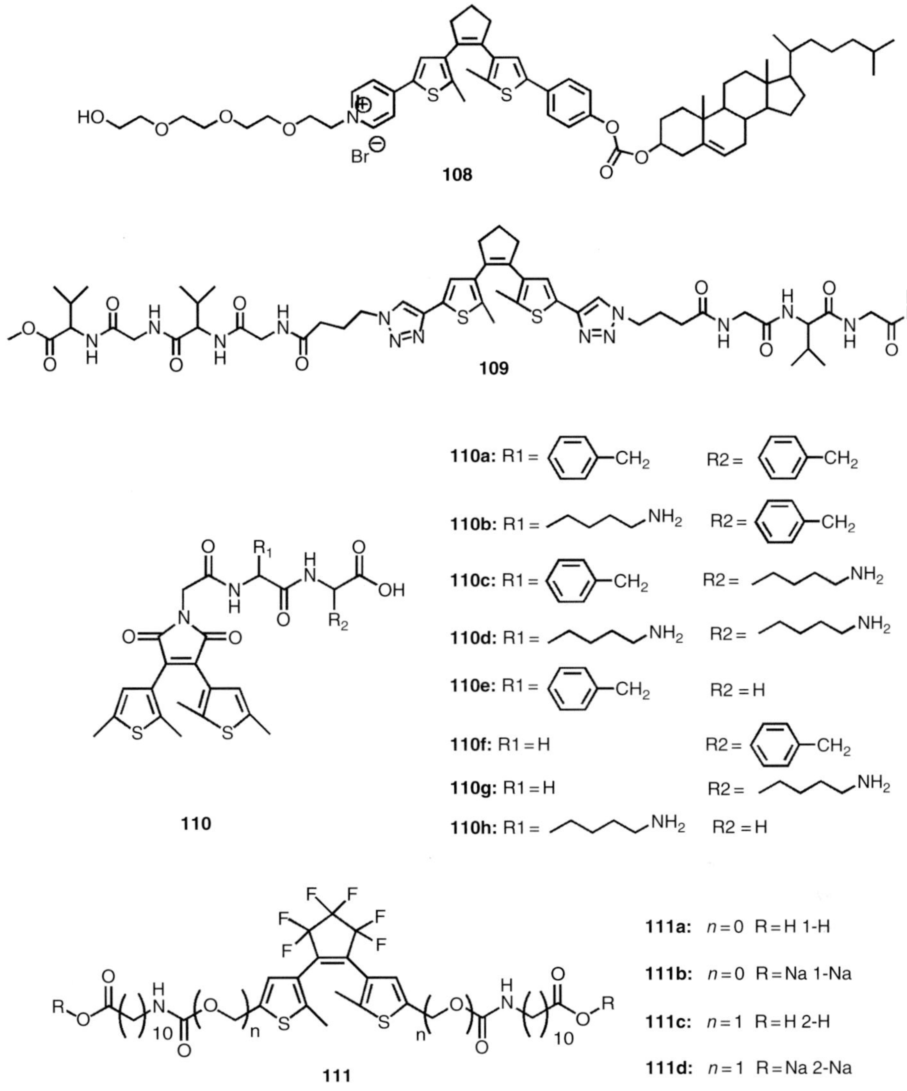

Figure 8.39 Chemical structures of functional photochromic diarylethene gels.

a dithienylethene-based rewritable hydrogelator **110** incorporated into a hydrogelating system based on a tripeptide motif [257]. The resulting hybrid system exhibits both a photochromic response and the ability to gelate water under acidic and neutral conditions. In addition, multi-addressable supramolecular gels based on linear amino acid and 1,2-bisthienylperfluorocyclopentenes covalently linked to two aggregative side arms derived from 11-aminoundecanoic acid **111** have been designed to self-assemble leading to the formation of supramolecular gels, which could be reversibly and independently revealed or suppressed upon acidity, temperature changes, and also light irradiation [258].

8.4.5 Naphthopyran Gels

Naphthopyran are known to display interesting photochromic properties based on their photoirradiation-induced fatigue-resistant reversible color change [259]. The closed form is colorless, whereas its photoisomerized open form is colored. The latter is thermally unstable and reverts back to its original form by an electrocyclization process. Pozzo's groups reported light- and pH-sensitive properties of organogels of 2*H*-chromene derivative **112** (Figure 8.40) [260]. The neutral carboxylic acid form was found to be readily soluble in polar organic solvents, which upon addition of NaOH turned to gel. Furthermore, on irradiation at 366 nm, a yellow color was developed and the gel started to flow upon inversion. In the dark, a colorless viscous solution was formed, which on heating followed by

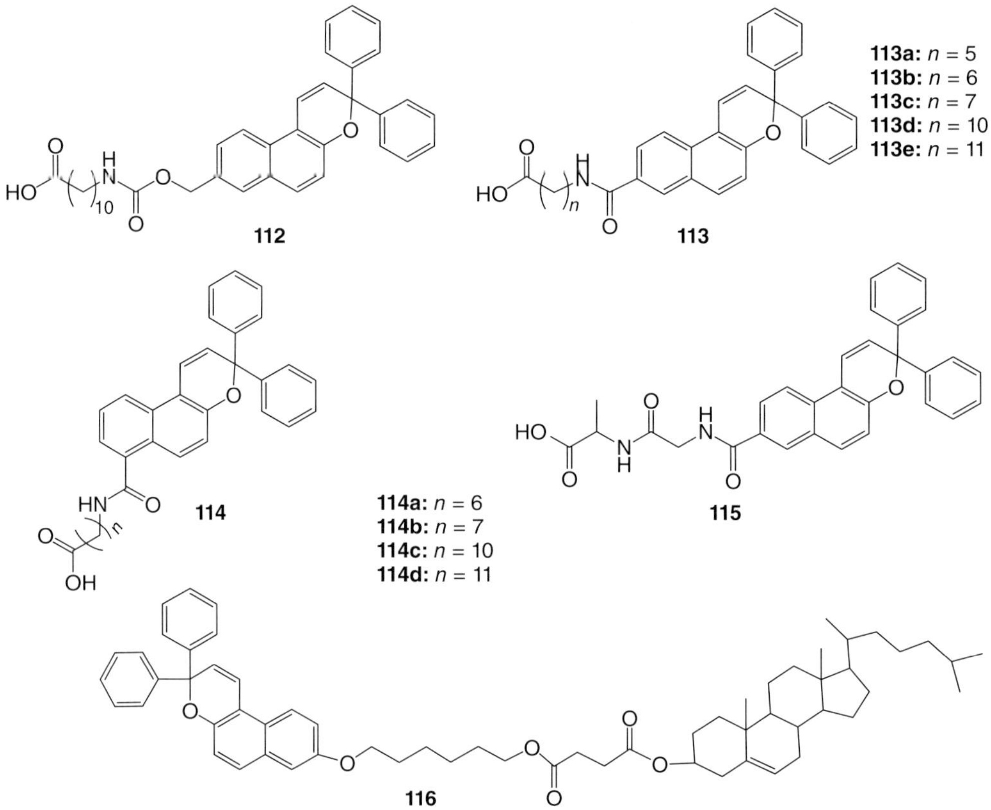

Figure 8.40 Chemical structures of photochromic naphthopyran gels.

cooling regenerated the original gel. These transitions were caused by light irradiation ring opening of the colorless cyclic form to the colored acyclic form, which partially disrupts the gel structure due to its incompatibility with the network. The acyclic form is, however, thermally unstable and returns to the cyclic form upon heating and cooling, resulting in the formation of a gel. A series of naphthopyran derivatives functionalized with *N*-acyl-1,ω-amino acids **113**, **114** and dipeptide **115** (Figure 8.40) also reported to form gels in presence of sodium salts [261]. A cholesterol-bridge-naphthopyran dyad **116** can readily self-assemble into gels under ultrasound radiation, and the formed gels easily transfer to solution by heat. It displayed the normal photochromism in both solution and gel states. The kinetic results confirm that the colored merocyanine in gels show a slower fading speed than that in solution due to the compact aggregation of **116** molecules in gels [262].

8.4.6 The Other Photochromic Gels

Some other system photochromic gels such as [2.2]paracyclophane-bridged imidazole dimmer, Schiff base, are also developed as shown in Figure 8.41.

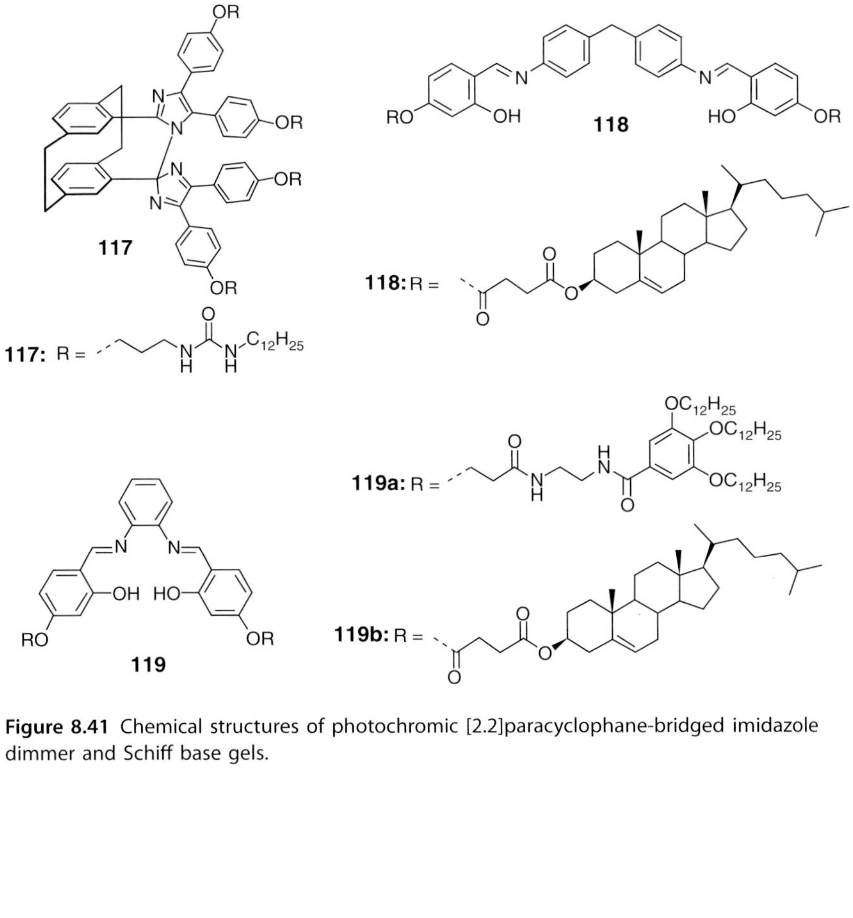

Figure 8.41 Chemical structures of photochromic [2.2]paracyclophane-bridged imidazole dimmer and Schiff base gels.

Photochromic [2.2]paracyclophane-bridged imidazole dimmer is a kind of instantaneous coloration upon exposure to UV light and rapid fading in the dark [263]. This behavior can be attributed to the photogenerated homolytic cleavage of the C–N bond between the imidazole rings, resulting in the formation of the colored triphenylimidazolyl radical. Abe's groups have prepared a novel organogelator **117** possessing [2.2]paracyclophane-bridged imidazole dimer as a photochromic unit. They demonstrated the inherently fast photochromic reactions in the gel phase. This system makes it possible to realize simultaneous morphological/physicochemical control only under light irradiation, which will create a new class of photo-oriented intelligent materials system [264].

The cholesterol-appended salicylideneaniline derivatives **118** and **119a** undergo isomerization of salicylideneaniline and keto–enol tautomerism under UV light, accompanied by gel–sol transition [265, 266]. Interestingly, replacement of the cholesterol group and ester linkage with 3,4,5-tridodecyloxybenzoic acid unit and amide linkage resulted in a gelator **119b** that showed thermochromic/nonphotochromic behavior in the self-assembled state. In addition, these molecules showed AIEE in the gel state due to J-aggregation and inhibition of nonradiative decay via intramolecular rotation [267].

Stilbenes are recognized as one of the thoroughly studied photochemical systems. Upon direct excitation of stilbenes, photoisomerization is confined entirely to the singlet excited state and competes with fluorescence on the *trans* side and photocyclization on the *cis* side. Stilbenes in the *cis* form are not thermally stable and return to the *trans* form in the dark [268]. Stilbenes are also suitable photochromic moieties for the design of "smart" gelators because of their photoresponsive conformational changes. A lot of stilbene gels such as appended cholesterol, surfactant, and triethoxysilane are investigated. For instance, Eastoe *et al.* have made a photoresponsive organogel system by refluxing stilbene-containing photosurfactant **120**. Upon UV-light irradiation, this gel transformed to sol phase with a spatial control and has been demonstrated within the sample as showed in Figure 8.42 [269].

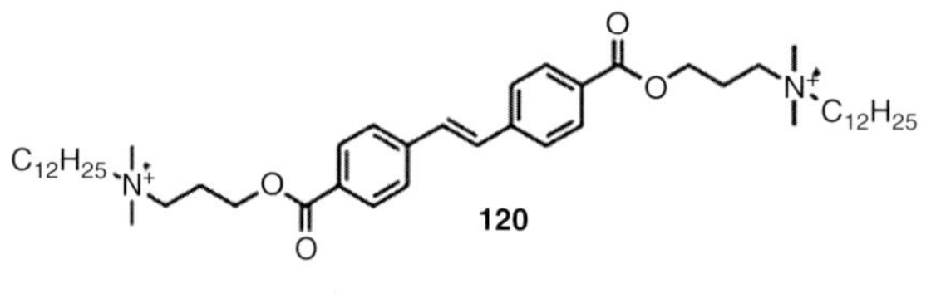

120

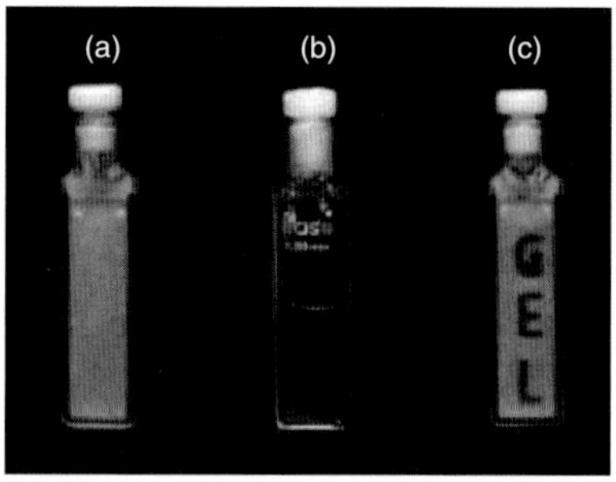

Figure 8.42 Photoinduced gel-to-sol transition of **120**–*N*,*N*′-dimethyldodecylamine organogel in toluene-d_8: (a) initial gel state, (b) after irradiation, and (c) after irradiation through a mask. (Adapted with permission from [269]. Copyright 2004, The Royal Society of Chemistry.)

Recently, Chen *et al.* reported a photochromic helicene gelators **121** based on gallamide-containing pseudoenantiomeric helicene pair bearing a (10*R*,11*R*)-dimethoxymethyldibenzosuberane core, which can self-assemble by intermolecular amide H-bonding and π–π stacking into bundled helical fibers with helical tunnels of complementary helicity [270]. The helicenes undergo excellent complementary photoswitchings of ternary logic at 280, 318, and 343 nm through (−)-gel–sol-(+)-gel interconversion as shown in Figure 8.43. Upon irradiation of pure (*M*)-**121**-containing gel by 318 nm light, the gel turned into a homogeneous solution after 60 min, with the overall dopant composition reaching (*M*)-**121**/(*P*)-**121′** = 60/40. Further irradiation of the homogeneous solution at 280 nm for 60 min led to an overall dopant composition of 25/75, where the complementary helical gel state started to form, and the dopant composition gradually shifted to pure (*P*)-**121′** after another 6 h of irradiation. Conversely, irradiation of the homogeneous solution [(*M*)-**121**/(*P*)-**121′** = 50/50] at 343 nm led to a dopant composition of 69/31, in which the original gel state began to be restored, and the dopant composition gradually returned to the initial (*M*)-**121** state [(*M*)-**121**/(*P*)-**121′** = 90/10] after another 8 h of irradiation. Thus, the helical superstructures of the bundled helical fibril tubes can be controlled in a complementary and reversible manner by exposing the gel materials to either 280 or 343 nm irradiation.

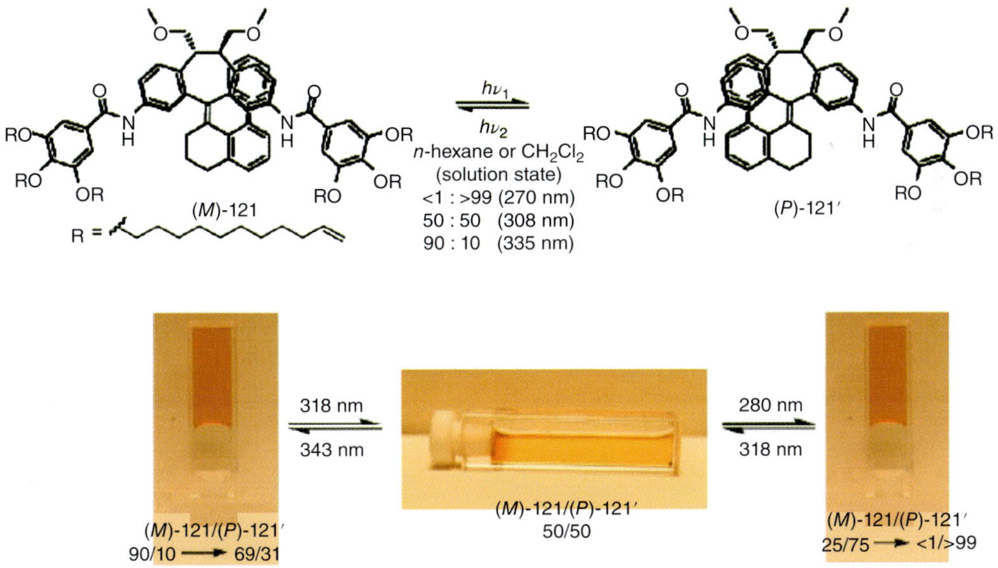

Figure 8.43 Photoisomerization profiles of helicenes (*M*)-**121** and (*P*)-**121′** and photographs of the gel at ambient temperature after irradiation of the original (*M*)-**121**-containing gel at 318 nm (middle) and then at 280 nm (right) or 343 nm (left). (Adapted with permission from [270]. Copyright 2013, American Chemical Society.)

8.4.7
Conclusion

In summary, we described recent developments of photochromic gels in the fields of stimuli-responsive soft materials. All kinds of photochromic compounds were almost induced to organogelators to prepare light stimuli-responsive gels. However, molecular design of photochromic gels is still not straightforward since incorporation of photochromic into gelator frameworks may perturb the self-assembly processes in a destructive way. Efforts for a deeper understanding of the processes and mechanism of gelation have helped scientists to design new and novel photochromic molecular systems that form gels comprised of exotic structures with controlled size, shape, and photoresponsive properties such as sol–gel phase transition, gel–precipitation transition, color or fluorescence change, and morphological change by photochromic reaction of these moieties upon light irradiation. Those photochromic gels may have potential applications in a number of areas including nanomaterials and delivery or modification agents for paints, inks, cleaning agents, cosmetics, polymers, drugs by photon controlled. More structural and functional photochromic gels are expected to be developed.

References

1. Dürr, H. and Bouas-Laurent, H. (eds) (2003) *Photochromism: Molecules and Systems*, Elsevier, Amsterdam.
2. Irie, M., Fukaminato, T., Matsuda, K., and Kobatake, S. (2014) Photochromism of diarylethene molecules and crystals: memories, switches, and actuators. *Chem. Rev.*, **114** (24), 12174–12277.
3. Irie, M. (1983) in *Molecular Models of Photoresponsiveness* (eds G. Montagnoli and B.F. Erlanger), Plenum Press, New York, pp. 291–312.
4. Irie, M. (1990) Photoresponsive polymers. in *Adv. Polym. Sci.*, vol. **94**, Springer-Verlag, Berlin Heidelberg, pp. 27–67.
5. Irie, M. (1993) Stimuli-responsive poly(*N*-isopropylacrylamide). Photo- and chemical-induced phase transitions. in *Adv. Polym. Sci.*, vol. **110**, Springer-Verlag, Berlin Heidelberg, pp. 49–65.
6. Irie, M. (2000) Diarylethenes for memories and switches. *Chem. Rev.*, **100** (5), 1685–1716.
7. Krongauz, V.A. (2003) in *Photochromism: Molecules and Systems* (eds H. Dürr and H. Bouas-Laurent), Elsevier, Amsterdam, pp. 793–821.
8. Ercole, F., Davis, T.P., and Evans, R.A. (2010) Photo-responsive systems and biomaterials: photochromic polymers, light-triggered self-assembly, surface modification, fluorescence modulation and beyond. *Polym. Chem.*, **1** (1), 37–54.
9. Irie, M., Fukaminato, T., Sasaki, T., Tamai, N., and Kawai, T. (2002) Organic chemistry: a digital fluorescent molecular photoswitch. *Nature*, **420** (6917), 759–760.
10. Fukaminato, T., Sasaki, T., Kawai, T., Tamai, N., and Irie, M. (2004) Digital photoswitching of fluorescence based on the photochromism of diarylethene derivatives at a single-molecule level. *J. Am. Chem. Soc.*, **126** (45), 14843–14849.
11. Fukaminato, T., Umemoto, T., Iwata, Y., Yokojima, S., Yoneyama, M., Nakamura, S., and Irie, M. (2007) Photochromism of diarylethene single molecules in

polymer matrices. *J. Am. Chem. Soc.*, **129** (18), 5932–5938.
12. Tsujioka, T., Sesumi, Y., Takagi, R., Masui, K., Yokojima, S., Uchida, K., and Nakamura, S. (2008) Selective metal deposition on photoswitchable molecular surfaces. *J. Am. Chem. Soc.*, **130** (32), 10740–10747.
13. Tsujioka, T. and Tsuji, K. (2012) Metal-vapor deposition modulation on soft polymer surfaces. *Appl. Phys. Exp.*, **5** (2), 021601-1–021601-3.
14. Tsujioka, T. (2013) Metal-vapor deposition modulation on polymer surfaces prepared by the coffee-ring effect. *Soft Matter*, **9** (24), 5681–5685.
15. Tsujioka, T., Takagi, R., and Shiozawa, T. (2010) Light-controlled metal deposition on photochromic polymer films. *J. Mater. Chem.*, **20** (43), 9623–9627.
16. Cho, H. and Kim, E. (2002) Highly fluorescent and photochromic diarylethene oligomer bridged by *p*-phenylenevinylene. *Macromolecules*, **35** (23), 8684–8687.
17. Lim, S.-J., An, B.-K., and Park, S.Y. (2005) Bistable photoswitching in the film of fluorescent photochromic polymer: enhanced fluorescence emission and its high contrast switching. *Macromolecules*, **38** (15), 6236–6239.
18. Hayasaka, H., Tamura, K., and Akagi, K. (2008) Dynamic switching of linearly polarized emission in liquid-crystallinity-embedded photoresponsive conjugated polymers. *Macromolecules*, **41** (7), 2341–2346.
19. Li, X. and Tian, H. (2005) High-content pendant photochromic copolymer with dithienylethene/fluorene 2:1 mole ratio. *Macromol. Chem. Phys.*, **206** (17), 1769–1777.
20. Harbron, E.J., Davis, C.M., Campbell, J.K., Allred, R.M., Kovary, M.T., and Economou, N.J. (2009) Photochromic dye-doped conjugated polymer nanoparticles: photomodulated emission and nanoenvironmental characterization. *J. Phys. Chem. C*, **113** (31), 13707–13714.
21. Davis, C.M., Childress, E.S., and Harbron, E.J. (2011) Ensemble and single-particle fluorescence photomodulation in diarylethene-doped conjugated polymer nanoparticles. *J. Phys. Chem. C*, **115** (39), 19065–19073.
22. Yang, T., Liu, Q., Pu, S., Dong, Z., Huang, C., and Li, F. (2012) Fluorophore–photochrome co-embedded polymer nanoparticles for photoswitchable fluorescence bioimaging. *Nano Res.*, **5** (7), 494–503.
23. Deniz, E., Kandoth, N., Fraix, A., Cardile, V., Graziano, A.C.E., Furno, D.L., Gref, R., Raymo, F.M., and Sortino, S. (2012) Photoinduced fluorescence activation and nitric oxide release with biocompatible polymer nanoparticles. *Chem. Eur. J.*, **18** (49), 15782–15787.
24. Zhu, M.-Q., Zhu, L., Han, J.J., Wu, W., Hurst, J.K., and Li, A.D.Q. (2006) Spiropyran-based photochromic polymer nanoparticles with optically switchable luminescence. *J. Am. Chem. Soc.*, **128** (13), 4303–4309.
25. Schenderlein, H., Voss, A., Stark, R.W., and Biesalski, M. (2013) Preparation and characterization of light-switchable polymer networks attached to solid substrates. *Langmuir*, **29** (14), 4525–4534.
26. Kawai, T., Kunitake, T., and Irie, M. (1999) Novel photochromic conducting polymer having diarylethene derivative in the main chain. *Chem. Lett.*, **28** (9), 905–906.
27. Kawai, T., Nakashima, Y., Kunitake, T., and Irie, M. (2005) Photon-mode modulation of fluorescence and electrical current with a photochromic conducting polymer. *Curr. Appl Phys.*, **5** (2), 139–142.
28. Kawai, T., Nakashima, Y., and Irie, M. (2005) A novel photoresponsive π-conjugated polymer based on diarylethene and its photoswitching effect in electrical conductivity. *Adv. Mater.*, **17** (3), 309–314.
29. Choi, H., Lee, H., Kang, Y., Kim, E., Kang, S.O., and Ko, J. (2005) Photochromism and electrical transport characteristics of a dyad and a polymer with diarylethene and quinoline units. *J. Org. Chem.*, **70** (21), 8291–8297.
30. Kim, E. and Lee, H.W. (2006) Photo-induced electrical switching through a

mainchain polymer. *J. Mater. Chem.*, **16** (14), 1384–1389.

31. Friedle, S. and Thomas, S.W. III, (2010) Controlling contact electrification with photochromic polymers. *Angew. Chem. Int. Ed.*, **49** (43), 7968–7971.
32. Pariani, G., Castagna, R., Dassa, G., Hermes, S., Vailati, C., Bianco, A., and Bertarelli, C. (2011) Diarylethene-based photochromic polyurethanes for multistate optical memories. *J. Mater. Chem.*, **21** (35), 13223–13231.
33. Pariani, G., Bertarelli, C., Dassa, G., Bianco, A., and Zerbi, G. (2011) Photochromic polyurethanes for rewritable CGHs in optical testing. *Opt. Express*, **19** (5), 4536–4541.
34. Bertarelli, C., Castagna, R., Pariani, G., and Binanco, A. (2011) Photochromic polymers as a versatile tool for devices based on switchable absorption and other optical properties. *Proc. SPIE*, **8113**, 81130S.
35. Sciascia, C., Castagna, R., Dekermenjian, M., Martel, R., Kandada, A.R.S., Di Fonzo, F., Bianco, A., Bertarelli, C., Meneghetti, M., and Lanzani, G. (2012) Light-controlled resistance modulation in a photochromic diarylethene-carbon nanotube blend. *J. Phys. Chem. C*, **116** (36), 19483–19489.
36. Toccafondi, C., Occhi, L., Cavalleri, O., Penco, A., Castagna, R., Bianco, A., Bertarelli, C., Comoretto, D., and Canepa, M. (2014) Photochromic and photomechanical responses of an amorphous diarylethene-based polymer: a spectroscopic ellipsometry investigation of ultrathin films. *J. Mater. Chem. C*, **2** (23), 4692–4698.
37. Huang, C.-Q., Wang, Y., Hong, C.-Y., and Pan, C.-Y. (2011) Spiropyran-based polymeric vesicles: preparation and photochromic properties. *Macromol. Rapid Commun.*, **32** (15), 1174–1179.
38. Nishi, H., Asahi, T., and Kobatake, S. (2009) Light-controllable surface plasmon resonance absorption of gold nanoparticles covered with photochromic diarylethene polymers. *J. Phys. Chem. C*, **113** (40), 17359–17366.
39. Nishi, H., Asahi, T., and Kobatake, S. (2011) Enhanced one-photon cycloreversion reaction of diarylethenes near individual gold nanoparticles. *J. Phys. Chem. C*, **115** (11), 4564–4570.
40. Nishi, H., Asahi, T., and Kobatake, S. (2011) Enhanced photocycloreversion reaction of diarylethene polymers attached to gold nanoparticles in the solid state. *J. Photochem. Photobio. A: Chem.*, **221** (2-3), 256–260.
41. Nishi, H., Asahi, T., and Kobatake, S. (2012) Plasmonic enhancement of gold nanoparticles on photocycloreversion reaction of diarylethene derivatives depending on particle size, distance from the particle surface, and irradiation wavelength. *Phys. Chem. Chem. Phys.*, **14** (14), 4898–4905.
42. Imao, S., Nishi, H., and Kobatake, S. (2013) Thermo- and photoresponsive reversible changes in localized surface plasmon resonance of gold nanoparticles covered by poly(*N*-isopropylacrylamide) with photochromic diarylethene end group. *J. Photochem. Photobiol. A: Chem.*, **252**, 37–45.
43. Rochon, P., Batalla, E., and Natansohn, A. (1995) Optically induced surface gratings on azoaromatic polymer films. *Appl. Phys. Lett.*, **66** (2), 136–138.
44. Kim, D.Y., Tripathy, S.K., Li, L., and Kumar, J. (1995) Laser-induced holographic surface relief gratings on nonlinear optical polymer films. *Appl. Phys. Lett.*, **66** (10), 1166–1168.
45. Natansohn, A. and Rochon, P. (2002) Photoinduced motions in azo-containing polymers. *Chem. Rev.*, **102** (11), 4139–4175.
46. Irie, M., Hirano, Y., Hashimoto, S., and Hayashi, K. (1981) Photoresponsive polymers. 2. Reversible solution viscosity change of polyamides having azobenzene residues in the main chain. *Macromolecules*, **14** (2), 262–267.
47. Bléger, D., Liebig, T., Thiermann, R., Maskos, M., Rabe, J.P., and Hecht, S. (2011) Light-orchestrated macromolecular "accordion": reversible photoinduced shrinking of rigid-rod polymers. *Angew. Chem. Int. Ed.*, **50** (52), 12559–12563.

48. Irie, M. and Schnabel, W. (1981) Photoresponsive polymers. On the dynamics of conformational changes of polyamides with backbone azobenzene groups. *Macromolecules*, **14** (5), 1246–1249.
49. Finkelmann, H., Nishikawa, E., Pereira, G.G., and Warner, M. (2001) A new opto-mechanical effect in solids. *Phys. Rev. Lett.*, **87** (1), 015501-1–015501-4.
50. Yu, Y., Nakano, M., and Ikeda, T. (2003) Directed bending of a polymer film by light. *Nature*, **425** (6954), 145.
51. Ikeda, T., Nakano, M., Yu, Y., Tsutsumi, O., and Kanazawa, A. (2003) Anisotropic bending and unbending behavior of azobenzene liquid-crystalline gels by light exposure. *Adv. Mater.*, **15** (3), 201–205.
52. Yu, Y., Nakano, M., Shishido, A., Shiono, T., and Ikeda, T. (2004) Effect of cross-linking density on photoinduced bending behavior of oriented liquid-crystalline network films containing azobenzene. *Chem. Mater.*, **16** (9), 1637–1643.
53. Ikeda, T., Mamiya, J., and Yu, Y. (2007) Photomechanics of liquid-crystalline elastomers and other polymers. *Angew. Chem. Int. Ed.*, **46** (4), 506–528.
54. Hosono, N., Kajitani, T., Fukushima, T., Ito, K., Sasaki, S., Takata, M., and Aida, T. (2010) Large-area three-dimensional molecular ordering of a polymer brush by one-step processing. *Science*, **330** (6005), 808–811.
55. Bléger, D., Yu, Z., and Hecht, S. (2011) Toward optomechanics: maximizing the photodeformation of individual molecules. *Chem. Commun.*, **47** (45), 12260–12266.
56. Wang, D.H., Lee, K.M., Yu, Z., Koerner, H., Vaia, R.A., White, T.J., and Tan, L.-S. (2011) Photomechanical response of glassy azobenzene polyimide networks. *Macromolecules*, **44** (10), 3840–3846.
57. Lee, K.M., Wang, D.H., Koerner, H., Vaia, R.A., Tan, L.-S., and White, T.J. (2012) Enhancement of photogenerated mechanical force in azobenzene-functionalized polyimides. *Angew. Chem. Int. Ed.*, **51** (17), 4117–4121.
58. Sun, X., Wang, W., Qiu, L., Guo, W., Yu, Y., and Peng, H. (2012) Unusual reversible photomechanical actuation in polymer/nanotube composites. *Angew. Chem. Int. Ed.*, **51** (34), 8520–8524.
59. Golden, J.H. (1961) Bi(anthracene-9,10-dimethylene) (tetrabenzo-[2,2]-paracyclophane). *J. Chem. Soc.*, 3741–3748.
60. Maeda, K. and Hayashi, T. (1970) The mechanism of photochromism, thermochromism and piezochromism of dimers of triarylimidazolyl. *Bull. Chem. Soc. Jpn.*, **43** (2), 429–438.
61. Kawano, M., Sano, T., Abe, J., and Ohashi, Y. (1999) The first in situ direct observation of the light-induced radical pair from a hexaarylbiimidazolyl derivative by X-ray crystallography. *J. Am. Chem. Soc.*, **121** (35), 8106–8107.
62. Ichimura, K. and Watanabe, S. (1976) Photocyclization of diphenylmaleonitrile in the crystalline state. *Bull. Chem. Soc. Jpn.*, **49** (8), 2220–2223.
63. Trozzolo, A.M., Leslie, T.M., Sarpotdar, A.S., Small, R.D., Ferraudi, G.J., DoMinh, T., and Hartless, R.L. (1979) Photochemistry of some three-membered heterocycles. *Pure Appl. Chem.*, **51** (2), 261–270.
64. Sixl, H. and Warta, R. (1985) Reaction mechanism of photochromic 2-(2′,4′-dinitrobenzyl)pyridine (DNBP) single crystals. *Chem. Phys.*, **94** (1–2), 147–155.
65. Schmidt, A., Kababya, S., Appel, M., Khatib, S., Botoshansky, M., and Eichen, Y. (1999) Measuring the temperature width of a first-order single crystal to single crystal phase transition using solid-state NMR: application to the polymorphism of 2-(2,4-dinitrobenzyl)-3-methylpyridine. *J. Am. Chem. Soc.*, **121** (49), 11291–11299.
66. Hadjoudis, E., Vittorakis, M., and Moustakali-Mavridis, I. (1987) Photochromism and thermochromism of schiff bases in the solid state and in rigid glasses. *Tetrahedron*, **43** (7), 1345–1360.
67. Amimoto, K., Kanatomi, H., Nagakari, A., Fukuda, H., Koyama, H., and Kawato, T. (2003) Deuterium isotope effect on the solid-state thermal

isomerization of photo-coloured cis-keto species of *N*-salicylideneaniline. *Chem. Commun.*, (7), 870–871.

68. Mori, Y., Ohashi, Y., and Maeda, K. (1989) Crystal structure and photochemical behavior of 2,2,4,6-tetraphenyldihydro-1,3,5-triazine and its inclusion compounds. *Bull. Chem. Soc. Jpn.*, **62** (10), 3171–3176.
69. Kobatake, S. and Irie, M. (2004) Single-crystalline photochromism of diarylethenes. *Bull. Chem. Soc. Jpn.*, **77** (2), 195–210.
70. Morimoto, M. and Irie, M. (2005) Photochromism of diarylethene single crystals: crystal structures and photochromic performance. *Chem. Commun.*, (31), 3895–3905.
71. Irie, M. (2008) Photochromism and molecular mechanical devices. *Bull. Chem. Soc. Jpn.*, **81** (8), 917–926.
72. Irie, M. (2010) Photochromism of diarylethene molecules and crystals. *Proc. Jpn. Acad., Ser. B*, **86** (5), 472–483.
73. Irie, M. (2010) Photochromism of diarylethene single molecules and single crystals. *Photochem. Photobiol. Sci.*, **9** (12), 1535–1542.
74. Irie, M., Lifka, T., Kobatake, S., and Kato, N. (2000) Photochromism of 1,2-bis(2-methyl-5-phenyl-3-thienyl)perfluorocyclopentene in a single-crystalline phase. *J. Am. Chem. Soc.*, **122** (20), 4871–4876.
75. Takami, S., Kobatake, S., Kawai, T., and Irie, M. (2003) Extraordinarily high thermal stability of the closed-ring isomer of 1,2-bis(5-methyl-2-phenylthiazol-4-yl)perfluorocyclopentene. *Chem. Lett.*, **32** (10), 892–893.
76. Hanazawa, M., Sumiya, R., Horikawa, Y., and Irie, M. (1992) Thermally irreversible photochromic systems. Reversible photocyclization of 1,2-bis(2-methylbenzo[*b*]thiophen-3-yl)perfluorocyclocoalkene derivatives. *J. Chem. Soc., Chem. Commun.*, (3), 206–207.
77. Uchida, K., Nakayama, Y., and Irie, M. (1990) Thermally irreversible photochromic systems. Reversible photocyclization of 1,2-bis(benzo[*b*]thiophen-3-yl)ethene derivatives. *Bull. Chem. Soc. Jpn.*, **63** (5), 1311–1315.
78. Jean-Ruel, H., Cooney, R.R., Gao, M., Lu, C., Kochman, M.A., Morrison, C.A., and Dwayne Miller, R.J. (2011) Femtosecond dynamics of the ring closing process of diarylethene: a case study of electrocyclic reactions in photochromic single crystals. *J. Phys. Chem. A*, **115** (45), 13158–13168.
79. Jean-Ruel, H., Gao, M., Kochman, M.A., Lu, C., Liu, L.C., Cooney, R.R., Morrison, C.A., and Dwayne Miller, R.J. (2013) Ring-closing reaction in diarylethene captured by femtosecond electron crystallography. *J. Phys. Chem. B*, **117** (49), 15894–15902.
80. Miyasaka, H., Araki, S., Tabata, A., Nobuto, T., Mataga, N., and Irie, M. (1994) Picosecond laser photolysis studies on photochromic reactions of 1,2-bis(2,4,5-trimethyl-3-thienyl)maleic anhydride in solutions. *Chem. Phys. Lett.*, **230** (3), 249–254.
81. Tamai, N., Saika, T., Shimidzu, T., and Irie, M. (1996) Femtosecond dynamics of a thiophene oligomer with a photoswitch by transient absorption spectroscopy. *J. Phys. Chem.*, **100** (12), 4689–4692.
82. Ern, J., Bens, A.T., Martin, H.-D., Mukamel, S., Schmid, D., Tretiak, S., Tsiper, E., and Kryschi, C. (1999) Reaction dynamics of photochromic dithienylethene derivatives. *Chem. Phys.*, **246** (1–3), 115–125.
83. Okabe, C., Nakabayashi, T., Nishi, N., Fukaminato, T., Kawai, T., Irie, M., and Sekiya, H. (2003) Picosecond time-resolved stokes and anti-stokes Raman studies on the photochromic reactions of diarylethene derivatives. *J. Phys. Chem. A*, **107** (28), 5384–5390.
84. Yamada, T., Kobatake, S., Muto, K., and Irie, M. (2000) X-ray crystallographic study on single-crystalline photochromism of bis(2,5-dimethyl-3-thienyl)perfluorocyclopentene. *J. Am. Chem. Soc.*, **122** (8), 1589–1592.

85. Yamada, T., Kobatake, S., and Irie, M. (2000) X-ray crystallographic study on single-crystalline photochromism of 1,2-bis(2,5-dimethyl-3-thienyl)perfluorocyclopentene. *Bull. Chem. Soc. Jpn.*, **73** (10), 2179–2184.
86. Kobatake, S., Uchida, K., Tsuchida, E., and Irie, M. (2002) Single-crystalline photochromism of diarylethenes: reactivity–structure relationship. *Chem. Commun.*, (23), 2804–2805.
87. Shibata, K., Muto, K., Kobatake, S., and Irie, M. (2002) Photocyclization/cycloreversion quantum yields of diarylethenes in single crystals. *J. Phys. Chem. A*, **106** (1), 209–214.
88. Morimoto, M., Kobatake, S., and Irie, M. (2002) Multi-colored photochromic crystals of diarylethene mixtures. *Adv. Mater.*, **14** (15), 1027–1029.
89. Morimoto, M., Kobatake, S., and Irie, M. (2003) Multicolor photochromism of two- and three-component diarylethene crystals. *J. Am. Chem. Soc.*, **125** (36), 11080–11087.
90. Kuroki, L., Takami, S., Shibata, K., and Irie, M. (2005) Photochromism of single crystals composed of dioxazolylethene and dithiazolylethene. *Chem. Commun.*, (48), 6005–6007.
91. Takami, S., Kuroki, L., and Irie, M. (2007) Photochromism of mixed crystals containing bisthienyl-, bisthiazolyl-, and bisoxazolylethene derivatives. *J. Am. Chem. Soc.*, **129** (23), 7319–7326.
92. Morimoto, M., Kobatake, S., and Irie, M. (2003) Photochromism of diarylethenes in nanolayers of a single crystal. *Photochem. Photobiol. Sci.*, **2** (11), 1088–1094.
93. Morimoto, M., Kobatake, S., and Irie, M. (2004) Crystal engineering of photochromic diarylethene single crystals. *Chem. Rec.*, **4** (1), 23–38.
94. Kobatake, S., Matsumoto, Y., and Irie, M. (2005) Conformational control of photochromic reactivity in a diarylethene single crystal. *Angew. Chem. Int. Ed.*, **44** (14), 2148–2151.
95. Kobatake, S., Takami, S., Muto, H., Ishikawa, T., and Irie, M. (2007) Rapid and reversible shape changes of molecular crystals on photoirradiation. *Nature*, **446** (7137), 778–781.
96. Kuroki, L., Takami, S., Yoza, K., Morimoto, M., and Irie, M. (2010) Photoinduced shape changes of diarylethene single crystals: correlation between shape changes and molecular packing. *Photochem. Photobiol. Sci.*, **9** (2), 221–225.
97. Terao, F., Morimoto, M., and Irie, M. (2012) Light-driven molecular-crystal actuators: rapid and reversible bending of rodlike mixed crystals of diarylethene derivatives. *Angew. Chem. Int. Ed.*, **51** (4), 901–904.
98. Morimoto, M. and Irie, M. (2010) A diarylethene cocrystal that converts light into mechanical work. *J. Am. Chem. Soc.*, **132** (40), 14172–14178.
99. Kitagawa, D., Nishi, H., and Kobatake, S. (2013) Photoinduced twisting of a photochromic diarylethene crystal. *Angew. Chem. Int. Ed.*, **52** (35), 9320–9322.
100. Uchida, K., Sukata, S., Matsuzawa, Y., Akazawa, M., de Jong, J.J.D., Katsonis, N., and Feringa, B.L. (2008) Photoresponsive rolling and bending of thin crystals of chiral diarylethenes. *Chem. Commun.*, (3), 326–328.
101. Kobatake, S., Hasegawa, H., and Miyamura, K. (2011) High-convertible photochromism of a diarylethene single crystal accompanying the crystal shape deformation. *Cryst. Growth Des.*, **11** (4), 1223–1229.
102. Koshima, H., Ojima, N., and Uchimoto, H. (2009) Mechanical motion of azobenzene crystals upon photoirradiation. *J. Am. Chem. Soc.*, **131** (20), 6890–6891.
103. Koshima, H., Takechi, K., Uchimoto, H., Shiro, M., and Hashizume, D. (2011) Photomechanical bending of salicylideneaniline crystals. *Chem. Commun.*, **47** (41), 11423–11425.
104. Koshima, H., Nakaya, H., Uchimoto, H., and Ojima, N. (2012) Photomechanical motion of furylfulgide crystals. *Chem. Lett.*, **41** (1), 107–109.
105. Al-Kaysi, R.O., Müller, A.M., and Bardeen, C.J. (2006) Photochemically driven shape changes of crystalline organic nanorods. *J. Am. Chem. Soc.*, **128** (50), 15938–15939.

106. Al-Kaysi, R.O. and Bardeen, C.J. (2007) Reversible photoinduced shape changes of crystalline organic nanorods. *Adv. Mater.*, **19** (9), 1276–1280.
107. Good, J.T., Burdett, J.J., and Bardeen, C.J. (2009) Using two-photon excitation to control bending motions in molecular-crystal nanorods. *Small*, **5** (24), 2902–2909.
108. Zhu, L., Al-Kaysi, R.O., Dillon, R.J., Tham, F.S., and Bardeen, C.J. (2011) Crystal structures and photophysical properties of 9-anthracene carboxylic acid derivatives for photomechanical applications. *Cryst. Growth Des.*, **11** (11), 4975–4983.
109. Zhu, L., Al-Kaysi, R.O., and Bardeen, C.J. (2011) Reversible photoinduced twisting of molecular crystal microribbons. *J. Am. Chem. Soc.*, **133** (32), 12569–12575.
110. Kim, T., Al-Muhanna, M.K., Al-Suwaidan, S.D., Al-Kaysi, R.O., and Bardeen, C.J. (2013) Photoinduced curling of organic molecular crystal nanowires. *Angew. Chem. Int. Ed.*, **52** (27), 6889–6893.
111. Kim, T., Zhu, L., Mueller, L.J., and Bardeen, C.J. (2014) Mechanism of photoinduced bending and twisting in crystalline microneedles and microribbons composed of 9-methylanthracene. *J. Am. Chem. Soc.*, **136** (18), 6617–6625.
112. Kim, T., Zhu, L., Mueller, L.J., and Bardeen, C.J. (2012) Dependence of the solid-state photomechanical response of 4-chlorocinnamic acid on crystal shape and size. *CrystEngComm*, **14** (22), 7792–7799.
113. Sun, J.-K., Li, W., Chen, C., Ren, C.-X., Pan, D.-M., and Zhang, J. (2013) Photoinduced bending of a large single crystal of a 1,2-bis(4-pyridyl)ethylene-based pyridinium salt powered by a [2+2] cycloaddition. *Angew. Chem. Int. Ed.*, **52** (26), 6653–6657.
114. Li, Q. (ed) (2012) *Liquid Crystals Beyond Displays: Chemistry, Physics, and Applications*, John Wiley & Sons, Inc., Hoboken, NJ.
115. Li, Q. (ed) (2011) *Self-organized Organic Semiconductors: From Materials to Device Applications*, John Wiley & Sons, Inc., Hoboken, NJ.
116. Li, Q. (ed) (2013) *Intelligent Stimuli Responsive Materials: From Well-Defined Nanostructures to Applications*, John Wiley & Sons, Inc., Hoboken, NJ.
117. Li, Q. (ed) (2014) *Nanoscience with Liquid Crystals: From Self-Organized Nanostructures to Applications*, Springer-Verlag, Heidelberg.
118. Li, Q. (ed) (2015) *Anisotropic Nanomaterials: Preparation, Properties, and Applications*, Springer-Verlag, Heidelberg.
119. de Gennes, P.G. and Prost, J. (1993) *The Physics of Liquid Crystals*, Oxford University Press, Oxford.
120. Collings, P.J. and Hird, M. (1997) *Introduction to Liquid Crystals: Chemistry and Physics*, Taylor & Francis, London.
121. Goodby, J.W., Collings, P.J., Kato, T., Tschierske, C., Glesson, H., and Raynes, P. (eds) (2014) *Handbook of Liquid Crystals*, 2nd edn, Wiley-VCH Verlag GmbH, Weinheim.
122. Chen, R.H. (2011) *Liquid Crystal Displays: Fundamental Physics and Technology*, John Wiley & Sons, Inc., Hoboken, NJ.
123. Woltman, S.J., Jay, G.D., and Crawford, G.P. (eds) (2007) *Liquid Crystals: Frontiers in Biomedical Applications*, World Scientific, New Jersey, USA.
124. Bushby, R.J., Kelly, S.M., and O'Neill, M. (eds) (2013) *Liquid Crystalline Semiconductors: Materials, Properties and Applications*, Springer, Dordrecht.
125. Tschierske, C. (ed) (2012) *Liquid Crystals: Materials Design and Self-Assembly*, Springer-Verlag, Berlin and Heidelberg.
126. Ramamoorthy, A. (ed) (2007) *Thermotropic Liquid Crystals: Recent Advances*, Springer, Dordrecht.
127. Kitzerow, H.-S. and Bahr, C. (eds) (2001) *Chirality in Liquid Crystals*, Springer-Verlag, New York.
128. Kumar, S. (2001) *Chemistry of Discotic Liquid Crystals: From Monomers to Polymers*, CRC Press, Boca Raton, FL.
129. Donald, A.M., Windle, A.H., and Hanna, S. (2006) *Liquid Crystalline*

Polymers, Cambridge University Press, Cambridge.

130. De Jeu, W.H. (ed) (2012) *Liquid Crystal Elastomers: Materials and Applications*, Springer-Verlag, Heidelberg.

131. Wang, Y. and Li, Q. (2012) Light-driven chiral molecular switches or motors in liquid crystals. *Adv. Mater.*, **24**, 1926–1945.

132. Yokoyama, Y. (2000) Fulgides for memories and switches. *Chem. Rev.*, **100**, 1717–1739.

133. Ichimura, K. (2000) Photoalignment of liquid-crystal systems. *Chem. Rev.*, **100**, 1847–1873.

134. Berkovic, G., Krongauz, V., and Weiss, V. (2000) Spiropyrans and spirooxazines for memories and switches. *Chem. Rev.*, **100**, 1741–1753.

135. Feringa, B.L., van Delden, R.A., Koumura, N., and Geertsema, E.M. (2000) Chiroptical molecular switches. *Chem. Rev.*, **100**, 1789–1816.

136. Zhang, J., Zou, Q., and Tian, H. (2013) Photochromic materials: more than meets the eye. *Adv. Mater.*, **25**, 378–399.

137. Tian, H. and Yang, S. (2004) Recent progresses on diarylethene based photochromic switches. *Chem. Soc. Rev.*, **33**, 85–97.

138. Orgiu, E. and Samori, P. (2014) Organic electronics marries photochromism: generation of multifunctional interfaces, materials, and devices. *Adv. Mater.*, **26**, 1827–1845.

139. Bisoyi, H.K. and Li, Q. (2014) Light-directing chiral liquid crystalline nanostructures: from 1D to 3D. *Acc. Chem. Res.*, **47**, 3184–3195.

140. Cabrera, I., Krongauz, V., and Ringsdorf, H. (1987) Photo- and thermochromic liquid crystal polysiloxanes. *Angew. Chem. Int. Ed.*, **26**, 1178–1180.

141. Shragina, L., Buchholtz, F., Yitzchaik, S., and Krongauz, V. (1990) Searching for photochromic liquid crystals spironaphthoxazine substituted with a mesogenic group. *Liq. Cryst.*, **7**, 643–655.

142. Yitzchaik, S., Carbrera, I., Buchholtz, F., and Krongauz, V. (1990) Photochromism of liquid-crystal polyacrylates containing spiropyran groups. *Macromolecules*, **23**, 707–713.

143. Natarajan, L.V., Bunning, T.J., Klei, H.E., Crane, R.L., and Adams, W.W. (1991) Novel photochromic cholesteric liquid crystal siloxane. *Macromolecules*, **24**, 6554–6556.

144. Hattori, H. and Uryu, T. (1999) Synthesis and characterization of polymerizable photochromic liquid crystals containing a spiro-oxazine group. *Liq. Cryst.*, **26**, 1085–1095.

145. Hattori, H. and Uryu, T. (2001) Photochromic chiral liquid crystalline systems containing spiro-oxazine with a chiral substituent I. Synthesis and characterization of compounds. *Liq. Cryst.*, **28**, 25–34.

146. Keum, S.-R., Ahn, S.-M., and Lee, S.-H. (2004) Synthesis and characterization of new liquid crystallines containing a mono-aza styrylated indolinobenzospiropyran. Part 6. *Dyes Pigm.*, **60**, 55–59.

147. Keum, S.-R., Cho, C.-H., and Choi, Y.-K. (2004) Synthesis and mesomorphic properties of liquid crystallines, alkyloxybenzoyloxy phenylated indolinobenzospiropyrans. Part 7. *Dyes Pigm.*, **60**, 147–153.

148. Jin, L.-M., Li, Y., Ma, J., and Li, Q. (2010) Synthesis of novel thermally reversible photochromic axially chiral spirooxazines. *Org. Lett.*, **12**, 3552–3555.

149. Romani, A., Chidichimo, G., Formoso, P., Manfredi, S., Favaro, G., and Mazzucato, U. (2002) Photochromic behavior of a spiro-indolino-oxazine in reverse-mode polymer dispersed liquid crystal films. *J. Phys. Chem. B*, **106**, 9490–9495.

150. Macchione, M., Filpo, G.D., Nicoletta, F.P., and Chidichimo, G. (2005) Photochromic reverse mode polymer dispersed liquid crystals. *Liq. Cryst.*, **32**, 315–318.

151. Li, Y., Wang, M., Wang, H., Urbas, A., and Li, Q. (2014) Rationally designed axially chiral switches with high helical twisting power. *Chem. Eur. J.*, **20**, 16286–16292.

152. Frigoli, M. and Mehl, G.H. (2003) A photochromic liquid crystal system. *ChemPhysChem*, **1**, 101–103.

153. Frigoli, M. and Mehl, G.H. (2004) Modulation of the absorption, fluorescence, and liquid-crystal properties of functionalized diarylethene derivatives. *Chem. Eur. J.*, **10**, 5243–5250.
154. Frigoli, M. and Mehl, G.H. (2004) The enhancement of photoswitching in a diarylethene derivative by the incorporation of cyanobiphenyl groups. *Chem. Commun.*, 818–819.
155. Frigoli, M. and Mehl, G.H. (2005) Multiple addressing in a hybrid biphotochromic system. *Angew. Chem. Int. Ed.*, **44**, 5048–5052.
156. Frigoli, M., Welch, C., and Mehl, G.H. (2004) Design of mesomorphic diarylethene-based photochromes. *J. Am. Chem. Soc.*, **126**, 15382–15383.
157. Chen, S.H., Chen, H.M.P., Geng, Y., Jacobs, S.D., Marshall, K.L., and Blanton, T.N. (2003) Novel glassy nematic liquid crystals for non-destructive rewritable optical memory and photonic switching. *Adv. Mater.*, **15**, 1061–1065.
158. Rameshbabu, K., Urbas, A., and Li, Q. (2011) Synthesis and characterization of thermally irreversible photochromic cholesteric liquid crystals. *J. Phys. Chem. B*, **115**, 3409–3415.
159. Li, Y., Urbas, A., and Li, Q. (2011) Synthesis and characterization of light-driven dithienylcyclopentene switches with axial chirality. *J. Org. Chem.*, **76**, 7148–7156.
160. Li, Y., Urbas, A., and Li, Q. (2012) Reversible light-directed red, green and blue reflections with thermal stability enabled by a self-organized helical superstructure. *J. Am. Chem. Soc.*, **134**, 9573–9576.
161. Li, Y., Xue, C., Wang, M., Urbas, A., and Li, Q. (2013) Photodynamic chiral molecular switches with thermal stability: from reflection wavelength tuning to handedness inversion of self-organized helical superstructures. *Angew. Chem. Int. Ed.*, **52**, 13703–13707.
162. Li, Y., Wang, M., Urbas, A., and Li, Q. (2013) Photoswitchable but thermally stable axially chiral dithienylperfluorocyclopentene dopant with high helical twisting power. *J. Mater. Chem. C*, **1**, 3917–3923.
163. Lemieux, R.P. (2005) Photoswitching of ferroelectric liquid crystals using photochromic dopants. *Soft Matter*, **1**, 348–354.
164. Maly, K.E., Wand, M.D., and Lemieux, R.P. (2002) Bistable ferroelectric liquid crystal photoswitch triggered by a dithienylethene dopant. *J. Am. Chem. Soc.*, **124**, 7898–7899.
165. Maly, K.E., Zhang, P., Wand, M.D., Buncel, E., and Lemieux, R.P. (2004) Reversible photocyclization of achiral dithienylperfluorocyclopentene dopants in a ferroelectric liquid crystal: bistable SSFLC photoswitching. *J. Mater. Chem.*, **14**, 2806–2812.
166. Ube, T. and Ikeda, T. (2014) Photomobile polymer materials with crosslinked liquid-crystalline structures: molecular design, fabrication, and function. *Angew. Chem. Int. Ed.*, **53**, 10290–10299.
167. Matharu, A.S., Jeeva, S., and Ramanujam, P.S. (2007) Liquid crystals for holographic optical data storage. *Chem. Soc. Rev.*, **36**, 1868–1880.
168. Hvilsted, S., Andruzzi, F., Kulinna, C., Siesler, H.W., and Ramanujam, P.S. (1995) Novel side-chain liquid crystalline polyester architecture for reversible optical storage. *Macromolecules*, **28**, 2172–2183.
169. Norikane, Y., Hirai, Y., and Yoshida, M. (2011) Photoinduced isothermal phase transitions of liquid-crystalline macrocyclic azobenzenes. *Chem. Commun.*, **47**, 1770–1772.
170. Rahman, M.L., Tschierske, C., Yusoff, M., and Silong, S. (2005) Synthesis and liquid crystalline properties of a disc-shaped molecule with azobenzene at the periphery. *Tetrahedron Lett.*, **46**, 2303–2306.
171. Tanaka, D., Ishiguru, H., Shimizu, Y., and Uchida, K. (2012) Thermal and photoinduced liquid crystalline phase transitions with a rod-disc alternative change in the molecular shape. *J. Mater. Chem.*, **22**, 25065–25071.
172. Westphal, E., Bechtold, I.H., and Gallardo, H. (2010) Synthesis and optical/thermal behavior of new azo

photopolymerizable discotic liquid crystals. *Macromolecules*, **43**, 1319–1328.
173. Prasad, V., Kang, S.-W., Suresh, K.A., Joshi, L., Wang, Q., and Kumar, S. (2005) Thermotropic uniaxial and biaxial nematic and smectic phases in bent-core mesogens. *J. Am. Chem. Soc.*, **127**, 17224–17227.
174. Mathews, M., Kang, S., Kumar, S., and Li, Q. (2011) Designing bent-core nematogens towards biaxial nematic liquid crystals. *Liq. Cryst.*, **38**, 31–40.
175. Mathews, M., Zola, R.S., Yang, D., and Li, Q. (2011) Thermally, photochemically and electrically switchable reflection colors from self-organized chiral bent-core liquid crystals. *J. Mater. Chem.*, **21**, 2098–2103.
176. Gimeno, N., Pintre, I., Martinez-Abadia, M., Serrano, J.L., Ros, M.B., Bent-core liquid crystal phases promoted by azo-containing molecules: from monomers to side-chain polymers. *RSC Adv.*, **4**, 19694-19702.
177. Folcia, C.L., Alonso, I., Ortega, J., Etxebarria, J., Pintre, I., and Ros, M.B. (2006) Achiral bent-core liquid crystals with azo and azoxy linkages: structural and nonlinear optical properties and photoisomerization. *Chem. Mater.*, **18**, 4617–4626.
178. Alaasar, M., Prehm, M., Brautzsch, M., and Tschierske, C. (2014) 4-Methylresorcinol based bent-core liquid crystals with azobenzene wings a new class of compounds with dark conglomerate phases. *J. Mater. Chem. C*, **2**, 5487–5501.
179. Alaasar, M., Prehm, M., and Tschierske, C. (2013) A new room temperature dark conglomerate mesophase formed by bent-core molecules combining 4-iodoresoorcinol with azobenzene units. *Chem. Commun.*, **49**, 11062–11064.
180. Alaasar, M., Prehm, M., May, K., Eremin, A., and Tschierske, C. (2014) 4-Cyanoresorcinol-based bent-core mesogens with azobenzene wings: emergence of sterically stabilized polar order in liquid crystalline phases. *Adv. Funct. Mater.*, **24**, 1703–1717.
181. Alaasar, M., Prehm, M., and Tschierske, C. (2014) New azobenzene containing bent-core liquid crystals based on disubstituted resorcinol. *Liq. Cryst.*, **41**, 126–136.
182. Choi, S.-W., Izumi, T., Hoshino, Y., Takanishi, Y., Ishikawa, K., Watanabe, J., and Takezoe, H. (2006) Circular-polarization-induced enantiomeric excess in liquid crystals of an achiral, bent-shaped mesogen. *Angew. Chem. Int. Ed.*, **45**, 1382–1385.
183. Wang, Y., Urbas, A., and Li, Q. (2012) Reversible visible-light tuning of self-organized helical superstructures enabled by unprecedented light-driven axially chiral molecular switches. *J. Am. Chem. Soc.*, **134**, 3342–3345.
184. Li, Q., Li, L., Kim, J., Park, H.-K., and Williams, J. (2005) Reversible photoresponsive chiral liquid crystals containing cholesteryl moiety and azobenzene linker. *Chem. Mater.*, **17**, 6018–6021.
185. Ji, M., Li, Y., White, T., Urbas, A., and Li, Q. (2010) Light-driven nanoscale chiral molecular switch: reversible dynamic full range color phototuning. *Chem. Commun.*, **46**, 3463–3465.
186. Li, Q., Li, Y., Ma, J., Yang, D.-K., White, T.J., and Bunning, T.J. (2011) Directing dynamic control of red, green and blue reflection enabled by a light-driven self-organized helical superstructure. *Adv. Mater.*, **23**, 5069–5073.
187. Green, L., Li, Y., White, T., Urbas, A., Bunning, T., and Li, Q. (2009) Light-driven chiral molecular switches with tetrahedral and axial chirality. *Org. Biomol. Chem.*, **7**, 3930–3933.
188. Han, J., Li, Y.-X., Pang, M.-L., Cheng, K.-G., Wang, Y.-M., Ma, Y.-X., and Meng, J.-B. (2007) A novel photochromic liquid crystal system based on biindenylidenedione derivatives. *New J. Chem.*, **31**, 543–548.
189. Abraham, S., Mallia, V.A., Ratheesh, K.V., Tamaoki, N., and Das, S. (2006) Reversible thermal and photochemical switching of liquid crystalline phases and luminescence in diphenylbutadiene-based mesogenic dimers. *J. Am. Chem. Soc.*, **128**, 7692–7698.
190. Lee, C.-H. and Yamamoto, T. (2001) Synthesis and characterization of a

new class of liquid-crystalline, highly luminescent molecules containing a 2,4,6-triphenyl-1,3,5-triazine unit. *Tetrahedron Lett.*, **42**, 3993–3996.

191. Yashuda, T., Ooi, H., Morita, J., Akama, Y., Minoura, K., Funahashi, M., Shimomura, T., and Kato, T. (2009) π-Conjugated oligothiophene-based polycatenar liquid crystals: self-organization and photoconductive, luminescent, and redox properties. *Adv. Funct. Mater.*, **19**, 411–419.
192. Gallardo, H., Conte, G., Tuzimoto, P.A., Behramand, B., Molin, F., Eccher, J., and Bechtold, I.H. (2012) New luminescent liquid crystals based on 2,1,3-benzothiadiazole and bent five-membered *N*-heterocyclic cores. *Liq. Cryst.*, **39**, 1099–1111.
193. Cristiano, R., Vieira, A.A., Ely, F., and Gallardo, H. (2006) Synthesis and characterization of luminescent hockey stick-shaped liquid crystalline compounds. *Liq. Cryst.*, **33**, 381–390.
194. Gallardo, H., Cristiano, R., Vieira, A.A., Filho, R.A.W.N., Srivastava, R.M., and Bechtold, I.H. (2008) Non-symmetrical luminescent 1,2,4-oxadiazole-based liquid crystals. *Liq. Cryst.*, **35**, 857–863.
195. Cristiano, R., Ely, F., and Gallardo, H. (2005) Light-emitting bent-shape liquid crystals. *Liq. Cryst.*, **32**, 15–25.
196. Cristiano, R., Santos, D.M.P.D.O., and Gallardo, H. (2005) Synthesis and characterization of low molecular mass luminescent liquid crystalline materials with 1,3,4-oxadiazole units. *Liq. Cryst.*, **32**, 7–14.
197. Binnemans, K. (2009) Luminescence of metallomesogens in the liquid crystal state. *J. Mater. Chem.*, **19**, 448–453.
198. Gimenez, R., Pinol, M., and Serrano, J.L. (2004) Luminescent liquid crystals derived from 9,10-bis(phenylethynyl)anthracene. *Chem. Mater.*, **16**, 1377–1383.
199. Terazzi, E., Torelli, S., Bernardinelli, G., Rivera, J.-P., Benech, J.-M., Bourgogne, C., Donnio, B., Guillon, D., Imbert, D., Bunzil, J.-C.G., Pinto, A., Jeannerat, D., and Piguet, C. (2005) Molecular control of macroscopic cubic, columnar, and lamellar organizations in luminescent lanthanide-containing thermotropic liquid crystals. *J. Am. Chem. Soc.*, **127**, 888–903.
200. Diring, S., Camerel, F., Donnio, B., Dintzer, T., Toffanin, S., Capelli, R., Muccini, M., and Ziessel, R. (2009) Luminescent ethynyl-pyrene liquid crystals and gels for optoelectronic devices. *J. Am. Chem. Soc.*, **131**, 18177–18185.
201. Chen, L., Li, Y., Fan, J., Bisoyi, H.K., Weitz, D.A., and Li, Q. (2014) Photoresponsive monodisperse cholesteric liquid crystalline microshells for tunable omnidirectional lasing enabled by a visible light-driven chiral molecular switch. *Adv. Opt. Mater.*, **2**, 845–848.
202. Fan, J., Li, Y., Bisoyi, H.K., Zola, R.S., Yang, D., Bunning, T., Weitz, D.A., and Li, Q. (2015) Light-directing omnidirectional circularly polarized reflection from liquid crystal droplets. *Angew. Chem. Int. Ed.*, **54**, 2160–2164.
203. Wang, L., Dong, H., Li, Y., Xue, C., Sun, L.D., Yan, C.H., and Li, Q. (2014) Reversible near-infrared light directed reflection in a self-organized helical superstructure loaded with upconversion nanoparticles. *J. Am. Chem. Soc.*, **136**, 4480–4483.
204. Wang, L., Dong, H., Li, Y., Liu, R., Wang, Y., Bisoyi, H.K., Sun, L.D., Yan, C.H., and Li, Q. (2015) Luminescence-driven, reversible handedness inversion of self-organized helical superstructure enabled by a novel near-infrared light nanotransducers. *Adv. Mater.*, **27**, 2065–2069.
205. Wang, L., Gutierrez-Cuevas, K., Bisoyi, H.K., Xiang, J., Singh, G., Zola, R.S., Kumar, S., Lavrentovich, O.D., Urbas, A., and Li, Q. (2015) NIR light-directing self-organized 3D photonic superstructures loaded with mesogen-functionalized anisotropic plasmonic nanorods. *Chem. Commun.*, **51**, 15039–15042.
206. Xue, C., Xiang, J., Nemati, H., Bisoyi, H.K., Gutierrez-Cuevas, K., Wang, L., Gao, M., Zhou, S., Lavrentovich, O.D., Yang, D., Urbas, A., and Li, Q. (2015) Light-driven reversible alignment switching of liquid crystal enabled by azo thiol grafted gold nanoparticles. *ChemPhysChem*, **16**, 1852–1856.

207. Wang, L. and Li, Q. (2016) Stimuli-directing self-organized 3D liquid crystalline nanostructures: from materials design to photonic applications. *Adv. Funct. Mater.*, **26**, 10–28.
208. Bisoyi, H.K. and Li, Q. (2016) Light-directed dynamic chirality inversion in functional self-organized helical superstructures. *Angew. Chem. Int. Ed.*, **55**, 2994–3010.
209. Abdallah, D.J. and Weiss, R.G. (2000) Organogels and low molecular mass organic gelators. *Adv. Mater.*, **12** (17), 1237–1247.
210. de Loos, M., Feringa, B.L., and van Esch, J.H. (2005) Design and application of self-assembled low molecular weight hydrogels. *Eur. J. Org. Chem.*, **17**, 3615–3631.
211. Neralagatta, M.S. and Uday, M. (2005) Supramolecular gels: functions and uses. *Chem. Soc. Rev.*, **34**, 821–836.
212. Steed, J.W. (2011) Supramolecular gel chemistry: developments over the last decade. *Chem. Commun.*, **47**, 1379–1383.
213. Duan, P.F., Cao, H., Zhang, L., and Liu, M.H. (2014) Gelation induced supramolecular chirality: chirality transfer, amplification and application. *Soft Matter*, **10**, 5428–5448.
214. George, M. and Weiss, R.G. (2006) Molecular organogels. Soft matter comprised of low-molecular-mass organic gelators and organic liquids. *Acc. Chem. Res.*, **39** (8), 489–497.
215. Yang, X.Y., Zhang, G.X., and Zhang, D.Q. (2012) Stimuli responsive gels based on low molecular weight gelators. *J. Mater. Chem.*, **22**, 38–50.
216. Qi, Z.H. and Schalley, C.A. (2012) Exploring macrocycles in functional supramolecular gels: from stimuli responsiveness to systems chemistry. *Acc. Chem. Res.*, **47** (7), 2222–2233.
217. Weiss, R.G. (2014) The past, present, and future of molecular gels. What is the status of the field, and where is it going? *J. Am. Chem. Soc.*, **136** (21), 7519–7530.
218. Babu, S.S., Praveen, V.K., and Ajayaghosh, A. (2014) Functional π-Gelators and their applications. *Chem. Rev.*, **114** (4), 1973–2129.
219. Jung, J.H., Ono, Y., Sakurai, K., Sano, M., and Shinkai, S. (2000) Novel vesicular aggregates of crown-appended cholesterol derivatives which act as gelators of organic solvents and as templates for silica transcription. *J. Am. Chem. Soc.*, **122**, 8648–8653.
220. Jung, J.H., Ono, Y., and Shinkai, S. (2000) Sol–gel polycondensation of tetraethoxysilane in a cholesterol-based organogel system results in chiral spiral silica. *Angew. Chem. Int. Ed.*, **39**, 1862–1865.
221. Tamai, N. and Miyasaka, H. (2000) Ultrafast dynamics of photochromic systems. *Chem. Rev.*, **100** (5), 1875–1890.
222. Murata, K., Aoki, M., Suzuki, T., Harada, T., Kawabata, H., Komri, T., Ohseto, F., Ueda, K., and Shinkai, S. (1994) Thermal and light control of the sol-gel phase transition in cholesterol-based organic gels. Novel helical aggregation modes as detected by circular dichroism and electron microscopic observation. *J. Am. Chem. Soc.*, **116**, 6664–6676.
223. Koumura, N., Kudo, M., and Tamaoki, N. (2004) Photocontrolled gel-to-sol-to-gel phase transitioning of meta-substituted azobenzene bisurethanes through the breaking and reforming of hydrogen bonds. *Langmuir*, **20** (23), 9897–9900.
224. Wang, C., Chen, Q., Sun, F., Zhang, D., Zhang, G., Huang, Y., Zhao, R., and Zhu, D. (2010) Multistimuli responsive organogels based on a new gelator featuring tetrathiafulvalene and azobenzene groups: reversible tuning of the gel–sol transition by redox reactions and light irradiation. *J. Am. Chem. Soc.*, **132** (9), 3092–3096.
225. Wu, Y., Wu, S., Tian, X., Wang, X., Wu, W., Zou, G., and Zhang, Q. (2011) Photoinduced reversible gel–sol transitions of dicholesterol-linked azobenzene derivatives through breaking and reforming of van der Waals interactions. *Soft Matter*, **7**, 716–721.

226. Wu, Y., Wu, S., Zou, G., and Zhang, Q. (2011) Solvent effects on structure, photoresponse and speed of gelation of a dicholesterol-linked azobenzene organogel. *Soft Matter*, **7**, 9177–9183.
227. Jiao, T., Wang, Y., Gao, F., Zhou, J., and Gao, F. (2012) Photoresponsive organogel and organized nanostructures of cholesterol imide derivatives with azobenzene substituent groups. *Prog. Nat. Sci. Mater. Int.*, **22** (1), 64–70.
228. Kobayashi, H., Friggeri, A., Koumoto, K., Amaike, M., Shinkai, S., and Reinhoudt, D.N. (2002) Molecular design of "super" hydrogelators: understanding the gelation process of azobenzene-based sugar derivatives in water. *Org. Lett.*, **4**, 1423–1426.
229. Ogawa, Y., Yoshiyama, C., and Kitaoka, T. (2012) Helical assembly of azobenzene-conjugated carbohydrate hydrogelators with specific affinity for lectins. *Langmuir*, **28** (9), 4404–4412.
230. Rajaganesh, R., Gopal, A., Das, T.M., and Ajayaghosh, A. (2012) Synthesis and properties of amphiphilic photoresponsive gelators for aromatic solvents. *Org. Lett.*, **14** (3), 748–751.
231. Mamiya, J.-I., Kanie, K., Hiyama, T., Ikeda, T., and Kato, T. (2002) A rodlike organogelator: fibrous aggregation of azobenzene derivatives with a syn-chiral carbonate moiety. *Chem. Commun.*, 1870–1871.
232. Moriyama, M., Mizoshita, N., Yokota, T., Kishimoto, K., and Kato, T. (2003) Photoresponsive anisotropic soft solids: liquid-crystalline physical gels based on a chiral photochromic gelator. *Adv. Mater.*, **15** (16), 1335–1338.
233. Moriyama, M., Mizoshita, N., and Kato, T. (2004) Photopatterning of discotic liquid-crystalline gels. *Polym. J.*, **36** (8), 661–664.
234. Moriyama, M., Mizoshita, N., and Kato, T. (2006) Reversible on–off photoswitching of hydrogen bonding for self-assembled fibers comprising physical gels. *Bull. Chem. Soc. Jpn.*, **79**, 962–964.
235. Yagai, S., Nakajima, T., Kishikawa, K., Kohmoto, S., Karatsu, T., and Kitamura, A. (2005) Hierarchical organization of photoresponsive hydrogen-bonded rosettes. *J. Am. Chem. Soc.*, **127** (31), 11134–11139.
236. Duan, P., Li, Y., Li, L., Deng, J., and Liu, M. (2011) Multiresponsive chiroptical switch of an azobenzene-containing lipid: solvent, temperature, and photoregulated supramolecular chirality. *J. Phys. Chem. B*, **115** (13), 3322–3329.
237. Minkin, V.I. (2004) Photo-, thermo-, solvato-, and electrochromic spiroheterocyclic compounds. *Chem. Rev.*, **104** (5), 2751–2776.
238. Hachisako, H., Nakayama, H., and Ihara, H. (1999) Determination of critical aggregation concentrations of self-assembling lipids in nonpolar organic media using spiropyrans as photochromic probes. *Chem. Lett.*, **28**, 1165–1166.
239. Sumiya, S., Shiraishi, Y., and Hirai, T. (2013) Spiropyran–cholesterol conjugate as a photoresponsive organogelator. *New J. Chem.*, **37**, 2642–2647.
240. Qiu, Z., Yu, H., Li, J., Wang, Y., and Zhang, Y. (2009) Spiropyran-linked dipeptide forms supramolecular hydrogel with dual responses to light and to ligand–receptor interaction. *Chem. Commun.*, 3342–3344.
241. Chen, Q., Feng, Y., Zhang, D., Zhang, G., Fan, Q., Sun, S., and Zhu, D. (2010) Light-triggered self-assembly of a spiropyran-functionalized dendron into nano-/micrometer-sized particles and photoresponsive organogel with switchable fluorescence. *Adv. Funct. Mater.*, **20** (1), 36–42.
242. Shumburo, A. and Biewer, M. (2002) Stabilization of an organic photochromic material by incorporation in an organogel. *Chem. Mater.*, **14** (9), 3745–3750.
243. Lee, H.-Y., Diehn, K.K., Sun, K., Chen, T., and Raghavan, S.R. (2011) Reversible photorheological fluids based on spiropyran-doped reverse micelles. *J. Am. Chem. Soc.*, **133** (22), 8461–8463.
244. Lokshin, V., Samat, A., and Metelitsa, A.V. (2002) Spirooxazines: synthesis, structure, spectral and photochromic properties. *Russ. Chem. Rev.*, **71**, 893–916.

245. Li, Y., Wong, K.M.-C., Tam, A.Y.-Y., Wu, L., and Yam, V.W.-W. (2010) Thermo- and acid-responsive photochromic spironaphthoxazine-containing organogelators. *Chem. Eur. J.*, **16** (29), 8690–8698.
246. Tian, H. and Wang, S. (2007) Photochromic bisthienylethene as multifunction switches. *Chem. Commun.*, 781–792.
247. Lucas, L.N., van Esch, J., Kellogg, R.M., and Feringa, B.L. (2001) Photocontrolled self-assembly of molecular switches. *Chem. Commun.*, 759–760.
248. de Jong, J.J.D., Lucas, L.N.R., Kellogg, M., van Esch, J.H., and Feringa, B.L. (2004) Reversible optical transcription of supramolecular chirality into molecular chirality. *Science*, **304**, 278–281.
249. de Jong, J.J.D., Hania, P.R., Pugžlys, A., Lucas, L.N., de Loos, M., Kellogg, R.M., Feringa, B.L., Duppen, K., and van Esch, J.H. (2005) Light-driven dynamic pattern formation. *Angew. Chem. Int. Ed.*, **44** (16), 2373–2376.
250. Akazawa, M., Uchida, K.J.J., de Jong, D., Areephong, J., Stuart, M., Caroli, G., Browne, W.R., and Feringa, B.L. (2008) Photoresponsive dithienylethene-urea-based organogels with "reversed" behavior. *Org. Biomol. Chem.*, **6**, 1544–1547.
251. de Jong, J.J.D., Tiemersma-Wegman, T.D., van Esch, J.H., and Feringa, B.L. (2005) Dynamic chiral selection and amplification using photoresponsive organogelators. *J. Am. Chem. Soc.*, **127** (40), 13804–13805.
252. Wang, S., Shen, W., Feng, Y., and Tian, H. (2006) Clickity-click: highly functionalized peptoid oligomers generated by sequential conjugation reactions on solid-phase support. *Chem. Commun.*, **4**, 1497–1502.
253. Rameshbabu, K., Zou, L., Kim, C., Urbas, A., and Li, Q. (2011) Self-organized photochromic dithienylcyclopentene organogels. *J. Mater. Chem.*, **21**, 15673–15677.
254. Xiao, S., Zou, Y., Yu, M., Yi, T., Zhou, Y., Li, F., and Huang, C. (2007) A photochromic fluorescent switch in an organogel system with non-destructive readout ability. *Chem. Commun.*, 4758–4760.
255. van Herpt, J.T., Areephong, J., Stuart, M.C.A., Browne, W.R., and Feringa, B.L. (2014) Light-controlled formation of vesicles and supramolecular organogels by a cholesterol-bearing amphiphilic molecular switch. *Chem. Eur. J.*, **20** (6), 1737–1742.
256. van Herpt, J.T., Stuart, M.C.A., Browne, W.R., and Feringa, B.L. (2014) A dithienylethene-based rewritable hydrogelator. *Chem. Eur. J.*, **20** (11), 3077–3083.
257. Jiang, Y., Zeng, F., Gong, R., Guo, Z., Chen, C., and Wan, X. (2013) A multi-stimuli responsive organogel based on a tetrapeptide–dithienylcyclopentene conjugate. *Soft Matter*, **9**, 7538–7544.
258. Sevez, G., Gan, J., Pan, J., Sallenave, X., Colin, A., Saadoui, H., Saleh, A., Vögtle, F., and Pozzo, J.-L. (2007) Multi-addressable supramolecular gels based on linear amino acid and bisthienylcyclopentene. *J. Phys. Org. Chem.*, **20** (11), 888–893.
259. Bouas-Laurent, H. and Dürr, H. (2001) Organic photochromism. *Pure Appl. Chem.*, **73**, 639–665.
260. Ahmed, S.A., Sallenave, X., Fages, F., Mieden-Gundert, G., Müller, W.M., Müller, U., Vögtle, F., and Pozzo, J.-L. (2002) Multiaddressable self-assembling organogelators based on 2H-chromene and N-acyl-1,ω-amino acid units. *Langmuir*, **18** (19), 7096–7101.
261. Katritzky, A.R., Sakhuja, R., Khelashvili, L., and Shanab, K. (2009) Gelation behavior of 2H-chromene N-acylamino acid conjugates. *J. Org. Chem.*, **74** (8), 3062–3065.
262. Sun, L., Wang, G., Liu, L., and Wang, A. (2014) Study on the controlled gel formation and photochromic properties of a new cholesterol-bridge-naphthopyran dyad. *Bull. Korean Chem. Soc.*, **35** (5), 1343–1348.
263. Mutoh, K. and Abe, J. (2011) Comprehensive understanding of structure-photosensitivity relationships of photochromic [2.2]paracyclophane-bridged imidazole dimmers. *J. Phys. Chem. A*, **115** (18), 4650–4656.

264. Takizawa, M., Kimoto, A., and Abe, J. (2011) Photochromic organogel based on [2.2]paracyclophane-bridged imidazole dimer with tetrapodal urea moieties. *Dyes Pigm.*, **89**, 254–259.
265. Xue, P., Lu, R., Chen, G., Zhang, Y., Nomoto, H., Takafuji, M., and Ihara, H. (2007) Functional organogel based on a salicylideneaniline derivative with enhanced fluorescence emission and photochromism. *Chem. Eur. J.*, **13** (29), 8231–8239.
266. Simalou, O., Zhao, X., Lu, R., Xue, P., Yang, X., and Zhang, X. (2009) Strategy to control the chromism and fluorescence emission of a perylene dye in composite organogel phases. *Langmuir*, **25** (19), 11255–11260.
267. Datta, S. and Bhattacharya, S. (2012) Evidence of aggregation induced emission enhancement and keto-enol-tautomerism in a gallic acid derived salicylideneaniline gel. *Chem. Commun.*, **48**, 877–879.
268. Görner, H. and Kuhn, H.J. (2007) in *Advances in Photochemistry*, vol. **19** (eds D.C. Neckers, D.H. Volman, and G. von Bünau), John Wiley & Sons, Inc., pp. 1–117.
269. Eastoe, J., Sáchez-Dominguez, M., Wyatt, P., and Heenan, R.K. (2004) A photo-responsive organogel. *Chem. Commun.*, 2608–2609.
270. Chen, C.T., Chen, C.-H., and Ong, T.-G. (2013) Complementary helicity interchange of optically switchable supramolecular-enantiomeric helicenes with (−)-Gel–Sol– (+)-Gel transition ternary togic. *J. Am. Chem. Soc.*, **135** (14), 5294–5297.

9
Photochromic Materials in Biochemistry

Danielle Wilson and Neil R. Branda

9.1
Introduction

The use of photoresponsive materials to reversibly control biological events is an exciting prospect to externally manipulate important processes in cells, tissues, and organisms with a high degree of spatial and temporal accuracy in a relatively noninvasive manner. Light is a convenient stimulus because it can be precisely tuned and focused giving the user the ability to alter a particular process "on command" with a high degree of efficiency and with minimal perturbation to the environment. These features have resulted in researchers developing new tools and techniques to investigate complex biological functions using light. The field has attracted considerable interest over the past several decades and photoresponsive molecules have been extensively used in biochemical settings for a range of tasks including controlling neurotransmitters, DNA hybridization, tissue imaging, and protein structure and dynamics. In this chapter, we discuss the photocontrol of biologically relevant systems using synthetically derived photochromic molecules. We will review some of the considerations and requirements when designing effective systems, and highlight a few selected examples to demonstrate their capacity to influence biological systems. Discussion of optogenetics or photocages will not be included, although these techniques also offer excellent means to achieve photocontrol over biological functions [1, 2]. This review is far from exhaustive. Instead, our intent is to present a general overview of recent progress made in the field of light-controlled biomolecules in a variety of contexts. These individual topics have been reviewed in detail elsewhere, and the reader is referred to the following references for earlier examples [1, 3–19].

Photochromic Materials: Preparation, Properties and Applications, First Edition.
Edited by He Tian and Junji Zhang.

9.2 Reversible Photochemical Switching of Biomaterial Function

Light is effectively used in nature as a signal to regulate many complex cellular functions ranging from vision to photosynthesis. A key feature that facilitates the harnessing of light energy is the presence of a photon-absorbing ligand or chromophore within particular proteins. The chromophores absorb light of a specific wavelength and convert the absorbed light energy into a chemical signal. There are a limited number of specialized protein families that naturally express light-responsive chromophores, a few of these include phytochromes, rhodopsins, xanthopsins, cryptochromes, phototropins, and light oxygen voltage domains [20]. The bulk of biological molecules are not inherently photoresponsive, although it would be desirable to be able to introduce such a chromophore into these systems so that light-induced manipulations can be extended to a wider range of biological processes. This can be achieved through the careful incorporation of an appropriate chromophore into a bioactive molecule, either through direct attachment to the target biomolecule itself or to another compound such as a ligand that is capable of interacting with the target. Through the appropriate tuning of the photoresponsive molecule, the steric and electronic changes it undergoes as it absorbs light can be exploited to influence a biological pathway remotely, and where and when the user wants. Today, there is a wide range of tools capable of modifying various biochemical functions [18, 21, 22]. Those relevant to artificial photochromic biosystems will be discussed in the following section.

9.3 General Design Strategies and Considerations

The design of artificial light-controlled biological systems using photochromic molecules typically relies on harnessing the light-induced structural changes that accompany the chromophore's photoisomerization process and exploiting these changes to influence biomolecule function. Unfortunately though, there is not one single and straightforward strategy applicable for a wide range of targets, as it is challenging to predict exactly how a chromophore will interact with a selected target. However, the techniques for designing these have evolved and expanded considerably and there are many options to aid in rationally designing a light-controlled biomolecule. For example, the availability of crystal structures of targets enables the prediction of binding interactions. Ultimately, the design requires some degree of a trial and error to prepare a derivative that can be optimally controlled. The most widely used strategies for designing light controlled biomolecules can be broadly classified into two principally different categories (Figure 9.1) [15], these include:

1) *Direct – Modification of a biomolecule with a photochromic molecule.* This approach involves the covalent attachment of a photochromic group to one or more sites on the biomolecule. For instance, proteins can be modified

through covalent attachment of an appropriate photochromic molecule to an amino acid group, at or in the vicinity of an active site or other critical area required for function. Other biomolecules such as nucleic acids can be modified in a similar manner through incorporation of a photochromic group onto a position on the phosphate backbone, nucleobase moiety, or ribose moiety.

2) *Indirect – Modification of an external molecule with a photochromic molecule.* For example, a ligand that can interact with the target biomolecule can result in the desired effect. This can be achieved through the incorporation of a photochromic molecule into the structure of the ligand, often an inhibitor, activator, or cofactor that can interact with the biomolecule. Changes in the properties of the ligand are then used to indirectly influence the activity of the biomolecule through differences in their affinity for a particular site, for example, an active site. Alternatively, another form of indirect control is by incorporating the biomolecule into a photoresponsive environment such as a micelle, a solution, or a surface that has been modified with a chromophore.

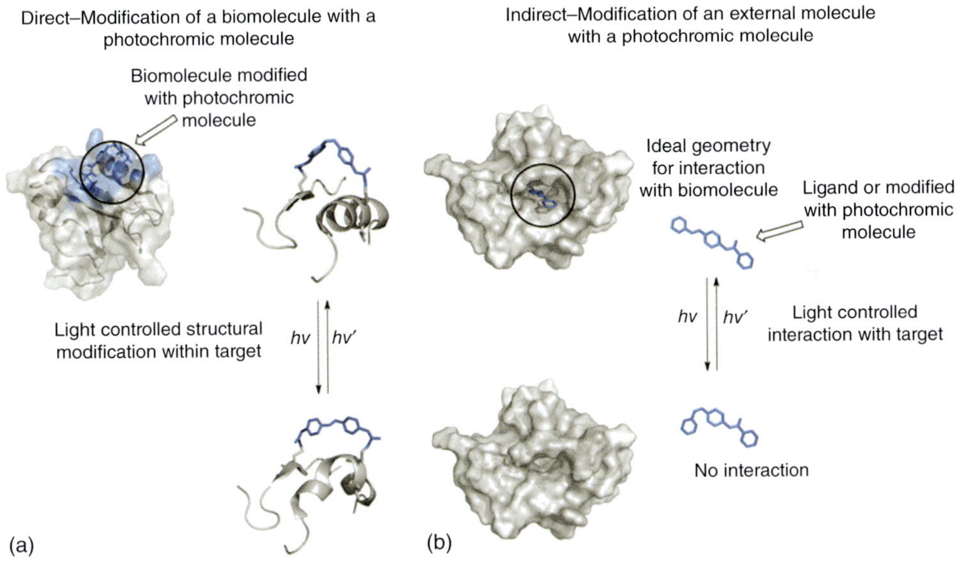

Figure 9.1 (a) A photochromic group is introduced to one or more sites within the biological target. The reversible structural changes that occur with isomerization influence the conformation and function of the macromolecule, either through alteration to the native structure (shown) or by changing the orientation of the photochromic group so that it can physically block or bind to a critical site (not shown). (b) The photochromic group is introduced into an external ligand. The structure and properties of the ligand are altered with isomerization and each isomer has a different affinity for the target. Images were generated using PyMOL [23].

9.3.1 Photoswitchable Tethers

The direct incorporation of a photochromic group into the biomolecule has been investigated for a number of different uses ranging from modulating enzyme function, protein conformation, activation of channels, and DNA hybridization [24–33]. This approach involves the modification of a site within the biomolecule with a photochromic group typically through a linkage to one or more sites on the photochromic group. The photoswitch is then used to control either the structural elements of the biomolecule itself, for example, by disrupting the native folded state, or by physically blocking a critical functional region through steric interactions. Ideally, in one photoisomeric state, the native structure of the biomolecule is retained so that it can perform normally, whereas in the other state, the biomolecule is perturbed in such a way that hinders normal function, thus allowing the user to initiate or interrupt a functional process simply with light illumination.

9.3.1.1 The Incorporation Method

A number of methods have been used to incorporate a photochromic group into the biomolecule of interest. The selection will depend on several factors, mainly the type of biomolecule and the ease with which it can be modified effectively, the location of the desired attachment site(s), and the type of distortion predicted to be necessary to influence the function. Some of the earliest attempts to influence the properties of proteins involved the random covalent attachment of photoswitchable groups to the side chain of amino acids in the native protein structure [28, 33]. Alternatively, a specific group capable of reacting with a chromophore at a later time can be introduced through mutation. This approach involves introduction of a reactive amino acid at a specific site within the molecule, typically cysteine, that is introduced through a mutation in the native structure. The photochromic molecule can also be selectively incorporated into a particular site during the synthesis of the biomolecule using photochromic amino acids. In the case of nucleic acids, photoresponsive-modified base pairs can be selectively introduced into desired positions during synthesis.

9.3.1.2 Considerations

There are a number of factors to consider when designing a light-controlled biomolecule that has an incorporated photochromic tether. Some of the challenges include the isolation of the biomolecule, selecting an attachment site, and the reaction conditions to induce covalent bond formation between a site on the biomolecule and a reactive moiety on the chromophore. In other words, the ease with which the biomolecule can be derivatized with a photochromic group and how well the biomolecule tolerates manipulation. The random attachment of photoresponsive groups to the biomolecule of interest requires that the chromophore contains a reactive site such as an acid or activated ester that can undergo a reaction with a residue such as a lysine, naturally present on the biomolecule of interest. Although perhaps convenient in some cases, this approach has several challenges that may limit its use including the incomplete reaction with the photochromic species, the generation of a mixture of products with differing numbers of photochromic groups attached, and the restricted control over the attachment site. Alternatively, site-specific or selective incorporation of a photochromic species allows more control over where the chromophore is placed. However, the identification of an optimal attachment site may involve trial and error, for example, the preparation of a set of several different amino acid substitutions in the vicinity of a binding site or other critical regulation site may be necessary. This does not guarantee success and there is always a possibility that incorporation of the photochromic species disrupts the structure and completely hinders its normal behavior, in which case photocontrol is not possible. Information on the structure of the biomolecule is useful for predicting which sites may be tolerable to incorporation of a photoswitch without interfering with the normal functioning of the system.

9.3.2 Photoswitchable Small Molecules

Biological macromolecules rely on interactions with effector molecules to signal changes and to trigger different processes required for efficient functioning. The affinity of an effector or ligand for a target system is dependent on its overall geometry and the sum of the electrostatic, hydrogen bonds, and van der Waals forces formed. Slight changes to the structure of the ligand can have a dramatic influence on its binding affinity for a particular target. Small molecule ligands with a photochromic group incorporated into their structure provide a unique handle to make changes to the structure of the ligand and take advantage of the specific nature often observed for ligand–target binding complex. They offer a number of attractive advantages as they can be synthesized independently and then later introduced into any organism without the need for manipulation to the biomolecule.

9.3.2.1 The Incorporation Method

Photochromic ligands are typically designed by rational modification of a natural effector molecule, but could alternatively mimic the structure of an inhibitory drug or intercalator. Light is then used to change the conformation of the photochromic species, from one that structurally and functionally mimics the effector molecule and is recognized by the target biomolecule to another that does not, allowing for modulation of activity.

9.3.2.2 Considerations

Although, in principle, this design concept is applicable to many different biological systems, it has several challenges and important considerations. The effector must include a recognizable group that can interact favorably with the target and must also include a photochromic species that will absorb light of a specific wavelength and undergo a change in its properties, most often its steric or electronic makeup. The photochromic group must also be placed in such a way that it will not significantly impede binding of the ligand to the target biomolecule in one of the isomeric forms. Rational design of the small molecule is critical, but overall it is difficult to predict how the presence of the photochromic group will influence the binding properties and whether or not a significant enough change in binding affinity will be observed upon photoswitching. An inherit disadvantage that can make the design of these systems challenging is that the chromophore-modified ligand may be too large and no longer mimic the activity of the original substrate in either photoisomeric form leading to significant reduction or even

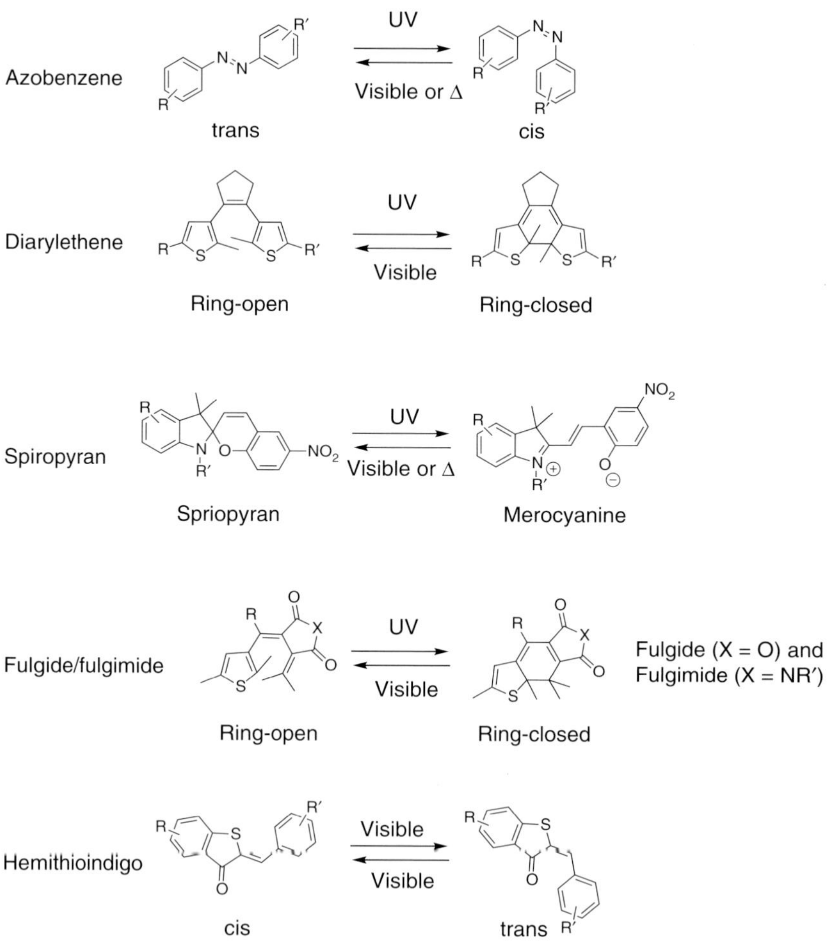

Figure 9.2 General classes of photochromic species that can be used to prepare reversibly light-controlled biological molecules.

abolishment of its affinity for the target in both photoisomers. Moreover, the photochromic group must possess decent photochemical conversion from one isomer to another, especially in cases where complete ON/OFF states are desired.

9.3.3
Chromophore Selection

There are several requirements to consider when selecting a photoswitch to use in a light-controlled biological system. Most importantly, the photoswitch should be able to efficiently absorb light in an energy range and of a power density that are tolerated by biological systems. In addition, the isomerization process should occur on an appropriate timescale to avoid prolonged irradiation. The changes that accompany isomerization must be significant enough to influence the activity of the target molecule. The photoswitch should be stable and resistant to degradation or side reactions in the biological environment. Depending on the application, the yield of the transformation from one isomer to the other may need to be as close to quantitative as possible and each isomer thermally stable, especially in cases where complete ON/OFF function is required. Alteration of the substituents on the photochromic backbone can optimize properties such as the excitation wavelength and water solubility. Several types of photoswitches are available for use in preparation of light-controlled biological molecules; some of these include spiropyrans, diarylethenes, azobenzenes, fulgides/fulgimides, and hemithioindigos (Figure 9.2) [34].

9.4
Selected Examples

9.4.1
Photoswitchable Enzymes

Enzymes play a critical role in the regulation of cellular processes. They are responsible for catalyzing a large number of diverse reactions and at speeds that would otherwise be slow or impossible in their absence. The ability to selectively regulate their function using an external light source is a desirable concept that could find use in signal amplification, sensors, medicine, and as a research tool to probe function and processes without disrupting other biomolecules. This section will review several recent examples from the literature that use a variety of methods to influence function of enzyme target.

9.4.1.1 Drug-Inspired Small Molecule Inhibitors

A number of researchers have successfully illustrated the concept of using an existing drug scaffold and then modifying it with a photoswitch to alter the drug's ability to interact with its target [35–44]. This concept generally involves using the changes that occur to the structure and overall geometry of the photoswitch or linkage distance in the case of two binding sites to regulate activity. This approach is appealing for cases where modification to the enzyme is not desired or perhaps not possible. Many enzymes have regions that bind specific small molecules such as inhibitors, activators, and cofactors, and modifying the small molecule with a photochromic species can allow for rapid light-induced changes, which can ultimately lead to differences in binding affinity for the enzyme. As mentioned in the previous section, the site where the photochromic group is incorporated into the small molecule is crucial, as enzymes usually bind to their substrates in a specific orientation through a number of critical interactions with the binding site. Therefore, the chromophore must be minimally invasive and attached in such a way that does not impede binding altogether in one of the conformations. The design of small molecule photoswitchable inhibitors is not always straightforward, and although not required, it is certainly helpful to have an idea of the nature of the binding site and critical interactions made between the residues of the binding site and the ligand. These critical interactions should be maintained in one isomer and then either removed or hindered upon photoswitching to the other, inactive isomer. The design requires a delicate balance of structural analysis of the binding site and the photoswitch.

Protease Inhibitors Proteasomes are part of an important enzyme class involved in a number of cellular processes ranging from regulation of the cell cycle, DNA construction, apoptosis, and immune responses [45]. The enzyme functions by cleaving peptide bonds in damaged or unnecessary proteins. Researchers have observed that when rates of protein synthesis become too high, often characteristic in some types of cancer cells, the proteasome cannot keep up with normal degradation and apoptosis is triggered. Therefore, this enzyme class is an important target for treating malignancies, as evidenced by several proteasome inhibitors available for treatment of cancers. A number of examples illustrating the photoregulation of proteasomes using azobenzene-derived inhibitors have been developed to date [35, 40, 41, 46, 47]. Recently, a photoswitchable protease inhibitor, inspired by the clinically approved chemotherapy drug bortezomib, was developed [42]. Bortezomib is a proteasome inhibitor used for treatment of multiple myeloma, although one of the severe drawbacks is the side effects due to nonspecific cytotoxicity toward healthy tissue [42]. The structures of the drug and the photoswitch-modified drug are shown in Figure 9.3. The design of photocontrolled proteasome inhibitor combines three structural elements: (i) an electrophilic boronic acid group for reversible covalent bond formation with the active site, (ii) an azobenzene group for photoswitchability, and (iii) a peptide moiety that is essential for active site recognition. The cocrystal structure of bortezomib bound to the active site of the proteasome reveals that the inhibitor binds deep within the active site in an extended linear conformation [48]. Therefore, the *trans*-azobenzene peptide can adopt a linear conformation similar to that observed for bortezomib, and this orientation allows for several hydrogen bonds to be made. This is in contrast to the bent cis-isomer that prevents it from interacting with the residues on the interior of the pocket. A two- to threefold

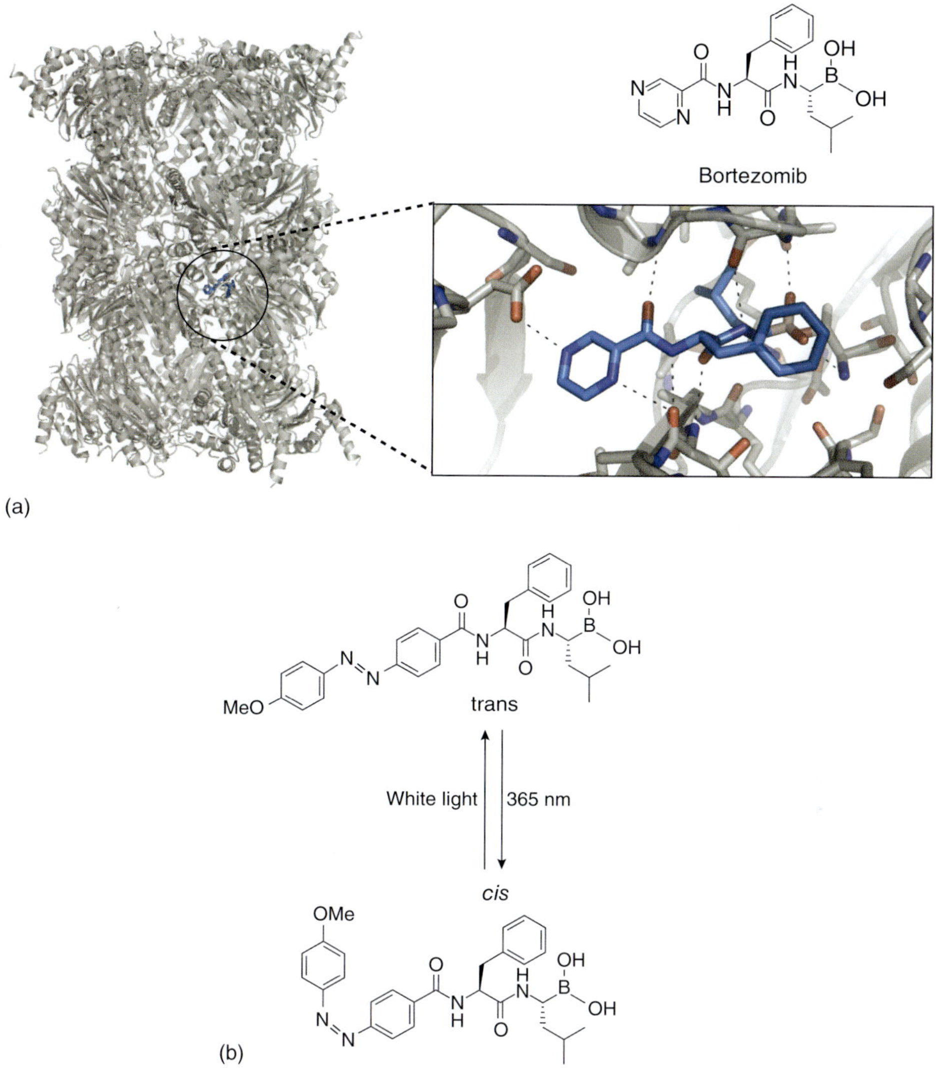

Figure 9.3 (a) Multiple myeloma drug bortezomib bound in a linear orientation in the active site of the 20s proteasome, PDB ID: 2F16 [48]. (b) Photochromic azobenzene-modified proteasome inhibitor, UV light converts the more active trans-isomer into the cis-isomer, and the isomerization can be reversed with white light [42]. Images were generated using PyMOL [23].

difference in activity is observed by photoconversion between the cis- and trans-isomers. The trans-isomer's being more active than the cis-counterpart is consistent with the hypothesis that the former would be able to fit into the tight gap of the active site, while the latter isomer would hinder binding.

Acetylcholinesterase Inhibitors The Trauner group demonstrated photocontrol of the acetylcholinesterase (AChE) using an azobenzene derivative based on the drug Tacrine, approved for early stage Alzheimer's disease [49]. AChE catalyzes the hydrolysis of the neurotransmitter acetylcholine. On the basis of the X-ray cocrystal structure of tacrine and AChE, this particular inhibitor appears to bind near the active site [50]. This differs from an earlier example described by Erlanger *et al.*, in which the trimethylammonium azobenzene targeted an allosteric site [38]. The *in vitro* activity studies measuring the rate of hydrolysis of acetylthiocholine, an acetylcholine analog, by AChE in the presence of the photochromic inhibitors indicated that the cis-isomer was an effective inhibitor (residual activity was 4% of the control). However, the trans-isomer also appeared to inhibit the enzyme to a fairly significant degree (17% of the activity remained). The photochromic inhibitor can effectively reduce the activity of the enzyme approximately fourfold by irradiation of the trans-isomer with UV light. This research group also investigated the effect of their photochromic inhibitor on the relaxation kinetics of acetylcholine-induced muscle constriction in mouse trachea. This study revealed a similar trend. In the presence of the more active cis-isomer, clearance rates of acetylcholine were reduced compared to the trans isomer (Figure 9.4).

9.4.1.2 Phosphoribosyl Isomerase Inhibitor with Two Binding Units

Another clever approach to creating small light-controlled molecules is using photochromic species that can interact with more than one binding site on the enzyme [37, 43, 51]. This strategy can give the user another means to control binding because not only is the geometry of the photoresponsive species important, but the distance and angle between the recognition sites are also influenced by photoisomerization, allowing for a unique method of manipulating its binding interaction with the enzyme. If designed properly, only one of the isomers will display an appropriate distance between the binding sites so that binding is favorable, while the other isomer will not have the appropriate distance leading to reduced binding. Recently, a series of dithienylethene-based Phosphoribosyl Isomerase A (PriA) inhibitors were designed (Figure 9.5) [44]. This enzyme catalyzes the isomerization of an aminoaldose to an amino ketose in the biosynthesis of tryptophan and histidine. The PriA enzyme has a symmetrical barrel-like scaffold and contains two binding sites on opposite sides of the barrel to accommodate binding of the phosphate groups of the aminoaldose substrate and product. The inhibitor design consists of a central dithienylethene scaffold with two pendant phosphate or phosphonate groups, the distance between the two groups and their relative orientation can change with light. In the ring-open form, the *meta*-substituted bisphosphate inhibitor displayed an IC_{50} of approximately 0.55 μM, eightfold more

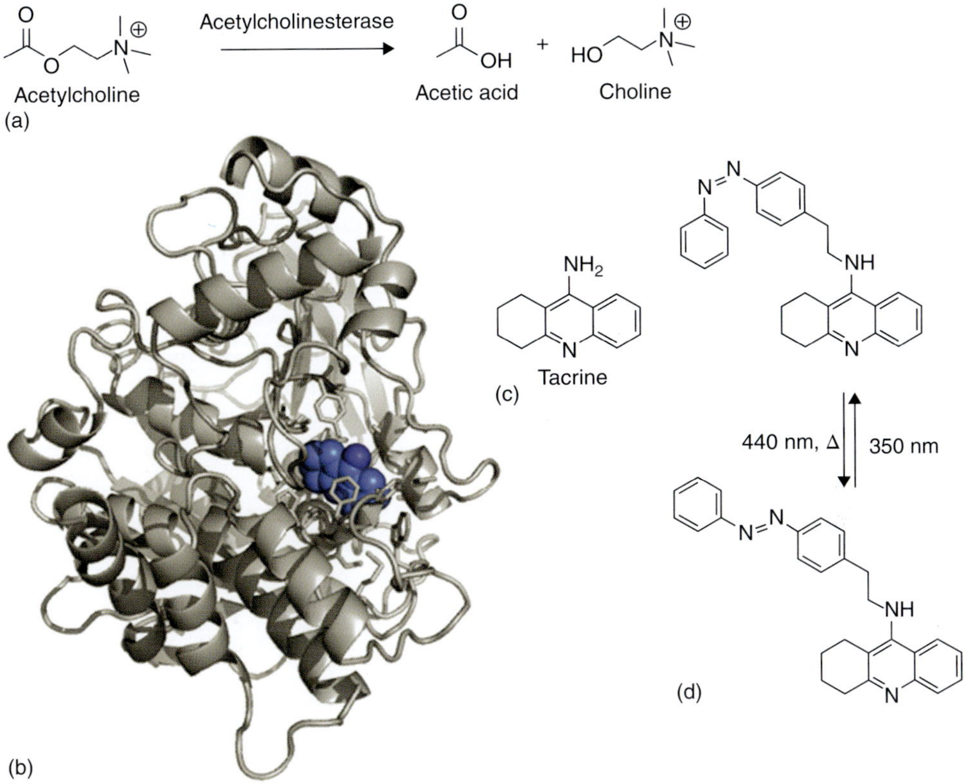

Figure 9.4 (a) Hydrolysis of acetylcholine catalyzed by acetylcholinesterase. (b) Tacrine inhibitor bound to acetylcholinesterase (PDB ID: 1ACJ) hinders acetylcholine binding. (c) Structure of tacrine. (d) Photochromic acetylcholinesterase inhibitors [49]. Image was generated using PyMOL [23].

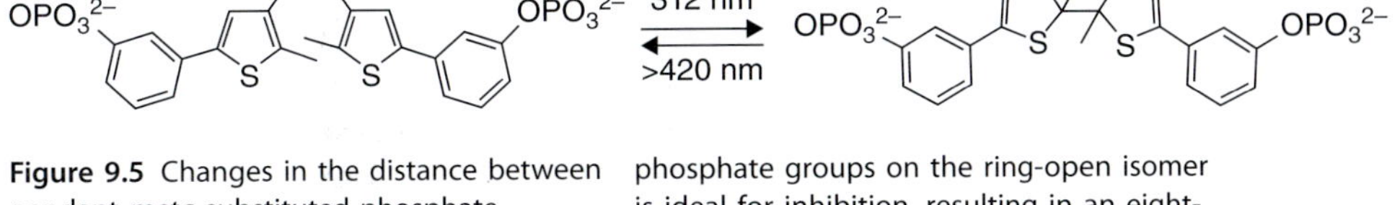

Figure 9.5 Changes in the distance between pendant *meta*-substituted phosphate groups on dithienylethene-modified PriA enzyme inhibitor. The distance between the phosphate groups on the ring-open isomer is ideal for inhibition, resulting in an eight-fold higher inhibitor potency compared to the ring-closed isomer [44].

potent than the ring-closed isomer. The decrease in affinity of the ring-closed isomer is attributed to the shortened distance between the two pendant phosphate groups of the dithienylethene. In contrast, the ortho- and para-substituted derivatives showed similar inhibitory activities regardless of isomerization state.

9.4.1.3 Direct Modification of Enzymes with Photochromic Groups

Not all enzymes have structures that can be readily targeted with small molecule effectors. For example, the binding site may not accommodate the increase in size that occurs after addition of a photoswitchable group. This consideration is especially important for enzymes with binding sites that are buried deep within the macromolecule. Therefore, an alternative strategy is to attach the photochromic group to the enzyme itself. It is important to consider where the photoswitch is to be incorporated; enzymes are flexible molecules and can often absorb small changes in structure induced by photoresponsive species. At the other extreme, attachment of a photochromic group should not alter the structure too dramatically as this could lead to permanent misfolding and loss of function of enzyme.

Site-Selective Introduction of Photochromic Species – Single Linkage One of the most common strategies used for the site-selective incorporation of a photochromic molecule is through the introduction of an orthogonal coupling group using site-directed mutagenesis. In this approach, typically an amino acid located at a particular site in the enzyme is mutated with a cysteine residue and can subsequently undergo reaction with a photochromic group at a later point. In this case, the photochromic group must contain thiol-reactive moiety such as a maleimide or haloacetamide so that attachment to the desired mutation site

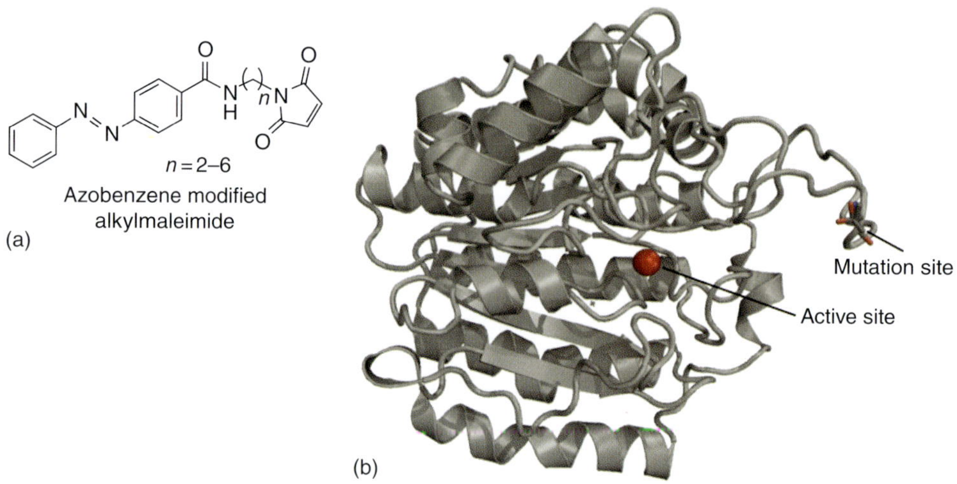

Figure 9.6 (a) Azobenzene-modified maleimide of varying alkyl chain linkage length [52]. (b) Histone deacetylase-like aminohydrolase (HDAH) enzyme, active site, and external loop of enzyme mutation site. Image is of wild-type PDB ID: 2VCG. Image was generated using PyMOL [23].

is possible. In a recent publication by the Meyer-Almes group (Figure 9.6), they described how they modified several alkylmaleimides of varying chain length with an azobenzene system, and then covalently attached the resulting photochromic assembly to a predetermined site on a mutant histone deacetylase-like aminohydrolase (HDAH) [52, 53]. The mutant variant of HDAH (M30S) contains a maleimide-reactive cysteine along the edge of an outward facing solvent loop, adjacent to the active site of the enzyme. This site is solvent accessible, facilitating the covalent attachment of the thiolate of cysteine to the maleimide double bond. The researchers found that when the alkyl linkage was two or three carbon atoms in length, the deacetylation rate could be influenced by light. There was an approximately twofold difference achieved between the activities of the two isomers. Although the attachment site of the photochromic species was located on a flexible loop away from the core of the enzyme, and would predictably cause minimal perturbation to the native protein structure, linkage lengths of four to six carbons exhibited much lower rates of substrate conversion in both cis- and trans-isomers, making photocontrol of activity in these cases not possible.

Site-Selective Introduction of Photochromic Species – Cross-Linking Successful light-induced control has also been demonstrated in enzyme systems by attaching a photoswitchable cross-linker [29]. This approach relies on the light-induced changes to the end-to-end distance of the photochromic molecule. In an example published by Gorostiza and coworkers, they prepared an azobenzene derivative containing a thiol-reactive sites on each end. They introduced a series of appropriately spaced cysteine mutations into the restriction enzyme for incorporation of the bifunctional azobenzene cross-linker. In one mutant variant, where the cysteine mutations were introduced close to the active site, they observed a nearly 16-fold change in enzymatic activity upon light-induced isomerization of the photoresponsive cross-linker. Overall, in contrast to single linkage sites, this approach typically results in larger structural changes. However, a potential drawback to this strategy is that it requires introduction of an appropriate cysteine into the desired biomolecule, followed by attachment of the photochromic moiety, which makes the system more complex and may limit *in vivo* applications.

9.4.2 Photoswitchable Peptides and Proteins

The introduction of photoswitches into proteins and peptides could allow for the reversible control of their structure and function, and as a result of the photoinduced isomerization, these changes can subsequently alter protein–protein interactions, protein–DNA interactions, as well as other intracellular signaling events, and can function as a convenient tool for studying protein-folding dynamics.

9.4.2.1 Peptide Cross-Linking

The conformational changes within a small-molecule photochromic species are often relatively insignificant in comparison to those in most proteins.

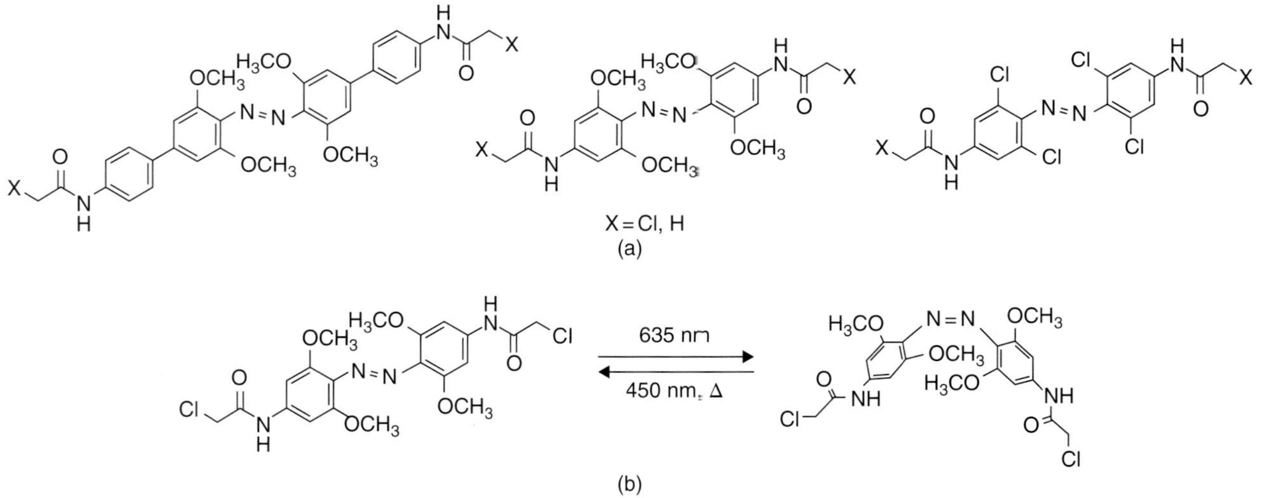

Figure 9.7 (a) Tetra-*ortho*-substituted azobenzene photoswitches that can be isomerized using red light. (b) Red light converts the trans-isomer to the cis-isomer, and this process can be reversed with blue light [58].

Light-induced changes in the photochromic group may be accommodated by a flexible protein backbone and therefore have little effect on the target's activity. Several studies have been conducted on smaller peptide sequences and simpler elements of protein secondary structure, such as α-helices and β-hairpins, using photoswitchable cross-linked groups [9, 12, 54–65]. The structural distortion can often be predicted to some degree by comparing the distance between the attachment points in the peptide and the distance between the linkage sites on the chromophore [61, 66].

There has been increasing motivation to develop photochromic systems that can be photoswitched using longer wavelengths of light to reduce tissue damage and increase the light penetration depth, often a concern for studies with biological systems. Recently, Woolley and coworkers designed a series of visible light-triggered photoswitchable peptide linkers with good photoconversion yields and thermal stabilities [58]. The ortho-positions of the azobenzene groups were substituted with bulky methoxy- or halogen-groups as shown in Figure 9.7. This substitution pattern distorts the planar ring system and results in a shift in the absorption spectrum of the trans-isomer, specifically the $n-\pi^*$ band, allowing for effective isomerization using red light, while maintaining good photochromic performance. Linking the chloroacetamide derivative to two appropriately spaced cysteine residues allowed them to control the helical content of a α-helical peptide, FK-11. The trans-isomer has an ideal end-to-end distance to support helix formation, whereas the distance is too long to be accommodated by the cis-isomer. These researchers demonstrated that irradiation of the peptide with red light produced the cis-isomer in nearly 98% yield. This transformation was accompanied by a decrease in helical content as evidenced by the changes in the circular dichroism spectra, which could be reversed by exposure of the peptide to blue light.

9.4.2.2 Cyclic Antimicrobial Peptide

Photochromic cyclic peptides have been exploited in the past using azobenzenes to control muscle fiber contraction [63]. Recently, Ulrich and Komarov prepared a cyclic peptidomimetic containing a dithienylethene group based on the antimicrobial peptide gramicidin S (Figure 9.8) [67].

The introduction of a photoresponsive dithienylethene group into the cyclic peptide allowed for changes in the ring size and overall geometry to be achieved reversibly with light. This research group predicted that the more flexible ring-open isomer would more closely resemble the conformation and amphiphilic properties of gramicidin S, which are important for its antibiotic activity. The site of attachment of the photochromic moiety was varied and they were able to demonstrate approximately 16-fold change in biological activity upon photoisomerization, which allowed for effective control over the growth of bacteria. As predicted, the ring-open form was distinctly more active than the closed form with all strains tested. Using molecular dynamic simulations, the authors calculated the distribution of nonpolar and polar surface charges. They found that the ring-open isomer, although less hydrophobic than Gramicidin S, is more hydrophobic than the ring-closed isomer and suggested that the difference

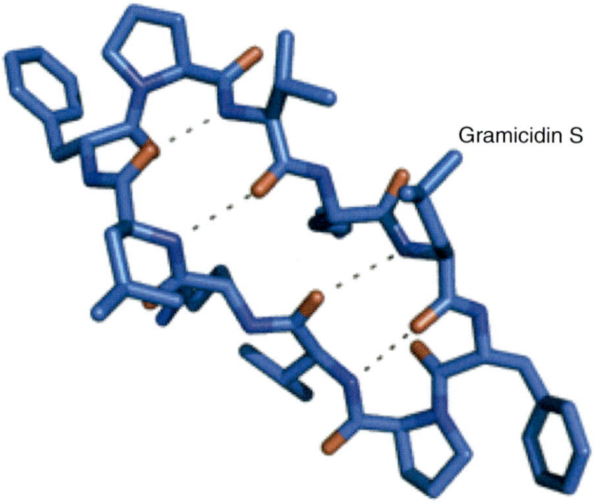

Figure 9.8 Antimicrobial peptide gramicidin S [67].

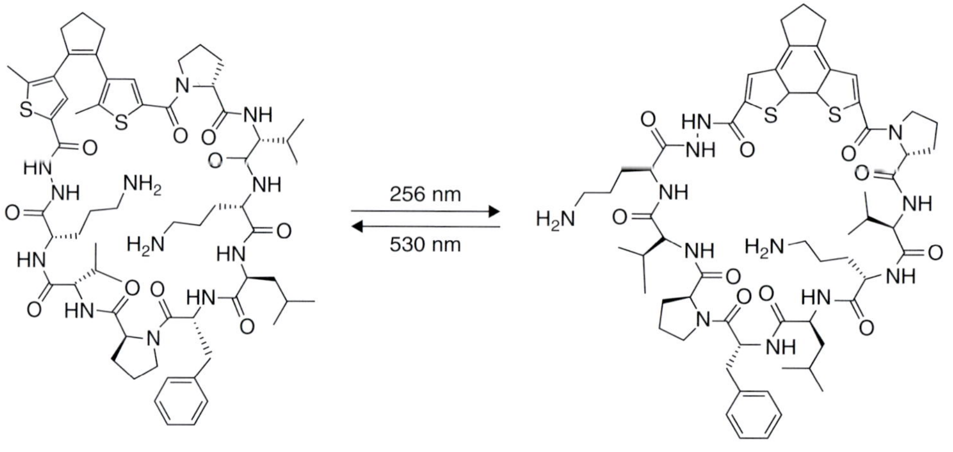

Figure 9.9 Photochromic dithienylethene-based cyclic peptide with light-dependent antimicrobial activity [67].

in activity could be attributable to changes in the amphipathic nature of the molecule that occur with photoswitching (Figure 9.9).

9.4.2.3 **Genetically Encoded Amino Acids**

Peptides or proteins can be modified with photochromic groups during their synthesis using an unnatural photoswitch-derived amino acid. For smaller peptides that can be made using solid phase synthesis, photochromic amino acids can be incorporated in place of a natural amino acid. More complex protein structures can be prepared using a modified amino acid that is incorporated into the protein during translation with the use of chemically modified amino acyl tRNAs and tRNA suppression technology. This method was developed by Schultz and has since been used to introduce photoswitchable amino acids at predetermined locations within the protein [32, 68, 69]. For example, a photochromic azobenzene-derived phenylalanine residue is usually selected as the photochromic counterpart (Figure 9.10). This hybrid photoswitchable amino acid is then coupled to pdCpA and ligated to a modified tRNA lacking two terminal nucleotides, followed by *in vitro* translation using mRNA. Recently, it has been demonstrated that the successful incorporation of a photochromic amino acid with a second reactive group can undergo addition to another nearby site on the protein [70]. This approach allows for the formation of another attachment site, or cross-linking to the protein, giving the user the opportunity to make larger scale changes to the protein backbone and secondary structure. It does, however, require the presence of a reactive amino acid nearby, such as a cysteine residue, so that cross-linking can be accomplished. Depending on the sequence of the protein, it may require mutation to introduce an appropriate reactive residue to increase the likelihood of cross-linking. The authors incorporated the modified azo-phenylalanine containing a reactive benzyl chloride to a helical portion of Calmodulin using a mutated tRNA-synthetase. In order to ensure cross-linkage formation, a cysteine residue was positioned nearby, at an appropriate residue spacing to accommodate the trans-isomer. The cross-linking efficiency with the nearby cysteine proved to be very high with no detectable uncross-linked protein. The photoinduced structural changes in the protein were measured using circular dichroism and indicated that as anticipated, the trans-isomer had a higher degree of helical content than the cis-isomer, which led to distortion of the helix.

9.4.2.4 Control of Motor Protein Function Using Site-Selective Mutation

In a recent study by Maruta *et al.*, a series of photochromic derivatives were synthesized and attached to predetermined sites of a critical loop region of the Kinesin Eg5 motor protein, allowing for control over the ATPase activity using light [71, 72]. They also investigated whether or not the photochromic groups in either isomeric state could block active site access and overall reduce the inhibitory effect of the known inhibitor, S-trityl cysteine.

The photochromic species were attached to the kinesin target by reaction of an alkyl iodide with several cysteine mutants introduced via site-directed mutagenesis. The native kinesin contains four cysteine residues, and to incorporate the cysteine-reactive photochromic species specifically, they prepared a version in which the intrinsic cysteines were substituted by other amino acids (Figure 9.11). The trityl-substituted derivative **1** displayed similar activities in both the cis- and trans-isomers and, therefore, photocontrol of activity was not possible. In the absence of treatment with S-trityl cysteine, the maleimide **2** and spiropyran **3** modified proteins did not show significant difference in activity between the two isoforms. However, in the presence of S-trityl cysteine, both the maleimide and spiropyran modified proteins attached to the D130C mutant exhibited light-dependent control over ATPase activity. In the case of

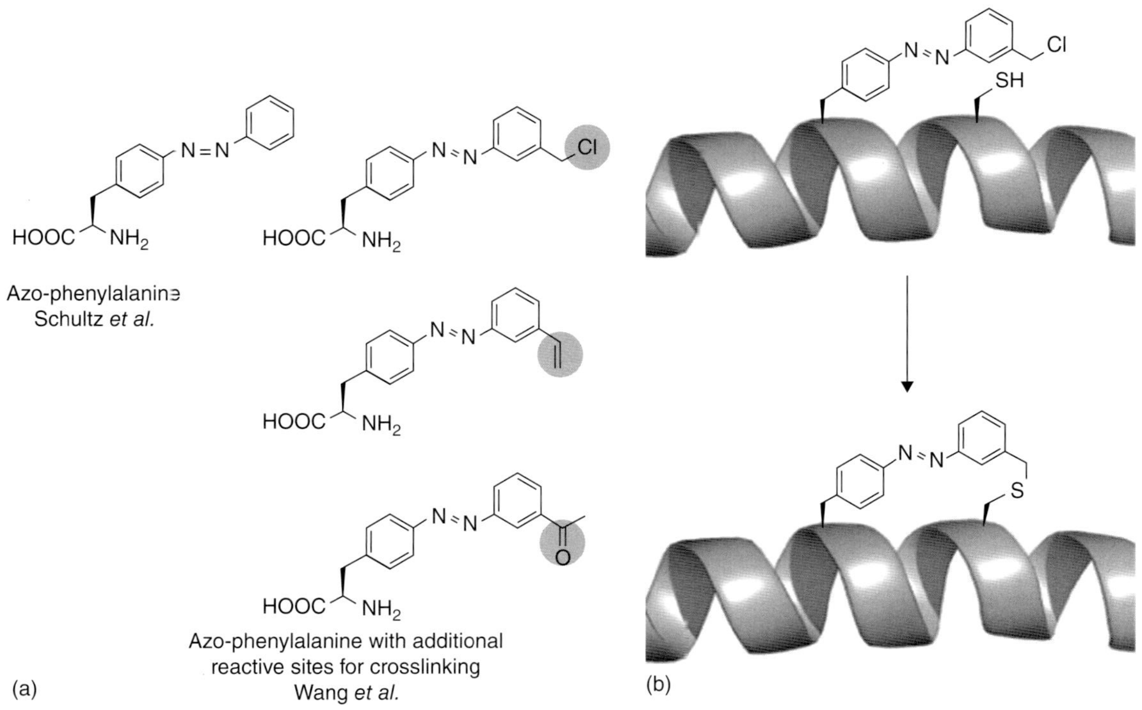

Figure 9.10 (a) Traditionally used azobenzene-modified amino acid azo-phenylalanine and novel azo-phenylalanine with an additional reactive benzyl chloride, alkene, and keto site. (b) Reaction with a nucleophile is possible to prepare cross-linked photochromic amino acids [70].

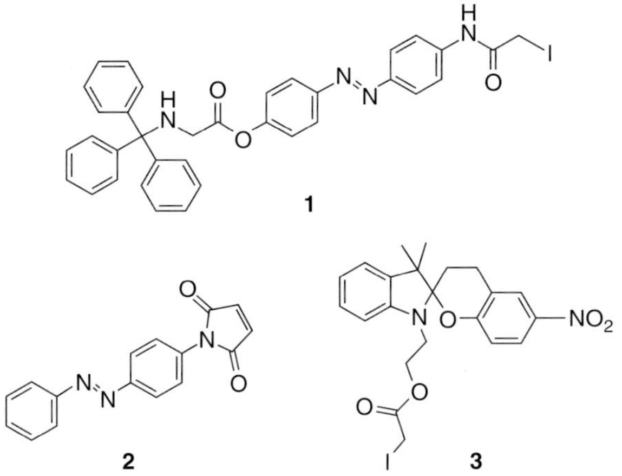

Figure 9.11 Structures of attached photochromic groups: (**1**) trityl-substituted azobenzene, (**2**) maleimide-substituted azobenzene, and (**3**) spiropyran [71].

the merocyanine isomer, the calculated K_i, V_{max}, and K_m values for the inhibitor indicated that it significantly reduced the activity of the S-trityl cysteine inhibitor, likely by blocking access to its binding site. A similar observation was made for the maleimide inhibitor. Using this approach they were able to successfully incorporate the photochromic groups into the selected cysteine residues and use light to influence the binding of the S-trityl cysteine inhibitor.

9.4.3 Photoswitchable Ion Channels and Receptors

Ion channels and receptors are made up of proteins that span the membrane of cells and function to transport ions or metabolites across cell walls. They are important for signaling and neuronal function. There has been much progress in the field of light-controlled neuronal signaling, and these advances have extended to living cells and organisms [1, 8, 73, 74]. The combination of synthetic photoswitches and naturally occurring receptor proteins has proved to be an effective means for rapidly and reversibly controlling channel function, and have aided in studying processes related to channel function. One of the methods commonly used to create light-controlled ion channels and receptors involve the covalent attachment of photochromic ligands to a site on the receptor protein using site-specific mutation of an amino acid to cysteine [75–79]. This approach involves genetic manipulation of the protein so that a reactive cysteine residue can be placed in a specific site on the receptor. A second method involves a small molecule ligand that contains a head group that resembles a known receptor agonist or antagonist [74, 79–85]. The affinity is light dependent, the conformation of one isomer interferes with binding and/or activation and the other permits it.

9.4.3.1 Photocontrol of Channel Activation and Desensitization with a Tethered Glutamate

In a recent publication, Isacoff *et al.* describe an ion channel that uses covalently tethered photochromic ligands to demonstrate the effects of rapid and reversible binding and dissociation events with a high degree of precision [86]. They designed a cysteine-reactive maleimide azobenzene derivative containing a pendant glutamate group to mimic binding of the agonist glutamate. This photochromic derivative was covalently attached to an engineered cysteine residue located near the glutamate-binding site. The azobenzene allowed for light-induced changes in the positioning of the pendant glutamate moiety, and when in the trans-conformation, the glutamate is facing away from the channel. Alternatively, in the cis-conformation, the glutamate faces toward the channel and can access the binding site. This mimics a high concentration of glutamate ligand, which allows for rapid light-induced changes in receptor binding. This technique provides unique study tool for measuring channel response to receptor-binding events. It has the advantage over traditional methods that it does not rely on addition of free agonist, often at high concentrations. This is especially exciting for use in cases where multiple binding events may be required for a response.

9.4.3.2 Photocontrol of Insulin Release Using a Small Molecular Sulfonylurea

Despite the abundance of impressive examples that illustrate how light-controlled tethers can manipulate the function of channels and receptors, these examples have a drawback that limits their use in therapeutic applications. These examples typically require genetic manipulation to one or more of the protein residues within the channel. As was described in earlier sections of this chapter, an option that avoids genetic modification of channels is the use of small molecule ligands with light-dependent affinity for a biological target. An example described by the Trauner group demonstrates the controlled increase in calcium and insulin levels in human and mouse islet cells using a photochromic inhibitor [87]. The inhibitor structure is based on the known drug, Glimepiride (Figure 9.12), used to treat glucose imbalances in diabetic patients. The drug behaves as an inhibitor targeting the activity of the K_{ATP} channel by binding to one of the subunits involved in ion efflux and ultimately production of insulin. In their photochromic inhibitor design, these researchers conserved the cyclohexyl-substituted phenylsulfonamide group of glimepiride and then replaced the alkyl pyrrolidinone group with a *p*-diethylaminoazobenzene. The trans-isomer could be readily converted to its cis-conformation with exposure to 400–500 nm light, and regeneration of the trans-isomer could be achieved within a few seconds of removing the light source. The IC_{50} values could not be measured directly due to the rapid thermal isomerization, but instead they were able to compare the activity of the two isomers by monitoring changes in calcium levels and insulin secretion. The cis-isomer had a higher EC_{50} value compared to Glimepiride but comparable, whereas the trans-isomer appeared to have little effect on these parameters.

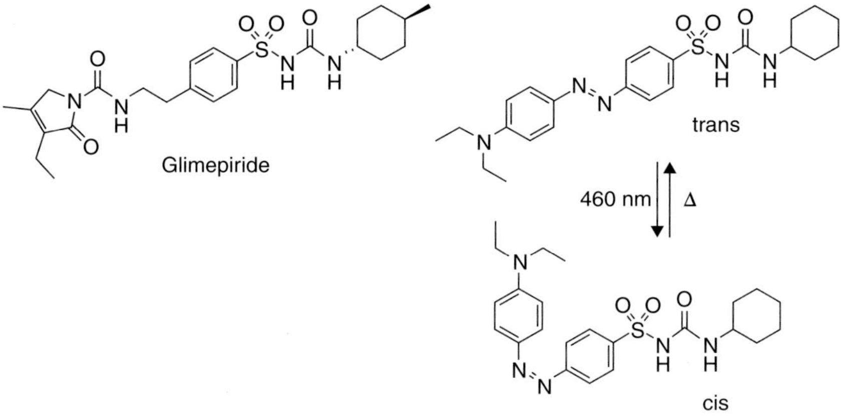

Figure 9.12 Diabetic drug Glimepiride and a photochromic version that show increased binding affinity for the K_{ATP} channel in the cis-isomer [87].

9.4.3.3 Photocontrol of Receptors Using Red Light

As was discussed in the previous section on photocontrol of proteins and peptides, the drawbacks that accompany the use of ultraviolet light to control channel function in living systems have prompted researchers to design photoswitches that can be isomerized using longer wavelengths of light. A number of visible light-activated photochromic systems have been prepared by addition of appropriate functional groups to an azobenzene core [58].

In a recent publication by Isacoff and Trauner [88], they described how they took an existing azobenzene-modified glutamate ligand, and included an electron-donating tertiary amine substituent (Figure 9.13). This modification resulted in a shift in the molecule's absorption spectrum and permitted the use of blue light instead of UV light for conversion of the trans- to the cis-isomer. The extended trans-isomer does not reach the binding cavity of glutamate, whereas the bent cis-form is able to bind effectively inducing activation of the channel. The photochromic species was attached to a genetically modified GluK2 receptor, L439C, by reacting the maleimide group with the mutated cysteine residue. Exposure

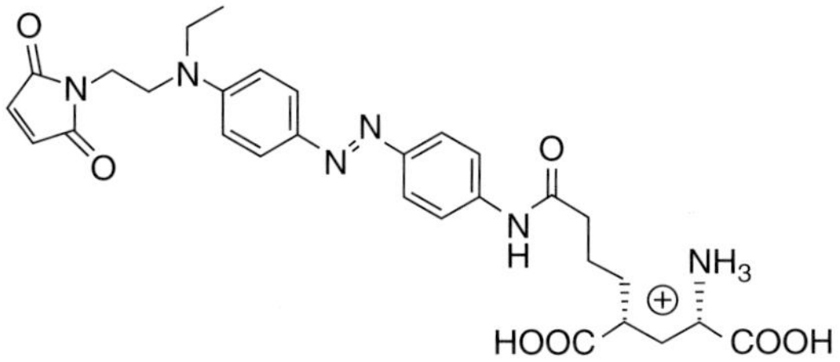

Figure 9.13 Structure of azobenzene-modified glutamate ligand [88].

with blue light resulted in channel opening, suggesting that the cis-form of the photoswitch binds to the receptor as predicted. Although the half-life of the cis-isomer is relatively short, 0.71 s, this can be used advantageous in some cases. Conversion to the cis-isomer with brief light exposure was sufficient to induce binding and subsequent channel activation and the system does not require deactivation with a second light source.

Woolley *et al.* describe a method to prepare a photoswitchable azobenzene that can be triggered using red light [89]. The photoresponsive derivative contains an *ortho*-tetrasubstituted azobenzene attached to a maleimide (Figure 9.14), which can later be covalently attached to the receptor protein via cysteine mutation introduced in the vicinity of the binding site of an ionotropic glutamate receptor. The ortho-substituted pattern is particularly appealing as it not only allows for visible light-induced photoswitching, but also the half-life of the cis-isomer is on the timescale of hours so that the photoinduced effects can be sustained without continuous irradiation. They demonstrate that channel opening can be induced by isomerization from the trans-isomer to the cis-isomer after exposure to high-intensity red light. The reverse process could be achieved with exposure to blue light.

9.4.4 Photoswitchable Nucleotides

Nucleic acids are critical for storing and encoding genetic information in all life forms, and from a biological perspective, the ability to control their structure and function with light could allow for control over gene expression [90]. Photocontrol of nucleotide hybridization or binding properties can be achieved by introducing a photochromic group into site on the nucleotide [25, 26, 91–94]. The photochromic species could be incorporated into the strand backbone in place of a nucleobase or ribose moiety or also by appending a photoswitch to an existing base pair. Photocontrol can also be achieved with external photochromic agents, for example, a ligand with light-dependent affinity or intercalation ability [95–98].

9.4.4.1 Spiropyran-Modified Oligonucleotide Backbones

In a report by Heckel, they describe the preparation of a series of spiropyran-modified oligonucleotides in which the spiropyran is directly positioned into the backbone of the nucleotide strand [99]. In this example, the spiropyran functions as a nucleobase analog, and the charged chromene unit of the planar merocyanine isomer is expected to intercalate into the base pairs and form favorable stacking interactions. The spiropyran, on the other hand, is more sterically demanding and leads to distortion of base pairing. The photochromic species was covalently attached to phosphate groups on the oligonucleotide strand via two linkage sites, one on the methyl groups on the indoline core and the other on the nitrogen

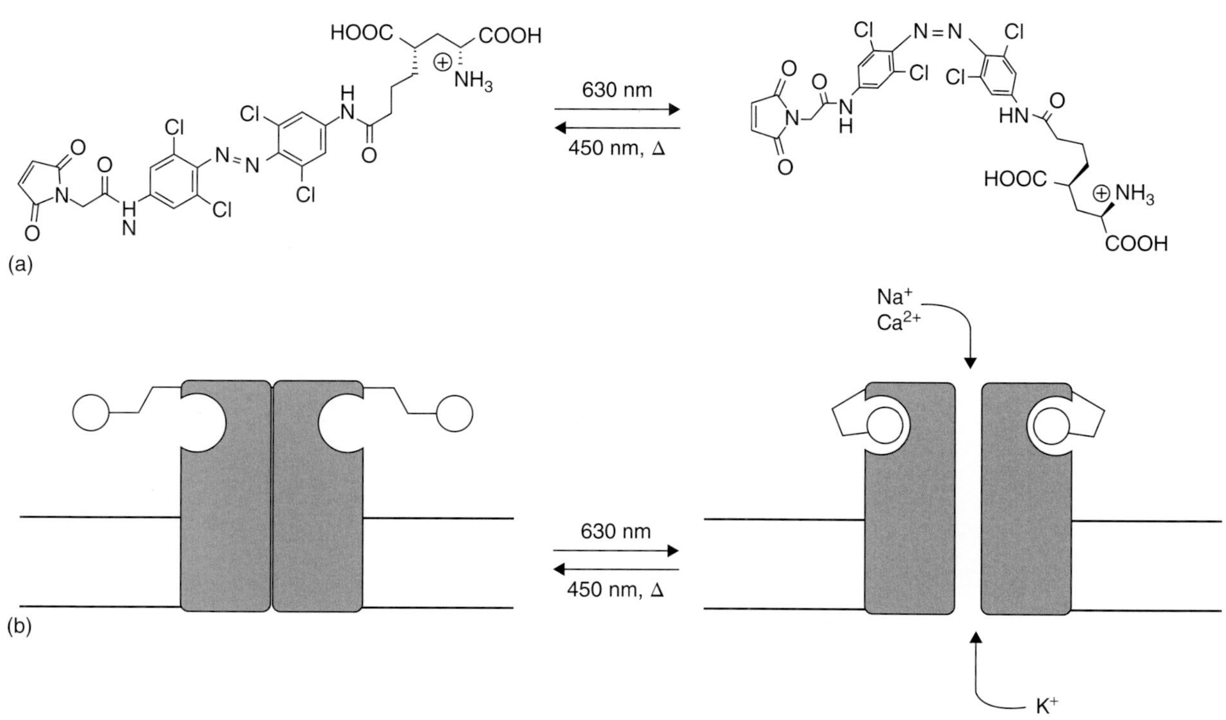

Figure 9.14 (a) Tetra-*ortho*-substituted azobenzene-derived glutamate tether for linkage to an ionotropic glutamate receptor, irradiation of the trans-isomer with 630 nm light produces the cis-isomer, and the reverse process is achieved with 450 nm light. (b) The tethered glutamate-azobenzene derivative allows for photocontrol of glutamate binding and channel activation by isomerization of azobenzene from the trans- to the cis-isomer [89].

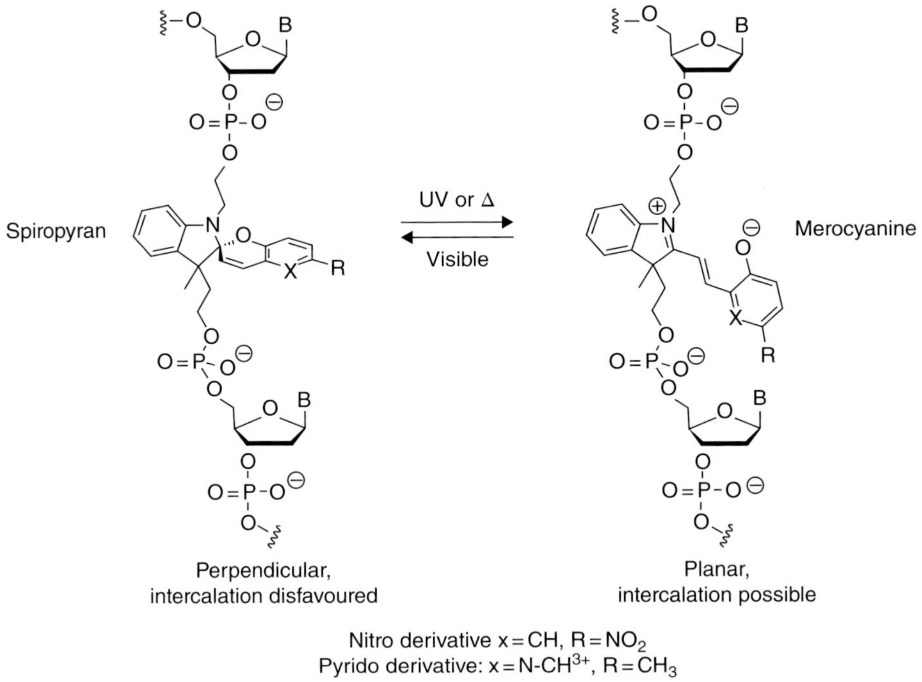

Figure 9.15 Oligonucleotide modified with a spiropyran group. The planar charged merocyanine allows for more efficient intercalation into the base pairs than the bulkier spiropyran form [99].

of the indoline (Figure 9.15). Preparation of this modified photoresponsive DNA strand was accomplished using solid phase synthesis. The yield for the photoconversion reaction of the spiropyran to the merocyanine was highly dependent on the chromene structure. They found that in aqueous buffer, the nitro-substituted derivative could not be isomerized to the merocyanine state with 365 nm light, although 25% conversion was possible when incubated at 25 °C. A second derivative containing a pyrido group redshifted the absorption spectrum so that photoisomerization to 35% merocyanine was possible with 365 nm light. This particular derivative showed greater thermal stability in the merocyanine state when incorporated into the DNA strand. The monomer alone quickly reverted from the merocyanine to the spiropyran form spontaneously within several minutes, however when incorporated into the DNA strand, the conversion occurred in several hours, suggesting that more favorable stacking interactions are achieved with the merocyanine group and the base pairs present.

9.4.4.2 Controlling RNA Duplex Hybridization with Light

In a recent study, the hybridization of RNA was controlled with light by incorporating an azobenzene group into the structure of a ribose unit on the oligonucleotide strand [100]. The nitrogenous base was substituted for an

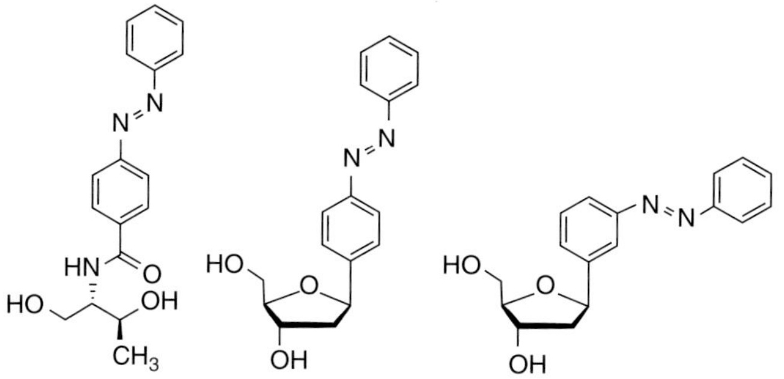

Figure 9.16 Structures of D-threoninol linker, *meta*- and *para*-substituted azobenzene ribose [100].

azobenzene derivative (Figure 9.16). The authors predict that the more planar *trans*-azobenzene can intercalate between adjacent base pairs of the duplex effectively, whereas the slightly more bulky *cis*-azobenzene isomer cannot lead to distortion of the helix. The synthesized monomers were incorporated into oligoribonucleotides using solid phase synthesis. They found that when the azobenzene-modified deoxyriboses were incorporated into the RNA duplex, the isomerization from the cis-isomer to the trans-isomer occurred more slowly, likely due to steric interactions with the base pairs of the oligonucleotide. Using melting temperature studies they compared the effects that different placement of the photoswitches within the oligonucleotide strand had on destabilization of the RNA duplex. In all cases the cis-isomer destabilized the duplex more so than the trans-isomer. The difference in melting temperatures between the two isomers depends on which nucleotide was located opposite the photoswitch.

9.4.4.3 **Diarylethene-Modified Oligonucleotides**

Jäschke and Cahová describe a photoswitch in which the nucleobase itself was involved in the isomerization process (Figure 9.17) [101]. In their design, they replace one of the thiophene rings of a dithienylethene with a deoxycytidine or deoxyuridine molecule, see Figure 9.17. This introduces several new structural and property elements, including electronics, flexibility, and bond hybridization of the nucleotide. In a previous work, they investigated the effects of incorporation of deoxyadenosine and guanosine into diarylethenes [19, 102]. The photochromic species were attached to several different oligonucleotide sequences, ranging from 15 to 19 residues long using a palladium-catalyzed cross-coupling reaction between an iodinated base pair and boronic acid-derived dithienylethene. They investigated the effect of each photoisomer on the ability of DNA polymerase to transcribe a modified template strand containing the photochromic deoxyuridine oligonucleotide. They found that the ring-closed isomer yielded higher amounts of RNA product than the ring-open isomer.

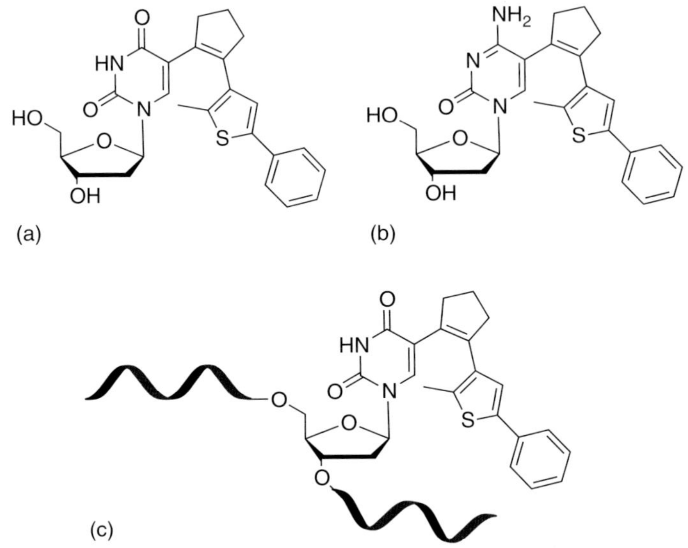

Figure 9.17 Structures of photochromic diarylethene-derived deoxyuridine (**a**), deoxycytidine (**b**), and the photoswitch-modified nucleotide sequence based on deoxyuridine (**c**) [101].

9.5 Summary

It is clear that photoresponsive small molecules can have a large impact on the operation of many complex biochemical systems, and we trust that this chapter provided an overview of some of the more recent examples. Given the success by several groups in using the light-induced changes in structure of biological ligands and modified macromolecules, we can expect more advances in the future, and potentially these concepts will be applied to biomedical use.

References

1. Fehrentz, T., Schönberger, M., and Trauner, D. (2011) Optochemical genetics. *Angew. Chem. Int. Ed.*, **50** (51), 12156–12182.
2. Gautier, A., Gauron, C., Volovitch, M., Bensimon, D., Jullien, L., and Vriz, S. (2014) How to control proteins with light in living systems. *Nat. Chem. Biol.*, **10** (7), 533–541.
3. Gorostiza, P. and Isacoff, E. (2007) Optical switches and triggers for the manipulation of ion channels and pores. *Mol. Biosyst.*, **3** (10), 686–704.
4. Kramer, R.H., Fortin, D.L., and Trauner, D. (2009) New photochemical tools for controlling neuronal activity. *Curr. Opin. Neurobiol.*, **19** (5), 544–552.
5. Banghart, M.R., Volgraf, M., and Trauner, D. (2006) Engineering light-gated ion channels. *Biochemistry*, **45** (51), 15129–15141.
6. Krauss, U., Drepper, T., and Jaeger, K.-E. (2011) Enlightened enzymes: strategies to create novel photoresponsive proteins. *Chem. Eur. J.*, **17** (9), 2552–2560.

7. Stoll, R.S. and Hecht, S. (2010) Artificial light-gated catalyst systems. *Angew. Chem. Int. Ed.*, **49** (30), 5054–5075.
8. Mourot, A., Tochitsky, I., and Kramer, R.H. (2013) Light at the end of the channel: optical manipulation of intrinsic neuronal excitability with chemical photoswitches. *Front. Mol. Neurosci.*, **6** (5), 1–15.
9. Beharry, A.A. and Woolley, G.A. (2011) Azobenzene photoswitches for biomolecules. *Chem. Soc. Rev.*, **40** (8), 4422–4437.
10. Sadovski, O., Beharry, A.A., Zhang, F., and Woolley, G.A. (2009) Spectral tuning of azobenzene photoswitches for biological applications. *Angew. Chem. Int. Ed.*, **48** (8), 1484–1486.
11. Woolley, G.A. (2012) Lighting up with azobenzenes. *Nat. Chem.*, **4** (2), 75–77.
12. Woolley, G.A. (2005) Photocontrolling peptide alpha helices. *Acc. Chem. Res.*, **38** (6), 486–493.
13. Ud-Dean, S.M.M. (2011) Kinetics of optical control of enzyme activity with photoswitchable inhibitors. *Interdiscip. Sci., Comput. Life Sci.*, **3** (2), 79–90.
14. Velema, W.A., Szymanski, W., and Feringa, B.L. (2014) Photopharmacology: beyond proof of principle. *J. Am. Chem. Soc.*, **136** (6), 2178–2191.
15. Szymanski, W., Beierle, J.M., Kistemaker, H.A.V., Velema, W.A., and Feringa, B.L. (2013) Reversible photocontrol of biological systems by the incorporation of molecular photoswitches. *Chem. Rev.*, **113** (8), 6114–6178.
16. Zhang, J., Wang, J., and Tian, H. (2014) Taking orders from light: progress in photochromic bio-materials. *Mater. Horiz.*, **1** (2), 169.
17. Willner, I. (1997) Photoswitchable biomaterials: en route to optobioelectronic systems. *Acc. Chem. Res.*, **30** (9), 347–356.
18. Givens, R. and Goeldner, M. (2005) *Dynamic Studies in Biology. Phototriggers, Photoswitches and Caged Biomolecules*, Wiley-VCH Verlag GmbH & Co KGaA, Weinheim.
19. Jäschke, A. (2012) Genetically encoded RNA photoswitches as tools for the control of gene expression. *FEBS Lett.*, **586** (15), 2106–2111.
20. van der Horst, M.A. and Hellingwerf, K.J. (2004) Photoreceptor proteins, "Star actors of modern times": a review of the functional dynamics in the structure of representative members of six different photoreceptor families. *Acc. Chem. Res.*, **37** (1), 13–20.
21. Cambridge, S. (2014) *Methods in Molecular Biology. Photoswitching Proteins*, vol. **1148**, Humana Press, New York.
22. Brieke, C., Rohrbach, F., Gottschalk, A., Mayer, G., and Heckel, A. (2012) Light-controlled tools. *Angew. Chem. Int. Ed.*, **51** (34), 8446–8476.
23. DeLano Scientific LLC (2006). Pymol. http://www.pymol.org.
24. Dong, S.-L., Löweneck, M., Schrader, T.E., Schreier, W.J., Zinth, W., Moroder, L., and Renner, C. (2006) A photocontrolled β-hairpin peptide. *Chem. Eur. J.*, **12** (4), 1114–1120.
25. Liang, X., Zhou, M., Kato, K., and Asanuma, H. (2012) Photoswitch nucleic acid catalytic activity by regulating topological structure with a universal supraphotoswitch. *ACS Synth. Biol.*, **2** (4), 194–202.
26. Liang, X., Nishioka, H., Takenaka, N., and Asanuma, H. (2008) A DNA nanomachine powered by light irradiation. *ChemBioChem*, **9** (5), 702–705.
27. Nishioka, H., Liang, X., Kato, T., and Asanuma, H. (2012) A photon-fueled DNA nanodevice that contains two different photoswitches. *Angew. Chem. Int. Ed.*, **51** (5), 1165–1168.
28. Willner, I., Rubin, S., and Riklin, A. (1991) Photoregulation of Papain activity through anchoring photochromic azo groups to the enzyme backbone. *J. Am. Chem. Soc.*, **113** (9), 3321–3325.
29. Schierling, B., Noël, A.-J., Wende, W., Hien, L.T., Volkov, E., Kubareva, E., Oretskaya, T., Kokkinidis, M., Römpp, A., Spengler, B., and Pingoud, A. (2010) Controlling the enzymatic activity of a restriction enzyme by light. *Proc. Natl. Acad. Sci. U.S.A.*, **107** (4), 1361–1366.
30. Nakayama, K., Endo, M., and Majima, T. (2005) A hydrophilic azobenzene-bearing amino acid for photochemical

control of a restriction enzyme BamHI. *Bioconjugate Chem.*, **16** (6), 1360–1366.

31. Harvey, J.H. and Trauner, D. (2008) Regulating enzymatic activity with a photoswitchable affinity label. *ChemBioChem*, **9** (2), 191–193.
32. Bose, M., Groff, D., Xie, J., Brustad, E., and Schultz, P.G. (2006) The incorporation of a photoisomerizable amino acid into proteins in E. coli. *J. Am. Chem. Soc.*, **128** (2), 388–389.
33. Aizawa, M., Namba, K., and Suzuki, S. (1977) Light-induced enzyme activity changes associated with the photoisomerization of bound spiropyran. *Arch. Biochem. Biophys.*, **182** (1), 305–310.
34. Feringa, B.L. (2001) *Molecular Switches*, vol. **1**, Wiley-VCH Verlag GmbH, Weinheim.
35. Pearson, D. and Abell, A.D. (2006) Photoswitch inhibitors of alpha-chymotrypsin--increased substitution and peptidic character in peptidomimetic boronate esters. *Org. Biomol. Chem.*, **4** (19), 3618–3625.
36. Falenczyk, C., Schiedel, M., Karaman, B., Rumpf, T., Kuzmanovic, N., Grøtli, M., Sippl, W., Jung, M., and König, B. (2014) Chromo-pharmacophores: photochromic diarylmaleimide inhibitors for sirtuins. *Chem. Sci.*, **5** (12), 4794–4799.
37. Chen, X., Wehle, S., Kuzmanovic, N., Merget, B., Holzgrabe, U., König, B., Sotriffer, C.A., and Decker, M. (2014) Acetylcholinesterase inhibitors with photoswitchable inhibition of β-amyloid aggregation. *ACS Chem. Neurosci.*, **5** (5), 377–389.
38. Bieth, J., Vratsanos, J.S.M., Wassermann, N.H., Cooper, A.G., and Erlanger, B.F. (1970) Photoregulation of biological activity by photochromic reagents. Inactivators of acetylcholinesterase. *Biochemistry*, **12** (16), 3023–3027.
39. Bieth, J., Vratsanos, S.M., Wassermann, N., and Erlanger, B.F. (1969) Photoregulation of biological activity by photochromic reagents II. Inhibitors of acetylcholinesterase. *Proc. Natl. Acad. Sci. U.S.A.*, **64** (3), 1103–1106.
40. Pearson, D., Alexander, N., and Abell, A.D. (2008) Improved photocontrol of α-chymotrypsin activity: peptidomimetic trifluoromethylketone photoswitch enzyme inhibitors. *Chem. Eur. J.*, **14** (24), 7358–7365.
41. Westmark, P.R., Kelly, J.P., and Smith, B.D. (1993) Photoregulation of enzyme activity. Photochromic, transition-state-analogue inhibitors of cysteine and serine proteases. *J. Am. Chem. Soc.*, **115** (9), 3416–3419.
42. Hansen, M.J., Velema, W.A., de Bruin, G., Overkleeft, H.S., Szymanski, W., and Feringa, B.L. (2014) Proteasome inhibitors with photocontrolled activity. *ChemBioChem*, **15** (14), 2053–2057.
43. Vomasta, D., Högner, C., Branda, N.R., and König, B. (2008) Regulation of human carbonic anhydrase I (hCAI) activity by using a photochromic inhibitor. *Angew. Chem. Int. Ed.*, **47** (40), 7644–7647.
44. Reisinger, B., Kuzmanovic, N., Löffler, P., Merkl, R., König, B., and Sterner, R. (2014) Exploiting protein symmetry to design light-controllable enzyme inhibitors. *Angew. Chem. Int. Ed.*, **53** (2), 595–598.
45. Rentsch, A., Landsberg, D., Brodmann, T., Bulow, L., Girbig, A.-K., and Kalesse, M. (2013) Synthesis and pharmacology of proteasome inhibitors. *Angew. Chem. Int. Ed.*, **52** (21), 5450–5488.
46. Abell, A.D., Jones, M.A., Neffe, A.T., Aitken, S.G., Cain, T.P., Payne, R.J., McNabb, S.B., Coxon, J.M., Stuart, B.G., Pearson, D., Lee, H.Y., and Morton, J.D. (2007) Investigation into the P3 binding domain of m-calpain using photoswitchable diazo- and triazene-dipeptide aldehydes: new anti-cataract agents. *J. Med. Chem.*, **50** (12), 2916–2920.
47. Harvey, A.J. and Abell, A.D. (2000) Azobenzene-containing, peptidyl α-ketoesters as photobiological switches of α-Chymotrypsin. *Tetrahedron*, **56** (50), 9763–9771.
48. Groll, M., Berkers, C.R., Ploegh, H.L., and Ovaa, H. (2006) Crystal structure of the boronic acid-based proteasome inhibitor bortezomib in complex with the yeast 20S proteasome. *Structure*, **14** (3), 451–456.

49. Broichhagen, J., Jurastow, I., Iwan, K., Kummer, W., and Trauner, D. (2014) Optical control of acetylcholinesterase with a tacrine switch. *Angew. Chem. Int. Ed.*, **53** (29), 7657–7660.
50. Harel, M., Schalkt, I., Bouett, F., Goeldnert, M., Hirtht, C., Axelsen, P.H., Silmanii, I., and Sussman, J.L. (1993) Quaternary ligand binding to aromatic residues in the active-site gorge of acetylcholinesterase. *Proc. Natl. Acad. Sci. U.S.A.*, **90** (19), 9031–9035.
51. Fujita, D., Murai, M., Nishioka, T., and Miyoshi, H. (2006) Light control of mitochondrial complex I activity by a photoresponsive inhibitor. *Biochemistry*, **45** (21), 6581–6586.
52. Horstmann, B., Korbus, M., Friedmann, T., Wolff, C., Thiele, C.M., and Meyer-Almes, F.-J. (2014) Synthesis of azobenzenealkylmaleimide probes to photocontrol the enzyme activity of a bacterial histone deacetylase-like amidohydrolase. *Bioorg. Chem.*, **57**, 155–161.
53. Korbus, M., Balasubramanian, G., Müller-Plathe, F., Kolmar, H., and Meyer-Almes, F.-J. (2014) Azobenzene switch with a long-lived cis-state to photocontrol the enzyme activity of a histone deacetylase-like amidohydrolase. *Biol. Chem.*, **395** (4), 401–412.
54. Zhang, F., Timm, K.A., Arndt, K.M., and Woolley, G.A. (2010) Photocontrol of coiled-coil proteins in living cells. *Angew. Chem. Int. Ed.*, **49** (23), 3943–3946.
55. Beharry, A.A., Chen, T., Al-abdul-wahid, M.S., Samanta, S., Davidov, K., Sadovski, O., Ali, A.M., Chen, S.B., Prosser, R.S., Chan, H.S., and Woolley, G.A. (2012) Quantitative analysis of the effects of photoswitchable distance constraints on the structure of a globular protein. *Biochemistry*, **51** (32), 6421–6431.
56. Zhang, F., Zarrine-Afsar, A., Al-Abdul-Wahid, M.S., Prosser, R.S., Davidson, A.R., and Woolley, G.A. (2009) Structure-based approach to the photocontrol of protein folding. *J. Am. Chem. Soc.*, **131** (6), 2283–2289.
57. Samanta, S., Qin, C., Lough, A.J., and Woolley, G.A. (2012) Bidirectional photocontrol of peptide conformation with a bridged azobenzene derivative. *Angew. Chem. Int. Ed.*, **51** (26), 6452–6455.
58. Samanta, S., Beharry, A.A., Sadovski, O., McCormick, T.M., Babalhavaeji, A., Tropepe, V., and Woolley, G.A. (2013) Photoswitching azo compounds in vivo with red light. *J. Am. Chem. Soc.*, **135** (26), 9777–9784.
59. Samanta, S., Qureshi, H.I., and Woolley, G.A. (2012) A bisazobenzene crosslinker that isomerizes with visible light. *Beilstein J. Org. Chem.*, **8**, 2184–2190.
60. Blanco-Lomas, M., Samanta, S., Campos, P.J., Woolley, G.A., and Sampedro, D. (2012) Reversible photocontrol of peptide conformation with a rhodopsin-like photoswitch. *J. Am. Chem. Soc.*, **134** (16), 6960–6963.
61. Ali, A.M. and Woolley, G.A. (2013) The effect of azobenzene cross-linker position on the degree of helical peptide photo-control. *Org. Biomol. Chem.*, **11** (32), 5325–5331.
62. Hoppmann, C., Seedorff, S., Richter, A., Fabian, H., Schmieder, P., Rück-Braun, K., and Beyermann, M. (2009) Light-directed protein binding of a biologically relevant β-sheet. *Angew. Chem. Int. Ed.*, **48** (36), 6636–6639.
63. Hoppmann, C., Schmieder, P., Domaing, P., Vogelreiter, G., Eichhorst, J., Wiesner, B., Morano, I., Rück-Braun, K., and Beyermann, M. (2011) Photocontrol of contracting muscle fibers. *Angew. Chem. Int. Ed.*, **50** (33), 7699–7702.
64. Kneissl, S., Loveridge, E.J., Williams, C., Crump, M.P., and Allemann, R.K. (2008) Photocontrollable peptide-based switches target the anti-apoptotic protein Bcl-xL. *ChemBioChem*, **9** (18), 3046–3054.
65. Revilla-López, G., Laurent, A.D., Perpète, E.A., Jacquemin, D., Torras, J., Assfeld, X., and Alemán, C. (2011) Key building block of photoresponsive biomimetic systems. *J. Phys. Chem. B*, **115** (5), 1232–1242.
66. Fujimoto, K., Maruyama, T., Okada, Y., Itou, T., and Inouye, M. (2013) Development of a new class of photochromic

peptides by using diarylethene-based non-natural amino acids. *Tetrahedron*, **69** (30), 6170–6175.

67. Babii, O., Afonin, S., Berditsch, M., Reißer, S., Mykhailiuk, P.K., Kubyshkin, V.S., Steinbrecher, T., Ulrich, A.S., and Komarov, I.V. (2014) Controlling biological activity with light: diarylethene-containing cyclic peptidomimetics. *Angew. Chem. Int. Ed.*, **53** (13), 3392–3395.
68. Nakayama, K., Endo, M., and Majima, T. (2004) Photochemical regulation of the activity of an endonuclease Bam HI using an azobenzene moiety incorporated site-selectively into the dimer interface. *Chem. Commun.*, (21), 2386–2387.
69. Muranaka, N., Hohsaka, T., and Sisido, M. (2002) Photoswitching of peroxidase activity by position-specific incorporation of a photoisomerizable non-natural amino acid into horseradish peroxidase. *FEBS Lett.*, **510** (1-2), 10–12.
70. Hoppmann, C., Lacey, V.K., Louie, G.V., Wei, J., Noel, J.P., and Wang, L. (2014) Genetically encoding photoswitchable click amino acids in Escherichia coli and mammalian cells. *Angew. Chem. Int. Ed.*, **53** (15), 3932–3936.
71. Ishikawa, K., Tamura, Y., and Maruta, S. (2014) Photocontrol of mitotic kinesin Eg5 facilitated by thiol-reactive photochromic molecules incorporated into the loop L5 functional loop. *J. Biochem.*, **155** (3), 195–206.
72. Ishikawa, K., Tohyama, K., Mitsuhashi, S., and Maruta, S. (2014) Photocontrol of the mitotic kinesin Eg5 using a novel S-trityl-L-cysteine analogue as a photochromic inhibitor. *J. Biochem.*, **155** (4), 257–263.
73. Polosukhina, A., Litt, J., Tochitsky, I., Nemargut, J., Sychev, Y., De Kouchkovsky, I., Huang, T., Borges, K., Trauner, D., Van Gelder, R.N., and Kramer, R.H. (2012) Photochemical restoration of visual responses in blind mice. *Neuron*, **75** (2), 271–282.
74. Stein, M., Middendorp, S.J., Carta, V., Pejo, E., Raines, D.E., Forman, S.A., Sigel, E., and Trauner, D. (2012) Azo-propofols: photochromic potentiators of GABA(A) receptors. *Angew. Chem. Int. Ed.*, **51** (42), 10500–10504.
75. Tochitsky, I., Banghart, M.R., Mourot, A., Yao, J.Z., Gaub, B., Kramer, R.H., and Trauner, D. (2012) Optochemical control of genetically engineered neuronal nicotinic acetylcholine receptors. *Nat. Chem.*, **4** (2), 105–111.
76. Sandoz, G., Levitz, J., Kramer, R.H., and Isacoff, E.Y. (2012) Optical control of endogenous proteins with a photoswitchable conditional subunit reveals a role for TREK1 in GABA(B) signaling. *Neuron*, **74** (6), 1005–1014.
77. Mourot, A. and Kramer, R.H. (2007) Staples, tape measures, and bungee cords: a variety of bifunctional reagents for understanding and controlling ion channels. *ACS Chem. Biol.*, **2** (7), 451–453.
78. Mourot, A., Kienzler, M.A., Banghart, M.R., Fehrentz, T., Huber, F.M.E., Stein, M., Kramer, R.H., and Trauner, D. (2011) Tuning photochromic ion channel blockers. *ACS Chem. Neurosci.*, **2** (9), 536–543.
79. Janovjak, H., Szobota, S., Wyart, C., Trauner, D., and Isacoff, E.Y. (2010) A light-gated, potassium-selective glutamate receptor for the optical inhibition of neuronal firing. *Nat. Neurosci.*, **13** (8), 1027–1032.
80. Fehrentz, T., Kuttruff, C.A., Huber, F.M.E., Kienzler, M.A., Mayer, P., and Trauner, D. (2012) Exploring the pharmacology and action spectra of photochromic open-channel blockers. *ChemBioChem*, **13** (12), 1746–1749.
81. Schönberger, M. and Trauner, D. (2014) A photochromic agonist for μ-opioid receptors. *Angew. Chem. Int. Ed.*, **53** (12), 3264–3267.
82. Schönberger, M., Althaus, M., Fronius, M., Clauss, W., and Trauner, D. (2014) Controlling epithelial sodium channels with light using photoswitchable amilorides. *Nat. Chem.*, **6** (8), 712–719.
83. Reiter, A., Skerra, A., Trauner, D., and Schiefner, A. (2013) A photoswitchable neurotransmitter analogue bound to its receptor. *Biochemistry*, **52** (50), 8972–8974.
84. Mourot, A., Fehrentz, T., Le Feuvre, Y., Smith, C.M., Herold, C., Dalkara, D.,

Nagy, F., Trauner, D., and Kramer, R.H. (2012) Rapid optical control of nociception with an ion-channel photoswitch. *Nat. Methods*, **9** (4), 396–402.

85. Bahamonde, M.I., Taura, J., Paoletta, S., Gakh, A.A., Chakraborty, S., Hernando, J., Fernández-Dueñas, V., Jacobson, K.A., Gorostiza, P., and Ciruela, F. (2014) Photomodulation of G protein-coupled adenosine receptors by a novel light-switchable ligand. *Bioconjugate Chem.*, **25** (10), 1847–1854.
86. Reiner, A. and Isacoff, E.Y. (2014) Tethered ligands reveal glutamate receptor desensitization depends on subunit occupancy. *Nat. Chem. Biol.*, **10** (4), 273–280.
87. Broichhagen, J., Schönberger, M., Cork, S.C., Frank, J.A., Marchetti, P., Bugliani, M., Shapiro, A.M.J., Trapp, S., Rutter, G.A., Hodson, D.J., and Trauner, D. (2014) Optical control of insulin release using a photoswitchable sulfonylurea. *Nat. Commun.*, **5**, 5116.
88. Kienzler, M.A., Reiner, A., Trautman, E., Yoo, S., Trauner, D., and Isacoff, E.Y. (2013) A red-shifted, fast-relaxing azobenzene photoswitch for visible light control of an ionotropic glutamate receptor. *J. Am. Chem. Soc.*, **135** (47), 17683–17686.
89. Rullo, A., Reiner, A., Reiter, A., Trauner, D., Isacoff, E.Y., and Woolley, G.A. (2014) Long wavelength optical control of glutamate receptor ion channels using a tetra-ortho-substituted azobenzene derivative. *Chem. Commun.*, **50** (93), 14613–14615.
90. Tang, X. and Dmochowski, I.J. (2007) Regulating gene expression with light-activated oligonucleotides. *Mol. Biosyst.*, **3** (2), 100–110.
91. Kamiya, Y. and Asanuma, H. (2014) Light-driven DNA nanomachine with a photoresponsive molecular engine. *Acc. Chem. Res.*, **47** (6), 1663–1672.
92. Barrois, S., Beyer, C., and Wagenknecht, H.-A. (2012) Covalent modification of 2′-deoxyuridine with two different molecular switches. *Synlett*, **23** (5), 711–716.
93. Zhou, M., Liang, X., Mochizuki, T., and Asanuma, H. (2010) A light-driven DNA nanomachine for the efficient photoswitching of RNA digestion. *Angew. Chem. Int. Ed.*, **49** (12), 2167–2170.
94. Asanuma, H., Ito, T., Yoshida, T., Liang, X., and Komiyama, M. (1999) Photoregulation of the formation and dissociation of a DNA duplex by using the cis-trans isomerization of azobenzene. *Angew. Chem. Int. Ed.*, **38** (16), 2393–2395.
95. Young, D.D. and Deiters, A. (2008) Light-regulated RNA-small molecule interactions. *ChemBioChem*, **9** (8), 1225–1228.
96. Fujimoto, K., Kajino, M., Sakaguchi, I., and Inouye, M. (2012) Photoswitchable, DNA-binding helical peptides assembled with two independently designed sequences for photoregulation and DNA recognition. *Chem. Eur. J.*, **18** (32), 9834–9840.
97. Ihmels, H., Mattay, J., May, F., and Thomas, L. (2013) Photoswitchable DNA-binding properties of a photochromic spirooxazine derivative. *Org. Biomol. Chem.*, **11** (31), 5184–5188.
98. Andersson, J., Li, S., Lincoln, P., and Andréasson, J. (2008) Photoswtiched DNA-binding of a photochromic spiropyran. *J. Am. Chem. Soc.*, **130** (36), 11836–11837.
99. Brieke, C. and Heckel, A. (2013) Spiropyran photoswitches in the context of DNA: synthesis and photochromic properties. *Chem. Eur. J.*, **19** (46), 15726–15734.
100. Goldau, T., Murayama, K., Brieke, C., Steinwand, S., Mondal, P., Biswas, M., Burghardt, I., Wachtveitl, J., Asanuma, H., and Heckel, A. (2015) Reversible photoswitching of RNA hybridization at room temperature with an azobenzene C-nucleoside. *Chem. Eur. J.*, **21** (7), 2845–2854.
101. Cahová, H. and Jäschke, A. (2013) Nucleoside-based diarylethene photoswitches and their facile incorporation into photoswitchable DNA. *Angew. Chem. Int. Ed.*, **52** (11), 3186–3190.
102. Singer, M., Nierth, A., and Jäschke, A. (2013) Photochromism of diarylethene-functionalized 7-deazaguanosines. *Eur. J. Org. Chem.*, **2013** (14), 2766–2769.

10
Industrial Applications and Perspectives

Junji Zhang and He Tian

10.1
Industrialization and Commercialization of Organic Photochromic Materials

Photochromism attracted scientific interest in the mid-nineteenth century, while systematic research of organic photochromic molecules started almost one century later in the 1950s. In 1966, Corning (USA) marketed photochromic ophthalmic lenses [1] that darkened reversibly in sunlight owing to silver halide crystals trapped within the matrix of the glass [2]. Since then millions of pairs of inorganic photochromic spectacles incorporating this technology have been prescribed. Photochromism thus forms the basis for what has become a global multimillion-dollar business, and a deep understanding of the phenomenon has been fundamental to the growth of the industries reliant on it. Organic photochromic materials, on the other hand, although dominating scientific research publications, has relatively less appeared in commercial industrial applications at their early age. However, with the fast development of polymer industrials (e.g., plastics), organic photochromic compounds have found their advantages in constructing commercial photochromic materials with greater robustness, lightness, as well as lower cost, which is essential for commercialization. Therefore, organic photochromic materials have become one of the booming fine chemical industry in the past two decades.

There are generally two types of organic photochromic materials: T-type and P-type. T-type refers to those that could undergo thermally decoloration, such as azobenzenes, spiropyrans, spirooxazines, naphthopyrans, and so on (Figure 10.1). In contrast, P-type photochromism is thermally irreversible, that is, all coloration and decoloration processes are driven only by light. Compounds such as dithienylethenes, fulgides, belong to P-type photochromes (Figure 10.2) [3]. While scientific research is mostly keen on the thermostable P-type photochromism, in industry, T-type photochromism (mostly spirooxazines and naphthopyrans) is the one that dominates.

Photochromic Materials: Preparation, Properties and Applications, First Edition.
Edited by He Tian and Junji Zhang.

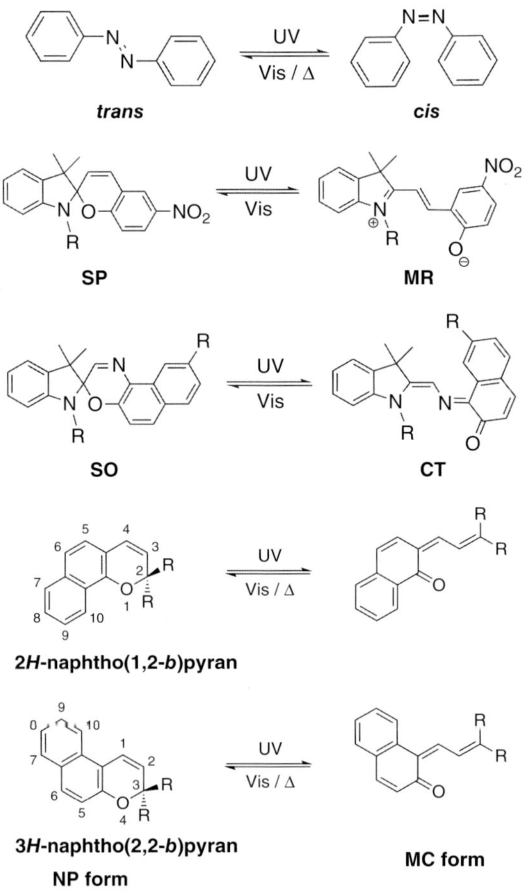

Figure 10.1 T-type photochromes: azobenzene, spiropyran, spirooxazine, and naphthopyran (from top to bottom).

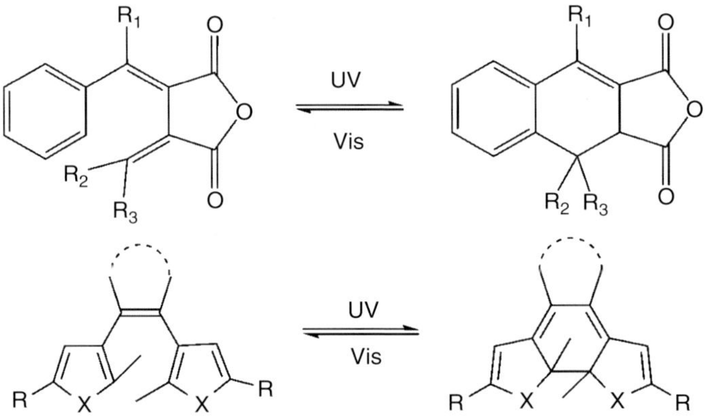

Figure 10.2 P-type photochromes: fulgide and diarylethene (from top to bottom).

10.1.1 Commercialized T-type Photochromic Materials

Conversion from scientific results to commercial products has to overcome formidable technical and commercial challenges. Not only the functional molecules must be practical and economical to manufacture, but also their photochromic performances have to satisfy an array of criteria:

- *Quick coloration rate*: Turning from colorless to colored forms rapidly for "on demand" photochromic operations, and the coloring should be sufficiently dark when light is applied.
- *Appropriate half-life of decoloration for desired applications*: For T-type photochromism, it is the thermobleaching half-life that matters. Take ophthalmic lenses, for example, the decoloration of lenses should be fast when the users are in-door, in case that the vision is not impaired when light levels fall. In addition, the lenses should be completely colorless when it is decolored.
- *High interconversion efficiency*: For coloration, the more colored form the less products are used, hence, more economical. For decoloration, the colored residue should be lower enough to avoid the remaining of the unwanted hues.
- *Good fatigue*: Decent photostability to ensure the recycling of the product (except for some disposable ones, e.g., security labels). Successful photochromic materials have intrinsically low fatigue properties and respond well to the presence of stabilities.
- *Thermostability*: This means the two photoisomers themselves show little change in photochromism with variation of temperatures.
- Good solubility and malleability.

Except for these above-mentioned criteria, still many complicated factors should be considered during industrial manufacturing. Take polymeric matrix for example, since T-type photochromic compounds often undergo substantial structural changes upon photoisomerization, the rigidity of the polymer matrix along with the polarity significantly affects the photochromic kinetics [4]. Generally, polymers with flexible chains and low glass transition temperatures (T_g) are the best candidate in photochromic industry. Levels of photochromic dye loadings (varying in the range of 0.01–0.30% w/w) for most efficient and durable photochromic behaviors are dependent on different media, applications, and dyes themselves. Yet one thing is the same, colorant in a high concentration will be detrimental to the performances of the products.

Also, a good start is half the battle. Thus, the optimization from the basic level should be focused on the molecular design. Recently, Momoda *et al.* [5] have reported that the introduction of electron-donating groups on the skeleton of a naphthopyran, a traditional photochromic lenses material, has led to the development of a bathochromic shift of the MC (Merocyanine) form absorption band toward the longer wavelength. Further introduction of methoxy groups to the *para*-position of the phenyl groups on C3 induced faster decoloration of the MC form (Figure 10.3). The comprehensive and detailed research on molecular synthesis and design would provide further guidance on development

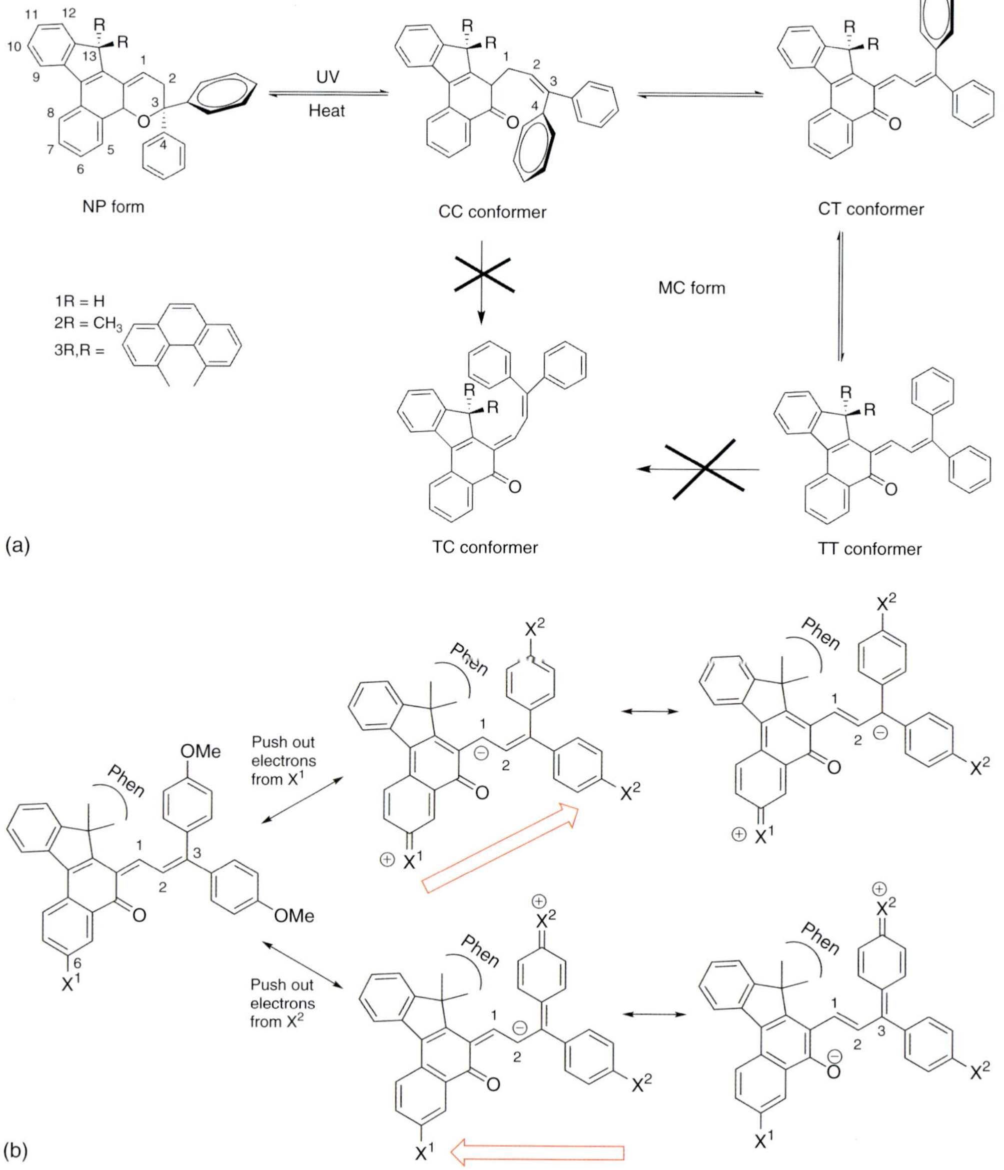

Figure 10.3 (a) Photochromism of 2*H*-indeno(2,3-*f*)naphtha(1,2-*b*)pyran. (b) Effect of electron-donating groups on C6 and C4′ of a phenyl group on C3 on resonance structures. (c) Effect of methoxy groups on C6, C11, and C4′ of a phenyl group on C3 on the resonance structures.

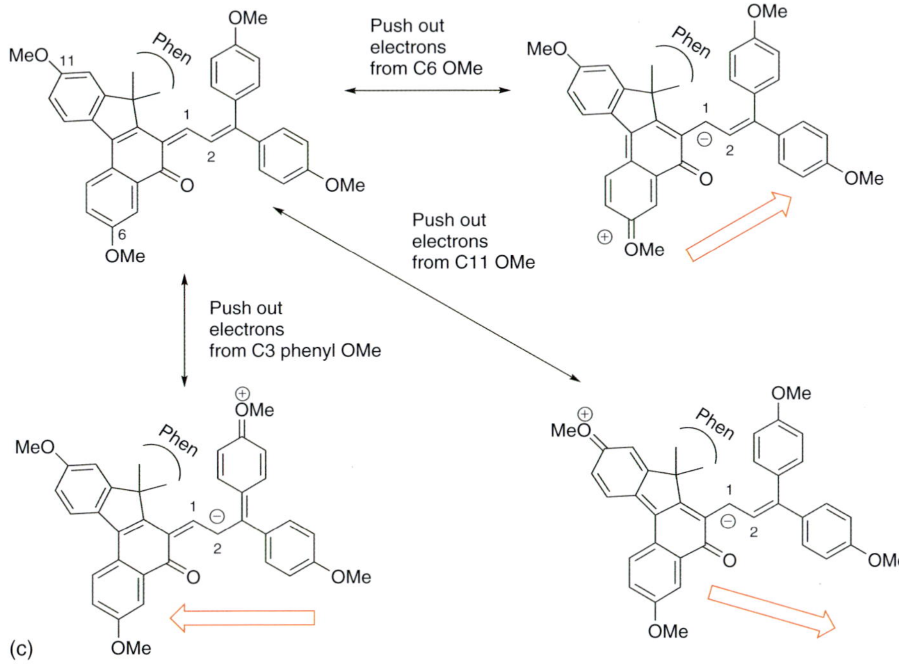

Figure 10.3 (*Continued*)

of novel photochromic molecules, which indeed will be profitable for daily industrialization.

At present, commercial T-type photochromic ophthalmic lenses have accounted for the greatest volume usage in photochromic industry. Generally, there are four manufacturing technologies for photochromic lenses productions [6]:

- *In mass*: Dyes are injection-molded in a thermoplastic or are dissolved in a monomer or resin system, which is then thermally or UV-cured into a semifinished lens that can be ground to the desired prescription.
- *Coating*: The colorants are coated onto the front or the back face of a lens in solution along with a resin by a technique like spin- or dip-coating.
- *Imbibition*: A photochromic layer about 0.2 mm thick is formed in the lens surface by allowing dye to diffuse into the polymer matrix.
- *Lamination*: A film containing the photochromic dyes is sandwiched between two sides of a lens.

Apart from ordinary sun glasses for daily life, recent innovation on commercial photochromic sun glasses for drivers has come out [7]. The hue of the glasses changes to different colors depending on whether they are exposed under natural sunlight or light filtered through the car windscreen. This two-stage photochromic coloration is achieved by using a combination of photochromic compounds with different spectral sensitivities.

Photochromic surface coatings, including photochromic inks [8], nail varnish [9], security printings [10], have become a more popular commercial reality for applications in esthetics, security, and authentications [11]. The simplest method to apply photochromic dyes to the coating formulations is to dissolve them in a commercially available solvent-based gloss varnish. As a long-term headache for coating industry, this method is only available for organic solvents (as photochromic dyes are mainly organic soluble). For more environmental-friendly water solutions, alternative approaches should be employed. One of the mostly used techniques nowadays is the microencapsulation. Photochromic dyes are encapsulated in a polymer, providing a photochromic particle with diameters of $10^{-5}-10^{-6}$ m, which can be dispersed homogeneously in water as those in conventional organic solvents.

Another widely available photochromic industry lies in photochromic plastics. Products such as drink bottles, hair clips, key fobs, and mobile phone accessories are from photochromic PVCs (Polyvinyl Chloride). Toy cars and dolls with photochromic magic are welcomed by children all around the world. Photochromic crash helmet visors for motorcyclists are available from several manufacturers, with adhesive light-responsive film being another option for riders [12]. Clever use of photochromism has been made in fishing line, which consists of a clear colorless polymeric fluorocarbon impregnated with photochromic dye [13]. Photochromic dyes have also been combined with other special effects colorants, such as pearlescents and metallics, to generate unusual looks, for example, in mobile phone covers.

Recent interest has been shown in using photochromic dyes as colorants in textiles. However, for the most commercially important hydrophobic polymers such as polyester and nylon, several problems are encountered with the application of T-type photochromic compounds [14–16]. The relative bulk of the photochromic colorants (compared with conventional disperse dyes) does not favor in diffusion into the fiber, and the polymers themselves tend to inhibit photochromism [14–16]. Although several obstacles exist, marketable photochromic textile has been still achieved by the application of novel methods, such as screen-printing pastes incorporating dye onto fabric [14–16]. Applications of commercial photochromic dyes as disperse dyes to polyesters fabric via solvent-based dyeing method and exhaust dyeing have also been reported and evaluated [17].

Other commercialized and potential applications of T-type photochromism are widely spread in a variety of fields, such as smart windows and agricultural films [11a, 18], personal care and cosmetics [8a, 19], and even in the field of battling fraud and counterfeiting including merchandise labels or even currency [20].

10.1.2 Commercialized P-Type Photochromic Materials

Major P-type photochromic materials include diarylethenes and fulgides, with a typical cyclohexadiene ring open and close motif. As thermoirreversible photochromic species, P-type photochromic materials are long time considered as potential all-optical circuitry for next-generation computing, switches/logic gates, memory technology, and ultra-density data storage devices [21]. Several all-optical molecular memory/data storage systems and photoactuators have been already demonstrated [22]. Despite all the effort being expended on this class, including the study of continuous microflow reactor techniques to improve selectivity and yield [23], it is not clear which, if any, of these avenues will bear fruit in the long term. Up to now, one family of the few commercially available P-type photochromism are fulgide derivatives, which are widely applied in chemical actinometry [24].

Although P-type photochromism is still in the stage of "potential" and "promising," optimism should be still held for this sprouting field. Owing to the potential impact that some of the technologies could have if they ever do eventually become commercial reality, P-type photochromism should have faith in their successful development and introduction that would signal the end of the dominance of T-type dyes in terms of industrial importance.

10.2
Perspectives for Organic Photochromic Materials

As Mr Hirshberg raised the term *photochromism* in the 1950s, during more than 60 years' development, the photochromic materials have found their popular applications from traditional dye industry to modern academic research and bounded for advanced future materials realm. In the recent decade, most efforts have been devoted to the construction of functional materials with photoswitchable fragments in various scientific research fields, ranging from chemistry, physics, and materials science to biology and nanotechnology. Azobenzenes, spiropyrans, diarylethenes, fulgides, and crowded ethenes, as well as nature-produced rhodopsin-based photoswitches are among the mostly used photoreversible "molecular gadgets" in building photoswitchable smart systems.

Trans–cis isomerization dyes, typically azobenzene and stilbenes, are one of the most traditional photochromic families. Apart from being traditional industrial dyes, the *trans–cis* photoisomerization induced remarkable geometric and dipole changes, which renders azobenzene and stilbene derivatives with an important role in photomodulated electro/optic devices, photoswitchable surface and nanostructures, and photo-triggered biological applications [22b, 25]. Nowadays, azobenzene derivatives are still the most popularly used photoswitches in scientific research and industrial applications. As a traditional T-type photochrome, azobenzenes suffer from several shortcomings as thermoreversibility (or thermo unstable), short wavelength absorption and excitation, strong overlap between two isomers, absorption bands, and related low conversion efficiency. Fortunately, recent research breakthroughs have promoted azobenzenes with several exciting optimizations: visible light-triggered isomerization [26], ultra-longer half-time for thermoreverse reaction, as well as separated isomer absorption bands and much higher conversion rate [27]. These molecular design and synthesis innovations rejuvenate azobenzene family and make them "never die and neither fade away."

Spirocyclic photochromes, taking spiropyran as an example, belong to another famous photochromic family. Although suffering from thermoinstability and low fatigue, spirocyclic photochrome still offers a unique feature of significantly increased dipole moment after photoisomerized from ring-closed spiropyran form to ring-open charge-separated zwitterionic merocyanine form [28]. Generally, this distinctive feature has been utilized in controlling the electrostatic interactions between spiropyran units and charged entities (e.g., charged nanoparticles and metal ions), both in bulk solution [29] and in modified surfaces [30]. The zwitterionic properties also facilitate a photoswitchable wettability of surfaces, which is commonly used in fabrication of hydrophilicity controllable smart surface materials [31]. Another smart property of spiropyran derivatives that catch peoples' eyes is their inherent photoacid nature [32]. The photoisomerization from open merocyanine isomer to spiropyran isomer induces a proton release from the phenol group on merocyanine moiety, according to different solvents and pK_a of phenol group (decided by different functional group modifications). In addition, due to the elongated conjugation of opened spirocyclic photoisomers, they are usually colored and fluorescent, which finds themselves valuable in photoswitchable bioimaging research [33]. More interestingly, spirocyclic photochromes have recently played an important role in mechanochromic polymer materials, where it is found that the open-closed isomerization of spirocyclic photochromic molecules could also be achieved by strain forces [34]. This promotes spirocyclic photoswitches as an eye-popper in advanced functional materials world.

Diarylethene and fulgide derivatives, famous for their thermoirreversibility (P-type) and high fatigue, perform negligible steric or geometrical changes while displaying large variations in conjugation length upon photoisomerization. This provides a wide utility in modulating optical and electronic coupling properties (e.g., energy transfer, donor–acceptor, conductivity, catalysis, metal complexation) of multicomponent functional systems attached with diarylethene and fulgide photochromes [35]. So far, a large variety of molecular data storage systems, molecular gates, molecular wires, nonlinear photoswitches, and photocontrolled catalyst based on diarylethene photoswitches have been reported [21]. For diarylethenes, the inherent chirality caused by two alkyl groups attached to the reaction carbon atom also attracts much interest due to its potential to be used in photochiral materials. Recently reported separation of para/anti-para isomers along with chiral isomers provides diarylethenes with more brilliant future in chiral environments, which is commonly existed in biological systems [36]. In addition, it is found that even the small geometrical change of diarylethenes would affect the biological functions when they are introduced into biosystems, which broaden the utility of diarylethene derivatives into even larger extents [37].

Overcrowded alkene, a fabulous photo-/thermo-triggered molecular rotor, has nowadays ascended as a shining star in fabrication of molecular dynamic system. With exquisitely joint driven by alternate light and heat, the overcrowded alkene would perform delicate rotation or even monodirectional rotation through the alkene linkage bond. This smart feature has now been widely used in molecular machines as nanocar, nanopropeller, and so on [38]. Moreover, as four chiral enantiomers would be generated with alternate applied light and heat in different sequences, overcrowded alkene photorotors have been also frequently used in photo-/thermo-controlled chiral synthesis and catalysis.

As mentioned in Section 10.1, photochromic molecules have been commercialized in various areas in our daily life. Yet compared to the enormous publications, these applications are just the tip of the iceberg. Especially, the routinely used photochromic materials are just based on their fundamental color changes, while more novel and complex functions of photochromic materials mentioned in the earlier chapters are still confined in papers and electronic publications. Thus, transform smart ideas and concepts into applications and make revolutionary materials are the ultimate goals for materials and chemistry scientists.

Photochromic polymers might be the most industrialized photochromic materials covering the commercials as photochromic glasses, photogels, coatings, optical waveguides, holographic recording media, UV sensors, nonlinear optics, and so on (see Sections 8.1 and 10.1). These photochromic polymers are usually produced by incorporating photochromores into the polymer backbones or side chains, which are always monofunctional and lack versatility. Block copolymer could be an excellent candidate for multifunctional photochromic polymers, as different functional molecules that modified monomers could be polymerized into one polymer chain with great versatility [39]. Supramolecular polymers, a sparkling new-concept polymer family based on noncovalent weak interactions (or supramolecular interactions, see Chapter 4), further extend the potential realm for multifunctional smart photochromic polymer materials [40]. Supramolecular polymers are constructed generally via holding functional monomers into macromolecular structures by highly directional noncovalent interactions such as hydrogen bonding, hydrophobic/hydrophilic effects, donor–acceptor interaction, π–π stacking, coordination, electrostatic interaction, and so on [41–43]. Theoretically, taking advantage of different functional monomers through different kinds of assembly, the supramolecular polymers would be omnipotent and carry out multiple tasks at the same time both individually and synergistically under different stimuli. Owing to their dynamic and reversible nature, photochromic supramolecular polymers have been long active in constructing photo-triggered gels and self-healing materials [44]. Although weaponed with various functional monomers, it is still a long way for photochromic supramolecular polymers to take challenge from traditional polymer materials. High-efficient assembly and operation, high synergy, and low self-interference between different monomers are major challenges lay ahead of chemists. For instance, the light signal for one photochrome should be exclusive as not to induce undesired excitation of other photochromes or photosensitive unites. Furthermore, to efficiently control the polymerization/assembly process of supramolecular polymers, deeper insights into the kinetics and thermodynamics theory and model study are urgently needed.

As life evolution started from the sea to the land, the original study of the photoswitchable molecular systems were also in the solution phase. Solution phase provides a homogeneous environment for free molecules, which is the basis for putting forward and proving new concepts as well as more complex models for construction and study. With the development of surface/interface science and device technology, photochromic molecules begin to leave the solution and embrace the new "solid land." However, we have to correct an easy-to-misunderstand point of view, that is, photochromic systems in solution are solely for basic science and concept studies. In fact, photochromic systems always find their popular application fields in solution environment, just as marine species evolution does not stop even after the emergence of life on land. One important application of photochromic systems in solution is photo-gated catalysis. As depicted in Chapter 5, issues as *in situ* switching, concentration effects, photocatalyst efficiency and robustness, reactivity differences along with higher spatial and temporal resolution are concluded as challenges remained to be improved. For sub-stoichiometric photoswitchable catalysis, in which catalytic amount of photoswitches are used, supramolecular strategy would be introduced to enhance the reaction activity by bringing the substrate closer via various supramolecular interactions.

Another important application of photochromic systems in solution is light-triggered biological process targeting and controlling, which is a rapidly burgeoning research interest among scientists in chemistry, biology, and medicine at present. Endowed with high degree of spatial/temporal resolution, noninvasive manner with low or negligible contamination, and controllable qualitative/quantitative regulations (by adjusting wavelength and intensity), light is rendered as one of the most convenient stimuli to precisely tune and control the biological processes on command with minimal perturbation of the physiological environment [45]. Green fluorescent proteins (GFPs), for example, are one of the most famous natural photochromes. Recently, artificial photoswitchable oligonucleotides, peptides, enzymes, and ion channels have showed their talents in various scientific research areas [46] (see Chapter 9). In addition, other applications such as photoswitchable bioimaging, targeting, and drug loading and release have long been the research interests among scientists and global companies.

Just as human body is a rather complicated natural system, for future perspective of *in vivo* photocontrolling/regulating application (in particular, photopharmacy and phototherapy), a lot of considerations should be satisfied and improved on modern photochromic systems. First of all, for photochromic molecule itself, several function and property issues should be considered. Due to the low issue permeability, biological harmfulness, and background fluorescence excitation, the traditional UV excitation is one major setback for common photochromic materials in biology. Although recently several solutions, such as upconversion nanoparticles (UCNPs), metal-to-ligand charge transfer (MLCT), and multiphoton excitation processes, have been proposed [47–49], they are more suitable for *in vitro* rather than for *in vivo* due to the use of toxic heavy transition metal ions. Recently, visible light (even red light) based photoswitches of azobenzene [26] and diarylethene [50] derivatives have been synthesized and reported, which give light to the biocompatible photoswitching in biological systems. Stability of photochromes in cellular environment is another important issue to be noted. The complex cellular environment contains a lot of redox substances (e.g., GSH (Glutathione), ROS (Reactive Oxygen Species)) [51], whether the photochrome could withstand the redox attack during their travel in human body is essential for their functionality and efficiency in application. For T-type photochromes, thermostability is a decisive parameter. This is not simply to say that the more the thermal stable, the better the performance. As a matter of fact, tuning the rate of the thermal relaxation due to different requirements constitutes one of the main advantages over using traditional approaches, as it opens the possibility of temporal control of drug activity, in addition to approaches based on the use of two different wavelengths of light (*vide supra*). Figure 10.4 presents possible therapeutic scenario that one can envision when exploiting thermal relaxation of a photoswitchable drug [46]. Second, for the photochromic functionalized systems, how to amplify or optimize the switching efficiency or contrast while doing limited or negligible harm to their inherent biological function is a key issue. To functionalize biological entities, careful modifications or even mutations are to carry out and more or less loss of biological activities is unavoidable. Thus, the choice of the modification sites

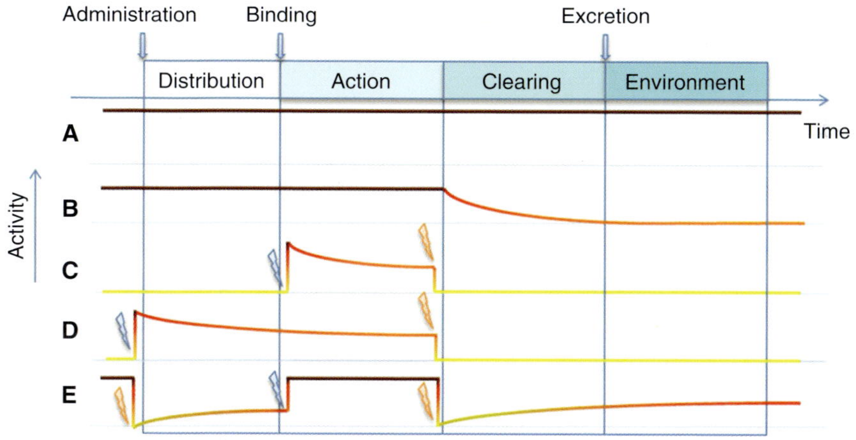

Figure 10.4 Possible therapeutic scenarios using photoswitchable bioactive compounds with a controlled half-life that show their activity in the thermodynamically unstable (A–C) or stable (D,E) state. (A) Irradiation prior to administration. (B) Irradiation at the point of action. (C) Multiple irradiation cycles of a drug with a short half-life for the on-state. (D) Irradiation prior to administration. (E) Irradiation prior to administration and while the drug is being cleared. (Adapted with permission [45]. Copyright 2014, American Chemical Society.)

and balance between activity and functionality should be beforehand tested and evaluated (see Chapter 9). Unlike small molecules, the working efficiency of photo-switches functionalized on bio-macromolecules should also be put emphasis on. Compared to the biomacromolecules, photochromic molecules are only small parts in this big machine. Thus, the molecule itself, the modification sites, and the macromolecule structures should be meticulously designed to achieve the maneuvering of the large biomacromolecules via small photochromic components. Last but not the least, in photochromic biological applications, the factor that concerned the most is the biotoxicity. For small organic molecules, the toxicity may stem from the molecule itself and the decomposition products by either metabolism or other effects. Although most small organic molecules show low biotoxicity when they are in low concentrations, when high dose should be used, the toxicity problem cannot be ignored anymore. Covalently linking photochromic molecules to biomolecules will "dilute" the toxicity to some extent, yet whether the modified biomolecules would affect the subsequent physiological processes is still unclear. For instance, questions like whether the photochrome-modified oligonucleotides would affect the subsequent gene transcription and expression should be greatly concerned when they are 1 day certified to be introduced in clinical therapies. A good substitution is the development of natural photoswitchable molecules, such as GFP. This should be the preference in resolving the toxicity concerns as they are originated from natural organisms. Despite these limitations, it should be mentioned that photochromic compounds have been successfully applied as prodrugs in therapies (mainly azobenzene derivatives), as exemplified by sulfasalazine and balsalazide, food/drug/cosmetic additives such as allura red AC, and diagnostic aids such as anazolene. These examples of successful applications of azobenzenes in medicinal contexts show that careful consideration of possible toxicity may allow the construction of biologically safe compounds [52].

Other potential applications in solution are based on binary coding nature of photochromic materials. Light-triggered logic operation, for example, is one of the hottest concept raised for future substitution of modern silicon semiconductor-computing devices (see Chapter 3). Although some consider molecular logic gates constructed in solutions (or "soft" logic gates) are merely concepts or even fantasy, we have reasons to believe that our "soft" photoactive materials will still perform real logic applications in certain fields. Actually, the word "computing" is not confined to modern computer and information engineering, the whole human body can be regarded as a complete computer. Different from semiconductor-based logic/memory operations in conventional computers, the logic/memory components in human "computers" are all based in solution phase. That is to say, we do not have to build "solid" computing systems via our photochromic molecules, while we can use our smart molecules to perform excellent logic operations in our bodies. Recent researches on logical sensing and operation in neuron systems offer a great opportunity for photochromic logic gates in future organ dysfunction recovering and therapy [53]. In addition, as a typical binary unit, photochromic compounds could be also employed in combination

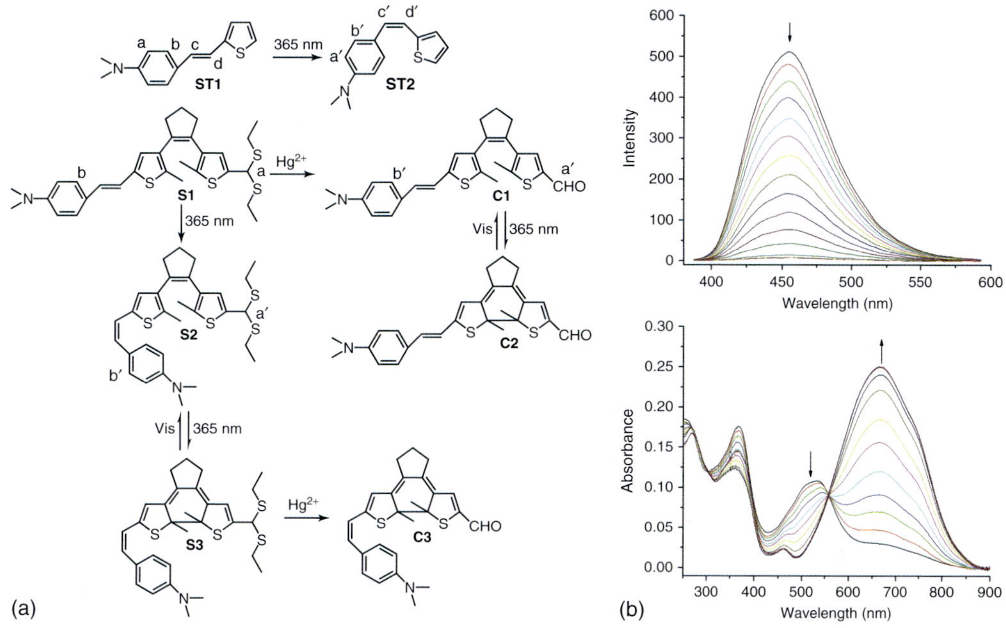

Figure 10.5 (a) Proposed sensing processes of compounds S1 and S3 with Hg^{2+} and the photochromic processes responding to light stimuli. (b) The fluorescence change upon detecting Hg^{2+} with open isomer (top) and the absorption change upon irradiation with UV and adding Hg^{2+} sequentially (bottom). (Adapted with permission [55]. Copyright 2013, Wiley-VCH Verlag GmbH & Co. KGaA, Weinheim.)

of "double-check" sensory systems. Traditional single-compound sensors can only check once on target molecules, which would easily cause false-positive, or even worse, false-negative signals. To tackle this problem, double-check sensors are demanded to improve the accuracy. Photochromic sensors usually possess alternations in several properties after photoswitching, therefore are potential candidates for double-check sensory systems. As shown in Figure 10.5, the reaction of mercury ion with dithioacetals on DTE (dithienylethenes) group caused the fluorescence quenching, which could be the signal 1 for mercury detection. After irradiation to its closed DTE isomer, the addition of mercury ion caused a further bathochromic shift in absorption spectra compared to the original closed isomer, which could be regarded as the signal 2 for mercury detection. The two signals responded to mercury ion based on two photoisomers improved the accuracy of the mercury ion detection by this photochromic sensor [55].

A step forward to more practical applications, we have to first give the photochromic molecules "a place to stand," similar to Archimedes' "*need for a place to stand to move the earth.*" Thus, two-dimensional construction of photochromic materials on surfaces and interfaces provides an opportunity to manipulate both chemical and physical property switches on nano-/micro or even macroscale surfaces. As described in Chapter 6, amounts of photochromic surficial materials on wettability controlling, photopatterning, photoelectron wiring, reversible functional target capturing, logic operation, cargo loading and releasing, photo-triggered conductance manipulation, and so on, have been presented, ranging from macroscale SAM (Self-assembled Monolayers) to microscale particles to even single-molecule scale molecular junctions. Apart from real applications, photochromic surficial materials also provide research models for further photoswitchable devices' fabrications (see Chapter 7). Thanks to the ever-developing methods for surface modification, more and more functional molecules could be conveniently and efficiently attached to the surface materials [54]. The surface modification could be regarded as "reaction and purification" process since the "wanted" molecules can be anchored on the surface specifically and other unreacted or coexisted impurities could be directly washed away. Although one should be careful about the unspecific adsorption of unwanted molecules to the surface, the modification process indeed saves time for tedious purification of traditional synthesis. Photochromic surficial materials are at present commonly used in various scientific research areas, in order to promote surficial materials into industrialization, scientists have long focused on the precise functionalization of the surface and even gone deep into the working mechanism and process of surface materials. Since the surface-confined molecule is not so free as they are in solutions, the assembly of the molecules on the surface determines the efficiency of the surface material. Surfaces with low density of molecules would be certainly inefficient, while overdensed assembly will also harm the performances of the molecules due to the steric hindrance between each molecule. Therefore, how to control the modification process to achieve the optimal surface density is one of the primary prerequisites. In addition, different from the reaction in homogeneous solvents, the surface modification takes place on the heterogeneous surface or interface. The reaction kinetics and thermodynamics should be further studied to determine universal and comprehensive preparation methods for all kinds of modifications. Another essential issue for photochromic surficial materials is the interactions between photochromic molecules and the supported surfaces along with the mechanism of photo-triggered interfacial energy/material flow. When attached to the surface, the interactions between photochromic molecules and surfaces would have different effects (either positive or negative) on the confined molecules. Effects such as plasmon-enhanced photochromism (positive) and excited-state quenching by gold surface (negative) have been successively reported (see Chapter 6). To avoid the negative effects, different linkage groups (flexible nonconjugated anchors, rigid conjugated anchors, multipod anchors, etc.) have been designed. Yet, systematic design and study of the molecule–surface interactions and mechanism are still rare.

The integration of dynamic photoswitches into three-dimensional architectures leads to another novel photochromic materials – photochromic crystal materials. This is a kind of ordered crystal that changes its shape under alternate light irradiation in the process of crystal-to-crystal chemical transformation.

The ordered assembly of crystal materials provides a strategy for amplification of micro-/nanoscale behaviors of individual molecule to macroscale mechanical performances. Up until now, photochromic crystals have been made into different shapes (rodlike, platelike, etc.) to perform mechanical works as photo-triggered molecular actuators, ribbons, and photopatterning for surface wettability [56] (see Section 8.2). Yet, most of the photochromic crystal materials are based on diarylethene derivatives due to their small geometric changes after isomerization. Crystal materials based on azobenzenes and spiropyrans are very rare. For diarylethene-based photochromic crystals, due to the hindered rotation of aryl moieties in the crystal lattices, the photoswitching behavior only works when both two requirements are met: (i) the diarylethene molecules are in photoactive antiparallel form; and (ii) the distance between two active carbon atoms should be less than 0.42 nm [35b]. Thus, precise structure design for these diarylethene-based photochromic crystal materials is rather necessary. Another crystalline-related photochromic materials are photochromic liquid crystals. By attaching photochromes with mesogenic groups or doping them with commercial liquid crystals, precisely light-modulated phase change, helix inversion, and related reflection color changes have been achieved in photochromic liquid crystals (see Section 8.3). As concluded in Section 8.3, the combination of photochromic molecules and LCs (Liquid Crystals) would probably cause the loss of one of these properties. Therefore, photochromes with inherent and separable chirality could be considered as candidates for fabricating novel photochromic LC materials.

Furthermore, to find their niche in future electronic information processing industry, the combination of photochromic materials with integrated circuits based on microelectromechanical systems offers advantages over above-mentioned crystalline molecular photoswitches due to the compatibility of microfabrication processes with complementary metal-oxide semiconductor (CMOS) technology on the scale of 1 μm to millimeters. Even the most anticipation of photochromic materials is the strategy that the fabrication of organic electro-optic devices would break through the limitations of size miniaturization of traditional Si-based electronics. In the past decades, amounts of photoswitchable organic electronic devices have been fabricated as transistors, luminescent displays, and so on [57] (see Chapter 7). More complex functional elements, such as logic and sensing, could be further integrated into these photoswitchable devices to perform long-expected molecular memory and computing operations. Compared to early solution models, photochromic devices can be regarded as a large step forward. Nonetheless, several key challenges still remain to be solved before photochromic devices achieve real industrialization. First, the electron/charge separation and transfer processes, which is considered as the fundamental and core for the organic electronic devices, need to be further unraveled. As in the photochromic electronic devices, the photoswitching endows the different photoisomers with different electronic states. Thus, how the molecules switch the charge/electron transfer processes and whether an injected charge or electron would affect the isomerization are still ambiguous. In contrast, the photoexcitation required for photoswitching would also be possible to interfere with the semiconductor devices. Second, since the performances of organic electronic devices are significantly dependent on the molecule-docked surfaces and interfaces, the assembly and interactions of photochromic molecules with surfaces/interfaces along with the inherent properties of surface materials should be another research focus. For example, the photoswitchable efficiency and even the photoreversibility and activity would be significantly affected when attached to the surface, according to molecule surface distances, extent of conjugations, intermolecular spaces, and so on (see Chapter 6). In addition, for some metal surfaces, the plasmon resonances excited by photochromic excitation light would be a nonignorable side effect (whether harmful or helpful). One of the typical cases is in the photoswitch molecular junctions. As the gap between two gold electrodes is in single-molecule level, it is still unclear which energy source, direct exciting light or excited plasmon resonance, would take the major responsibility for the switching behavior. Third, the role of molecular assembly during the device fabrications still requires deep investigation. In photochromic organic electronic device researches, especially in photoswitchable transistors, ordered intermolecular assembly is believed as one of the decisive factors for device efficiency, reversibility, and stability. Meticulous molecule design and careful fabrication would provide optimized devices. Thus, continuous design and synthesis of precisely modified photochromic molecules would not only improve the molecule inherent property itself but also affect the assembly nature when used for device fabrication. This would be the ever-completed mission for photochromic chemists. Fourth, for more complex system, precise control and signal readout should be improved. Take logic gate for instance, in solution models various signals (absorption, fluorescence, pH, etc.) could be used as inputs/outputs, yet in real device barely light and electronic signals can be used as outputs due to their accuracy in operations. Apart from that, the logic operation cycles should be wasteless, that is, inputs such as ions and acids/bases are not suitable for applications in photochromic logic devices. All-photonic logic gates reported recently [58] are one of the promising models for device fabrication, while the broad absorption and emission bands along with the ratio contrast should be still further improved for practical applications.

All in all, in the past two decades, photochromic materials have evolved from one-dimensional (solution phase) to two-dimensional (surface/interface), three-dimensional (bulk crystals, polymers, gels), to even more complicated electronic devices. Although photochromism is not such a large research area compared to other electro-optic materials as supramolecular materials, sensors, and photovoltaics, it still finds itself extremely active in both scientific research and industrial world. Using clean energy light as a driving force, photochromic materials bear advantages of environmental friendliness, high spatio and temporal resolution, reversible regulation, remote control, no-waste production, and so on. To further improve photochromic materials and expand their applications in daily life, several prospects have been raised in the above perspectives, which, shamefully, are still far from comprehensive and instructive. To boost the world of photochromism, we scientists should work together, on the one hand, to develop new types of photochromic molecules with novel structures and characteristics to enrich the photochromic family and, on the other hand, open our minds to find new land to maximize the potential of our pretty molecules for industrial innovations. We also envision that with the fast development of theoretical and computing chemistry, more accurate and universal calculating methods and models will be established for peering the nature of photochemistry and revolutionizing the molecule design and performance estimation.

References

1. Van Gemert, B. and Kish, D. G. (1999) ChemiChromics USA '99 : Conference and Exhibition on High-Tech Colors & Functional Materials, Conference Proceedings, New Orleans.
2. Hoffmann, H.J. (1999) *Organic Photochromic and Thermochromic Compounds*, vol. **2**, Plenum, New York, p. 822.
3. (a) Feringa, B.L. (2001) *Molecular Switches*, Wiley-VCH Verlag GmbH, Weinheim; (b) Raymo, F.M. and Tomasulo, M. (2005) Electron and energy transfer modulation with photochromic switches. *Chem. Soc. Rev.*, **34**, 327; (c) Yildiz, I., Deniz, E., and Raymo, F.M. (2009) Fluorescence modulation with photochromic switches in nanostructured constructs. *Chem. Soc. Rev.*, **38** (7), 1859–1867.
4. Such, G., Evans, R.A., Yee, L.H., and Davis, T.P. (2003) *J. Macromol. Sci.*, **C43** (4), 547–579.
5. Momoda, J., Izumi, S., and Yokoyama, Y. (2015) Substituent effects on the photochromic properties of 3, 3-diphenyspiro [benzofluorenopyran-cyclopentaphenanthrene]s. *Dyes Pigm.*, **119**, 95–107.
6. Crano, J.C., Kwak, W.S., and Welch, C.N. (1992) *Applied Photochromic Polymer Systems*, Blackie, Glasgow.
7. (a) Ambler, D.M., Balch, T.A., and Yamasaki, N.L.S. (2005) Lens which provides active response to light in ultraviolet–visible spectral region, for eyewear e.g. sunglasses, has at least two photochromic materials being activated by ultraviolet light and visible light, respectively. US Patent 6926405-B2, filed Jun. 06, 2003 and issued Aug. 09, 2005; (b)Younger Optics (September 2009) http://www.drivewearlens.com (accessed 22 January 2016).
8. (a)Carroll, G.H. (2002) Composition for producing photochromatic tattoo, comprising non-pigmented carrier and photochromic compounds which transform from colorless to intensely colored on exposure to ultraviolet rays to make tattoo visible. US Patent 6470891-B2, filed Dec. 13, 1999 and issued Oct. 29, 2002; (b) Bamfield, P. (2001) *Chromic Phenomena*, The Royal Society of Chemistry, Cambridge.
9. (a) Suzuki, Y. (2002) *Kinosei Inki no Saishin Gijutsu*, 63; (b) Suzuki, Y. (2002) *Chem. Abstr.*, **137**, 280630c.
10. van Renesse, R.L. (1998) *Optical Document Security*, 2nd edn, Artech House, Boston, MA and London, p. 201.
11. (a) Higgins, S. (2003) *Chem. Brit.*, **39**, 26; (b) Maesri, M., Pina, F., and Balzani, V. (2001) *Molecular Switches*, Wiley-VCH Verlag GmbH & Co. KGaA, Weinheim, pp. 309–337; (c) Guglielmetti, R. (1990) *Photochromism Molecules and Systems*, Elsevier, Amsterdam, p. 855.
12. (a) Barrack, J. (2002) Visor Photochromic, WO0218986; (b) Akuma Helmets (September 2009) http://www.akumahelmets.com/products_visor_sunshift.php (accessed 22 January 2016).
13. Pure Fishing (September 2009) http://www.berkley-fishing.co.uk (accessed 22 January 2016).
14. (a) Lee, S.-J., Son, Y.-A., Suh, H.-J., Lee, D.-N., and Kim, S.-H. (2006) Preliminary exhaustion studies of spiroxazine dyes on polyamide fibers and their photochromic properties. *Dyes Pigm.*, **69** (1), 18–21; (b) Billah, S.M.R., Christie, R.M., and Shamey, R. (2008) Direct coloration

of textiles with photochromic dyes. Part 1: application of spiroindolinonaphthoxazines as disperse dyes to polyester, nylon and acrylic fabrics. *Color. Technol.*, **124** (4), 223–228; (c) Billah, S.M.R., Christie, R.M., and Morgan, K.M. (2008) Direct coloration of textiles with photochromic dyes. Part 2: the effect of solvents on the colour change of photochromic textiles. *Color. Technol.*, **124** (4), 229–233.

15. (a) Son, Y.-A., Park, Y.-M., Park, S.-Y., Shin, C.-J., and Kim, S.-H. (2007) Exhaustion studies of spiroxazine dye having reactive anchor on polyamide fibers and its photochromic properties. *Dyes Pigm.*, **73** (1), 76–80; (b) Shah, M.R.B., Christie, R.M., Morgan, K.M., and Shamey, R. (2005) Photochromic protein substrates. *Mol. Cryst. Liq. Cryst.*, **431** (1), 535–539; (c) Little, A.F. and Christie, R.M. (2010) Textile applications of photochromic dyes. Part 1: establishment of a methodology for evaluation of photochromic textiles using traditional colour measurement instrumentation. *Color. Technol.*, **126** (3), 157–163.
16. (a) Little, A.F. and Christie, R.M. (2010) Textile applications of photochromic dyes. Part 2: factors affecting the photocoloration of textiles screen-printed with commercial photochromic dyes. *Color. Technol.*, **126** (3), 164–170; (b) Little, A.F. and Christie, R.M. (2011) Textile applications of photochromic dyes. Part 3: factors affecting the technical performance of textiles screen-printed with commercial photochromic dyes. *Color. Technol.*, **127** (5), 275–281.
17. (a) Aldib, M. and Christie, R.M. (2011) Textile applications of photochromic dyes. Part 4: application of commercial photochromic dyes as disperse dyes to polyester by exhaust dyeing. *Color. Technol.*, **127** (5), 282–287; (b) Aldib, M. and Christie, R.M. (2013) Textile applications of photochromic dyes. Part 5: application of commercial photochromic dyes to polyester fabric by a solvent-based dyeing method. *Color. Technol.*, **129** (2), 131–143.
18. Granqvist, C.G., Lansåker, P.C., Mlyuka, N.R., Niklasson, G.A., and Avendaño, E. (2009) Progress in chromogenics: new results for electrochromic and thermochromic materials and devices. *Sol. Energy Mater. Sol. Cells*, **93** (12), 2032–2039.
19. (a) Blin, X. and Simon, J.C. (2004) Cosmetic composition containing oil phase and naphthopyran photochromic dye, useful e.g. as make up, provides rapid and reversible color change on exposure to sunlight. FR Patent 2845899-A1, filed Oct.18, 2002 and issued Apr. 23, 2004; (b) Gonzalez, A.D., Pechko, A.H., and Kalafsky, R.E. (2003) Topical composition having reversible visible change in color in response to change in light, for e.g. sunscreen product, comprises photochromic material(s) and vehicle acceptable for topical application to skin or hair. US Patent 6531118-B1, filed Dec.11, 2001 and issued Mar. 11, 2003; (c) Forest, S.E., Krummen, M.S., and Yelton, J.A. (2002) Indicator used in skin protective kit for alerting user when sunscreen needs to be reapplied, comprises photochromic molecule, and carrier, and indicator is applied directly to skin without separate adhesive layer. WO Patent 200203949-A2, filed Jun. 27, 2001 and issued Jan. 11, 2002; (d) Carroll, G.H. (2002) Photochromic tanning lotion for application to person's skin, contains photochromic compound(s) mixed in appropriate amount in the lotion. US Patent 6361763-B1, filed Mar. 27, 2001 and issued Mar. 26, 2002.
20. (a) McCallien, D.W.J. and Bezer, M. (2000) Method of marking liquids e.g. fuels, involves addition of photochromic compound having specific color intensity after exposure with UV-light source. GB Patent 2344599A, filed Mar. 06, 1999 and issued Jun. 14, 2000; (b) Wright, P. (1989) Woven security label – contains colourless photochromic yarn which reversibly becomes coloured under UV irradiation. EP Patent 328320 A1, filed Feb. 03, 1989 and issued Aug. 16, 1989; (c) Committee on Next-Generation Currency Design of the Commission on Engineering and Technical Systems of the National Research Council (1993) *Counterfeit Deterrent Features for*

the Next-Generation Currency Design, National Academies Press, Washington, DC, p. 39.

21. Zhang, J.J., Zou, Q., and Tian, H. (2013) Photochromic materials: more than meets the eye. *Adv. Mater.*, **25** (3), 378–399.
22. (a) Terao, F., Morimoto, M., and Irie, M. (2012) Light-driven molecular-crystal actuators: rapid and reversible bending of rodlike mixed crystals of diarylethene derivatives. *Angew. Chem. Int. Ed.*, **124** (4), 925–928; (b) Yu, H. and Ikeda, T. (2011) Photocontrollable liquid-crystalline actuators. *Adv. Mater.*, **23** (19), 2149–2180.
23. Ushiogi, Y., Hase, T., Iinuma, Y., Taketa, A., and Yoshida, J.-I. (2007) Synthesis of photochromic diarylethenes using a microflow system. *Chem. Commun.*, **28**, 2947–2949.
24. Kuhn, H.J., Braslavsky, S.E., and Schmidt, R. (2004) Chemical actinometry (IUPAC technical report). *Pure Appl. Chem.*, **76** (12), 2105–2146.
25. (a) Russew, M.M. and Hecht, S. (2010) Photoswitches: from molecules to materials. *Adv. Mater.*, **22** (31), 3348–3360; (b) Lee, S., Kang, H.S., and Park, J.K. (2012) Directional photofluidization lithography: micro/nanostructural evolution by photofluidic motions of azobenzene materials. *Adv. Mater.*, **24** (16), 2069–2103; (c) Beharry, A.A. and Woolley, G.A. (2011) Azobenzene photoswitches for biomolecules. *Chem. Soc. Rev.*, **40** (8), 4422–4437; (d) Bandara, H.M.D. and Burdette, S.C. (2012) Photoisomerization in different classes of azobenzene. *Chem. Soc. Rev.*, **41** (5), 1809–1825; (e) Matharu, A.S., Jeeva, S., and Ramanujam, P.S. (2007) Liquid crystals for holographic optical data storage. *Chem. Soc. Rev.*, **36** (12), 1868–1880; (f) Yagai, S. and Kitamura, A. (2008) Recent advances in photoresponsive supramolecular self-assemblies. *Chem. Soc. Rev.*, **37** (8), 1520–1529.
26. (a) Samanta, S., Babalhavaeji, A., Dong, M.X., and Woolley, G.A. (2013) Photoswitching of ortho-substituted azonium ions by red light in whole blood. *Angew. Chem. Int. Ed.*, **52** (52), 14127–14130; (b) Beharry, A.A., Sadovski, O., and Woolley, G.A. (2011) Azobenzene photoswitching without ultraviolet light. *J. Am. Chem. Soc.*, **133** (49), 19684–19687; (c) Bleger, D., Schwarz, J., Brouwer, A.M., and Hecht, S. (2012) o-Fluoroazobenzenes as readily synthesized photoswitches offering nearly quantitative two-way isomerization with visible light. *J. Am. Chem. Soc.*, **134** (51), 20597–20600; (d) Samanta, S., Beharry, A.A., Sadovski, O., McCormick, T.M., Babalhavaeji, A., Tropepe, V., and Woolley, G.A. (2013) Photoswitching azo compounds in vivo with red light. *J. Am. Chem. Soc.*, **135** (26), 9777–9784.
27. Weston, C.E., Richardson, R.D., Haycock, P.R., White, A.J.P., and Fuchter, M.J. (2014) Arylazopyrazoles: azoheteroarene photoswitches offering quantitative isomerization and long thermal half-lives. *J. Am. Chem. Soc.*, **136** (34), 11878–11881.
28. Minkin, V.I. (2004) Photo-, thermo-, solvato-, and electrochromic spiroheterocyclic compounds. *Chem. Rev.*, **104** (5), 2751–2776.
29. Liu, D., Chen, W., Sun, K., Deng, K., Zhang, W., Wang, Z., and Jiang, X. (2011) Resettable, multi-readout logic gates based on controllably reversible aggregation of gold nanoparticles. *Angew. Chem. Int. Ed.*, **50** (18), 4103–4107.
30. (a) Balogh, D., Tel-Vered, R., Freeman, R., and Willner, I. (2011) Photochemically and electrochemically triggered Au nanoparticles "sponges". *J. Am. Chem. Soc.*, **133** (17), 6533–6536; (b) Niazov, T., Shlyahovsky, B., and Willner, I. (2007) Photoswitchable electrocatalysis and catalyzed chemiluminescence using photoisomerizable monolayer-functionalized surfaces and Pt nanoparticles. *J. Am. Chem. Soc.*, **129** (20), 6374–6375.
31. (a) Chen, L., Wang, W., Su, B., Wen, Y., Li, C., Zhou, Y., Li, M., Shi, X., Du, H., Song, Y., and Jiang, L. (2014) A Light-responsive release platform by controlling the wetting behavior of hydrophobic surface. *ACS Nano*, **8** (1), 744–751; (b) Riskin, M., Gutkin, V., Felner, I., and Willner, I. (2008) Photochemical and electrochemical encoding

of erasable magnetic patterns. *Angew. Chem. Int. Ed.*, **47** (23), 4416–4420.

32. (a) Shi, Z., Peng, P., Strohecker, D., and Liao, Y. (2011) Long-lived photoacid based upon a photochromic reaction. *J. Am. Chem. Soc.*, **133** (37), 14699–14703; (b) Tatum, L.A., Foy, J.T., and Aprahamian, I. (2014) Waste management of chemically activated switches: using a photoacid to eliminate accumulation of side products. *J. Am. Chem. Soc.*, **136** (50), 17438–17441; (c) Maity, C., Hendriksen, W.E., van Esch, J.H., and Eelkema, R. (2015) Spatial structuring of a supramolecular hydrogel by using a visible-light triggered catalyst. *Angew. Chem. Int. Ed.*, **54** (3), 998–1001.
33. Tian, Z. and Li, A.D.Q. (2013) Photoswitching-enabled novel optical imaging: innovative solutions for real-world challenges in fluorescence detections. *Acc. Chem. Res.*, **46** (2), 269–279.
34. Davis, D.A., Hamilton, A., Yang, J.L., Cremar, L.D., Van Gough, D., Potisek, S.L., Ong, M.T., Braun, P.V., Martinez, T.J., White, S.R., Moore, J.S., and Sottos, N.R. (2009) Force-induced activation of covalent bonds in mechanoresponsive polymeric materials. *Nature*, **459** (7243), 68–72.
35. (a) Irie, M. (2000) Diarylethenes for memories and switches. *Chem. Rev.*, **100** (5), 1685–1716; (b) Irie, M., Fulcaminato, T., Matsuda, K., and Kobatake, S. (2014) Photochromism of diarylethene molecules and crystals: memories, switches, and actuators. *Chem. Rev.*, **114** (24), 12174–12277.
36. (a) Li, W.L., Jiao, C.H., Li, X., Xie, Y.S., Nakatani, K., Tian, H., and Zhu, W.H. (2014) Separation of photoactive conformers based on hindered diarylethenes: efficient modulation in photocyclization quantum yields. *Angew. Chem. Int. Ed.*, **53** (18), 4603–4607; (b) Li, W.L., Li, X., Xie, Y.S., Wu, Y., Li, M.Q., Wu, X.Y., Zhu, W.H., and Tian, H. (2015) Enantiospecific photoresponse of sterically hindered diarylethenes for chiroptical switches and photomemories. *Sci. Rep.*, **5**, 09186.
37. (a) Vomasta, D., Hogner, C., Branda, N.R., and Konig, B. (2008) Regulation of human carbonic anhydrase I (hCAI) activity by using a photochromic inhibitor. *Angew. Chem. Int. Ed.*, **47** (40), 7644–7647; (b) Babii, O., Afonin, S., Berditsch, M., Reisser, S., Mykhailiuk, P.K., Kubyshkin, V.S., Steinbrecher, T., Ulrich, A.S., and Komarov, I.V. (2014) Controlling biological activity with light: diarylethene-containing cyclic peptidomimetics. *Angew. Chem. Int. Ed.*, **53** (13), 3392–3395; (c) Fujimoto, K., Kajino, M., Sakaguchi, I., and Inouye, M. (2012) Photoswitchable, DNA-binding helical peptides assembled with two independently designed sequences for photoregulation and DNA recognition. *Chem. Eur. J.*, **18** (32), 9834–9840.
38. (a) Koumura, N., Zijlstra, R.W.J., van Delden, R.A., Harada, N., and Feringa, B.L. (1999) Light-driven monodirectional molecular rotor. *Nature*, **401** (6749), 152–155; (b) Conyard, J., Addison, K., Heisler, I.A., Cnossen, A., Browne, W.R., Feringa, B.L., and Meech, S.R. (2012) Ultrafast dynamics in the power stroke of a molecular rotary motor. *Nat. Chem.*, **4** (7), 547–551; (c) Fletcher, S.P., Dumur, F., Pollard, M.M., and Feringa, B.L. (2005) A reversible, unidirectional molecular rotary motor driven by chemical energy. *Science*, **310** (5745), 80–82; (d) van Delden, R.A., ter Wiel, M.K.J., Pollard, M.M., Vicario, J., Koumura, N., and Feringa, B.L. (2005) Unidirectional molecular motor on a gold surface. *Nature*, **437** (7063), 1337–1340; (e) Eelkema, R., Pollard, M.M., Vicario, J., Katsonis, N., Ramon, B.S., Bastiaansen, C.W.M., Broer, D.J., and Feringa, B.L. (2006) Molecular machines: nanomotor rotates microscale objects. *Nature*, **440** (7081), 163; (f) Ruangsupapichat, N., Pollard, M.M., Harutyunyan, S.R., and Feringa, B.L. (2011) Reversing the direction in a light-driven rotary molecular motor. *Nat. Chem.*, **3** (1), 53–60; (g) Wang, J.B. and Feringa, B.L. (2011) Dynamic control of chiral space in a catalytic asymmetric reaction using a molecular motor. *Science*, **331** (6023), 1429–1432; (h) Kudernac, T., Ruangsupapichat, N.,

Parschau, M., Macia, B., Katsonis, N., Harutyunyan, S.R., Ernst, K.H., and Feringa, B.L. (2011) Electrically driven directional motion of a four-wheeled molecule on a metal surface. *Nature*, **479** (7372), 208–211; (i) Qu, D.H. and Feringa, B.L. (2010) Controlling molecular rotary motion with a self-complexing lock. *Angew. Chem. Int. Ed.*, **49** (6), 1107–1110.

39. (a) Habault, D., Zhang, H., and Zhao, Y. (2013) Light-triggered self-healing and shape-memory polymers. *Chem. Soc. Rev.*, **42** (17), 7244–7256; (b) Gohy, J.-F. and Zhao, Y. (2013) Photo-responsive block copolymer micelles: design and behavior. *Chem. Soc. Rev.*, **42** (17), 7117–7129.
40. (a) Ma, X. and Tian, H. (2014) Stimuli-responsive supramolecular polymers in aqueous solution. *Acc. Chem. Res.*, **47** (7), 1971–1981; (b) Dong, S., Zheng, B., Wang, F., and Huang, F. (2014) Supramolecular polymers constructed from macrocycle-based host–guest molecular recognition motifs. *Acc. Chem. Res.*, **47** (7), 1982–1994.
41. (a) Brunsveld, L., Folmer, J.B., Meijer, E.W., and Sijbesma, R.P. (2001) Supramolecular polymers. *Chem. Rev.*, **101** (12), 4071–4098; (b) De Greef, T.F.A., Smulder, M.M.J., Wolffs, M., Schenning, A.P.H.J., Sijbesma, R.P., and Meijer, E.W. (2009) Supramolecular polymerization. *Chem. Rev.*, **109** (11), 5687–5754; (c) Qu, D.-H., Wang, Q.-C., Zhang, Q.-W., Ma, X., and Tian, H. (2015) Photoresponsive host–guest functional systems. *Chem. Rev.*, **115** (15), 7543–7588.
42. Aida, T., Meijer, E.W., and Stupp, S.I. (2012) Functional supramolecular polymers. *Science*, **335** (6070), 813–817.
43. Yan, X., Wang, F., Zheng, B., and Huang, F. (2012) Stimuli-responsive supramolecular polymeric materials. *Chem. Soc. Rev.*, **41** (18), 6042–6065.
44. (a) Yan, X., Xu, D., Chi, X., Chen, J., Dong, S., Ding, X., Yu, Y., and Huang, F. (2012) A multiresponsive, shape-persistent, and elastic supramolecular polymer network gel constructed by orthogonal self-assembly. *Adv. Mater.*, **24** (3), 362–369; (b) Zhang, M., Xu, D., Yan, X., Chen, J., Dong, S., Zheng, B., and Huang, F. (2012) Self-healing supramolecular gels formed by crown ether based host–guest interactions. *Angew. Chem. Int. Ed.*, **124** (28), 7117–7121.
45. Velema, W.A., Szymanski, W., and Feringa, B.L. (2014) Photopharmacology: beyond proof of principle. *J. Am. Chem. Soc.*, **136** (6), 2178–2191.
46. (a) Broichhagen, J., Frank, J.A., and Trauner, D. (2015) A roadmap to success in photopharmacology. *Acc. Chem. Res.*, **48** (7), 1947–1960; (b) Szymanski, W., Yilmaz, D., Kocer, A., and Feringa, B.L. (2013) Bright ion channels and lipid bilayers. *Acc. Chem. Res.*, **46** (12), 2910–2923; (c) Szymanski, W., Beierle, J.M., Kistemaker, H.A.V., Velema, W.A., and Feringa, B.L. (2013) Reversible photocontrol of biological systems by the incorporation of molecular photoswitches. *Chem. Rev.*, **113** (8), 6114–6178.
47. (a) Boyer, J.C., Carling, C.J., Chua, S.Y., Wilson, D., Johnsen, B., Baillie, D., and Branda, N.R. (2012) Photomodulation of fluorescent upconverting nanoparticle markers in live organisms by using molecular switches. *Chem. Eur. J.*, **18** (11), 3122–3126; (b) Boyer, J.C., Carling, C.J., Gates, B.D., and Branda, N.R. (2010) Two-way photoswitching using one type of near-infrared light, upconverting nanoparticles, and changing only the light intensity. *J. Am. Chem. Soc.*, **132** (44), 15766–15772; (c) Zhou, Z.G., Hu, H., Yang, H., Yi, T., Huang, K.W., Yu, M.X., Li, F.Y., and Huang, C.H. (2008) Up-conversion luminescent switch based on photochromic diarylethene and rare-earth nanophosphors. *Chem. Commun.*, **39**, 4786–4788.
48. (a) Ko, C.C. and Yam, V.W.W. (2010) Transition metal complexes with photochromic ligands—photosensitization and photoswitchable properties. *J. Mater. Chem.*, **20** (11), 2063–2070; (b) Tan, W.J., Zhou, J., Li, F.Y., Yi, T., and Tian, H. (2011) Visible light-triggered photoswitchable diarylethene-based iridium(III) complexes for imaging living cells. *Chem. Asian J.*, **6** (5), 1263–1268; (c) Lee, P.H.M., Ko, C.C., Zhu, N., and

Yam, V.W.W. (2007) Metal coordination-assisted near-infrared photochromic behavior: a large perturbation on absorption wavelength properties of N, N-donor ligands containing diarylethene derivatives by coordination to the rhenium (I) metal center. *J. Am. Chem. Soc.*, **129** (19), 6058–6059; (d) Chan, J.C.H., Lam, W.H., Wong, H.L., Zhu, N., Wong, W.T., and Yam, V.W.W. (2011) Diarylethene-containing cyclometalated platinum (II) complexes: tunable photochromism via metal coordination and rational ligand design. *J. Am. Chem. Soc.*, **133** (32), 12690–12705.

49. (a) Zhu, M.Q., Zhang, G.F., Li, C., Aldred, M.P., Chang, E., Drezek, R.A., and Li, A.D.Q. (2011) Reversible two-photon photoswitching and two-photon imaging of immunofunctionalized nanoparticles targeted to cancer cells. *J. Am. Chem. Soc.*, **133** (2), 365–372; (b) Feng, Y., Yan, Y., Wang, S., Zhu, W., Qian, S., and Tian, H. (2006) Photochromic thiophene oligomers based on bisthienylethene: syntheses, photochromic and two-photon properties. *J. Mater. Chem.*, **16** (37), 3685–3692.

50. (a) Fukaminato, T., Hirose, T., Doi, T., Hazama, M., Matsuda, K., and Irie, M. (2014) Molecular design strategy toward diarylethenes that photoswitch with visible light. *J. Am. Chem. Soc.*, **136** (49), 17145–17154; (b) Hu, F., Cao, M., Ma, X., Liu, S.H., and Yin, J. (2015) Visible-light-dependent photocyclization: design, synthesis, and properties of a cyanine-based dithienylethene. *J. Org. Chem.*, **80** (15), 7830–7835.

51. Boulegue, C., Loweneck, M., Renner, C., and Moroder, L. (2007) Redox potential of azobenzene as an amino acid residue in peptides. *ChemBioChem*, **8** (6), 591–594.

52. (a) McGirt, L.Y., Vasagar, K., Gober, L.M., Saini, S.S., and Beck, L.A. (2006) Successful treatment of recalcitrant chronic idiopathic urticaria with sulfasalazine. *Arch. Dermatol.*, **142** (10), 1337–1342; (b) Kruis, W., Schreiber, S., Theuer, D., Brandes, J.W., Schutz, E., Howaldt, S., Krakamp, B., Hamling, J., Monnikes, H., Koop, I., Stolte, M., Pallant, D., and Ewald, U. (2001) Low dose balsalazide (1.5 g twice daily) and mesalazine (0.5 g three times daily) maintained remission of ulcerative colitis but high dose balsalazide (3.0 g twice daily) was superior in preventing relapses. *Gut*, **49** (6), 783–789; (c) Abramsson-Zetterberg, L. and Ilback, N.G. (2013) The synthetic food colouring agent Allura Red AC (E129) is not genotoxic in a flow cytometry-based micronucleus assay in vivo. *Food Chem. Toxicol.*, **59**, 86–89; (d) Taylor, S.H. and Shillingford, J.P. (1959) Clinical applications of coomassie blue. *Br. Heart J.*, **21** (4), 497–504.

53. (a) Bonnet, J., Yin, P., Ortiz, M.E., Subsoontron, P., and Endy, D. (2013) Amplifying genetic logic gates. *Science*, **340** (6132), 599–603; (b) Stanton, B.C., Nielsen, A.A.K., Tamsir, A., Clancy, K., Peterson, T., and Voigt, C.A. (2014) Genomic mining of prokaryotic repressors for orthogonal logic gates. *Nat. Chem. Biol.*, **10** (2), 99–105; (c) Gaber, R., Lebar, T., Majerle, A., Ster, B., Dobnikar, A., Bencina, M., and Jerala, R. (2014) Designable DNA-binding domains enable construction of logic circuits in mammalian cells. *Nat. Chem. Biol.*, **10** (3), 203–208; (d) Siuti, P., Yazbek, J., and Lu, T.K. (2013) Synthetic circuits integrating logic and memory in living cells. *Nat. Biotechnol.*, **31** (5), 448–452; (e) Regot, S., Macia, J., Conde, N., Furukawa, K., Kjellen, J., Peeters, T., Hohmann, S., de Nadal, E., Posas, F., and Sole, R. (2011) Distributed biological computation with multicellular engineered networks. *Nature*, **469** (7329), 207–211; (f) Tamsir, A., Tabor, J., and Voigt, C.A. (2011) Robust multicellular computing using genetically encoded NOR gates and chemical 'wires'. *Nature*, **469** (7329), 212–215.

54. (a) Chinwangso, P., Jamison, A.C., and Lee, T.R. (2011) Multidentate adsorbates for self-assembled monolayer films. *Acc. Chem. Res.*, **44** (7), 511–519; (b) Gooding, J.J. and Ciampi, S. (2011) The molecular level modification of surfaces: from self-assembled monolayers to complex molecular assemblies. *Chem. Soc. Rev.*, **40** (5), 2704–2718.

55. Jin, J., Li, X., Zhang, J.J., Zhao, P., and Tian, H. (2013) Rational design of double-check mercury ion chemosensors based on photochromic compounds. *Isr. J. Chem.*, **53** (5), 288–293.
56. Abendroth, J.M., Bushuyev, O.S., Weiss, P.S., and Barrett, C.J. (2015) Controlling motion at the nanoscale: rise of the molecular machines. *ACS Nano*, **9** (8), 7746–7768.
57. Orgiu, E. and Samori, P. (2014) 25th anniversary article: organic electronics marries photochromism: generation of multifunctional interfaces, materials, and devices. *Adv. Mater.*, **26** (12), 1827–1845.
58. Andreasson, J., Pischel, U., Straight, S., Moore, T.A., Moore, A.L., and Gust, D. (2011) All-photonic multifunctional molecular logic device. *J. Am. Chem. Soc.*, **133** (30), 11641–11648.

Index

Photochromic Materials: Preparation, Properties and Applications, First Edition.
Edited by He Tian and Junji Zhang.

o

p